PROCESS CONTROL

PROCESS CONTROL

K. KRISHNASWAMY

B.E. (Electrical Engg.), M.Tech. (Measurements)

Vice Principal

Kongu Engineering College

Erode, Tamil Nadu

PUBLISHING FOR ONE WORLD

NEW AGE INTERNATIONAL (P) LIMITED, PUBLISHERS

New Delhi • Bangalore • Chennai • Cochin • Guwahati • Hyderabad
Jalandhar • Kolkata • Lucknow • Mumbai • Ranchi

Visit us at **www.newagepublishers.com**

Published by New Age International (P) Ltd., Publishers
First Edition: 2006
Second Edition: 2009
Reprint: 2012

Branches:
- No. 37/10, 8th Cross (Near Hanuman Temple), Azad Nagar, Chamrajpet, **Bangalore** - 560 018. Tel.: (080) 26756823, Telefax: 26756820, E-mail: bangalore@newagepublishers.com
- 26, Damodaran Street, T. Nagar, **Chennai** - 600 017. Tel.: (044) 24353401, Telefax: 24351463 E-mail: chennai@newagepublishers.com
- CC-39/1016, Carrier Station Road, Ernakulam South, **Cochin** - 682 016. Tel.: (0484) 2377004, Telefax: 4051303 E-mail: cochin@newagepublishers.com
- Hemsen Complex, Mohd. Shah Road, Paltan Bazar, Near Starline Hotel, **Guwahati** - 781 008.Tel.: (0361) 2513881 Telefax: 2543669, E-mail: guwahati@newagepublishers.com
- No. 105, 1st Floor, Madhiray Kaveri Tower, 3-2-19, Azam Jahi Road, Nimboliadda, **Hyderabad** - 500 027. Tel.: (040) 24652456, Telefax: 24652457, E-mail:hyderabad@newagepublishers.com
- RDB Chambers (Formerly Lotus Cinema)106A, 1st Floor, S.N. Banerjee Road, **Kolkata** - 700 014. Tel.: (033) 22273773, Telefax: 22275247, E-mail:kolkata@newagepublishers.com
- 16-A, Jopling Road, **Lucknow** - 226 001. Tel.: (0522) 2209578, 4045297, Telefax: 2204098 E-mail: lucknow@newagepublishers.com
- 142C, Victor House, Ground Floor, N.M. Joshi Marg, Lower Parel, **Mumbai** - 400 013. Tel.: (022) 24927869 Telefax: 24915415, E-mail: mumbai@newagepublishers.com
- 22, Golden House, Daryaganj, **New Delhi** - 110 002. Tel.: (011) 23262370, 23262368, Telefax: 43551305 E-mail: sales@newagepublishers.com

ISBN : 978-81-224-2736-3

₹ 299.00

C-11-10-5960

Printed in India at Saras Graphics, Rai, Haryana.
Typeset at In-house, Delhi.

PUBLISHING FOR ONE WORLD
NEW AGE INTERNATIONAL (P) LIMITED, PUBLISHERS
4835/24, Ansari Road, Daryaganj, New Delhi-110002
Visit us at **www.newagepublishers.com**

FOREWORD

Process Control is a principal part of recent industrial revolution. The advantages of automatic process control have been universally recognized. Automatic process control is used primarily because it results in economy of operation of industrial processes that more than pays for the expense of the control equipment. Elimination of human error is another positive contribution of use of automatic process control. Understanding of the principle of automatic process control is necessary in modern engineering and should therefore be a most important part of the span of knowledge of the engineer. It is all the more important for those who are pursuing Electronics and Instrumentation, Instrumentation and Control, and Chemical Engineering.

Prof. K. Krishnaswamy, with his vast industrial experience and his academic eminence, has brought out the various aspects of Process Control with numerous illustrations. To deal with the subject in depth, he has divided the whole book into five chapters. At the end of each chapter, a number of problems are worked out and a set of thought-provoking questions are given. The book will be of immense use both as a textbook and as a reference for the students. In addition, it will also serve as a reference book for the process control engineers in the field.

It really gives me an immense pleasure to introduce to the students and professionals the outcome of the excellent and dedicated effort of our senior professor in the form of 'Process Control' book.

Dr. A. M. NATARAJAN
Principal
Kongu Engineering College
Perundurai, Erode
Tamil Nadu

PREFACE TO THE SECOND EDITION

Three years have passed since the first edition of this book 'Process Control' was published. During this time, the book has been widely used in the industry and the response has been most gratifying. The book found overwhelming responses from both students and teaching faculty of engineering courses. The first edition covered all the topics demanded by universities from Tamil Nadu area. There is a popular demand now from students and faculty of other technical universities in India for adding the following two chapters to the book so that it can be used as a textbook covering their entire syllabus.

1. Signal Conditioning
2. Digital Controllers-Computer Based Control

The author felt, while bringing out the first edition, that the topic 'Signal conditioning' belonged to 'Sensors and transducers' or 'Transducer Engineering' subject. Also the topic was covered substantially in the books meant for that subject by different authors. Since there is a popular request now, the topic has been added elaborately in chapter 6, though the suitable place might be in the beginning of the book.

The topic 'Digital Controllers—Computer-Based Control', being very wide, requires to be taught in at least two to three separate courses. Some of them may include Microprocessors and microcontrollers, Computer control of process and computer network, and Distributed control systems. The author was really afraid of bringing all materials belonging to this topic in a very condensed form in the first edition. Due to popular demand now, the author has attempted to give very brief ideas in the newly added chapter 7 of this second edition. It covers mainly the Computer based controllers, Distributed control system, Supervisory control and data acquisition system, and Control system networks. Field buses are covered to some extent. May this new chapter confirm the trust which my readers have entrusted in me.

The arrangement of the book however remained unaltered as far as the first five chapters are concerned. Only the last two chapters are added to cover the entire gamut of process control subject so that students find all required information in a single volume.

My deepest appreciation is expressed to Ms. L. Kalpana who typed the original manuscript of chapters 6 & 7 with great care and dedication.

I hope that this book will be very useful to the reader in all respects. I would welcome constructive criticism of the book and will be grateful to the reader for suggestions for its further improvement.

K. KRISHNASWAMY

PREFACE TO THE FIRST EDITION

The purpose of this book is to present to the engineering students and to the beginning engineers in process industries the important principles of automatic control, beginning with mathematical modeling of processes and carrying on into the generalized behaviour of closed loop systems. Controller characteristics, controller tuning, control systems with multiple loops and final control elements are covered sufficiently. Selected unit operations in chemical industries and power plants are covered to make the subject easily understandable.

The book is intended primarily for the undergraduate in Engineering with some knowledge of the elements of calculus, differential equations, Laplace transforms, fluid mechanics and thermodynamics. The syllabi of various technical universities including Anna University, Chennai, is kept in mind while preparing the contents of the book. I have tried to cover the topics in a simple and clear way to the readers. The book is the culmination of my 30 years of industrial experience and 6 years of academic experience in this field. It is unavoidable that my descriptions are influenced by my personal experience and preference.

The material in this book is so arranged that the basic principles used are thoroughly understood before their application in the process industries is encountered. Chapter 1 deals with mathematical modeling of processes in addition to introduction of process control. Chapter 2 covers controller characteristics and tuning of controllers including evaluation criteria. Chapter 3 describes some of the control systems with multiple loops. Chapter 4 attempts to bring out the details of final control elements, whereas Chapter 5 concentrated on selected unit operations in chemical and power industries.

I have also used some of the available printed materials which are listed in the bibliography. Although carefully prepared and reviewed, this book may contain errors of typography, calculations or statement. Any report of these errors is welcome by the author.

K. KRISHNASWAMY

ACKNOWLEDGEMENTS

I wish to acknowledge the contributions in terms of material, assistance and encouragement made by colleagues at Kongu Engineering College. Special thanks are due to the staff members of Electronics and Instrumentation Engineering Department of Kongu Engineering College. Dr. A. M. Natarajan, Principal, Kongu Engineering College deserves much credit for his continuous encouragement, comments and suggestions. Thanks are due to the office bearers and trustees of Kongu Vellalar Institute of Technology Trust, Perundurai who have given active support and moral encouragement.

I am vastly indebted to many people who have helped and inspired me in various ways. My gratitude goes to those colleagues who worked with me in Rourkela Steel Plant, Steel Authority of India Ltd. for strengthening my resolve in so many direct and indirect ways. Mr. K. S. Bhagawan, Mr. S. C. Thiagarajan, Mr. R. Mishra and Mr. S. Ramakrishnan may not have realized what an influence their generous presence, teachings, and friendship have had in shaping this book.

I owe a special debt to Mr. J. Jegadeeswara Raja who typed the original manuscript with great care, artistic taste, skill, and dedication, unparalleled in my own experience.

Finally, to my parents, wife and daughter Ms. Bharati goes my gratitude for their love, support and dedication.

K. KRISHNASWAMY

CONTENTS

PROCESS CONTROL INTRODUCTION

1.1 PROCESS

A process denotes an operation or series of operations on fluid or solid materials during which the materials are placed in a more useful state.

The objective of a process is to convert certain raw materials (input feedstock) into desired products (output) using available sources of energy in the most economical way.

A unit process may involve either a change of chemical state or a change in physical state. Many external and internal conditions affect the performance of a process. These conditions may be expressed in terms of process variables such as temperature, pressure, flow, liquid level, dimension, weight, volume etc.

A process must satisfy several requirements imposed by its designers and the general technical, economic and social conditions in the presence of everchanging external influences (disturbances). The requirements include safety of men and machine, environmental regulations, production specifications, operational constraints and economics.

By process it is meant either a unit process like an alkylation reactor or unit operation like evaporator, distillation column or storage vessel.

A process can be described either by an ordinary differential equation (lumped parameter system) or by partial differential equation (distributed parameter system).

1.2 CONTROL (SYSTEM)

The term control means methods to force parameters in the environment to have specific values. This can be as simple as making the temperature in a room stay at 25°C or as complex as manufacturing an integrated circuit or guiding a spacecraft to Jupiter. In general, all of the elements necessary to accomplish the control objective are described by the term control system.

The basic strategy by which a control system operates is quite logical and natural. In fact, the same strategy is employed in living organisms to maintain temperature, fluid flow rate, and a host of other biological functions. This is natural process control.

The technology of artificial control was first developed with a human as an integral part of the control action. When it is learnt how to use machines, electronics, and computers to replace the human function, the term 'automatic control' came into use.

1.3 PROCESS CONTROL

The process may be controlled by measuring a variable representing the desired state of the product and automatically adjusting one of the other variables of the process. In process control, the basic objective is to regulate the value of some quantity. To regulate means to maintain that quantity at some desired value (reference value or set point) regardless of external influences.

During the first industrial revolution the workdone by human muscles was gradually replaced by the power of machines. Process control opened the door to the second industrial revolution, where the routine functions of the human mind and the need for the continuous presence of human observers were also taken care of by machines. In true sense process control had made optimization, and thereby, the beginning of the third industrial revolution possible. Here the traditional goal of maximizing the quantity production is gradually replaced by the goal of maximizing the quality and durability of the produced goods, while minimizing the consumption of energy and raw materials and maximizing recycling and reuse. Optimized process control is the moto.

1.4 AUTOMATIC PROCESS CONTROL

Automatic control is the maintenance of a desired value of a quantity or condition by measuring the existing value, comparing it to the desired value, and employing the difference to initiate action for reducing this difference. Thus automatic control requires a closed loop of action and reaction operating without human aid.

1.4.1 Variables of Automatic Process Control

(*a*) Set point variable : is the one that is set by the operator, master controller or computer as a desired value for a 'controlled variable'. It is also called sometimes as 'reference value'.

(*b*) Controlled variable : is the one that must be maintained precisely at the set point. Typically, the variable chosen to represent the state of the system is termed the 'controlled variable'. Examples of controlled variables are temperature, pressure, flow rate, level, vaccum pressure, concentration, density etc.

(*c*) Manipulated variable : is the one that can be changed in order to maintain the controlled variable at the set point value. In other words, the variable chosen to control the system's state is termed the 'manipulated variable'. It is also called sometimes as 'controlling variable'. Examples of manipulated variables are coolant flow, fuel flow, feed water flow etc.

(*d*) Load variables : are those variables that cause disturbances in the process. They are also called as load disturbances. The load variable may change either continuously or sporadically with some function of time. Sometimes it is fixed and not a function of time. Examples are feed rate, feed composition, steam header pressure, coolant temperature etc.

The load variables are uncontrolled independent variables, which, when they change, will upset the control system, and their effects can only be corrected in a 'feedback' manner. This means that a change in load variables is not responded until they have upset the controlled variable.

1.4.2 Process Degree of Freedom

The state of a process or the configuration of a system is determined when each of its degrees of freedom is specified. Consider, for example, a ball placed on a billiard table. In order to specify its position, we would require three coordinates : One north-south coordinate, one east-west coordinate, and the height. However, the height is not arbitrary because it is given by the height of the table surface above a reference plane. Consequently the ball has two degrees of freedom.

Mathematically, the number of degrees of freedom is defined as

$$n = n_v - n_e \tag{1.1}$$

where n = number of degrees of freedom of a system

n_v = number of variables that describe the system

n_e = number of defining equations of the system or number of independent relationships that exist among the various variables.

In the example of the billiard ball there are three variables of position ($n_v = 3$), one defining equation ($n_e = 1$), and therefore two degrees of freedom ($n = n_u - n_e = 3 - 1 = 2$). It is easy to see intuitively that a train has only one degree of freedom because only its speed can be varied, while boats have two and air planes have three [Refer Fig. 1.1 and Fig. 1.2].

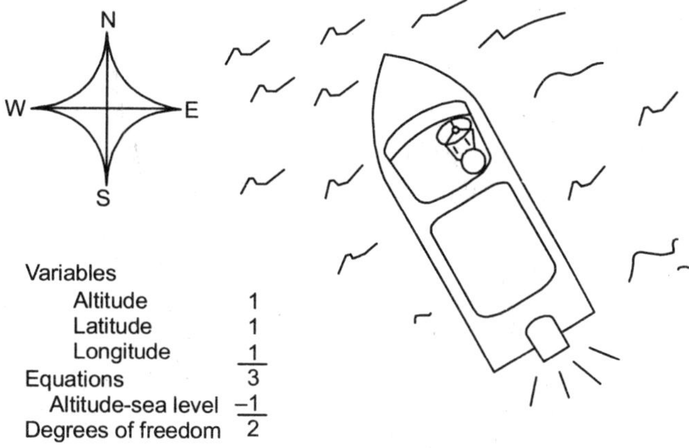

Fig. 1.1 Degrees of freedom of a boat

When looking at industrial processes, the determination of degrees of freedom becomes more complex and cannot always be determined intuitively. 'The degrees of freedom of a process represents the maximum number of independently acting controllers that can be placed on that process.' In other words, 'the number of independently acting automatic controllers on a system or process may not exceed the number of degrees of freedom.' System variables and parameters must be carefully distinguished. The weight of water in a tank, specific heat of water etc are parameters not variables. The inlet temperature, outlet temperature, water flow rate, heat input rate etc are variables.

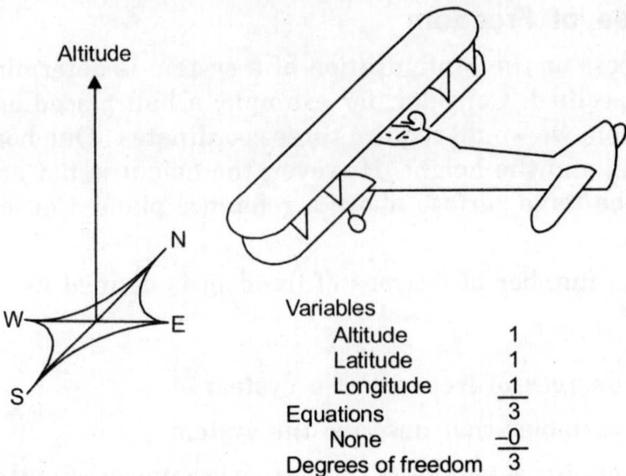

Fig. 1.2 Degrees of freedom of an airplane.

For chemical processes involving separation, distillation, or fractionation where heterogeneous equilibrium exists and where each component is present in each phase, a modification of the rule may be derived. It is known as Gibb's phase rule,

$$n = n_c - n_p + 2 \tag{1.2}$$

where, n = number of chemical degrees of freedom

n_c = number of components

n_p = number of phases.

The above equation applies only to the chemical states of the process, and the number 2 in the equation represents temperature and pressure. For an isothermal process

$$n = n_c - n_p + 1$$

and for a constant-pressure process

$$n = n_c - n_p + 1$$

For example, consider a steam boiler producing saturated steam (Refer Fig. 1.3) . The number of components is one (water), and the number of phases are two (liquid and gas). Therefore the number of degrees of freedom are

$$n = 1 - 2 + 2 = 1$$

and either temperature or pressure (but not both) may be selected as the independent variable. For a boiler producing superheated steam the number of degrees of freedom are two, and both temperature and pressure must be controlled.

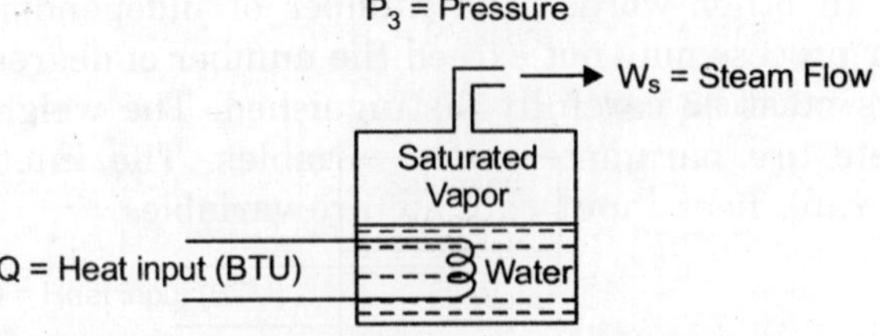

Fig. 1.3 The number of degrees of freedom of a saturated steam boiler is one

1.4.2.1 Water Heater

Consider the heat exchanger, a water-heating process shown in Fig. 1.4. There are four variables namely :

u = inlet temperature

c = outlet temperature

w = water flow rate

m = heat input rate and hence $n_v = 4$

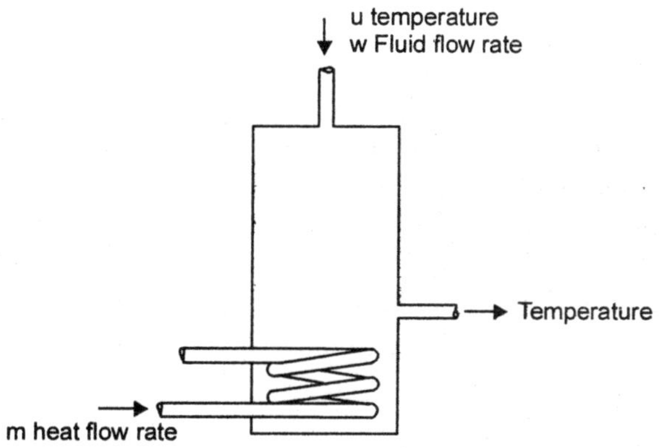

Fig. 1.4 A water-heating process

There is one defining equation obtained from conservation of energy (first law of thermodynamics). Therefore $n_e = 1$. Hence the number of degrees of freedom in this case are $n = n_u - n_e = 4 - 1 = 3$.

Whenever the heat input is through steam, the heat exchanger can be called as steam heater also.

1.4.2.2 Liquid to Liquid Heat Exchanger

Consider the Liquid to Liquid heat exchanger shown in Fig. 1.5. There are six variables as listed ($n_v = 6$). Here also, the defining equation is only one obtained from conservation of energy (first law of thermodynamics) ie n_e =1. Hence the number of degrees of freedom in this case are

$$n = n_v - n_e = 6 - 1 = 5$$

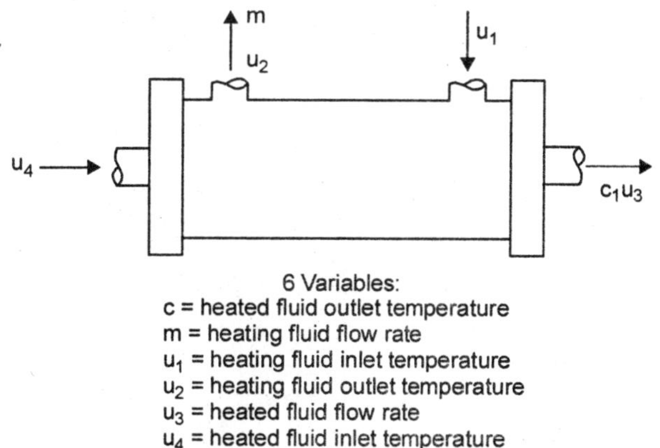

Fig. 1.5 The degrees of freedom of a liquid-to-liquid heat exchanger

This means that if five variables are held constant, this will result in a constant state for the sixth variable, the outlet temperature (c). Therefore, the maximum number of automatic controllers that can be placed on this process is five. Usually one would not consider this option of using five controllers, but one might use only a single control loop.

One would select the 'controlled variable (*c*)' to be the process property which is most important, because it has the most impact on plant productivity, safety, or product quality. One would select the 'manipulated variable (*m*)' to be that process input variable which has the most direct influence on the controlled variable (*c*) which in this case the flow rate of the heating fluid. The other 'load variables (u_1 to u_4)' are uncontrolled independent variables, which, when they change, will upset the control system, and their effects can only be corrected in a 'feedback' manner. This means that a change in load variables is not responded to until they have upset the controlled variable (*c*).

1.4.2.3 Binary Distillation

When the process is more involved, such as in the case of binary distillation, the calculation of the degrees of freedom also becomes more involved. Fig. 1.6 lists 14 variables of this process, but they are not all independent. There being two components and two phases at the bottom, feed and overhead, Gibb's law states that only two of the three variables (Pressure, temperature, and Composition) are independent. $n = n_c - n_p + 2 = 2 - 2 + 2 = 2$ which is true for bottom, overhead and feed.] Therefore the number of independent variables is only eleven. The number of defining equations is two [The conservation of mass and Energy], and, therefore the number of degrees of freedom for this process is $n = n_u - n_e = 11 - 2 = 9$. Consequently, not more than nine automatic controllers can be placed on this process.

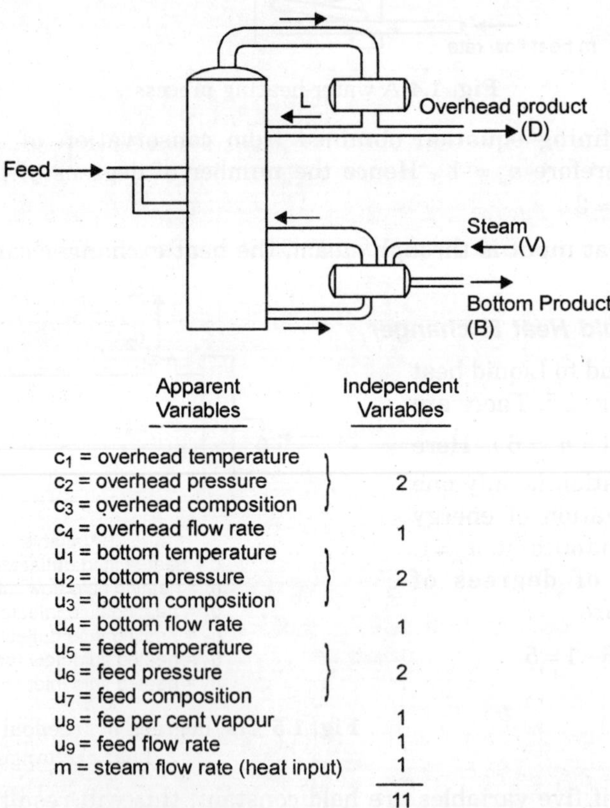

Apparent Variables	Independent Variables
c_1 = overhead temperature c_2 = overhead pressure c_3 = overhead composition	2
c_4 = overhead flow rate	1
u_1 = bottom temperature u_2 = bottom pressure u_3 = bottom composition	2
u_4 = bottom flow rate	1
u_5 = feed temperature u_6 = feed pressure u_7 = feed composition	2
u_8 = fee per cent vapour	1
u_9 = feed flow rate	1
m = steam flow rate (heat input)	1
	11

Fig. 1.6 In a Binary distillation process the number of independent variables is 11 and the number of degrees of freedom is 9

1.5 NEED FOR AUTOMATIC PROCESS CONTROL IN INDUSTRY

The first use of automatic control seems to have been the flyball governor on Watt's steam engine during 1775. This device was employed to regulate the speed of the engine by manipulating the steam flow by means of a valve.

Nowadays automatic process control devices are used in almost every phase of industrial operations. They are commonly employed in the following areas.

1. Processing industries such as petroleum, chemical, steel, power and food for the control of assembly operations, work flow, heat treating, and similar variables.
2. Goods manufacturers such as automobile parts, refrigerators, and electronic equipment like television sets, radio etc for the control of assembly operations, work flow, heat treating, and similar operations.
3. Transportation systems such as railways, airplanes, free missiles and ships.
4. Power machines such as machine tools, compressors and pumps, prime movers, and electric power-supply units for the control of position, speed and power.

Automatic control devices are used because their application results in economical behaviour of the system under control or because they are required for humanitarian purposes. Some of the needs and advantages of a process automation are listed below.

(*a*) Increase in productivity (increase in quantity or number of products) : Helps to increase the efficiency of both men and machine.

(*b*) Improvement in quality of products by meeting the product specifications overcoming operational constraints.

(*c*) Improvement in consistency of the product dimensions, performance and the length of service.

(*d*) Economical improvement by way of savings in processing raw materials, savings in energy, effective utilization of capital and human labour etc.

(*e*) Minimise/suppress the influence of external disturbances on the process.

(*f*) Ensure the stability of the process.

(*g*) Optimize the performance of the process.

(*h*) Meet environmental regulations.

All these factors generally lead to an increase in quantity produced with improved consistent quality at less cost taking the safety of men and machine into account and keeping the healthy environment.

1.6 MATHEMATICAL MODELING OF PROCESSES

To analyse the behaviour of a process, a mathematical representation of the physical and chemical phenomenon taking place in it is essential and this representation constitutes the mathematical model. The activities leading to the construction of the model is called modeling.

The main uses of mathematical modeling are

1. To improve understanding of the process
2. To optimize process design and hence operating conditions.
3. To design a control strategy for the process

4. To train operating personnel.
5. The model-based control action is 'intelligent' and helps in achieving uniformity, disturbance rejection, and set point tracking, all of which translate into better process economics.

1.6.1 Modeling Approaches

To have a simple description of how the process reacts to various inputs, mathematical model is required for the control engineers to develop a suitable control strategy.

The three dominant modeling approaches are :

1. Transfer function
2. Time series and
3. Non linear phenomenological

Transfer function models are based on open-loop laplace transform descriptions of the process response to a step input, and have been the traditional control modeling approach. Their familiarity and the simplicity of the resulting model based control are advantages which offset their limitations of linear and simplistic dynamic modeling.

Time series models represent the open-loop response of the process with a vector of impulses which are empirically determined and consist of 30 or so elements. The precision of the performance of the modeled dynamic process is an advantage which offsets the limiting assumption that the process is linear, as well as the need for using matrix / vector algebra. This is the most common modeling approach in the industrial use of model-based control today.

Non linear phenomenological models are design-type simulators. For markedly nonlinear or non stationary process applications, their control intelligence can offset the disadvantages of their modeling and computational complexity.

In all cases, control is only as good as the modeled representation of the process. Initializing the controller with the model which has been validated by process testing is the first and foremost implementation step.

1.6.2 Laplace Transforms

Laplace transforms can be used in process control for :

1. Solution of differential equations (Linear)
2. Analysis of linear control system
3. Prediction of transient response for different inputs.

The Laplace transform equation is defined as follows :

$$L\big(f(t)\big) = \int_0^\infty f(t).e^{-st}dt. = F(s) \tag{1.3}$$

Example :

Find out the transient response for $u(t)$ = Unit step at $t > 0$ if the input (u) and output (y) relation of a process is governed by the following equation.

$$\frac{d^3y}{dt^3}+6\frac{d^2y}{dt^2}+11\frac{dy}{dt}+6y=4\ \frac{du}{dt}+2x \tag{1.4}$$

and at $t = 0$, $y\ (0) = y^{\prime}(0) = y^{\prime\prime}(0)=0$ and $\frac{du}{dt}=0$

Steps involved

1. Take Laplace transform
2. Factorise using partial fraction decomposition
3. Take inverse Laplace transform.

Step 1 : Take Laplace transform (Note Zero initial conditions)

$$S^3\ Y(s) + 6S^2\ Y(s) + 11\ S\ Y(s) + 6\ Y(s) = 4\ s\ U(s) + 2\ U(s) \tag{1.5}$$

$$Y(s) = \frac{4s+2}{s^3+6s^2+11s+6}U(s) = G(s).U(s) \tag{1.6}$$

$$U(s) = 1/S \qquad [U(t) = \text{Unit step at } t > 0\ ] \tag{1.7}$$

$$\text{Transfer function} = \text{G(s)} = \frac{Y(s)}{U(s)}=\frac{4s+2}{s^3+6s^2+11s+6} \tag{1.8}$$

Step 2 : Factor denominator of Y(s)

$$Y(s) = \frac{4s+2}{s^3+6s^2+11s+6}x\frac{1}{s} \tag{1.9}$$

Denominator of $Y(s) = S\ \left(s^3+6s^2+11s+6\right)$

$$= S\ (S + 1)\ (S + 2)\ (S + 3)$$

Using partial fraction decomposition

$$\frac{4s+2}{s(s+1)(s+2)(s+3)} = \frac{a1}{s}+\frac{a2}{s+1}+\frac{a3}{s+2}+\frac{a4}{s+3}$$

Multiply by S, Set $S = 0$

$$\left.\frac{4s+2}{(s+1)(s+2)(s+3)}\right| S = 0 = a_1+s[\frac{a_2}{s+1}+\frac{a_3}{s+2}+\frac{a_4}{s+3}]$$

$$\frac{2}{1\times2\times3} = a_1=\frac{1}{3}$$

For a_2, Multiply by $(S + 1)$, Set $S = -1$ etc for a_3 and a_4

$$a_2 = 1,\ a_3 = -3,\ a_4 = 5/3$$

Step 3 : Take inverse of laplace transform

$$y(t) = \frac{1}{3} + e^{-t} - 3e^{-2t} + \frac{5}{3}e^{-3t} \tag{1.10}$$

$$t \to \infty \qquad y(t) \to \frac{1}{3}$$

1.6.3 State Variables and State equations

In order to characterize a processing systems such as tank heater, batch reactor, distillation column, heat exchanger etc and their behaviour we need :

1. A Set of fundamental dependent quantities whose values will describe the natural state of a given system.
2. A Set of equations in the variables above which will describe how the natural state of the given system changes with time.

For most of the processing systems of interest there are only three such fundamental quantities.

1. Mass 2. Energy and 3. Momentum

Quite often, though, the fundamental dependent variables cannot be measured directly and conveniently. In such cases we select other variables which can be measured conveniently, and when grouped approximately they determine the value of the fundamental variables. Thus mass, energy and momentum can be characterized by variables such as temperature, pressure, flowrate, density and concentration. These characterizing variables are called 'Sstate variables' and their values define the 'state' of a processing system.

The equations that relate the state variables (dependent variables) to the various independent variables are derived from application of the 'conservation principle' on the 'fundamental quantities' and are called state equations.

The 'principle of conservation' of a quantity S, for the considered period of time, states that:

$$A_a = F_{in} - F_{out} + A_g - A_c \tag{1.11}$$

where A_a = Accumulation of S within the system

A_g = Amount of S generated within the system

A_c = Amount of S consumed within the system

F_{in} = Flow of S into the System

F_{out} = Flow of S out of the system.

The quantity S can be any of the following fundamental quantities :

1. Total mass
2. Mass of individual components
3. Total energy
4. Momentum

[It should be remembered that for the physical and chemical processes, the total mass and total energy cannot be generated from nothing ; neither do they disappear.]

Let us review now the forms used most often for the balance equations. Consider the system shown in Fig. 1.7. We have :

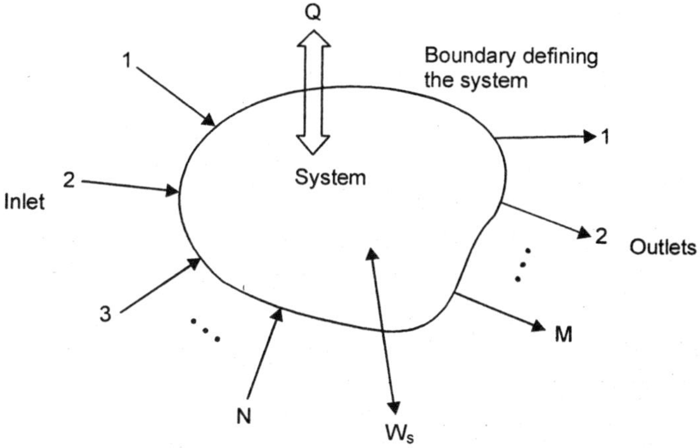

Fig. 1.7 A general system and its interactions with the external world

Total mass balance

$$\frac{d(\rho V)}{dt} = \sum_{i=inlet} \rho_i F_i - \sum_{j=outlet} \rho_j F_j \tag{1.12}$$

Mass balance on component A

$$\frac{d(n_A)}{dt} = \frac{d(C_A V)}{dt} = \sum_{i=inlet} C_{Ai} F_i - \sum_{j=outlet} C_{Aj} F_j \pm rV \tag{1.13}$$

Total energy balance

$$\frac{dE}{dt} = \frac{d(U+K+P)}{dt} = \sum_{i=inlet} \rho_i F_i h_i - \sum_{j=outlet} \rho_j F_j h_j \pm Q \pm W_S \tag{1.14}$$

where ρ = density of the material in the system

ρ_i = density of the material in the ith inlet stream

ρ_j = density of the material in the jth outlet stream.

V = Total volume of the system.

F_i = Volumetric flowrate of the ith inlet stream

F_j = Volumetric flowrate of the jth outlet stream

n_A = number of moles of component A in the system

C_A = molar concentration (moles/volume) of A in the system

C_{Ai} = molar concentration of A in the ith inlet stream

C_{Aj} = molar concentration of A in the jth outlet stream

r = reaction rate per unit volume for component A in the system

h_i = specific enthalpy of the material in the ith inlet stream

h_j = specific enthalpy of the material in the jth outlet stream

U, K, P = Internal, Kinetic and potential energies of the system respectively.

Q = Amount of Heat exchanged between the system and its surroundings per unit time.

W_s = Shaft work exchanged between the system and its surroundings per unit time.

By convention, a quantity is considered positive if it flows into the system and negative if it flows out. The state equations with the associated state variables constitute the 'mathematical model' of a process, which yields the dynamic or static behaviour of the process. The application of the conservation principle as defined by the above equations will yield a set of differential equations with the fundamental quantities as the dependent variables and time as the independent variable. The solution of the fundamental quantities, or equivalently, the state variables, change with time ; that is, it will determine the 'dynamic behaviour' of the process.

If the state variables donot change with time, the process is said to be at 'steady state'. In this case, the rate of accumulation of a fundamental quantity S per unit time is zero, and the resulting balances yield a set of algebraic equations.

1.6.4 Additional Elements of the Mathematical Models

In addition to the balance equations, we need other relationships to express thermodynamic equilibria, reaction rates, transport rates for heat, mass, momentum and the mathematical modeling of various chemical and/or physical processes can be classified as follows :

Transport rate equations

They are needed to describe the rate of mass, energy, and momentum transfer between a system and its surroundings.

Kinetic rate equations

They are needed to describe the rates of chemical reactions taking place in a system.

Reaction and Phase equilibria relationships :

These are needed to describe the equilibrium situations reached during a chemical reaction or by two or more phases.

Equations of state

Equations of state are needed to describe the relationship among the intensive variables describing the thermodynamic state of a system. The ideal gas law and the van der Waals equation are two typical equations of state for gaseous systems.

1.6.5 Dead Time

It cannot be assumed that whenever a change takes place in one of the input variables of a system its effect is instantaneously observed in the state variables and the outputs. Whenever an input variable of a system changes, there is a time interval (short or long) during which no effect is observed on the outputs of the system. The time interval is called 'dead time' or 'transportation lag', or 'pure delay', or 'distance-velocity lag'. The dead time is an important element in the mathematical modeling of processes and has a serious impact on the design of effective controllers. The presence of dead time can very easily destabilize the dynamic behaviour of a controlled system.

1.7 FIRST ORDER PROCESS SYSTEMS

A first order system is one whose output y(t) is modeled by a first order differential equation. Thus in the case of linear (or linearised) system, we have

$$a_1 \frac{dy}{dt} + a_0 y = b\,f(t) \tag{1.15}$$

where $f(t)$ is the input (forcing function).

If $a_0 \neq 0$, then the above equation yields

$$\frac{a_1}{a_0}\frac{dy}{dt} + y = \frac{b}{a_0} f(t)$$

$$\tau_P \frac{dy}{dt} + y = K_P f(t) \tag{1.16}$$

Where $\tau_P \left(= \frac{a_1}{a_0} \right)$ is known as the 'time constant' of the process and $K_P \left(= \frac{b}{a_0} \right)$ is called the 'steady-state gain' or 'static gain' or simply the 'gain' of the process.

If $y(t)$ and $f(t)$ are interms of deviation variables around a steady state, the initial conditions are $y(0) = 0$ and $f(0) = 0$. Then the transfer function of a first order process is given by,

$$G(s) = \frac{\bar{y}(s)}{\bar{f}(s)} = \frac{K_P}{\tau_P s + 1} \tag{1.17}$$

A first order process with a transfer function given above is also known as 'first-order lag', 'linear lag' or 'exponential transfer lag'.

If, on the other hand, $a_0 = 0$, then we get

$\frac{dy}{dt} = \frac{b}{a_1} f(t) = K'_P f(t)$ which gives a transfer function,

$$G(s) = \frac{\bar{y}(s)}{\bar{f}(s)} = \frac{K'_P}{s} \tag{1.18}$$

In such case, the process is called 'purely capacitive' or 'pure integrator'.

1.7.1 Processes Modeled as First Order Systems

The first order processes are characterized by :

1. Their capacity to store material, energy, or momentum.
2. The resistance associated with the flow of mass, energy, or momentum in reaching the capacity.

Thus the dynamic response of tanks that have the capacity to store liquids or gases can be modeled as first order. The resistance is associated with the pumps, valves, weirs and pipes which are attached to the inflowing or outflowing liquids or gases. Similarly the temperature response of solid, liquid, or gaseous systems which can store thermal energy (thermal capacity, C_p) is modeled as first order. For such systems the resistance is associated with the transfer of heat through walls, liquids, or gases. In other words, a process that possesses a capacity to store mass or energy and thus act as a buffer between inflowing and outflowing streams will be modeled as a first order system.

1.7.2 First Order Level Process—Liquid Storage Tank

Consider the tank shown in Fig. 1.8

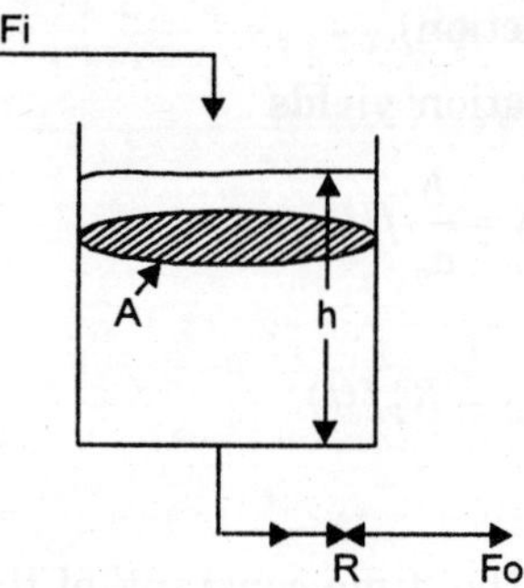

Fig. 1.8 Systems with capacity for mass storage—First-order lag

F_i = Inlet volumetric flow rate (m^3/sec)

F_0 = Outlet volumetric flow rate (m^3/sec)

A = Cross sectional area (m^2)

R = Resistance to outlet flow caused by pipe, valve or wier etc.

h = Liquid head in the tank

Assumption : F_o is linearly related to 'h' through 'R'.

$$F_o = \frac{h}{R} = \frac{\text{Driving Force of Flow}}{\text{Resistance to Flow}} \tag{1.19}$$

At any point of time, the tank has the capacity to store mass. The total mass balance gives

Input flow rate–Output flow rate = Rate of accumulation

$$Fi - F_o = \frac{d(Ah)}{dt} = A.\frac{dh}{dt} = F_i - \frac{h}{R}$$

or

$$AR\frac{dh}{dt} + h = RF_i \tag{1.20}$$

Where A = Cross-sectional area of the tank

At steady state,

$$h_s = R\ F_{is} \tag{1.21}$$

From equations (1.20) and (1.21), we get the following equation in terms of deviation variables.

$$AR\frac{dh'}{dt} + h' = RF'_i$$

(1.22)

Where $h' = h - h_s$, and $F'_i = F_i - F_{is}$

Let $\tau_P = AR$ = Time constant of the process.

$K_p = R$ = Steady – state gain of the process.

Then the transfer function is given by :

$$G(s) = \frac{\overline{h'}(s)}{\overline{F'}(s)} = \frac{K_P}{\tau_P s + 1} \tag{1.23}$$

The cross-sectional area of the tank 'A' is a measure of its capacitance to store mass. Thus the larger the value of 'A', the larger the storage capacity of the tank. Since $\tau_P = AR$, for the tank.

Time constant = Storage capacitance × Resistance to flow.

From Equation (1.19), we get

$$\frac{\overline{h'}(s)}{\overline{F_0}'(s)} = R \tag{1.24}$$

$$G(s) = \frac{\overline{F_0}'(s)}{\overline{F'}(s)} = \frac{1}{\tau_P s + 1} \qquad (\because K_p = R) \tag{1.25}$$

The term $\frac{1}{\tau_P s + 1}$ is called first order lag and is typical of first order processes.

1.7.3 Dynamic Response of a First-Order Lag System

The transfer function for such a system is given by the equation {Refer Eqns. 1.17 and 1.23}

$$G(s) = \frac{\overline{y}(s)}{\overline{f}(s)} = \frac{K_P}{\tau_P s + 1}$$

Let us examine how it responds to a unit step change in $f(t)$. Since $\overline{f}(s) = \frac{1}{s}$, from the above equation we get

$$\overline{y}(s) = \frac{K_P}{s(\tau_P s + 1)} = \frac{K_P}{s} - \frac{K_P \tau_P}{\tau_P s + 1} \tag{1.26}$$

Inverting equation (1.26), we get

$$y(t) = K_P(1 - e^{-\frac{t}{\tau_P}})$$

If the step change in $f(t)$ were of magnitude A, the response would be

$$y(t) = A.\ K_p(1 - e^{-\frac{t}{\tau_P}}) \tag{1.27}$$

Fig. 1.9 shows how $y(t)$ changes with time. The Plot is in terms if dimensionless coordinates $\frac{y(t)}{AK_P}$ versus $\frac{t}{\tau_P}$ and as such can be used to determine the response of any typical first order system, independently of the particular values of A, K_P and τ_P .

Several features of the plot in Fig. 1.9 are characteristic of the response of first order systems and these features are :

1. A first order lag process is 'self-regulating' and it reaches a new steady state. Refering to the tank level system, when the inlet flowrate increases by unit step, the liquid level goes up. As the liquid level goes up, the hydrostatic pressure increases, which inturn increases the output flow rate F_0 [Refer Eqn. 1.19]. This action works towards the restoration of an equilibrium state (steady state).

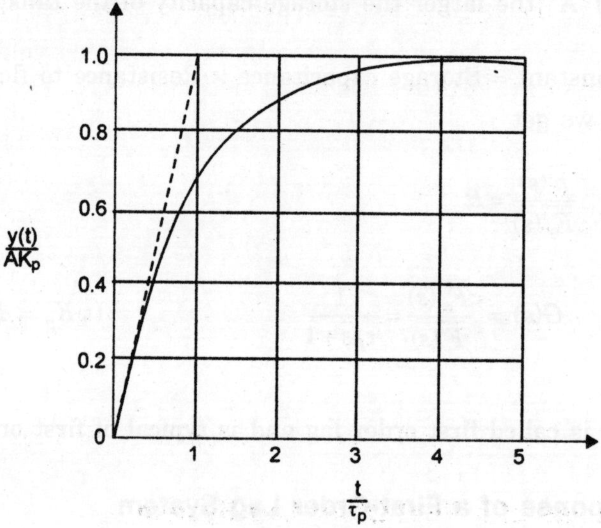

Fig. 1.9 Dimensionless response of first-order lag to step input change

2. The slope of the response at $t = 0$ is equal to 1.

$$\left.\frac{d[y(t)/AK_P]}{d\left(\frac{t}{\tau_P}\right)}\right|_{t=0} \quad (e^{-\frac{t}{\tau_P}})_{t=0} = 1$$

This implies that if the initial rate of change of $y(t)$ were to be maintained, the response should reach its final value in one time constant (see the dotted line in the Fig. 1.19). The corollary conclusions are :

The Smaller the value of the time constant τ_P, the steeper the initial response of the system.

The time constant τ_P of a process is a measure of the time necessary for the process to adjust to a change in its input'.

3. The value of the response $y(t)$ reaches 63.2 % of its final value when the time elapsed is equal to one time constant, τ_p. Subsequently we have $y(t)$ values for $2\tau_p$ = 86.5 %, $3\tau_p$ =95%. And $4\tau_p$ =98%. Thus, after four time constants, the response has essentially reached its ultimate value.
4. The ultimate value of the response (its value at the new steady state) is equal to K_p for a unit step change in the input, or $A.K_p$ for a step size A. This is easily seen from the equation (1.27), which yields $y -> A.K_p$ as $t -> \infty$. This characteristic explains the name 'steady state' or 'static gain' given to the parameter K_p since for any step change Δ (input) in the input, the resulting change in the output steady state is given by

$$\Delta \text{ (output)} = K_p \, \Delta \text{ (input)} \tag{1.28}$$

It is clear from equation 1.28 that, to effect the same change in the output, we need:

(*i*) A small change in the input if K_p is large (very sensitive system)

(*ii*) A large change in the input if K_p is small.

5. The effect of time constant τ_p (determined by the cross sectional area of the tank, A) and static gain K_p (determined by the resistance to the flow of the liquid, R) in the response of first order lag systems is given in Fig. 1.10.

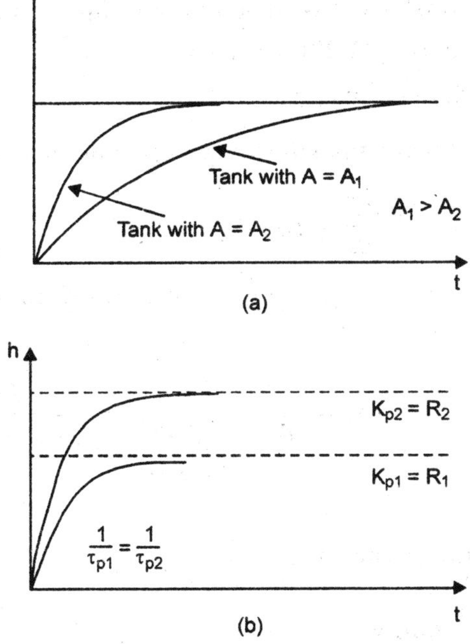

Fig. 1.10 Effect of (*a*) time constant and (*b*) static gain, in the response of first-order lag systems

1.7.4 First Order System with a Capacity for Energy Storage (*Stirred Tank Heater—A Thermal System*)

The liquid in a tank is heated with saturated steam, which flows through a coiled tube immersed in the liquid. (Refer Fig. 1.11). The energy balance at transient state is given by the equation (1.29).

$$V\rho\, C_P \frac{dT}{dt} = Q = UA_t (T_{st} - T) \qquad (1.29)$$

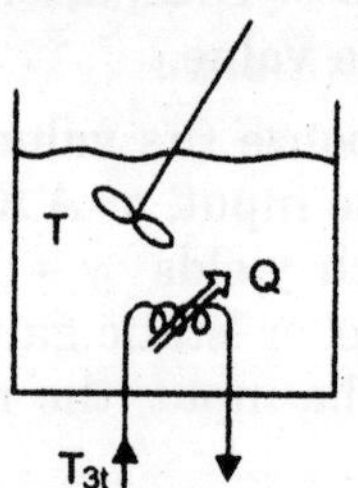

Fig. 1.11 System with capacity for energy storage

where V = Volume of liquid in the tank in m^3

ρ = Liquid's density in Kg/m^3

C_P = heat capacity of the liquid in $KJ/Kg\ K$

U = overall heat transfer coefficient between steam and liquid in KW/m^2K

A_t = Total heat transfer Area (surface area of the coil immersed in the liquid) in m^2

T_{st} = Temperature of the saturated steam in K.

Q = Rate of heat transfer between steam and liquid in KW.

At steady state the equation (1.29) becomes

$$UA_t (T_{sts} - T_s) = 0 \qquad (1.30)$$

Subtract (1.30) from (1.29) and take the following equation in terms of deviation variables :

$$V\rho\, C_P \frac{dT'}{dt} = UA_t (T_{st}' - T') \qquad (1.31)$$

where $T' = T - T_s$ and $T_{st}' = T_{st} - T_{st.s}$. The Laplace transform of equation (1.31) will yield the following transfer function:

$$G(s) = \frac{\overline{T}'(s)}{\overline{T}_{st}'(s)} = \frac{1}{\frac{V\rho C_P}{UA_t} s + 1} = \frac{K_P}{\tau_P s + 1} \qquad (1.32)$$

where τ_P = Time constant of the process = $\dfrac{V\rho C_P}{UA_t}$

K_P = Steady State gain =1.

The equation (1.32) clearly demonstrates that this is a first order lag system. The system possesses capacity to store thermal energy and a resistance to the flow of heat characterized by U. The capacity to store thermal energy is measured by the value of the term $V\rho\, C_P$. The resistance to the flow of heat from the steam to the liquid is expressed by the term $\dfrac{1}{UA_t}$. Therefore, we notice that the time constant of this system is given by the same equation as that of the tank level system discussed earlier under (1.7.2) :

$$\text{Time Constant} = \tau_P = \frac{V\rho C_P}{UA_t} = (\text{Storage capacitance}) \times (\text{Resistance to flow})$$

1.7.5 First Order Thermal Process (Mercury Thermometer)

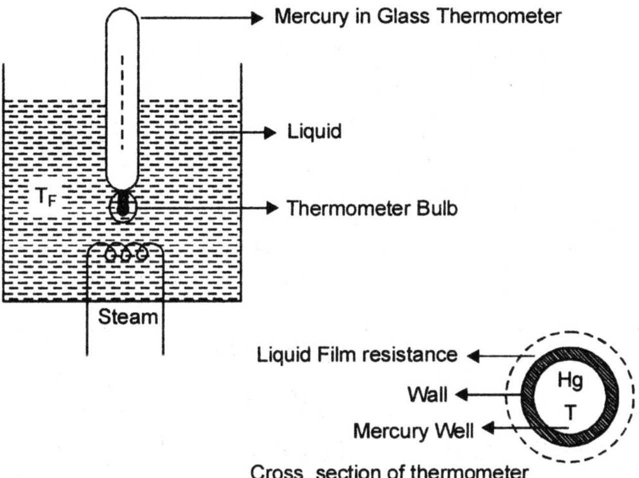

Fig. 1.12 Temperature measurement system

Consider a mercury-in-glass thermometer placed in a liquid tank to measure the temperature of the liquid which is heated by steam through a coil system. The temperature of the liquid (T_F) varies with time as shown in Fig. 1.12. T is the temperature of the mercury in the well of the Thermometer. The following assumptions are made to determine the transfer function relating the variation of the thermometer reading (T) for change in the temperature of the liquid (T_F).

1. The expansion or contraction of the glass walled well containing mercury is negligible (that means the resistance offered by glass wall for heat transfer is negligible)
2. The liquid film surrounding the bulb is the only resistance to the heat transfer.
3. The mercury assumes isothermal condition throughout.

Applying unsteady state heat balance for the bulb, we get

Input heat rate – Output heat rate = Rate of heat accumulation

$$U\,A\,(T_F - T) - 0 = M\,C_P \frac{dT}{dt}$$

$$U\,A\,(T_F - T) = M\,C_P \frac{dT}{dt} \qquad (1.33)$$

where,

A = Surface area of the bulb for heat transfer in m^2

M = Mass of mercury in the bulb, kg

CP = Heat capacity of the mercury in kJ/kgK

U = Film heat transfer coefficient kW/m^2K

At steady state, the equation (1.33) can be rewritten as

$$U\,A\,(T_{FS} - T_S) = 0 \tag{1.34}$$

Subtracting Eqn (1.34) from Eqn (1.33)

$$U\,A\,[(T_F - T_{FS}) - (T - T_S)] = M\,C_p \frac{d(T - T_S)}{dt}$$

Defining the deviation variables,

$$T_F - T_{FS} = \overline{T}_F \text{ and } T - T_S = \overline{T} \text{ and}$$

Substituting in the above eqn. we get,

$$U\,A\,(\overline{T}_F - \overline{T}) = M\,C_p \frac{d\overline{T}}{dt}$$

$$\overline{T}_F - \overline{T} = \frac{M\,C_p}{UA} \cdot \frac{d\overline{T}}{dt} \tag{1.35}$$

Defining time constant τ_P for the Thermometer,

$$\tau_P = \frac{M\,C_p}{UA}$$

Equation (1.35) can be rewritten as

$$\overline{T}_F - \overline{T} = \tau_P \cdot \frac{d\overline{T}}{dt} \tag{1.36}$$

Taking Laplace transform, we get

$$\overline{T}_{F(S)} - \overline{T}_{(S)} = \tau_P \text{ s } \overline{T}_{(S)}$$

$$\overline{T}_{F(S)} = \overline{T}_{(S)}[1 + \tau_P s]$$

The transfer function $G(s) = \dfrac{\overline{T}_{(S)}}{\overline{T}_{F(S)}} = \dfrac{1}{1 + \tau_P s}$ and hence is a first order lag system.

1.7.6 First–Order Pressure Process

The pressure control is similar to that of tank level control in that the flow equations must generally be linearised to obtain the system transfer function. Here the following two types of pressure processes will be dealt with : 1. Gas storage tank and 2. Process with inlet and outlet resistances.

1.7.6.1 Gas Storage Tank (Refer Fig. 1.13)

It is assumed that gas at an inlet pressure of P_i (N/m^2) flows into a storage tank of volume V (m^3) at a rate of W_i (Kg /sec). The resistance of the inlet pipe is R (N sec/KgKm^2). ρ is the density of the gas in Kg /m^3 in the container and will vary with pressure P in the tank. Temperature is assumed constant. Making a mass balance,

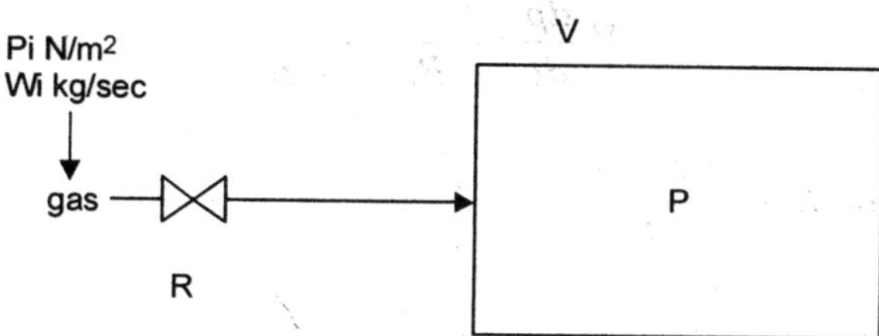

Fig. 1.13 Gas storage tank

$$Wi = \frac{d(V\rho)}{dt} = V\frac{d\rho}{dt} \tag{1.37}$$

If the gas is an ideal gas, $\rho = \dfrac{P}{R_g T}$

where R_g is the gas constant (Nm/KgK)

Also $W_i = \dfrac{P_i - P}{R}$

Hence equation (1.37) becomes

$$P_i - P = \left(\frac{RV}{R_g T}\right)\frac{dP}{dT} = \tau_P \frac{dP}{dt} \tag{1.38}$$

$$\left\{\text{where } \tau_P = \frac{RV}{RgT} = \left(\frac{\text{N Sec}}{\text{KgKm}^2}\right) \times \text{m}^3 \times \frac{\text{KgK}}{\text{Nm}} \times \frac{1}{\text{K}} = \text{sec}\right\}$$

Taking Laplace transform of equation (1.38),

$$P_{i(S)} = P_{(S)}\,[\tau_P s + 1], \;\therefore\; G(s) = \frac{P_{(S)}}{P_{i(S)}} = \frac{1}{\tau_P s + 1} \tag{1.39}$$

1.7.6.2 Pressure System with Two Resistances

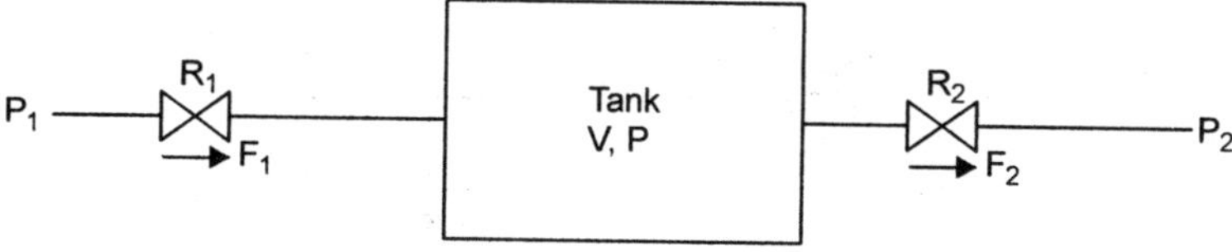

Fig. 1.14 A Pressure system with two resistances

Consider a pressure system with tank of volume V and varying pressure P at constant temperature. F_1 is inlet flow through resistance R_1 with source pressure P_1. F_2 is output flow through resistance R_2 and flowing out at pressure P_2. As the flows into and out of the tank are both influenced by the tank pressure, both flow resistances affect the time constant. Making a mass balance,

Accumulation in the tank = Input flow rate (F_1) – Output flow rate (F_2)

$$V.\frac{dp}{dt} = \frac{P_1 - P}{R_1} - \frac{P - P_2}{R_2} \tag{1.40}$$

where $R_1 = \dfrac{d(P_1 - P)}{dF_1}$ and $R_2 = \dfrac{d(P - P_2)}{dF_2}$

$$V.\frac{dp}{dt} + P\left(\frac{1}{R_1} + \frac{1}{R_2}\right) = \frac{P_1}{R_1} + \frac{P_2}{R_2}$$

$$V.\frac{dp}{dt} + P\left(\frac{R_1 + R_2}{R_1 R_2}\right) = \frac{P_1}{R_1} + \frac{P_2}{R_2}$$

$$V.\left(\frac{R_1 R_2}{R_1 + R_2}\right)\frac{dp}{dt} + P = \left(\frac{R_1 R_2}{R_1 + R_2}\right)\frac{P_1}{R_1} + \left(\frac{R_1 R_2}{R_1 + R_2}\right)\frac{P_2}{R_2}$$

$$\tau_P \frac{dP}{dt} + P = K_1 P_1 + K_2 P_2 \tag{1.41}$$

where $K_1 = \dfrac{R_2}{R_1 + R_2}$, $K_2 = \dfrac{R_1}{R_1 + R_2}$, $\tau_P = \dfrac{VR_1 R_2}{R_1 + R_2}$

Taking Laplace transform of equation (1.41),

$$\tau_P s\, P_{(S)} + P_{(S)} = K_1 P_{1(S)} + K_2 P_{2(S)}$$

$$P_{(S)}(1 + \tau_P s) = K_1 P_{1(S)} + K_2 P_{2(S)}$$

$$P_{(S)} = \frac{K_1}{1 + \tau_P s} P_{1(S)} + \frac{K_2}{1 + \tau_P s} P_{2(S)} \tag{1.42}$$

Equation (1.42) can be represented in a block diagram as below :

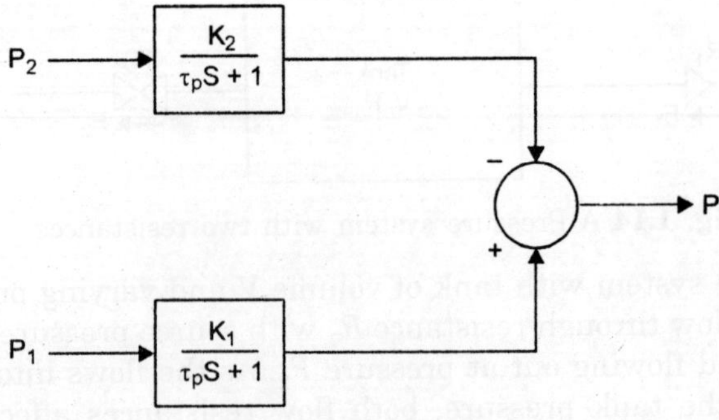

Fig. 1.15 Block Diagram of a pressure system with two resistances

The output pressure P (Tank pressure) depends on both P_1 and P_2. Either of them could be the controlled variable or the load variable or both could be load variables if the controller acts to change R_1 or R_2.

Note: If there are several inlets and outlets the system is still first order one. A first order system with multiple resistances (Refer Fig. 1.16) will have the following binding equation.

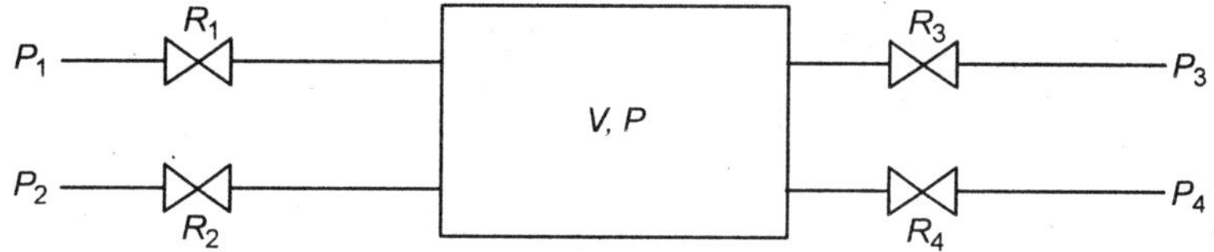

Fig. 1.16 Pressure System with several resistances

$$\tau_P \frac{dP}{dt} + P = K_1P_1 + K_2P_2 + K_3P_3 + K_4P_4 \tag{1.43}$$

where

$$\tau_P = \frac{V}{\Sigma\left(\frac{1}{R}\right)}$$

$$K_1 = \frac{1}{R_1\sum\left(\frac{1}{R}\right)}, \quad K_2 = \frac{1}{R_2\sum\left(\frac{1}{R}\right)} \text{ etc.}$$

1.7.7 Liquid Level Process with Constant Flow Outlet (*Pure Capacitive System or Integrating Process*)

Consider the liquid storage tank system discussed in section 1.7.2 with the following difference.

The effluent flow rate F_o is determined by a constant-displacement pump and not by the hydrostatic pressure of the liquid level h. (Refer Fig. 1.17)

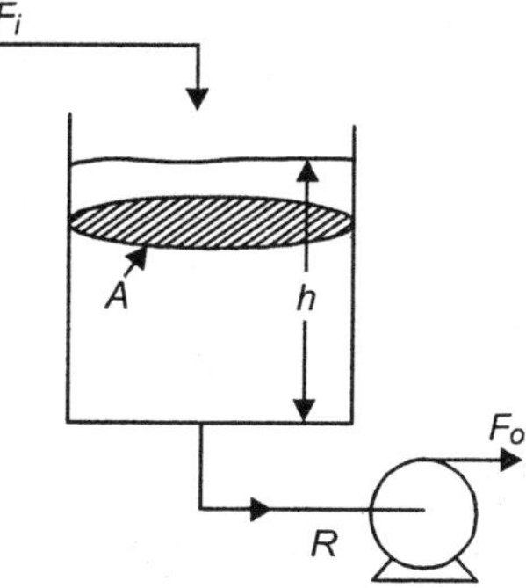

Fig. 1.17 Systems with capacity for mass storage – pure capacitive

In such case the total balance around the tank yields

$$A\frac{dh}{dt} = F_i - F_o \tag{1.44}$$

At steady state $$0 = F_{is} - F_o \tag{1.45}$$

Subtracting equation (1.45) from (1.44) and taking the equation in terms of deviation variables we get,

$$A\frac{dh'}{dt} = F_i'$$

Which yields the following transfer function

$$G(s) = \frac{\overline{h'}(s)}{\overline{F_i'}(s)} = \frac{1}{AS} \tag{1.46}$$

The 's' in the denominator denotes an integrating process (pure capacitive process).

1.7.8 Dynamic Response of a Pure Capacitive Process

The transfer function for such process is given by the equation (1.18)

$$G(s) = \frac{\overline{y}(s)}{\overline{f}(s)} = \frac{K'p}{s} \tag{1.18}$$

When $f(t)$ undergoes a unit step change, $y(t)$ changes with time as below :

$$f(t) = 1 \quad \text{for } t > 0$$

$$\overline{f}(s) = \frac{1}{s}$$

Therefore the equation (1.18) yields

$$\overline{y}(s) = \frac{K'p}{s^2}$$

and after inversion we get,

$$y(t) = K'pt$$

The output grows linearly with time in an unbounded fashion as in Fig. 1.18. Thus

$$y(t) \to \infty \quad \text{as} \quad t \to \infty$$

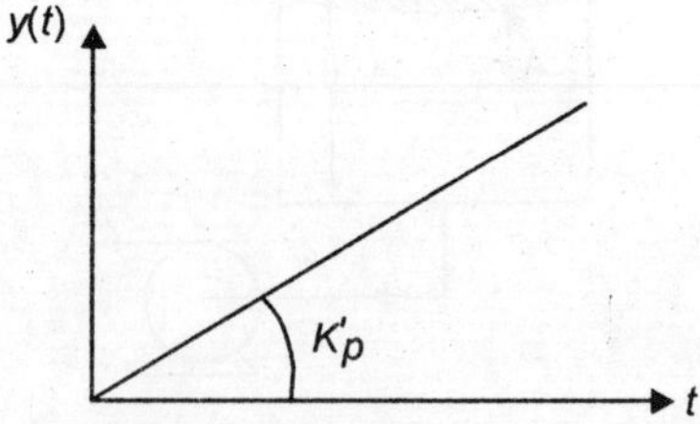

Fig. 1.18 Unbounded response of pure capacitive process

Such response, characteristic of a pure capacitive process, lends the name 'pure integrator' because it behaves as if there were an integrator between its input and output. A pure capacitive process will cause serious control problems, because it cannot balance itself. In the tank discussed under section 1.7.7, we can adjust manually the speed of the constant – displacement pump, so as to balance the flow coming in and thus keep the level constant. But any small change in the flow rate of the inlet steam will make the tank flood or run dry. This attribute is known as 'non-self-regulation'.

Processes with integrating action mostly commonly encountered in a chemical process are tanks with liquids, vessels with gases, inventory systems for raw materials or products, and so on.

1.7.9 Electrical Analogy of First Order Process

Fig 1.19 gives a circuit containing one capacitor – one resistor combination.

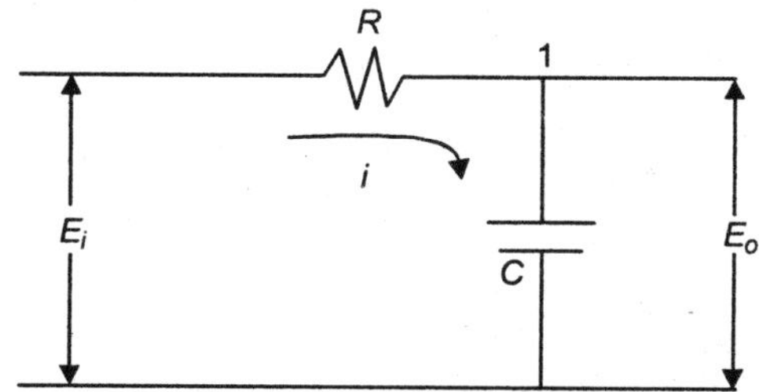

Fig 1.19 Electrical analogy of I order process

Applying Kirchoff's law,

$$E_0 = E_i - iR$$

$$= E_i - C\left(\frac{dE_0}{dt}\right)R$$

or $$RC\left(\frac{dE_0}{dt}\right) + E_0 = E_i$$

Taking Laplace transform,

$$\frac{E_{O(S)}}{E_{i(S)}} = \frac{1}{RCS+1} = \frac{1}{\tau_P s+1}$$

Where $\tau_P = RC$ has units of time.

1.7.10 First Order Systems with Variable Time Constant and Variable Gain

So far in our discussions we assumed that the coefficients of the first order differential equation [Equation 1.15] were constant. This led to the conclusion that the time constant τ_P and steady state gain K_p of the process were constant. But this is not true for a large number of components in a chemical process. As a matter of fact, in a chemical plant, we will more often encounter processes with variable time constants and gains than not.

1.7.10.1 Tank Level Control System with Variable Time Constant and Gain

For the tank system discussed in section 1.7.2, assume that the effluent flow rate, F_0, is not a linear function of the liquid level, but is given by the following relationship which holds for turbulent flow:

$$F_o = \beta\sqrt{h} \text{ where } \beta = \text{constant}$$

Then the material balance yields the following nonlinear equation :

$$A\frac{dh}{dt} + \beta\sqrt{h} = F_i$$

Linearise this equation around a steady state and put it in terms of deviation variables :

$$A\frac{dh'}{dt} + \frac{\beta}{2\sqrt{hs}}h' = F_i' \tag{1.47}$$

or

$$\tau_P\frac{dh'}{dt} + h' = K_P F_i'$$

Where $\tau_P = 2A\frac{\sqrt{hs}}{\beta}$ and $K_P = 2\frac{\sqrt{hs}}{\beta}$. We notice that both the time constant τ_P and the steady-state gain KP depend upon steady state value of the liquid level, h_s. Since we can vary the value of h, by varying the steady-state value of the inlet flow rate F_{is}, we conclude that the system has variable time constant and static gain.

1.7.10.2 Heater with Variable Time Constant

Let us consider the heater system discussed in section 1.7.4. The time constant and the static gain were found to be

$$\tau_P = \frac{V\rho C_P}{U\,A_t} \text{ and } K_P = 1$$

The overall heat transfer coefficient, U, does not remain the same for a long period of operation. Corrosion, dirt, or various other solids deposited on the internal or external surfaces of the heating coil result in a gradual decrease of the heat transfer coefficient. This, in turn, will cause the time constant of the system to vary. This example is characteristic of what can happen to even simple first-order systems.

1.7.10.3 Handling of First-order systems with variable time constant and static gains

We can assume that such systems possess constant time constants and static gains for a certain limited period of time only. At the end of such a period we will change the values of τ_p and K_p and consider that we have a new first order system with new but constant τ_p and K_p, which will be changed again at the end of the next period. Such an 'adaptive procedure' can be used successfully if the time constant and the static gain of a process change slowly, in which case the time period of relatively constant values is rather long.

1.7.11 Response of First Order Systems to Different Disturbances (*Different Forcing Functions*)

The five standard disturbances or forcing functions considered normally for discussions are 1.) Step, 2) Ramp, 3) Impulse, 4) Pulse and 5) Sinusoidal. The general first order transfer function.

$$G(s) = \frac{K_P}{\tau_P s + 1} \text{ is considered for the purpose.}$$

The dynamic response of the system for a unit step change is already discussed in the section 1.7.3. Hence the response for the remaining four forcing functions will be discussed here.

1.7.11.1 Ramp Forcing Function

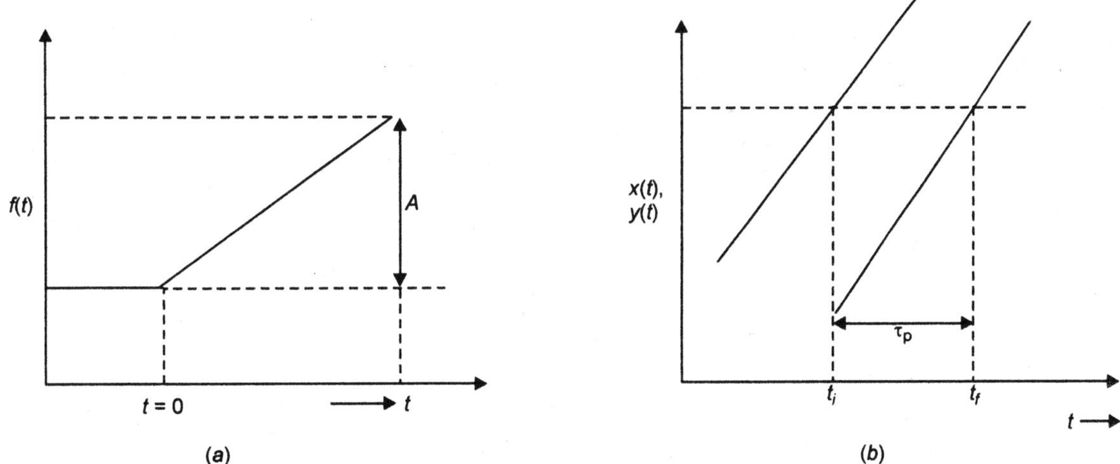

Fig. 1.20 Response of first order system for a Ramp disturbance

In a storage tank, if the inlet flow rate is gradually increasing, then it is called a ramp change. If the rate of change is constant it is called linear ramp. We will discuss here the response of a first order system with a linear ramp forcing function in the input. (Refer Fig. 1.20)

The function,

$$f(t) = 0 \text{ when } t < 0$$

$$= M\, t \text{ when } t \geq 0$$

Slope,

$$M = \frac{A}{t}$$

$$\text{Laplace transform of } f(t) = \frac{A}{s^2} = X(s)$$

$$\text{The transfer function, } G(s) = \frac{Y(s)}{X(s)} = \frac{1}{\tau_P s + 1}$$

$$Y(s) = \frac{\left(\frac{A}{\tau_P}\right)}{s^2(s+\frac{1}{\tau_P})} = \frac{C_1}{s+\frac{1}{\tau_P}} + \frac{C_2}{s} + \frac{C_3}{s^2} \quad (1.48)$$

$$\frac{A}{\tau_P} = C_1 s^2 + C_2 s(s+\frac{1}{\tau_P}) + C_3(s+\frac{1}{\tau_P})$$

Putting $s = 0$, we get $\frac{A}{\tau_P} = \frac{C_3}{\tau_P}$ or $C_3 = A$

Putting $s = -\frac{1}{\tau_P}$, we get $\frac{A}{\tau_P} = \frac{C_1}{\tau_P{}^2}$ or $C_1 = A\tau_P$

Differentiating equation (1.48) wrt 's', we get

$$0 = 2\,C_1 s + C_2\left[s(1) + \left(s + \frac{1}{\tau_P}\right)(1)\right] + C_3$$

$$0 = 2\,C_1 s + 2C_2 s + \frac{C_2}{\tau_P} + C_3$$

Putting $s = 0$, we get

$$\frac{C_2}{\tau_P} + C_3 = 0 \text{ and } \frac{C_2}{\tau_P} = -C_3 \text{ or } C_2 = -A\tau_P$$

$$\therefore \quad Y(s) = \frac{A\tau_P}{s+\frac{1}{\tau_P}} - \frac{A\tau_P}{s} + \frac{A}{s^2} \quad (1.49)$$

Taking inverse Laplace transform, we get

$$y(t) = A\ \tau_P e^{-\frac{t}{\tau_P}} - A\tau_P + At$$

$$y(t) = A(t-\tau_P) + A\tau_P e^{-\frac{t}{\tau_P}} \quad (1.50)$$

At steady state, $t \to \infty$

$$y(t) = A(t-\tau_P) \text{ or } \frac{y(t)}{A} = (t-\tau_P) \quad (1.51)$$

$$x(t) = At \quad \text{or} \quad \frac{x(t)}{A} = t \quad (1.52)$$

Dynamic Error : It is defined as the difference between input variable $x(t)$ and the response $y(t)$ at steady state.

$$\text{Dynamic Error} = x(t) - y(t) \text{ at steady state}$$
$$= At - A(t - \tau_P) = A\tau_P$$

Where 'A' is the slope of the ramp change.

Time Lag : It is defined as the time taken by the response to reach the value of the input.

At this condition,
$$\frac{x(t)}{A} = \frac{y(t)}{A}$$
$$\frac{x(t)}{A} = t_i \text{ and } \frac{y(t)}{A} = t_f - \tau_P$$

$\therefore$
$$t_i = t_f - \tau_P$$
$$\text{Time lag} = t_f - t_i = \tau_P \quad (1.53)$$

Therefore the time lag is only the time constant for ramp or linear forcing function for the first order control system.

1.7.11.2 Impulse Forcing Function

If the inlet flow rate to a storage tank is suddenly increased to a value 'A' for a fraction of a second (time almost equal to zero) and brought back to its original value, it is an impulse disturbance. Fig. 1.21 gives an impulse disturbance and the response.

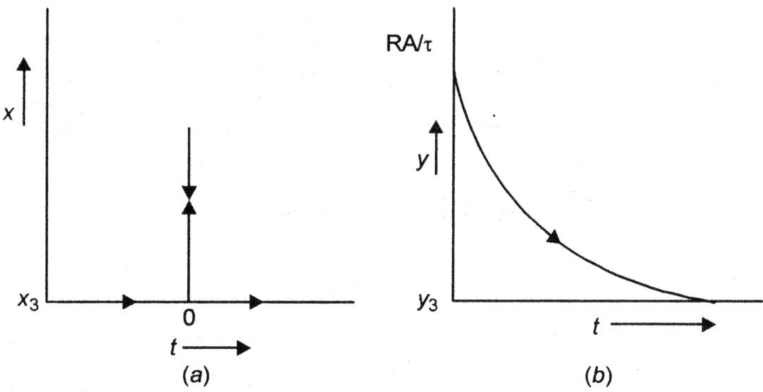

Fig. 1.21 Impulse response of first order system

For an impulse,
$$x(s) = A$$
$$G(s) = \frac{Y(s)}{X(s)} = \frac{1}{\tau_P s + 1}$$

$\therefore$
$$Y(s) = \frac{A}{\tau_P + 1}$$

Then
$$y(t) = \frac{A}{\tau_P} e^{\frac{-t}{\tau_P}} \quad (1.54)$$

At $t = 0$, $\quad y(t) = \dfrac{A}{\tau_P}$

And at $t = \infty$ $\quad y(t) = 0$

$$\frac{dy}{dt} = -\frac{A}{\tau_P^2} e^{-\frac{t}{\tau_P}}$$

$$= \frac{A}{\tau_P^2} \text{ at t =0}$$

$$= 0 \text{ at } t = \infty$$

1.7.11.3 Pulse Forcing Function

If the inlet flow rate is suddenly increased to a new value, maintained at that value for a short period of time 'δ' and decreased back to the original value, it is called a pulse disturbance. If the pulse duration δ is zero, then it is equivalent to impulse. Fig. 1.22 gives a pulse disturbance and the response of the first order system to it.

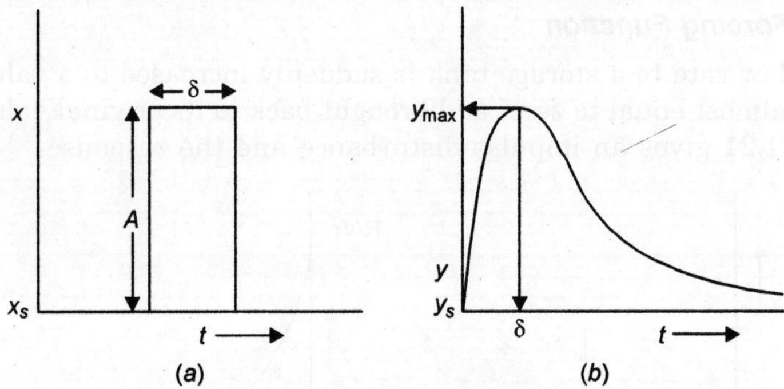

Fig. 1.22 Pulse response

For a pulse, $\quad X(s) = \left(\dfrac{A}{s}\right)\left(1 - e^{-\delta s}\right)$

and hence $\quad Y(s) = \left(\dfrac{A}{s}\right)\dfrac{\left(1 - e^{-\delta s}\right)}{\tau_P s + 1}$

$$Y(s) = \left(\frac{A}{s(\tau_P s + 1)}\right) - \frac{A\,e^{-\delta s}}{s(\tau_P s + 1)} \qquad (1.55)$$

The first part $\left(\dfrac{A}{s(\tau_P s + 1)}\right)$ is for a step increase of A from $t = 0$ to $t = \infty$ and the second part $\dfrac{A\,e^{-\delta s}}{s(\tau_P s + 1)}$ is decrease of A at $t \geq \delta$.

Inverting the equation (1.55), we get

$$y(t) = A\left[1-e^{-\frac{t}{\tau_P}}\right] \quad \text{for } 0 \le t < \delta \tag{1.56}$$

$$y(t) = A\left[1-e^{-\frac{t}{\tau_P}}\right] - A\left[1-e^{-\frac{(t-\delta)}{\tau_P}}\right] \quad \text{for } t \ge \delta \tag{1.57}$$

Hence in the response curve, $Y_{\max}$ occurs at t = δ and from equations (1.56) and (1.57), we get

$$Y_{\max} = A\left(1-e^{-\frac{\delta}{\tau_P}}\right) \tag{1.58}$$

1.7.11.4 Sinusoidal Response of First Order System

Fig. 1.23 shows the sinusoidal disturbance and the response of the first order system to such a disturbance.

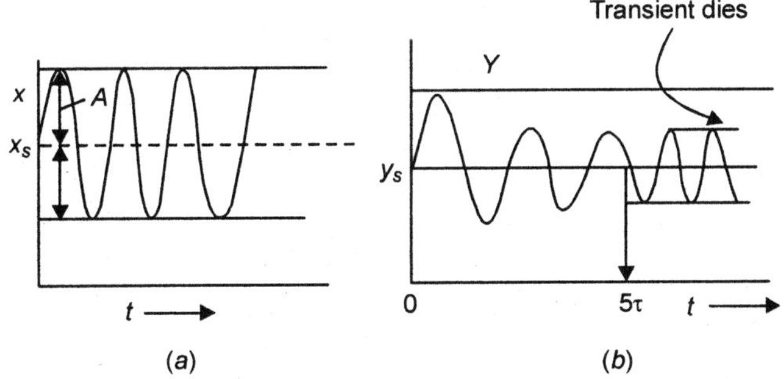

Fig. 1.23 Sinusoidal response of a first-order system

$$f(t) = 0 \text{ when } t < 0$$

$$= A \sin \omega t \qquad \text{when } t \ge 0$$

$$L\{f(t)\} = \frac{A\omega}{s^2+\omega^2}$$

$$X(s) = \frac{A\omega}{s^2+\omega^2}$$

$$G(s) = \frac{Y(s)}{X(s)} = \frac{1}{\tau_P s+1}$$

$$\therefore \quad Y(s) = \frac{A\omega/\tau_P}{\left(s^2+\omega^2\right)\left(s+\frac{1}{\tau_P}\right)}$$

Solving by partial fraction method,

$$Y(s) = \frac{A\omega/\tau_P}{(s^2+\omega^2)\left(s+\frac{1}{\tau_P}\right)} = \frac{C_1}{s+\frac{1}{\tau_P}} + \frac{C_2 s}{s^2+\omega^2} + \frac{C_3}{s^2+\omega^2} \tag{1.59}$$

Cross multiplying the above equation by $\left(s+\frac{1}{\tau_P}\right)(s^2+\omega^2)$

$$\frac{A\omega}{\tau_P} = C_1(s^2+\omega^2) + C_2 s\left(s+\frac{1}{\tau_P}\right) + C_3\left(s+\frac{1}{\tau_P}\right) \tag{1.60}$$

Putting $s = -\frac{1}{\tau_P}$,
$$\frac{A\omega}{\tau_P} = C_1\left[\omega^2 + \frac{1}{\tau_P^2}\right]$$

$$\therefore \quad C_1 = \frac{A\omega\tau_P}{1+\omega^2\tau_P^2}$$

Putting $s = 0$,
$$\frac{A\omega}{\tau_P} = C_1\omega^2 + \frac{C_3}{\tau_P}$$

$$\frac{A\omega}{\tau_P} = \left(\frac{A\omega\tau_P}{1+\omega^2\tau_P^2}\right)\omega^2 + \frac{C_3}{\tau_P}$$

$$\therefore \quad C_3 = \frac{A\omega}{1+\omega^2\tau_P^2}$$

Differentiating the equation (1.60) wrt 's', we get

$$0 = 2sC_1 + 2sC_2 + \frac{C_2}{\tau_P} + C_3$$

Putting $s = 0$,
$$C_2 = -C_3\tau_P = -\frac{A\omega\tau_P}{1+\omega^2\tau_P^2}$$

Substituting the values of C_1, C_2 and C_3 in equation (1.59), we get,

$$Y(s) = \left(\frac{A\omega\tau_P}{1+\omega^2\tau_P^2}\right)\left(\frac{1}{s+\frac{1}{\tau_P}}\right) - \left(\frac{A\omega\tau_P}{1+\omega^2\tau_P^2}\right)\left(\frac{s}{s^2+\omega^2}\right) + \left(\frac{A\omega}{1+\omega^2\tau_P^2}\right)\left(\frac{1}{s^2+\omega^2}\right) \tag{1.61}$$

Taking the inverse Laplace transform, we get,

$$y(t) = \frac{A\omega\tau_P}{1+\omega^2\tau_P^{\,2}} e^{-\frac{t}{\tau_P}} - \frac{A\omega\tau_P}{1+\omega^2\tau_P^{\,2}} \cos \omega t + \frac{A}{1+\omega^2\tau_P^{\,2}} \sin \omega t \tag{1.62}$$

using the trigonometric relation

$$p \cos \theta + q \sin \theta = r \sin(\theta + \phi) \tag{1.63}$$

where $r = \sqrt{p^2 + q^2}$ and $\phi = \tan^{-1}\left(\frac{p}{q}\right)$

Comparing equations (1.62) and (1.63), we get

$$p = -\frac{A\omega\tau_P}{1+\omega^2\tau_P^{\,2}},\ q = \frac{A}{1+\omega^2\tau_P^{\,2}} \text{ and } \theta = \omega t$$

$$\therefore \quad r = \sqrt{\left(\frac{-A\omega\tau_P}{1+\omega^2\tau_P^{\,2}}\right)^2 + \left(\frac{A}{1+\omega^2\tau_P^{\,2}}\right)^2} = \frac{A}{\sqrt{1+\omega^2\tau_P^{\,2}}}$$

$$\phi = \tan^{-1}\left(\frac{-A\omega\tau_P/1+\omega^2\tau_P^{\,2}}{A/(1+\omega^2\tau_P^{\,2})}\right) = \tan^{-1}(-\omega\tau_P)$$

$$\therefore \quad y(t) = \frac{A\omega\tau_P}{1+\omega^2\tau_P^{\,2}} e^{-\frac{t}{\tau_P}} + \frac{A}{\sqrt{1+\omega^2\tau_P^{\,2}}} \sin(\omega t + \phi) \tag{1.64}$$

At steady state, $t \to \infty$, the first term in the equation (1.64) becomes zero and hence,

$$y(t) = \frac{A}{\sqrt{1+\omega^2\tau_P^{\,2}}} \sin(\omega t + \phi) \tag{1.65}$$

where $\phi = \tan^{-1}(-\omega\tau_P)$ and is called 'Phase angle'. In Fig 1.23 (*b*) the transient dies after about $5\,\tau_P$ and we have a steady sine wave of radian frequency ω.

Amplitude Ratio

The ratio of the output or response amplitude of variation to the input amplitude of variation is called as 'Amplitude Ratio'.

$$x(t) = A \sin \omega t$$

$$y(t) = \frac{A}{\sqrt{1+\omega^2\tau_P^{\,2}}} \sin(\omega t + \phi)$$

$$\text{Amplitude Ratio} = \frac{\text{Output Amplitude of variation}}{\text{Input Amplitude of variation}}$$

$$= \frac{\dfrac{A}{\sqrt{1+\omega^2\tau_P{}^2}}}{A} = \frac{1}{\sqrt{1+\omega^2\tau_P{}^2}} \tag{1.66}$$

$$\text{Phase Angle } (\phi) = \tan^{-1}(-\omega\tau_P) \tag{1.67}$$

1.8 SECOND ORDER PROCESS SYSTEMS

Systems with First-order dynamic behaviour are not the only ones encountered in a chemical process. An output may change, under the influence of an input, in a drastically different way from that of a first-order system, following higher order dynamics. In this section we will discuss about the second order systems and in section 1.9 the systems with higher than second order dynamics will be discussed.

A second order system is one whose output, $y(t)$, is described by the solution of a second-order differential equation. In other words, the processes which obey a second order differential equation are termed second order systems. For example, the following equation describes a second-order linear system :

$$a_2 \frac{d^2y}{dt} + a_1 \frac{dy}{dt} + a_0 y = bf(t) \tag{1.68}$$

If $a_0 \neq 0$, the equation (1.68) can be written as

$$\tau^2 \frac{d^2y}{dt} + 2\zeta\tau \frac{dy}{dt} + y = K_P f(t) \tag{1.69}$$

Where $\tau^2 = \dfrac{a_2}{a_0}$, $2\,\zeta\tau = \dfrac{a_1}{a_0}$ and $K_P = \dfrac{b}{a_0}$

Equation (1.69) is in the standard form of a second-order system, where

τ = 'natural period' of oscillation of the system

ζ = damping factor

K_P = Steady state, or static, or simply gain of the system.

If equation (1.69) is interms of deviation variables, the initial conditions are zero and its Laplace transform yields the following standard transfer function for a second-order system :

$$G(s) = \frac{Y(s)}{X(s)} = \frac{\bar{y}(s)}{\bar{f}(s)} = \frac{K_P}{\tau^2 s^2 + 2\zeta\tau s + 1} \tag{1.70}$$

Systems with second-order dynamics can be classified as follows :

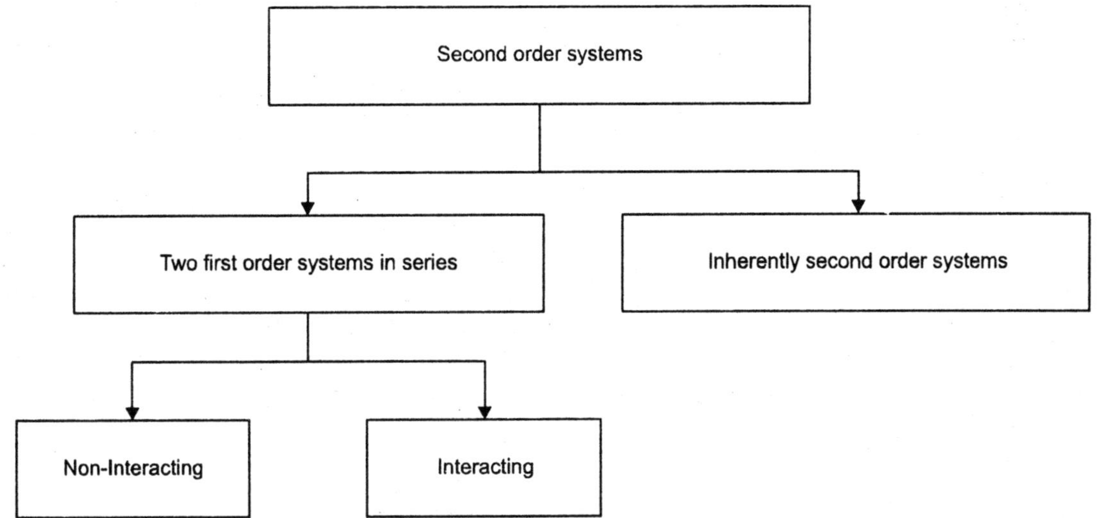

1.8.1 Dynamic Response of a Second-Order System

The analysis of the dynamic response of a second-order system to a unit step input will provide all the fundamental dynamic features of a second-order system.

For a unit step change in the input $f(t)$, equation (1.70) yields

$$Y(s) = \frac{K_P}{s(\tau^2 s^2 + 2\zeta\tau s + 1)} \tag{1.71}$$

The two poles of the second-order transfer function are given by the roots of the characteristic polynomial,

$$\tau^2 s^2 + \zeta\tau s = 0$$

and they are

$$p_1 = -\frac{\zeta}{\tau} + \frac{\sqrt{\zeta^2 - 1}}{\tau} \quad \text{and} \quad p_2 = -\frac{\zeta}{\tau} - \frac{\sqrt{\zeta^2 - 1}}{\tau}$$

Therefore the equation (1.71) becomes,

$$Y(s) = \frac{K_P / \tau^2}{s(s - p_1)(s - p_2)} \tag{1.72}$$

and the form of the response $y(t)$ will depend on the location of the two poles, p_1 and p_2, in the complex plane. Thus we can distinguish three cases.

Case-A : When $\zeta > 1$, we have two distinct and real poles.

Case-B : When $\zeta = 1$, we have two equal poles (multiple poles)

Case-C : When $\zeta < 1$, we have two complex conjugate poles.

Let us examine each case separately.

Case A : Over Damped Response When $\zeta > 1$

In this case the inversion of the equation (1.72) by partial-fractions yields

$$y(t) = K_P\left[1-e^{-\frac{\zeta t}{\tau}}\left(\cosh\sqrt{\zeta^2-1}\frac{t}{\tau}+\frac{\zeta}{\sqrt{\zeta^2-1}}\sinh\sqrt{\zeta^2-1}\frac{t}{\tau}\right)\right] \quad (1.73)$$

Where cosh(.) and sinh(.) are the hyperbolic trigonometric functions defined by

$$\sinh\alpha = \frac{e^{\alpha}-e^{-\alpha}}{2} \quad \text{and} \quad \cosh\alpha = \frac{e^{\alpha}+e^{-\alpha}}{2}$$

The response has been plotted in fig 1.24 (a) for various values of ζ, $\zeta > 1$. It is known as 'overdamped response' and resembles a little the response of a first order system to a unit step input. But when compared to a first-order response we notice that the system initially delays to respond and then its response is rather sluggish. It becomes more sluggish as ζ increases (i.e., as the system becomes more heavily overdamped.) Finally, we notice that as time goes on, the response approaches its ultimate value asymptotically. As it was the case with first-order systems, the gain is given by,

$$K_P = \frac{\Delta(\text{output steady state})}{\Delta(\text{input steady state})}$$

Overdamped are the responses of multicapacity processes, which result from the combination of first-order systems in series as we will see in section (1.8.2).

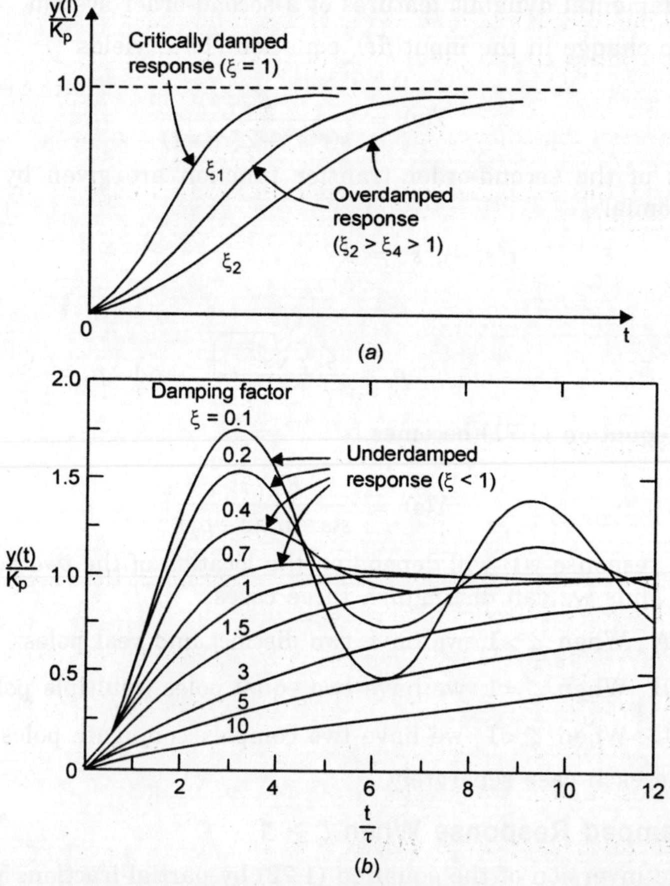

Fig. 1.24 Dimensionless response of second-order system to input step change

Case B: Critically damped response, When $\zeta = 1$

In this case, the inversion of equation (1.71) gives the result,

$$y(t) = K_P\left[1-\left(1+\frac{t}{\tau}\right)e^{-\frac{t}{\tau}}\right] \tag{1.74}$$

The response is also shown in Fig 1.24(a). We notice that a second-order system with critical damping approaches its ultimate value faster than does an over damped system.

Case C: Under damped Response, When $\zeta < 1$

The inversion of equation (1.71) yields

$$y(t) = K_P\left[1-\frac{1}{\sqrt{1-\zeta^2}}\,e^{-\frac{\zeta t}{\tau}}\,Sin(\omega t+\phi)\right] \tag{1.75}$$

where $$\omega = \frac{\sqrt{1-\zeta^2}}{\tau} \text{ and } \varphi = \tan^{-1}\left[\frac{\sqrt{1-\zeta^2}}{\zeta}\right]$$

The response has been plotted in Fig. 1.24(b) for various values of damping factor, ζ. From the plots we can observe the following :

1. The underdamped response is initially faster than the criticalled damped or over damped responses, which are characterized as sluggish.
2. Although the under damped response is initially faster and reaches its ultimate value quickly, it does not stay there, but it starts oscillating with progressively decreasing amplitude. This oscillatory behaviour makes an underdamped response quite distinct from all previous ones.
3. The oscillatory behaviour becomes more pronounced with smaller values of the damping factor, ζ.

It must be emphasized that almost all the underdamped responses in a chemical plant are caused by the interaction of the controllers with the process units they control. Therefore, it is a type of response that we will encounter very often, and it is wise to become well acquainted with its characteristics.

1.8.2 Characteristics of an Underdamped response

The underdamped response shown in Fig. 1.25 can be used as reference in order to define the various terms employed to describe an underdamped response.

1. Overshoot : Overshoot is the ratio between the maximum amount by which the response exceeds its ultimate value (A) and the ultimate value of the response. The overshoot is a function of ζ, and it can be shown that it is given by the following expression :

$$\text{Overshoot} = \frac{A}{B} = \exp\left(\frac{-\pi\zeta}{\sqrt{1-\zeta^2}}\right) \tag{1.76}$$

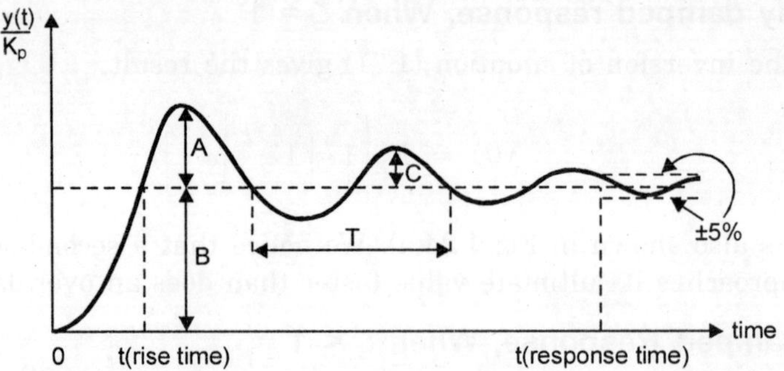

Fig. 1.25 Characteristics of an underdamped response.

Fig. 1.26 shows the plot of overshoot versus ζ given by equation (1.76). We notice that the overshoot increases with decreasing ζ, while as ζ approaches 1 the overshoot approaches zero (critically damped response).

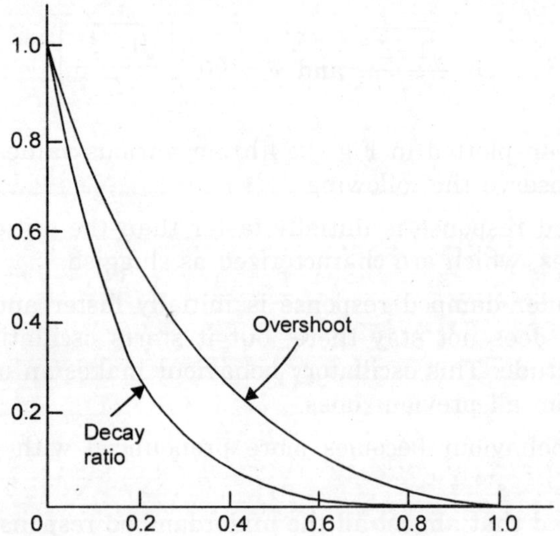

Fig. 1.26 Effect of damping factor on overshoot and decay ratio.

2. Decay Ratio : The decay ratio is the ratio of the amounts above the ultimate value of two successive peaks. The decay ratio can be shown to be related to the damping factor ζ through the equation.

$$\text{Decay Ratio} = \frac{C}{A} = \exp\left(\frac{-2\pi\zeta}{\sqrt{1-\zeta^2}}\right) = (\text{Overshoot})^2 \tag{1.77}$$

The equation (1.77) has also been plotted in Fig. 1.26.

3. Period of Oscillation : The time elapsed between two successive peaks is called the period of oscillation (T). The radian frequency (rad /time) of the oscillations of an underdamped response is given by the equation

$$\omega = \text{radian frequency} = \frac{\sqrt{1-\zeta^2}}{\tau} \tag{1.78}$$

$$\omega = 2\pi f = 2\pi \frac{1}{T} = \frac{\sqrt{1-\zeta^2}}{\tau}$$

$$\therefore \quad T = \frac{2\pi\tau}{\sqrt{1-\zeta^2}} \tag{1.79}$$

4. Natural Period of Oscillation

A second-order system with $\zeta = 0$ is a system free of any damping. Its transfer function is

$$G(s) = \frac{K_P}{\tau^2 s^2 + 1} = \frac{K_P / \tau^2}{\left(s - j\frac{1}{\tau}\right)\left(s + j\frac{1}{\tau}\right)} \tag{1.80}$$

It has two purely imaginary poles (on the imaginary axis) and it will oscillate continuously with a constant amplitude and a natural frequency.

$$\omega n = \frac{1}{\tau}$$

The corresponding cyclical period T_n is given by

$$T_n = 2\pi\tau \tag{1.81}$$

It is the property of the parameter τ that gave its name, the natural period of oscillation.

5. Response Time : The response of an underdamped system will reach its ultimate value in an oscillatory manner as $t \to \infty$. For practical purposes, it has been agreed to consider that the response reached its final value when it came within ± 5% of its final value and stayed there. The time needed for the response to reach this situation is known as the response time (Refer Fig. 1.25).

6. Rise Time : This term is used to characterize the speed with which an underdamped system responds. It is defined as the time required for the response to reach its final value for the first time (Refer Fig. 1.25). From Fig. 1.24(*b*) we notice that the smaller the value of ζ, the shorter the rise time (i.e., The faster the response of the system), but at the same time the larger the value of the overshoot.

Note : Good understanding of the underdamped behaviour of a second-order system will help tremendously in the design of efficient controllers. The values of τ and ζ are to be selected while designing a controller in such a way that the overshoot is small, the rise time short, the decay ratio small, and the response time short. It may not be possible to achieve all these objectives for the same values of τ and ζ, but an acceptable compromise is always possible.

1.8.3 Second Order Systems : Two First Order System in series : Non-Interacting

When material or energy flows through a single capacity, we get a first-order system. If on the other hand, mass or energy flows through a series of two capacities, the behaviour of the system is described by second-order dynamics. In Fig. 1.27 tank-1 feeds tank-2 and thus it affects its dynamic behaviour, whereas the opposite is not true. Such a system is characteristic of a large class of 'non interacting capacities or non interacting first order systems in series'.

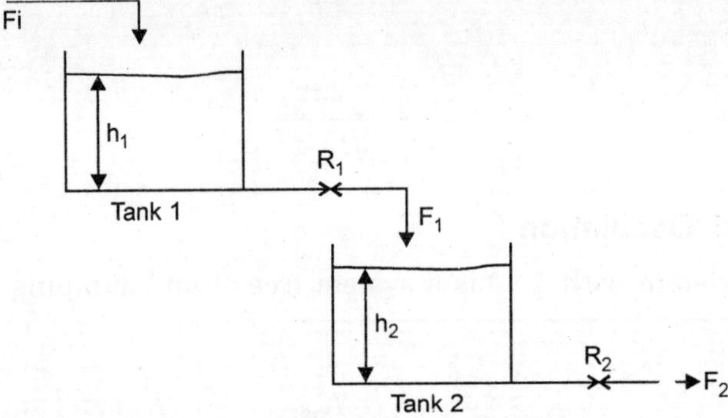

Fig. 1.27 Noninteracting tanks

Multicapacity processes do not have to involve more than one physical processing unit. It is quite possible that all capacities are associated with the same processing unit. For example, the distillation column is a multicapacity process. There is a series of liquid flow elements which are non-interacting.

When a system is composed of two non interacting capacities, it is described by a set of two differential equations of the general form :

$$\tau_{P1}\frac{dy_1}{dt}+y_1=K_{P1}f_1(t) \qquad \text{first capacity} \tag{1.82}$$

$$\tau_{P2}\frac{dy_2}{dt}+y_2=K_{P2}f_1(t) \qquad \text{second capacity} \tag{1.83}$$

$\bar{f}_1(s)$ → $G_1(s)$ → $\bar{y}_1(s)$ → $G_2(s)$ → $\bar{y}_2(s)$

(a)

$\bar{f}_1(s)$ → $G_1(s)$ → $\bar{y}_1(s)$ → $G_2(s)$ → $\bar{y}_2(s)$ ... $\bar{y}_{N-1}(s)$ → $G_N(s)$ → $\bar{y}_N(s)$

(b)

Fig. 1.28 Noninteracting capacities in series

In other words, the first system affects the second by its output, but it is not affected by the second system [Fig. 1.28(a)]. Equation (1.82) can be solved first and then we can solve equation (1.83). This sequential solution is characteristic of non interacting capacities in series. The corresponding transfer functions are

$$G_1(s)=\frac{Y_1(s)}{F_1(s)}=\frac{K_{P1}}{\tau_{P1}s+1} \quad \text{and} \quad G_2(s)=\frac{Y_2(s)}{Y_1(s)}=\frac{K_{P2}}{\tau_{P2}s+1}$$

The overall transfer function between the external input $f_1(t)$ and $y_2(t)$ is

$$G_0(s)=\frac{Y_2(s)}{F_1(s)}=\frac{Y_1(s)}{F_1(s)},\frac{Y_2(s)}{Y_1(s)}=G_1(s)\,G_2(s)$$

$$G_0(s) = \frac{K_{P1}}{\tau_{P1}s+1} \cdot \frac{K_{P2}}{\tau_{P2}s+1} \tag{1.84}$$

$$G_0(s) = \frac{K_P'}{(\tau')^2 s^2 + 2\zeta'\tau' s + 1} \tag{1.85}$$

Where $(\tau^{\text{l}})^2 = \tau_{P1}\tau_{P2}$, $2\zeta'\tau' = \tau_{P1} + \tau_{P2}$ $K_P' = K_{P1}.K_{P2}$

Equation (1.85) very clearly indicates that the overall response of the system is second order. From equation (1.84) we also notice that the two poles of the overall transfer function are real and distinct :

$$p_1 = -\frac{1}{\tau_{P1}}, \qquad p_2 = -\frac{1}{\tau_{P2}}$$

If the time constant τ_{P1} and τ_{P2} are equal, we have two equal poles. Therefore, non-interacting systems always result in an overdamped and critically damped second order system and never in an underdamped system.

For the case of N noninteracting systems [Fig 1.28(*b*)] it is easy to show that the overall transfer function is given by

$$G_0(s) = G_1(s)G_2(s)............G_N(s) = \frac{K_{P1}K_{P2}..........................K_{PN}}{(\tau_{P1}s+1)(\tau_{P2}s+1).....................(\tau_{PN}s+1)} \tag{1.86}$$

System in Fig. 1.27 is a system with two noninteracting tanks in series. The transfer functions for the two tanks are :

$$G_1(s) = \frac{K_{P1}}{\tau_{P1}s+1} \quad \text{and} \quad G_2(s) = \frac{K_{P2}}{\tau_{P2}s+1}$$

As we have already seen under first order systems, since $F_1' = \frac{h_1'}{R_1}, K_{P1} = R_1, K_{P2} = R_2,$

$$\tau_{P1} = A_1R_1 \text{ and } \tau_{P2} = A_2R_2$$

We can easily find that the overall transfer function is

$$G_0(s) = G_1(s).G_2(s) \quad = \frac{H_2(s)}{F_i(s)} = \frac{K_{P2}}{(\tau_{P1}s+1)(\tau_{P2}s+1)}$$

$$= \frac{R_2}{\tau_P\tau_{P2}s^2 + (\tau_{P1}+\tau_{P2})s+1} \tag{1.87}$$

1.8.4 Second Order Systems : Two First Order Systems in Series Interacting

In contrast to noninteracting systems, in the system shown in Fig 1.29, tank1 affects the dynamic behaviour of tank2, and vice versa, because the flow rate F_1 depends on the difference between liquid levels h_1 and h_2. This system represents interacting capacities or interacting first-order systems in series.

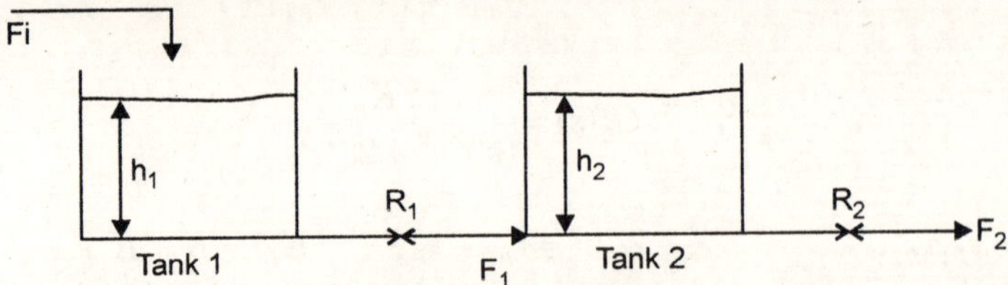

Fig. 1.29. Interacting tanks

The stirred tank heater, a multicapacity process with capacity to store mass and energy, is another example of such systems. It is easy to show that these two capacities interact when the inlet flowrate changes. Thus, a change in the inlet flowrate affects the liquid level in the tank, which inturn affects the temperature of the liquid. Consequently, the temperature response to an inlet flowrate change exhibits second-order overdamped characteristics. It is to be noted that the two capacities do not interact when the inlet temperature changes. Therefore, the temperature response to inlet temperature changes exhibits first-order characteristics. Also the multiple capacities need not correspond to physically different units, but could be present within the same processing unit.

Let us analyse the characteristics of an interacting system with two interacting tanks (Refer Fig. 1.29). The mass balances yield,

$$A_1 \frac{dh_1}{dt} = F_i - F_1 \qquad \text{for tank 1}$$

$$A_2 \frac{dh_2}{dt} = F_1 - F_2 \qquad \text{for tank 2}$$

Assuming linear resistances to flow we get

$$F_1 = \frac{h_1 - h_2}{R_1} \quad \text{and} \quad F_2 = \frac{h_2}{R_2}$$

Substituting F_1 and F_2 in the above equation, we get

$$A_1 R_1 \frac{dh_1}{dt} + h_1 - h_2 = R_1 F_i \tag{1.88}$$

$$A_2 R_2 \frac{dh_2}{dt} + (1 + \frac{R_2}{R_1})h_2 - \frac{R_2}{R_1} h_1 = 0 \tag{1.89}$$

The steady-state equivalents of equations (1.88) and (1.89) are

$$h_{1S} - h_{2S} = R_1 F_{iS} \tag{1.90}$$

$$\left(1 + \frac{R_2}{R_1}\right) R_{2S} - \frac{R_2}{R_1} h_{1S} = 0 \tag{1.91}$$

Subtracting (1.90) from (1.88) and (1.91) from (1.89) and introducing the deviation variables, we get

$$A_1R_1\frac{dh_1'}{dt}+h_1'-h_2'=R_1F_i' \tag{1.92}$$

$$A_2R_2\frac{dh_2'}{dt}+(1+\frac{R_2}{R_1})h_2'-\frac{R_2}{R_1}h_1'=0 \tag{1.93}$$

Where $h_1'=h_1-h_{1S}$, $h_2'=h_1-h_{2S}$, and $F_i'=F_i-F_{iS}$

Taking Laplace transforms of equations (1.92) and (1.93), we get

$$\left(A_1R_1s+1\right)\overline{h'_{1(S)}}-\overline{h'_{2(S)}}=R_1\overline{F'_{i(S)}}$$

$$-\frac{R_2}{R_1}\overline{h'_{1(S)}}+\left[A_2R_2s+\left(1+\frac{R_2}{R_1}\right)\right]h'_{2S}=0$$

Solving the above equations with respect to $\overline{h'_{1(S)}}$ and $\overline{h'_{2(S)}}$, We get

$$\frac{H_1(s)}{Fi(s)}=\frac{\overline{h'_{1(S)}}}{\overline{F'_{i(S)}}}=\frac{\tau_{P2}R_1s+(R_1+R_2)}{\tau_{P1}\tau_{P2}s^2+(\tau_{P1}+\tau_{P2}+A_1R_2)s+1} \tag{1.94}$$

$$G_0(s)=\frac{H_2(s)}{Fi(s)}=\frac{\overline{h'_{2(S)}}}{\overline{F'_{i(S)}}}=\frac{R_2}{\tau_{P1}\tau_{P2}s^2+(\tau_{P1}+\tau_{P2}+A_1R_2)s+1} \tag{1.95}$$

Where $\tau_{P1}=A_1R_1$ and $\tau_{p2}=A_2R_2$ are the time constants of the two tanks. Equations (1.94) and (1.95) indicate that the responses of both tanks follow second-order dynamics. Comparing equation (1.95) for the interacting tanks with equation (1.87) for non interacting tanks, we notice that they differ only in the coefficient of S in the denominator by the term A_1R_2. This term may be thought of as the 'interaction factor' and indicates the degree of interaction between the two tanks. The larger the value of A_1R_2, the larger the interaction between two tanks.

From Equation (1.95) it is easily found that the two poles of the transfer function are given by

$$p_{1,2}=\frac{-\left(\tau_{P1}+\tau_{P2}+A_1R_2\right)\pm\sqrt{\left(\tau_{P1}+\tau_{P2}+A_1R_2\right)^2-4\tau_{P1}\tau_{P2}}}{2\tau_{P1}\tau_{P2}} \tag{1.96}$$

But $\left(\tau_{P1}+\tau_{P2}+A_1R_2\right)^2-4\tau_{P1}\tau_{P2}>0$

Therefore, p_1 and p_2 are distinct and real poles. Consequently, the 'response of interacting capacities is always overdamped'.

Since the response is overdamped with poles p_1 and p_2 given by equation (1.96), then equation (1.96) can be written as follows :

$$\frac{H_2(s)}{F_i(s)} = \frac{R_2 / \tau_{P1}\tau_{P2}}{(s-p_1)(s-p_2)} = \frac{(\tau_1\tau_2)R_2 / \tau_{P1}\tau_{P2}}{(\tau_1 s+1)(\tau_2 s+1)} \tag{1.97}$$

Where $\tau_1 = -\frac{1}{p_1}$ and $\tau_2 = -\frac{1}{p_2}$. Equation (1.97) implies that two interacting capacities can be viewed as noninteracting capacities but with modified effective time constants. Thus, whereas initially the two interacting tanks had effective time constants τ_{P1} and τ_{P2}, when they are viewed as noninteracting, they have different time constants τ_1 and τ_2.

1.8.5 Inherently Second Order Processes

Inherently second-order process can exhibit underdamped behaviour, and consequently it cannot be decomposed into two first-order systems in series (Interacting or noninteracting) with physical significance. They occur rather rarely in a chemical process, and they are associated with the motion of liquid masses or the mechanical translation of solid parts, possessing : (1) Inertia to motion, (2) Resistance to motion and (3) Capacitance to store mechanical energy. Since resistance and capacitance are characteristic of the first order systems, we conclude that the inherently second-order systems are characterized by their inertia to motion.

Newton's law applied on a given system yields

Balance of forces on the system = mass of system × acceleration

$$F = m \cdot a \tag{1.98}$$

Since acceleration $a = \frac{d(\text{velocity})}{dt} = \frac{dv}{dt}$

And velocity $v = \frac{d(\text{spatial displacement})}{dt} = \frac{dx}{dt}$

We conclude that

Balance of forces on the system = mass of the system × $\frac{d^2(\text{spatial displacement})}{dt^2}$

$$F = m\frac{d^2x}{dt^2} \tag{1.99}$$

1.8.5.1 U-Tube Manometer

Consider the simple U-tube manometer shown in Fig. 1.30. When the pressures at the top of the two legs are equal, the two liquid levels are at rest at the same horizontal plane. Let us assume that there is suddenly a pressure difference $\Delta p = p_1 - p_2$ is imposed on the two legs of the manometer. To know the dynamic response of the levels in the two legs, we have to proceed as below.

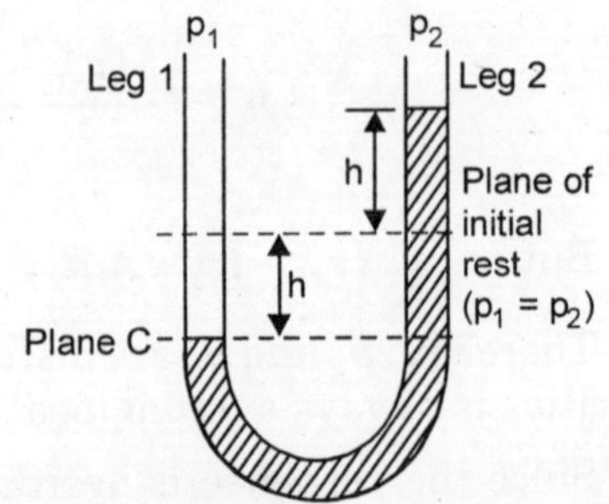

Fig. 1.30 Manometer

Let us apply Newton's law on the plane 'C' of the manometer. [Refer Eqn (1.98)]

{(Force due to pressure p_1 on leg1) – (force due to pressure p_2 on leg2)
– (Force due to liquid level difference in the two legs) – (Force due to fluid friction)}
= (Mass of liquid in the tube) × (Acceleration)

OR

$$\left\{p_1A_1 - p_2A_2 - \rho\frac{g}{g_C}A_2(2h) - (\text{force due to fluid friction})\right\} = \frac{m}{g_C}\cdot\frac{dv}{dt} \qquad (1.100)$$

where

p_1, p_2 = pressures at the top of legs 1 and 2 respectively.

A_1, A_2 = cross-sectional areas of legs 1 and legs 2 respectively. (Many times $A_1 = A_2 = A$)

ρ = density of liquid in manometer

g = acceleration due to gravity

g_C = conversion constant.

m = mass of liquid in the manometer

v = average velocity of the liquid in the tube

h = deviation of liquid level from the initial plane of rest.

L = length of liquid in the manometer tubes.

Poiseuille's equation for laminar flow in a pipe can be used to relate the force due to fluid friction with the flow velocity. Thus we have (Poiseuille's Equation)

$$\text{Volumetric flow rate :} \quad A\frac{dh}{dt} = \frac{\pi R^4}{8\mu}\cdot\frac{\Delta p}{L} \qquad (1.101)$$

where R = radius of the pipe through which liquid flows.

μ = viscosity of the flowing fluid.

L = length of the pipe.

ΔP = pressure drop due to fluid friction along the tube of length 'L'

Therefore, applying poiseuille's equation to the flow of liquid in the manometer we get,

$$\text{Force due to fluid friction} = \frac{\Delta p \pi R^2}{g_C} = A\frac{8\mu L}{R^2 g_C}\cdot\frac{dh}{dt} \qquad (1.102)$$

where $\Delta p = p_1 - p_2$. Recall also that the fluid velocity and acceleration are given by,

$$v = \frac{dh}{dt} \quad \text{and} \quad \frac{dv}{dt} = \frac{d^2h}{dt^2} \qquad (1.103)$$

Substituting Eqn. (1.102) from (1.103) in eqn. (1.100), we get

$$\Delta pA - \frac{2\rho gA}{g_C}h - \frac{8\mu LA}{R^2 g_C}\cdot\frac{dh}{dt} = \frac{pAL}{g_C}\cdot\frac{d^2h}{dt^2}$$

Dividing both sides by $2\rho g\ A/g_C$, we get

$$\left(\frac{L}{2g}\right)\frac{d^2h}{dt^2}+\left(\frac{4\mu L}{\rho g R^2}\right)\frac{dh}{dt}+h=\left(\frac{g_C}{2\rho g}\right)\Delta p \qquad (1.104)$$

By defining $\tau^2=\frac{L}{2g}$, $2\zeta\tau=\frac{4\mu L}{\rho g R^2}$ and $K_P=\frac{g_C}{2\rho g}$, the equation (1.104) becomes

$$\tau^2\frac{d^2h}{dt^2}+2\zeta\tau\frac{dh}{dt}+h=K_P.\Delta p \qquad (1.105)$$

Therefore, the transfer function between h and Δp is

$$\frac{H(s)}{\Delta p(s)}=\frac{\overline{h}(s)}{\overline{\Delta p}(s)}=\frac{K_P}{\tau^2 s^2+2\zeta\tau s+1}$$

Both equations (1.105) and (1.106) indicate the inherent second-order dynamics of the manometer.

Externally Mounted Level Indicator : [Refer Fig. 1.31]

Quite often we use the externally mounted displacement type transmitter for measuring liquid levels. The system of the tank-displacer chamber has many similarities with the manometer. The cross-sectional areas of the two legs are unequal and the Δp (external) pressure difference is caused by a change in the liquid level of the main tank. Therefore, we expect that the response of the level in the displacement chamber, h_m, will follow second-order dynamics with respect to a change in the liquid level of the tank, h :

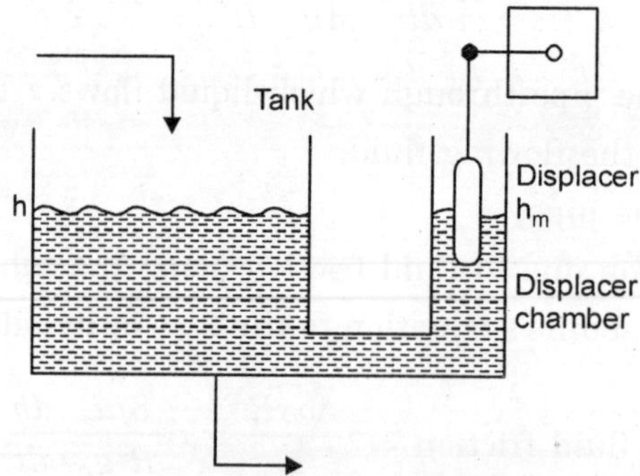

Fig. 1.31 Externally mounted level indicator.

$$\frac{H_m(s)}{H(s)}=\frac{\overline{h_m}(s)}{\overline{h}(s)}=\frac{K_{Pm}}{\tau_m{}^2 s^2+2\zeta_m\tau_m s+1} \qquad (1.107)$$

1.8.5.2 Variable Capacitance Differential Pressure Transducer

The variable capacitance differential transducer is a very popular device which is used to sense and transmit pressure differences. Fig. 1.32 shows a schematic of such a device. A pressure signal is transferred through an isolating diaphragm and fill liquid in a sealed capillary system with a differential -pressure sensing element (a) attached at the other end of the capillary (b). Here, the pressure is transmitted through a second isolating diaphragm and fill liquid (silicone oil), to a sensing diaphragm. A reference pressure will balance the sensing diaphragm on the other side of this diaphragm. The position of the sensing diaphragm is detected by capacitor plates on both sides of the diaphragm. A change in pressure p_1 of a processing unit will make the pressure p_2 change at the end of the capillary tube.

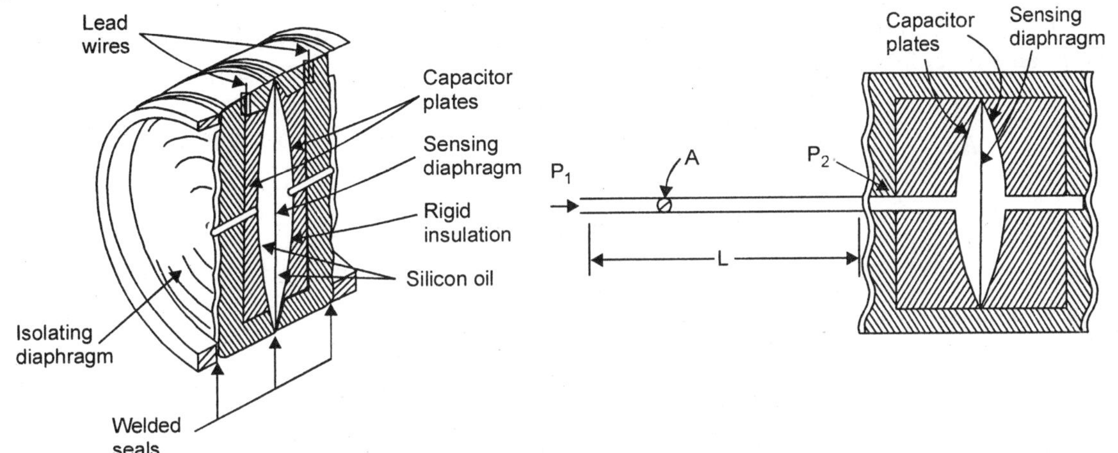

Fig. 1.32 Variable capacitance differential pressure transducer
(a) Differential pressure sensing element (b) Pressure signal transmission system

A force balance around the capillary will yield

{Force due to pressure p_1 of the process exercised at the end 1 of the capillary} {Force due to the pressure p_2 exercised at the end 2 of the capillary}

= (mass) × (acceleration)

OR

$$p_1A - p_2A = \frac{AL\rho}{g_C}\cdot\frac{d^2x}{dt^2} \tag{1.108}$$

where A = cross-sectional area of the capillary

l = length of the capillary tube

ρ = density of the liquid in the capillary tube

x = fluid displacement in the capillary tube = displacement of diaphragm

The force p_2A at the end of the capillary is balanced by two forces :

p_2A = Resistance exerted the diaphragm which acts like a spring + Viscous friction force exercised by the fluid

$$p_2A = Kx + C\frac{dx}{dt} \tag{1.109}$$

where K = hooke's constant for the diaphragm

C = dampling coefficient of the viscous liquid in front of the diaphragm

Substituting p_2A from equation (1.109) in equation (1.108), and rearranging, we get

$$\left(\frac{AL\rho}{Kg_C}\right)\frac{d^2x}{dt^2}+\frac{C}{K}.\frac{dx}{dt}+x=\frac{A}{K}p_1 \tag{1.110}$$

Equation (1.110) clearly indicates that the response of the device (i.e., the diaphragm displacement, x) follows second-order dynamics to any changes in process pressure p_1. If we define $\tau^2=\frac{AL\rho}{Kg_C}$, $2\zeta\tau=\frac{C}{K}$ and $K_P=\frac{A}{K}$ and take transfer function, we get

$$G(s) = \frac{X(s)}{p_1(s)}=\frac{\bar{x}(s)}{p_1(s)}=\frac{K_P}{\tau^2s^2+2\zeta\tau s+1} \tag{1.111}$$

1.8.5.3 Pneumatic Valve

The Pneumatic valve is the most commonly used final control element. It is a system that exhibits inherent second order dynamics.

Consider a typical pneumatic valve shown in Fig. 1.33. The position of the stem (or, equivalently, of the plug at the end of the stem) will determine the size of the opening for flow and consequently the quantity of the flow (flow rate). The position of the stem is determined by the balance of all forces acting on it. These forces are :

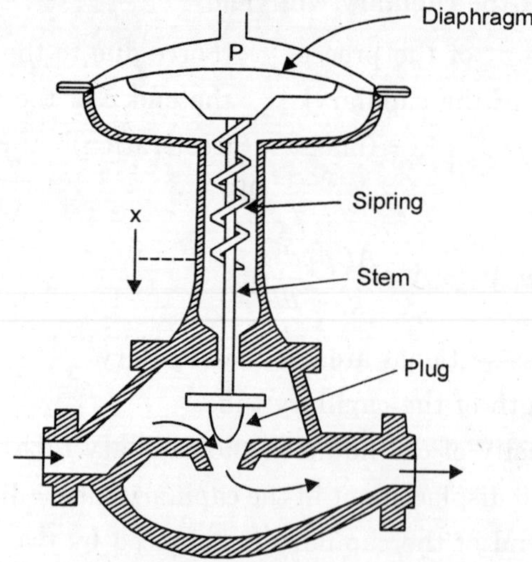

Fig. 1.33 Pneumatic valve

pA = Force exerted by the compressed air at the top of the diaphragm; pressure 'p' is the signal that opens or closes the valve and 'A' is the area of the diaphragm.

Kx = Force exerted by the spring attached to the stem and the diaphragm; 'K' is the Hooke's constant for the spring and 'x' is the displacement; it acts upward.

$C\frac{dx}{dt}$ = Frictional force exerted upward and resulting from the close contact of the stem with valuve packing; C is the friction coefficient between stem and packing.

Applying Newton's law, we get

$$pA - Kx - C\frac{dx}{dt} = \left(\frac{M}{g_C}\right)\frac{d^2x}{dt} \quad \text{OR} \quad \left(\frac{M}{Kg_C}\right)\frac{d^2x}{dt^2} + \frac{C}{K}\frac{dx}{dt} + x = \frac{A}{K}p$$

Defining $\tau^2 = \frac{M}{Kg_C}$, $2\zeta\tau = \frac{C}{K}$, and $K_P = \frac{A}{K}$,

We get $$\tau^2\frac{d^2x}{dt^2} + 2\zeta\tau\frac{dx}{dt} + x = K_P p$$

The last equation indicates that the stem position 'x' follows inherent second-order dynamics. The transfer function is

$$G(s) = \frac{X(s)}{P(s)} = \frac{\bar{x}(s)}{\bar{p}(s)} = \frac{\frac{A}{K}}{\left(\frac{M}{Kg_C}\right)s^2 + \frac{C}{K}s + 1}$$

$$G(s) = \frac{K_P}{\tau^2 s^2 + 2\zeta\tau s + 1} \tag{1.112}$$

Note : Usually, $M << KgC$ and as a result, the dynamics of a pneumatic value can be approximated by that of first-order system.

1.8.5.4 Damped Oscillator (Oscillatory Element)

The oscillatory element is shown in Fig. 1.34. Although it is not encountered in ordinary liquid, gas and thermal processes, it is typical of many measuring instruments such as the bourdan-tube pressure gauge. Consider the mass, spring and damping system shown in Fig. 1.34.

Newton's Second law of motion gives

$$M\frac{d^2x}{dt^2} = -B\frac{dx}{dt} - Kx + F(t) \tag{1.113}$$

where M = mass in Kg

B = damping coefficient in N.S/m

K = spring gradient or modulus of the spring in N/m

F = force acting on the mass M in N

x = displacement in m

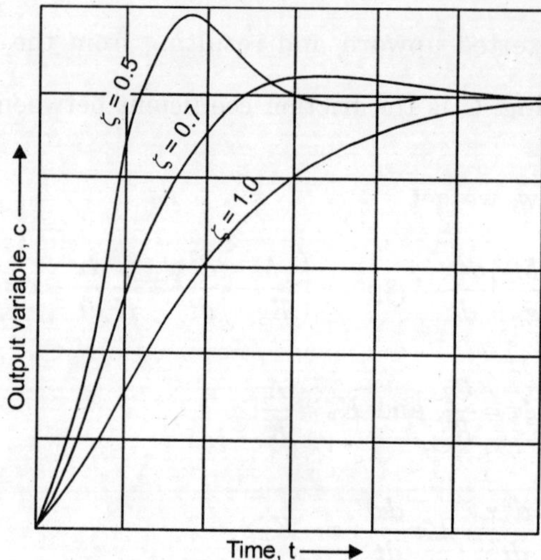

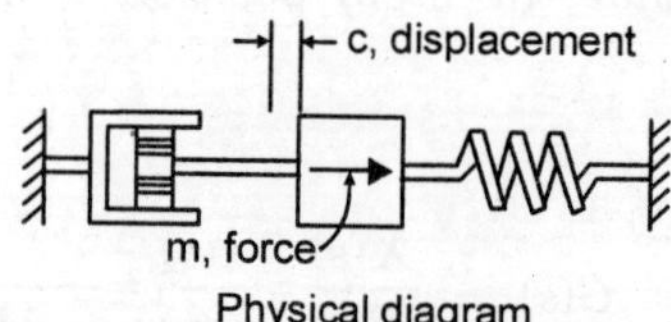

m → $\dfrac{1/K}{T^2s^2 + 2\xi Ts + 1}$ → c

Block diagram

Fig. 1.34 The oscillatory element

i.e., net force acting on the mass = (mass) × (acceleration)

= rate of change of momentum

Dividing the equation (1.113) on both sides by K and rearranging gives,

$$\frac{M}{K}.\frac{d^2x}{dt}+\frac{B}{K}.\frac{dx}{dt}+Kx=\frac{F}{K} \qquad (1.114)$$

Defining $\tau^2 = \dfrac{M}{K}$, $2\zeta\tau = \dfrac{B}{K}$, and $K_P = \dfrac{1}{K}$, we get

$$\tau^2\frac{d^2x}{dt^2}+2\zeta\tau\frac{dx}{dt}+x=K_P F(t)$$

The above equation indicates that the displacement 'x' follows inherent second-order dynamics. The transfer function is

$$G(s) = \frac{X(s)}{F(s)}=\frac{K_P}{\tau^2s^2+2\zeta\tau s+1} \qquad (1.115)$$

The response of an oscillatory system element is shown in Fig. 1.34 for a step change in input variable. The system is underdamped for $\zeta < 1$, critically damped for $\zeta = 1$ and overdamped for $\zeta > 1$.

1.9 DYNAMIC BEHAVIOUR OF HIGHER-ORDER SYSTEMS

Systems with higher than second-order dynamics are not unknown in chemical processes. Three classes of higher-order systems are most often encountered.

1 N first-order processes in series (multicapacity processes)

2 Processes with dead time

3 Processes with inverse response.

1.9.1 N Capacities in Series

In section 1.8 we found that two capacities in series, interacting or noninteracting, give rise to second-order system. If we extend the same procedure to N capacities (first-order systems) in series, we find that the overall response is of Nth order ; that is, the denominator of the overall transfer function is an Nth -order polynomial,

$$a_N s^N + a_{N-1} s^{N-1} + \dots\dots\dots\dots + a_1 s + a_0$$

The overall transfer function for the N capacities which are nonreacting is :

$$G_0(s) = G_1(s).G_2(s)\dots\dots G_N(s) = \frac{K_1 K_2 \dots\dots\dots\dots K_N}{(\tau_1 s+1)(\tau_2 s+1)\dots\dots\dots(\tau_N s+1)} \qquad (1.116)$$

For interacting capacities the overall transfer function is more complex.

The following general conclusions can be made regarding the basic characteristics of the capacities in series for a step change in the input.

N Noninteracting Capacities in Series

a) The response has the characteristics of an overdamped system.

b) Increasing number of capacities in series increases the sluggishness of the response.

N Interacting Capacities in Series

Interaction increases the sluggishness of the overall response.

Hence a process with N capacities in series will necessitate a controller that will not only keep the final output of a desired value but will also try to improve the speed of the system's response.

1.9.1.1 Jacekted Coolers as Multicapacity Processes

Consider the batch cooler shown in Fig 1.35(a). The content of the tank is a mixture of components A and B is being cooled by constant flow of cold water circulating through the jacket. We can identify the following three capacities in series.

1. Heat Capacity of the mixture in the tank
2. Heat capacity of the tank's wall
3. Heat capacity of the coolant in the jacket.

It is easy to show that these capacities interact.

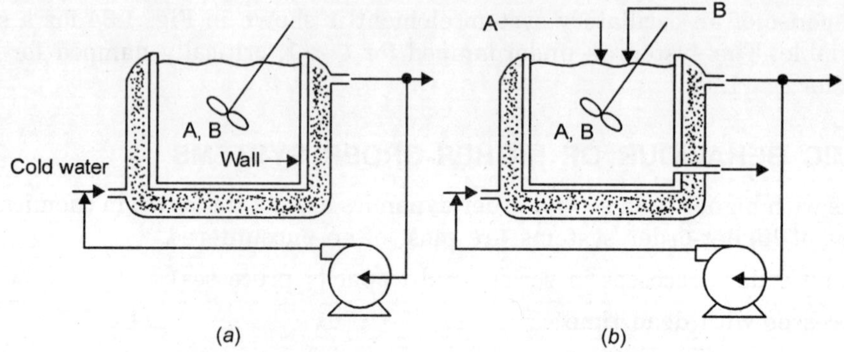

Fig. 1.35 Jacketed coolers (*a*) batch ; (*b*) continuous flow

For the jacketed continuous flow cooler of Fig. 1.35(*b*), we have more interacting capacities

1. Total material capacity of the tank
2. Tank's capacity for component *A*
3. Heat capacity of the tank's content
4. Heat capacity of the tank's wall
5. Heat capacity of the cold water in the jacket.

Here again, all the five capacities are interacting. It is expected that the response of the coolers to input changes will be overdamped and rather sluggish.

1.9.1.2 Staged Processes as Multicapacity Systems

Distillation and Gas absorption columns are very often encountered in chemical processes for the separation of a mixture into its components. Both systems have a number of trays. Each tray has material and heat capacities. Therefore, each column with *N* trays can be considered as a system with 2*N* capacities in series. From the physics of distillation and absorption it is easy to see that the 2*N* capacities interact.

Therefore, a step change in the liquid flow rate of the solvent at the top of the absorption column produces a very delayed, sluggish response for the content of solvent in the valuable component 'A' (Refer Fig. 1.36). This is because the input change has to travel through a large number of interacting capacities in series.

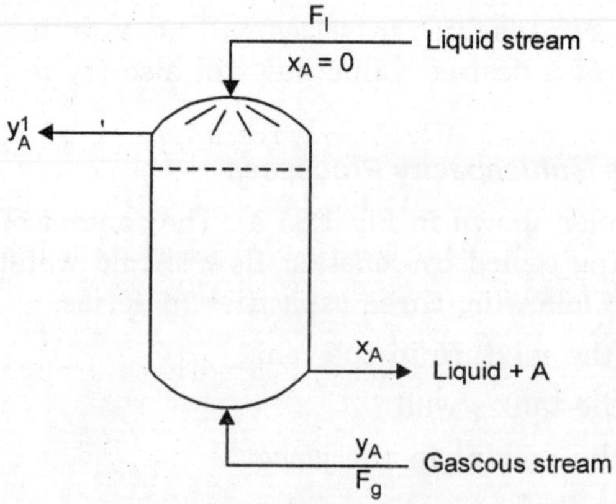

Fig. 1.36 Absorption column

Similarly, a step change in the reflux ratio of a distillation column (See Fig. 1.37) will quickly have an effect on the composition of the overhead product while the composition of the bottoms stream will respond very sluggishly (delayed and slow)

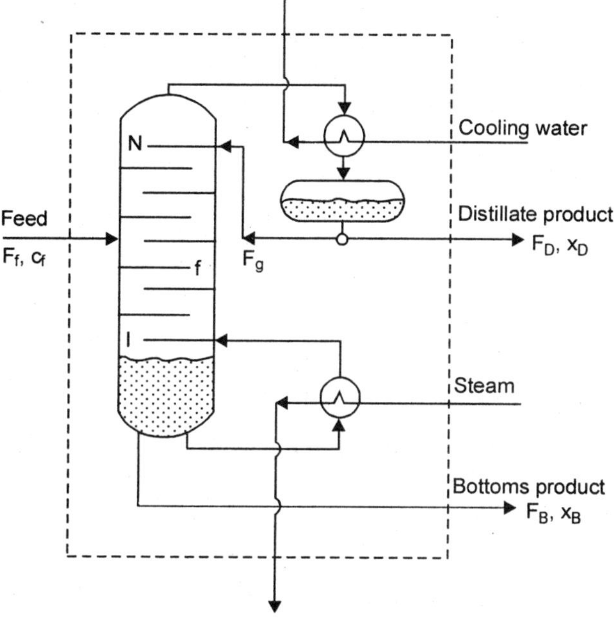

Fig. 1.37 Binary distillation column

Finally, a step change in the steam flow rate of the reboiler will have almost an immediate effect on the composition of the bottoms stream. On the contrary, the effect on the composition of the overhead product will be delayed and slow.

1.9.2 Dynamic Systems with Dead Time

So far we have assumed that there is no dead time between an input and output whenever a change took place in the input variable. In other words, whenever a change took place in the input variable, its effect was instantaneously observed in the behaviour of the output variable. But this is not true and contrary to our physical experience. Virtually all physical processes will involve sometime delay between the input and the output.

Consider a first-order system with a deadtime t_d between the input $f(t)$ and the output $y(t)$. We can represent such system by a series of two systems as shown in Fig. 1.38(a), that is, a first-order system in series with a dead time. For the first-order system we have the following transfer function :

$$\frac{L[y(t)]}{L[f(t)]} = \frac{\bar{y}(s)}{\bar{f}(s)} = \frac{Y(s)}{F(s)} = \frac{K_P}{\tau_P s + 1}$$

While for the dead time we have,

$$\frac{L[y(t-t_d)]}{L[y(t)]} = e^{-t_d s}$$

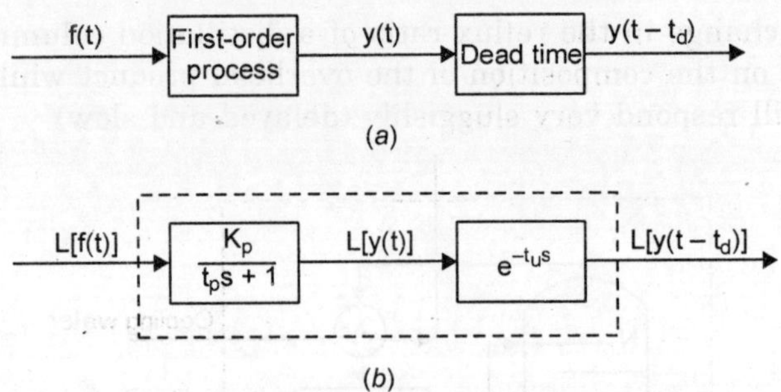

Fig. 1.38 (*a*) Process with dead time ; (*b*) Block diagram

Therefore, the transfer function between the input $f(t)$ and the delayed output $y(t - t_d)$ is given by : [Refer Fig 1.38(*b*)]

$$\frac{L[y(t-t_d)]}{L[f(t)]} = \frac{K_p e^{-t_d s}}{\tau_p s + 1} \tag{1.117}$$

Similarly, the transfer function for a second-order system with delay is given by :

$$\frac{L[y(t-t_d)\}}{L[f(t)]} = \frac{K_p e^{-t_d s}}{\tau^2 s^2 + 2\zeta \tau s + 1} \tag{1.118}$$

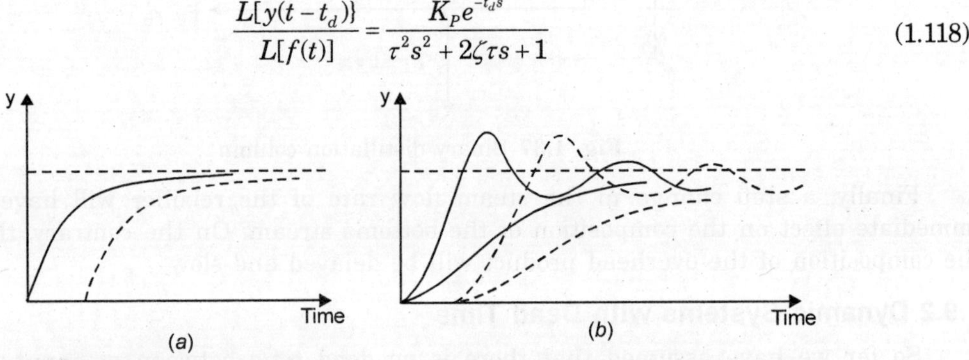

Fig. 1.39 Response of time-delayed systems to step input change
(*a*) First -order (*b*) Second-order

Fig. 1.39 shows the response of first-order and second-order systems with deadtime to a step change in the input. Processes with deadtime are difficult to control because the output does not contain information about current events.

1.9.3 Dynamic Systems with Inverse Response

The dynamic behaviour of certain processes is in such a way that initially the response is in the opposite direction to where it eventually ends up. Such behaviour is called 'Inverse response' or 'Nonminimum phase response' and it is exhibited by a small number of processing units. An example of such a process is boiler water level control system.

1.9.3.1 Boiler Drum Level Control : (Ref Fig. 1.40)

Proper boiler operation requires that the level of water in the steam drum be maintained within a certain band. The water level in the steam drum is related to, but is not a direct indicator of, the quantity of water in the drum. At each boiler load there is a different volume in the water that is occupied by steam bubbles. Thus, as load is increased there are more steam bubbles, and this causes the water to 'swell', or rise, rather than fall, because of the added water usage. Therefore if the drum volume is kept constant, the corresponding mass of water is minimum at high boiler loads and maximum at low boiler loads. The control of feed water therefore needs to respond to load changes and to maintain water by constantly adjusting the mass of water stored in the system.

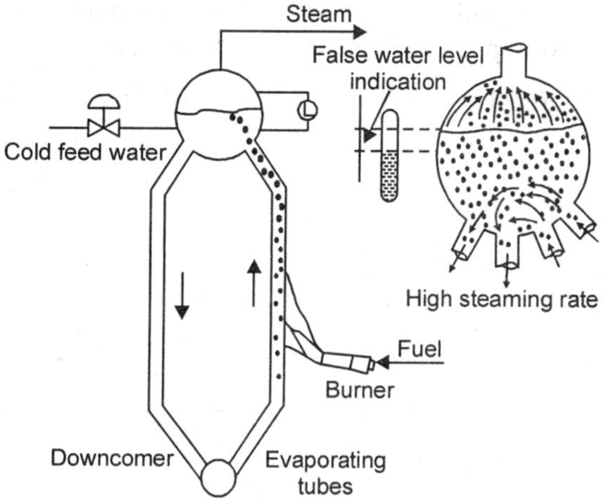

Fig. 1.40 Partial vaporization in the evaporating tubes causes drum level to shrink when feed water flow increases and when pressure rises. On the other hand, an increase in the demand for steam causes the level to "Swell"

Feed water is always colder than the saturated water in the drum. Some steam is then necessarily condensed when contacted by the feed water. As a consequence, a sudden increase in feed water flow (step change in feed water flow) tends to collapse some bubbles in the drum and temporarily reduce their formation in the evaporating tubes. Then, although the mass of the liquid in the system has increased, the apparent liquid level in the drum falls(shrinks). Equilibrium is restored within seconds, and the level will begin to rise. Nonetheless, the initial reaction to a change in feed water flow tends to be in the wrong (reverse) direction. This property is called 'Inverse Response' and it causes an effective delay in control action, making control more difficult. Such behaviour is the net result of two opposing effects and can be explained as follows referring to Fig 1.41.

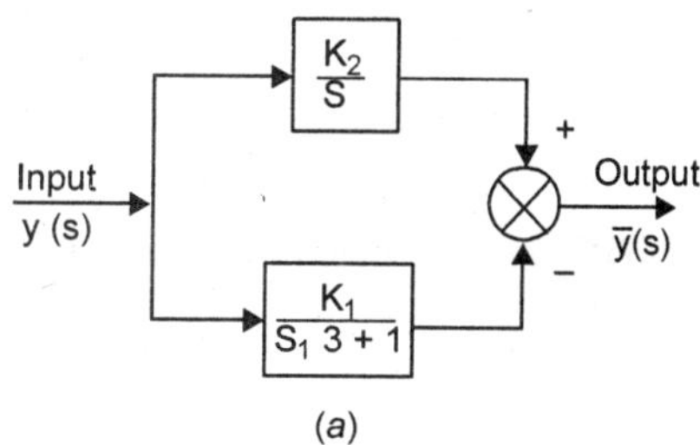

(*a*)

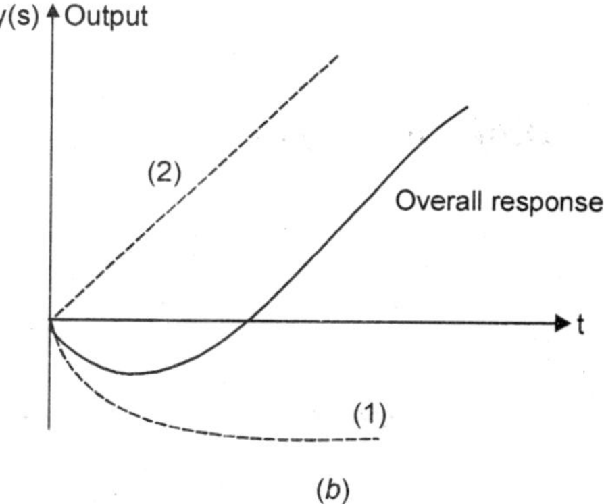

(*b*)

Fig. 1.41 Block diagram of liquid level in a boiler system and its inverse response

1. The cold feedwater causes a temperature drop which decreases the volume of the entrained steam bubbles. This leads to a decrease of the liquid level of the boiler water, following first-order behaviour [Curve 1 in Fig. 1.41(*b*)]

   That is, $-K_1/(\tau_1 s+1)$
2. With constant heat supply, the steam production remain constant and consequently the liquid level of the boiler water will start increasing in an integral form (pure capacity), leading to a pure capacitive response, K_2/s [curve 2 in Fig. 1.41(*b*)].
3. The result of the two opposing effects is given by [Refer Fig 1.41(*a*)],

$$\frac{K_2}{s} - \frac{K_1}{\tau_1 s+1} = \frac{(K_2\tau_1 - K_1)s + K_2}{s(\tau_1 s+1)} \tag{1.119}$$

   and for $K_2\tau_1 < K_1$ the second term $[-K_1/(\tau_1 s+1)]$ dominates initially and we take the inverse response. If the condition above is not satisfied, we don't have inverse response.

When $K_2\tau_1 < K_1$, then from equation (1.119) we notice that the transfer function has a positive zero, at the point s = $-K_2/(K_2\tau_1 - K_1) > 0$

Table 1.1 Systems with inverse response

1. Pure capacitive minus first-order response

$$G(s) = \frac{K_2}{s} - \frac{K_1}{\tau_1 s+1} = \frac{(K_2\tau_1 - K_1)s + K_2}{s(\tau_1 s+1)}$$

For $K_2\tau_1 < K_1$ zero $= -K_2(K_2\tau_1 - K_1) > 0$

2. Difference between two first-order responses

$$G(s) = \frac{K_1}{\tau_1 s+1} - \frac{K_2}{\tau_2 s+1} = \frac{(K_1\tau_2 - K_2\tau_1)s + (K_1 - K_2)}{(\tau_1 s+1)(\tau_2 s+1)}$$

For $\frac{\tau_1}{\tau_2} > \frac{K_1}{K_2} > 1$ zero $= -(K_1 - K_2)/(K_1\tau_2 - K_2\tau_1) > 0$

3. Difference between two first-order responses with dead time

$$G(s) = \frac{K_1 e^{-t_1 s}}{\tau_1 s+1} - \frac{K_2 e^{-t_2 s}}{\tau_2 s+1}$$

For $K_1 > K_2$ and $t_1 > t_2 \geq 0$.

4. Second order minus first order response

$$G(s) = \frac{K_1}{\tau^2 s^2 + 2\zeta\tau s + 1} - \frac{K_2}{\tau_2 s+1}$$

For $K_1 > K_2$.

5. Different between two second-order responses

$$G(s) = \frac{K_1}{\tau_1^2 s^2 + 2\zeta_1\tau_1 s + 1} - \frac{K_2}{\tau_2^2 s^2 + 2\zeta_2\tau_2 s + 1}$$

For $\frac{\tau_1^2}{\tau_2^2} > \frac{K_1}{K_2} > 1$

6. Different between two second-order responses with dead time

$$G(s) = \frac{K_1 e^{-t_1 s}}{\tau_1^2 s^2 + 2\zeta_1\tau_1 s + 1} - \frac{K_2 e^{-t_2 s}}{\tau_2^2 s^2 + 2\zeta_2\tau_2 s + 1}$$

For $K_1 > K_2$ and $t_1 > t_2 \geq 0$.

The above drum level control example demonstrates that the inverse response is the result of two opposing effects. Table 1.1 shows several such opposing effects between first or second-order systems. In all cases we notice that when the system possesses an inverse response, its transfer function has a positive zero. In general, the transfer function of a system with inverse response is given by :

$$G(s) = \frac{b_m s^m + b_{m-1} s^{m-1} + \ldots\ldots\ldots\ldots + b_1 s + b_0}{a_n s^n + a_{n-1} s^{n-1} + \ldots\ldots\ldots\ldots + a_1 s + a_0} \qquad (1.120)$$

Where one of the roots of the numerator (i.e.,one of the zeros of the transfer function) has positive real part.

Systems with inverse response are particularly difficult to control and require special attention.

1.9.3.2 Inverse response from two opposing First-order Systems

Fig. 1.42 shows another possibility of inverse response. Two opposing effects result from two different first-order processes, yielding an overall response equal to

$$\bar{y}(s) = \left(\frac{K_1}{\tau_1 s + 1} - \frac{K_2}{\tau_2 s + 1}\right)\bar{f}(s)$$

or

$$\bar{y}(s) = \left(\frac{(K_1\tau_2 - K_2\tau_1) + (K_1 - K_2)}{(\tau_1 s + 1)(\tau_2 s + 1)}\right)\bar{f}(s) \qquad (1.121)$$

We have inverse response when :

Initially (at $t = 0_1$) process 2, which reacts faster than process1 (i.e., $K_2/\tau_2 > K_1/\tau_1$), dominates the response of the overall system, but

Ultimately process 1 reaches a higher steady-state value than process 2 (i.e., $K_1 > K_2$), and forces the response of the overall system in the opposite direction. [Refer Fig. 1.42(*b*)].

Note that when $\tau_1/\tau_2 > K_1/K_2 > 1$ the process exhibits inverse response, and we find that the system's transfer function has a positive zero.

$$Z = -\frac{K_1 - K_2}{K_1\tau_2 - K_2\tau_1} > 0$$

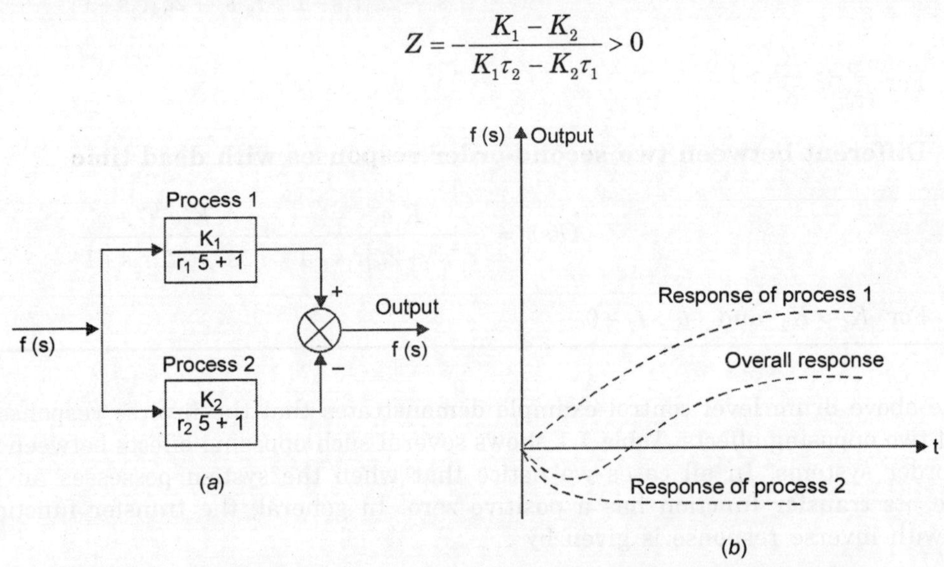

Fig. 1.42 (*a*) Block diagram of two opposing first-order systems ; (*b*) the resulting inverse response

1.10 BATCH AND CONTINUOUS PROCESS

A process in which the materials or work are stationary at one physical location while being treated is termed a 'Batch Process'. Batch processes are most often of the thermal type where materials are placed in a vessel or furnace and the system is controlled for a cycle of temperatures under controlled pressure for a period of time. Batch or Hood annealing of steel rolled coils, steel melting in bessemer converters, coke making in coke ovens, furnaces in foundries, batch reactors in chemical plants etc are some of the familiar batch processes. Idlies making in kitchen is one of the simplest example of batch process.

Batch processes are nearly always defined by temperature, pressure, or associated conditions such as composition. The degrees of freedom are usually well-defined. The purpose of such processes is to produce one or more products at (*a*) a given composition, (*b*) a maximum amount, and (c) best economy (employing least materials, energy, and time) . In short production rate (quantity), quality and economics are all to be taken care of.

The product composition desired is that at the end of the processing period and thus cannot be measured during the process. Consequently it is necessary to manipulate the variables of the process in such a manner that the behaviour of the process insures obtaining the desired composition. Maximum production and best economy result when the variables of the process are properly manipulated. A process computer may be used to insure a relationship among variables providing best operation.

A process in which the materials or work flows more or less continuously through a plant apparatus while being manufactured or treated is termed a 'continuous process'. Heating and rolling of steel ingots or billets, production of sinter, continuous annealing of metal sheets, production of steam and hence power, continuously stirred tank reactors (CSTR) etc are some of the familiar continuous processes.

Continuous processes possesses a number of degrees of freedom given by the number of variables and defining relations for the system. These variables are generally the temperature, pressure, flow-rate, and composition of each of the entering and leaving materials. Usually the purpose of the process is to produce one or more product at (a) a given composition, (b) a given or maximum flow rate, and (c) best economy (employing least materials, energy, personnel time, and equipment). Product composition is best insured by measuring product composition and controlling it by manipulating one of the degrees of freedom of the process. Fixed product flow rate usually requires flow controllers at several points of entering and leaving materials. Best economy is accomplished by maintaining all process variables in a predetermined relation such that the highest efficiency, least waste or some other criteria are satisfied.

1.10.1 Batch Process Control

If the process reaction rate is slow, and lags and dead time are small, a two-position (on-off) controller is generally satisfactory. If the process reaction rate is large or if lags are not small, it may be necessary to employ proportional control in order to avoid excessive cycling of the controlled variable. Offset in the final value may be eliminated by means of integral response.

Let us take a batch process where the temperature (controlled variable C) is raised slowly to the set point (v) and maintained for a particular period of time. The manipulated variable (m) is adjusted by the control valve and hence the valve position is indirectly represented for manipulated variable. Fig. 1.42(a) shows the temperature (C) and the corresponding valve position (m) with proportional control. When the processing begins, the valve opens fully to raise the temperature. When the temperature reaches the lower edge of the proportional band, the valve begins to close. The temperature cycles about the setpoint before becoming stable. With most processes a gradually decreasing valve setting is required in order to balance energy (offset). The gradual closing of the valve can be accomplished only by a corresponding deviation of the variable from the setpoint. As long as proportional sensitivity is high, offset will also be small.

Offset may be eliminated by means of integral response. The action of the proportional-integral controller is shown in Fig. 1.42(b). The offset of temperature is nearly eliminated, and the temperature is maintained closely to the setpoint after it has stabilished. Notice that the initial overshoot of temperature, when approaching the set point for the first time, is much larger than with only proportional control. The large initial overshoot is due to the action of integral response. During the heating up period the temperature is, of course, below the set point. The normal action of integral response is to shift the proportional band to urge the temperature toward the set point. Since the valve is already fully open, integral response simply shifts the proportional band all the way.

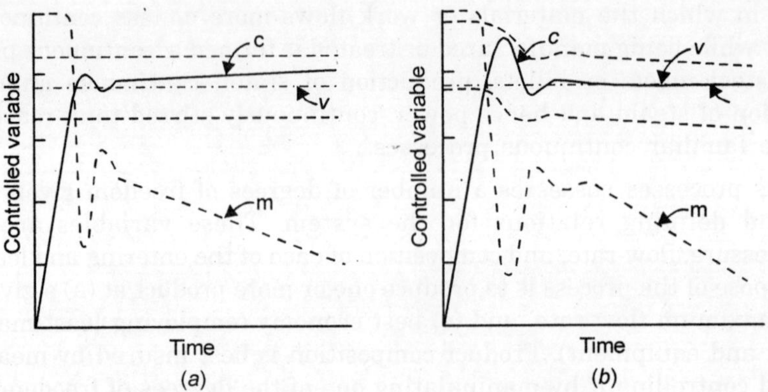

Fig. 1.42 Overshoot in batch-process control

Elimination of overshoot

Overshoot is prevented in one of several ways. First, many electric controllers with integral response incorporate a 'rate of approach setter'. This device limits the integral action until the first approach to the set point has been made. Second, a controller which incorporates a proportional derivative unit followed by a proportional -integral unit also may be adjusted to prevent overshoot. In the latter case the derivative response provides 'anticipation' by making the integral controller think that the controlled variable has approached the set point as illustrated in Fig. 1.43. The sensitivity of the first controller may be adjusted so that with a given setting of derivative time the overshoot is eliminated.

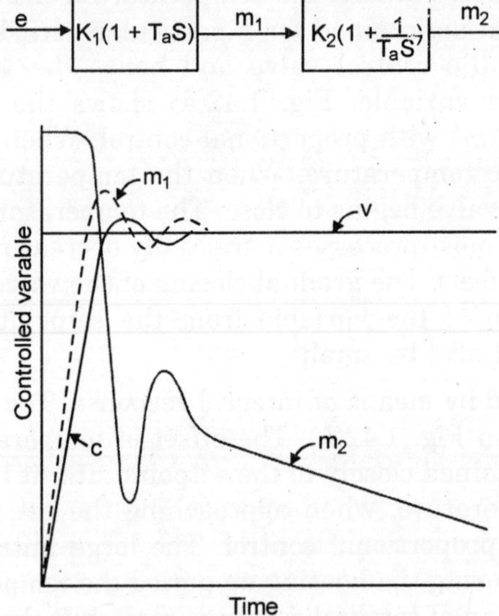

Fig. 1.43 Series-control elements

Time-variable control

In the processing of metals and chemicals it is sometimes necessary to vary the controlled variable over a definite time schedule. For example, in annealing steel, a schedule may be required such that the furnace temperature is raised at a desired rate to 1200°C, held or soaked for 8 hrs and lowered to 300°C in 6 hrs before allowing for natural cooling. This time schedule must be incorporated into the automatic control system. Time-variable control operates on the principle that the set point is moved through the desired time schedule. The control system functions to maintain the controlled variable (Temperature here) as close to the moving set point as required. Very often it is required to change the controlled variable at a steady rate in such batch processes involving controlled rates of heating or cooling.

1.10.2 Continuous-Process Control

The problems of continuous -process control are caused by load changes. Flow rate of the materials is almost always important in continuous processing, particularly where quantitative reactions are involved as in blending . On the other hand the supply of material is sometimes not constant and a flow control is necessary as shown in Fig. 1.44. The resulting flow of materials is nearly constant inspite of large changes of head in the tank. It is necessary to size the tank only large enough so that the longest charging period will not run the tank over, and the longest off-time of the inlet will not allow the tank to run dry. In order to keep load changes at a minimum, a short tank of large cross-section area (large capacitance) should be employed.

In many applications, particularly in continuous processing, the outflow of one unit becomes the feed to a succeeding unit. In order to obtain stability of operation in the plant, it is sometimes important that fluctuation of outflow and inflow be reduced to a minimum, thereby maintaining all feeds relatively constant. The storage capacity of the vessel may be utilized to proportion out flow against changes of inflow. The vessel thereby serves as a surge tank for absorbing fluctuations in flow rates. Averaging control gets its name since the outflow is 'averaged' against level, and the level is controlled between upper and lower limits rather than at a single point.

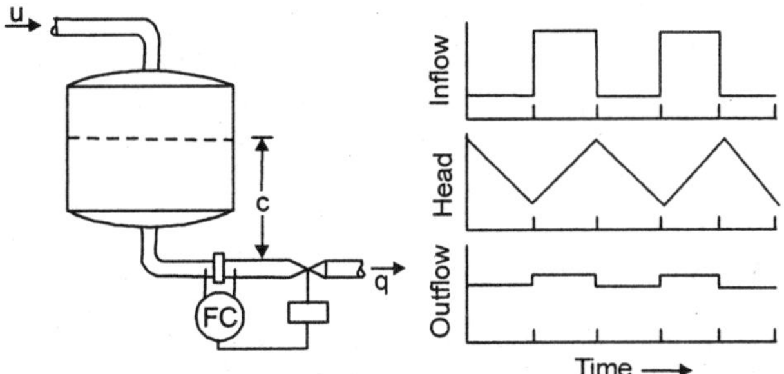

Fig. 1.44 Storage-vessel control

For the vessel in Fig 1.45 with control of outflow (outlet valve at a fixed setting),

$$A\frac{dc}{dt} = u - q \qquad (1.122)$$

Where A = cross-section area of tank. When the vessel is under controlled pressure, as is generally the case, the outflow is then constant if the pressure differential at the valve is constant. Obviously if there is no control, the tank would ultimately fill or run dry for changes in inflow. If control of level is added,

$$-q = K_C\,(u - c) \qquad (1.123)$$

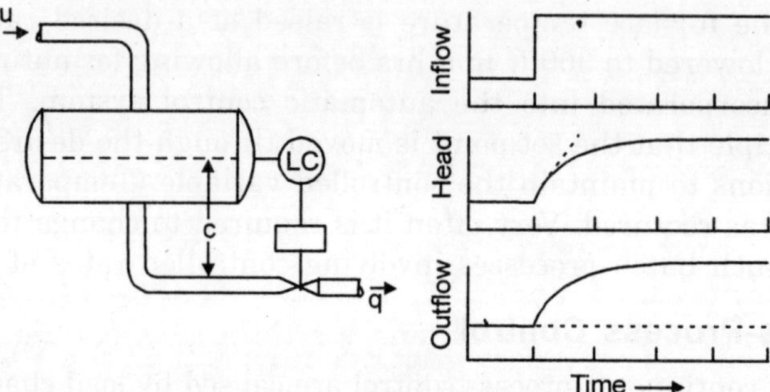

Fig. 1.45 Surge vessel control

where K_C is the proportional sensitivity of the proportional controller. Combining equations (1.122) and (1.123) after taking laplace transform, we get

$$\frac{Q_{(S)}}{U_{(S)}} = \frac{1}{\tau_S + 1} \tag{1.124}$$

and

$$\frac{C_{(S)}}{U_{(S)}} = \frac{1}{K_C} \cdot \frac{1}{\tau_S + 1} \tag{1.125}$$

where $\tau = \dfrac{A}{K_c}$

The inflow, head and outflow are also shown in Fig. 1.45. (Dotted curves are without level control). The time constant 'τ' may be selected by adjusting the proportional sensitivity of the controller so as to 'spread out' the inflow change over a period of about four time constants. In order to absorb changes of inflow the time constant 'τ' should be made as large as possible (proportional sensitivity K_e as small as possible) without completely filling or emptying the tank. Sometimes a proportional-integral controller is used with integral time set to a large value so that the outflow is further increased and the head in the tank maintained near the set point.

1.11 SELF REGULATION

A significant characteristic of some processes is the tendency to adopt a specific value of the controlled variable for nominal load with no control operations. Such a property of the process is called 'self-regulation'.

Consider the behaviour of the variable 'x' shown in Fig. 1.46. Notice that at time $t = t_0$ the constant value of 'x' is disturbed by some external factors, but that as time progresses the value of 'x' returns to its initial value and stays there. If 'x' is a process variable such as temperature, pressure, concentration, or flow rate, we say that the process is 'stable' or 'self-regulating' and needs no external intervention for its stabilization. It is clear that no control mechanism is needed to force 'x' to return to its initial value.

In contrast to the behaviour described above, the variable 'y' shown in Fig. 1.47 does not return to its initial value after it is disturbed by external influences. Processes whose variables follow the pattern indicated by 'y' in Fig. 1.47 (curves A, B, C) are called unstable processes and require external control for the stabilization of their behaviour.

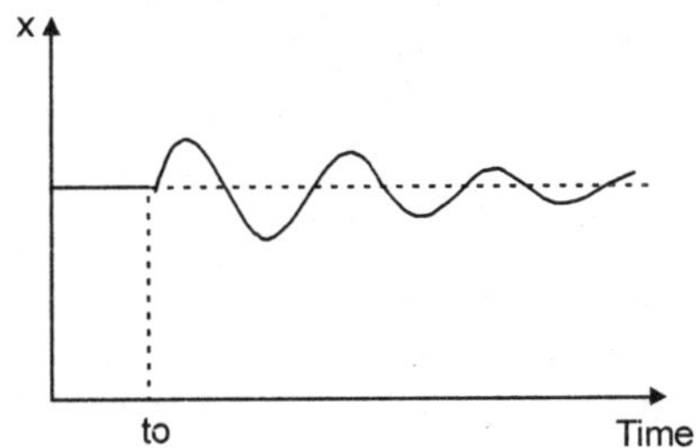

Fig. 1.46 Response of a stable system

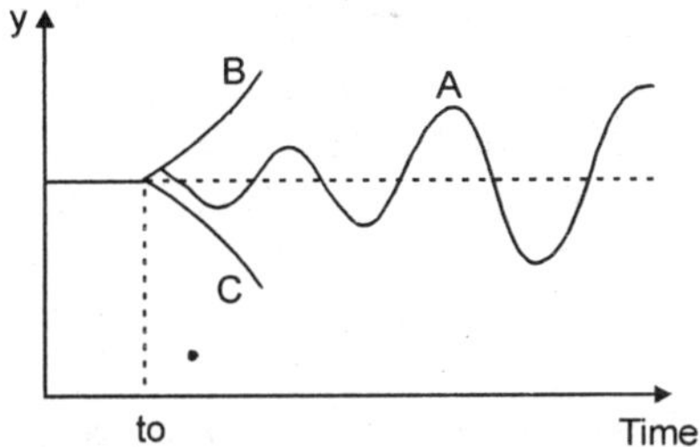

Fig 1.47 Alternative responses of unstable systems

1.11.1 Control of Temperature by Process Control

Consider the control of liquid temperature in a tank as shown in Fig. 1.48. The controlled variable is the liquid temperature 'T_L'. This temperature depends on many parameters in the process, for example, the input flow rate in a pipe 'A', the output flowrate via pipe 'B', the ambient temperature 'T_A', the steam temperature 'T_S', inlet liquid temperature 'T_O' and the steam flow rate 'Q_S'.

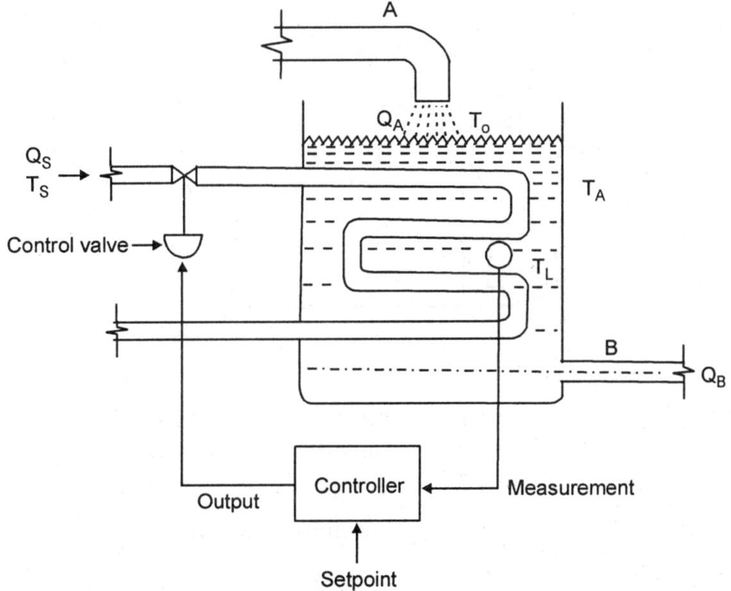

Fig. 1.48 Control of Temperature by process control

In this case, the steam flow rate is the 'Controlling parameter' chosen to provide control over the variable (liquid temperature). If one of the other parameters changes, a change in temperature results. To bring the temperature back to the set point value, we only change the steam flow rate, that is, heat input to the process.

The process explained above has 'self-regulation' as per the following argument. Suppose we fix the steam valve at 50 % opening so that no changes in valve position are possible when the process is going on. The liquid heats up until the energy carried away by the liquid equals that energy from steam flow. If the load changes, a new temperature is adopted (because the system temperature is not controlled). The process is self-regulating, however, because the temperature will not 'run away', but stabilises at some value under given conditions.

[An example of a process without self-regulation is tank from which liquid is pumped out at a fixed rate. Assume that the influx just matches with the outlet rate. Then the liquid in the tank is fixed at some nominal level. If the influx increases slightly, however, the level rises until the tank overflows. No self-regulation of the level is provided.]

1.11.2 CSTR with Cooling Jacket

Stable (self regulating) and unstable operations of a process can be well understood with the help of continuous stirred tank reactor (CSTR) operation. Consider a CSTR in which an irreversible exothermic reaction A → B takes place. The heat of reaction is removed by a coolant medium that flows through a jacket around the reactor. [Fig. 1.49]. As is known from the analysis of a CSTR system, the curve that describes the amount of heat released by the exothermic reaction is sigmoidal function of the temperature '*T*' in the reactor (curve '*A*' in Fig. 1.50). On the other hand, the heat removed by the coolant is the linear function of the temperature '*T*' (Line '*B*' in Fig. 1.50) consequently when the CSTR is at steady state, the heat produced by the reaction should be equal to the heat removed by the coolant. This requirement yields the steady states P_1, P_2 and P_3 at the intersection of curves A and B of Fig. 1.50. Steady states P_1 and P_3 are called 'stable', whereas P_2 is unstable. To understand the concept of 'Self-regulation' (stability), let us consider steady state P_2.

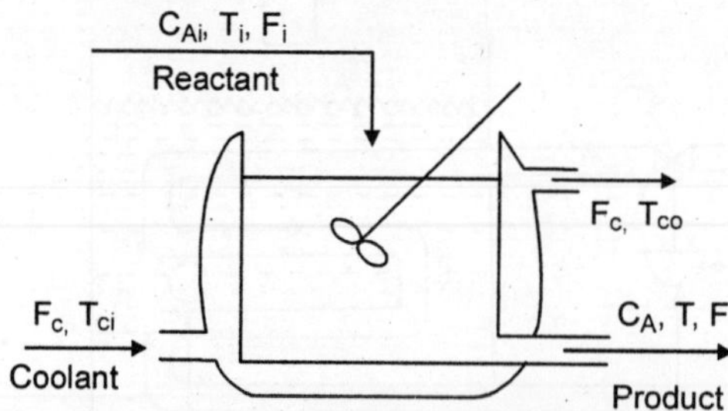

Fig 1.49 CSTR with cooling jacket

Assume that we are able to start the reactor at the temperature T_2 and the concentration C_{A2} that corresponds to this temperature. Consider that the temperature of the feed T_i increases. This will cause an increase in the temperature of the reacting mixture, say T_2'. At T_2' the heat released by the reaction (Q_2') is more than the heat removed by the coolant (Q_2''), thus leading to higher temperatures in the reactor and consequent to increased rates of reaction. Increased rates of reaction produce larger amounts of heat released by the exothermic reaction, which in turn lead to higher temperatures, and so on. Therefore, we see that an increase in T_i takes the reactor temperature away from steady state P_2 and that the temperature will eventually reach the value of the steady state P_3. (i.e., T_3). [Fig. 1.51(*a*)]. Similarly if T_i were to decrease, the temperature of the reactor would take off from P_2 and end up at P_1 (i.e., T_1) [Fig. 1.51(*b*)]. By contrast, if we were operating at steady state P_3 or P_1 and we perturbed the operation of the reactor, it would return naturally back to point P_3 or P_1 from which it started [Fig 1.51(*c*) and (*d*)].

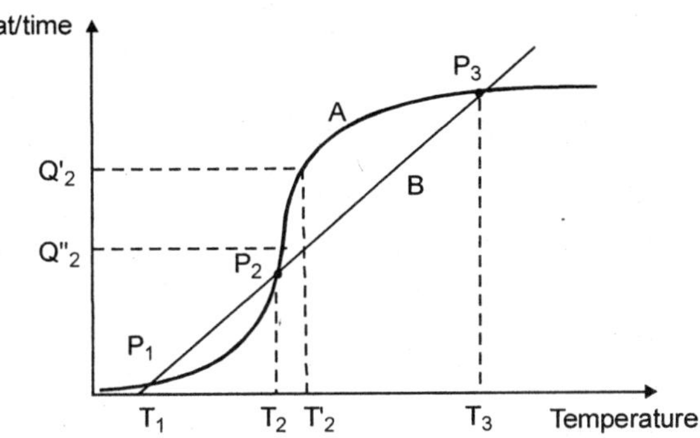

Fig 1.50 The three steady states of CSTR.

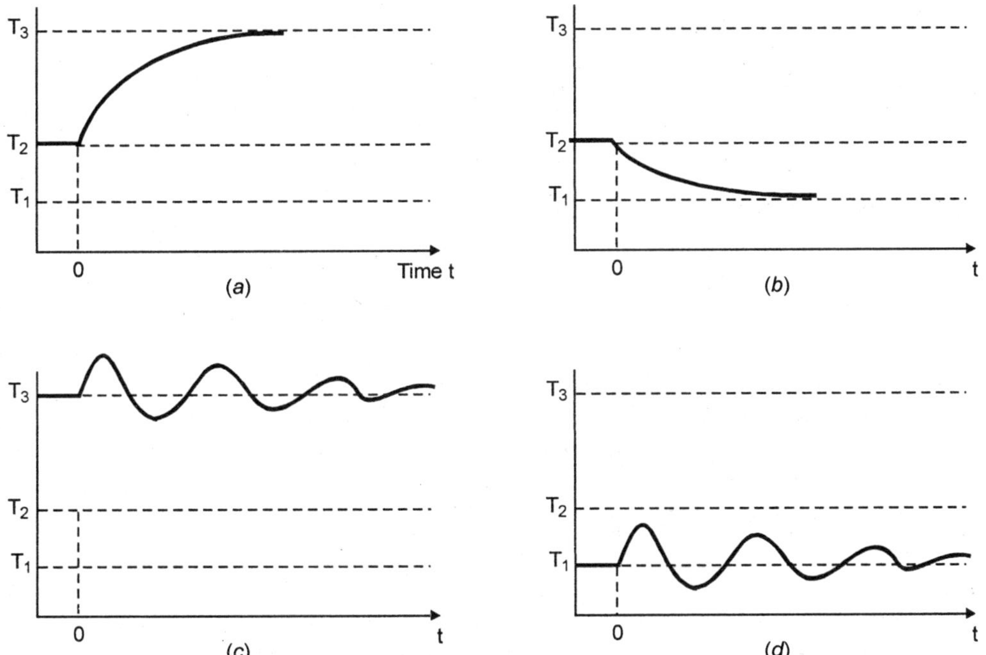

Fig. 1.51 Dynamic response of a CSTR
(*a*) and (*b*) indicate the instability of the middle steady state, while (*c*) and (*d*) demonstrates the stability of the other two.

For want of higher yield and safety reasons, sometimes we would like to operate the CSTR at the middle unstable steady state (P_2). In that case we need a controller that will ensure the stability of the operation at the middle steady state (P_2) in the presence of external disturbances that tend to take the system away from the desired point.

1.12 SERVO AND REGULATOR OPERATION

Servomechanisms and Regulators are used to control the process either via automatic controllers or as a self contained unit. They are physically doing the job of adjusting the manipulated variable to have the controlled variable at around set point. A controller automatically adjusts one of the inputs to the process in response to a signal fed back from the process output.

1.12.1 Servo Operation

If the purpose of the control system is to make the process follow changes in the set point as closely as possible, such an operation is called servo operation. Changes in load variables such as uncontrolled flows, temperature and pressure cause large errors than the set point changes (normally in batch processes). In such cases servo operation is necessary. Though the set point changes quite slowly and steadily, the errors from load changes may be as large as the errors caused by the change of set point. In such cases also servo operation may be considered.

A type of control system in common use, which has a slightly different objective from process control is called 'Servomechanism'. In this case the objective is to force some parameter to vary in a specific manner. This may be called 'a tracking control system' or 'a following control system'. Instead of regulating a variable value to a set point, the servomechanism forces the controlled variable value to follow variation of the reference value. This kind of automatic control is characterized mainly by a variable desired value, but a fixed load. Power-steering of automobiles, ship- sheering mechanism, robot arm movements etc are some of the examples.

1.12.2 Regulator Operation

In many of the process control applications, the purpose of control system is to keep the output (controlled variable) almost constant in spite of changes in load. Mostly in continuous processes the set point remains constant for longer time. Such an operation is called 'Regulator Operation'.

The set point generated and the actual value from sensors are given to a controller. The controller compares both the signals, generates error signal which is utilized to generate a final signal as controller output. The controller output is finally utilized to physically change the values of manipulated variable to achieve stability. The above action is achieved with the help of final control elements. They are operatable either with electrical, pneumatic or with hydraulic signals.

The system that serves good for servo operation will generally not be the best for regulator operation. Large capacity or inertia helps to minimize error here whereas it makes the system sluggish in case of servo operation. If the setpoint changes quite slowly and steadily, but the error from the load changes is as large as if caused by sudden change of setpoint, the responses of both regulator and servo operations may be considered for such systems.

Self-Contained Regulator Systems

The common feature of a self-contained regulator is that the sensor, the controller, and the final control element (in other words, the whole control loop) are combined into a single unit. Temperature regulators, Flow regulators, Pressure regulators, Differential pressure regualtors, Level regulators etc are such self-contained regulator systems. They are low cost ones and can be mounted directly on the process lines or furnaces. Regulators are mostly self contained mechanical devices, while controllers require an external energy (Pneumatic, Hydraulic or Electric). They are called regulators and not controllers because they are mechanical (mostly) and self-contained, requiring no external energy source.

1.13 PROBLEMS AND SOLUTIONS

Problem 1.1

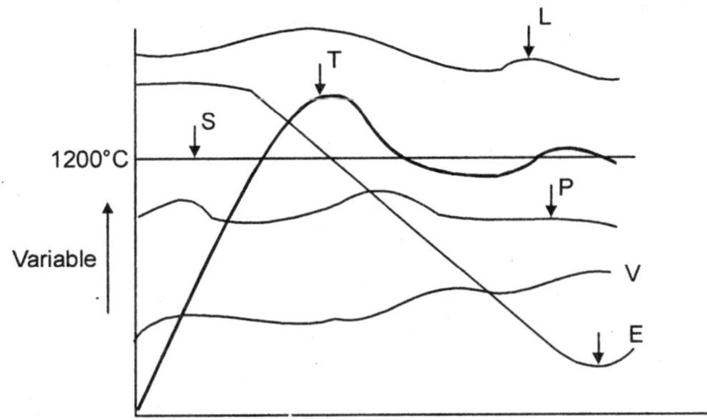

Fig 1.52 Variables of automatic process control

The various parameters of a certain process for controlling temperature is plotted as shown in Fig. 1.52. Identify the following variables :

(i) Controlled variable

(ii) Setpoint variable with its value.

(iii) Manipulated variable

(iv) Load variables.

Solution :

(*i*) Setpoint variable is '*S*' and its value = 1200°C [Remains constant through out the cycle].

(*ii*) Controlled variable is '*T*' [started raising from ambient temperature and reached the set point and stabilizes afterwards]

(*iii*) Manipulated variable is '*E*' [*E* is energy input to raise the temperature. Remains at maximum till controlled variable reaches setpoint and then decreases to the minimum and stays there to maintain the temperature].

(*iv*) Load variables are *V*, *P* and *L*.

Problem 1.2

Fig. 1.53 shows a steam heater where the process fluid is heated with the help of steam. Find out the number of degrees of freedom of this heating system.

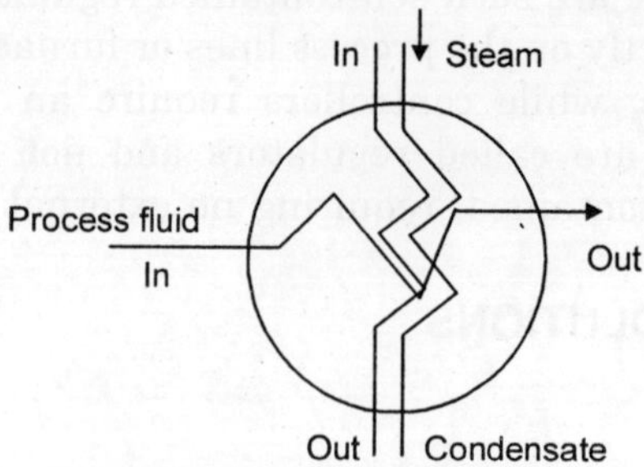

Fig. 1.53 Steam heater

Solution :

The number of system variables involved :

1. Inlet process fluid temperature T_i
2. Process fluid flowrate F_i
3. Outlet process fluid temperature T_O
4. Steam flow rate F_S.

The system-defining equation according to the first law of thermodynamics (law of conservation of energy) is given by :

$$H_S F_S = C_P F_i (T_O - T_i)$$

where H_S = Latent heat of steam

C_P = Specific heat of process fluid are parameters.

From the above we can conclude,

The number of variables that describe the system $n_V = 4$.

The number of system defining equation $n_e = 1$.

∴ Number of degrees of freedom

$$n = n_V - n_e = 4 - 1 = 3.$$

$$\therefore \quad n = 3$$

Hence a maximum of three automatic controllers can be used in this system.

Problem 1.3

A binary mixture (benzene and toluene) is to be distilled at atmospheric pressure as shown in Fig. 1.54. The variables are :

1. F_O–Overhead flow

2. T_O–Overhead temperature

3. E_i–Heat input rate.

Calculate the degrees of freedom. Also state how many number of automatic controllers may be used.

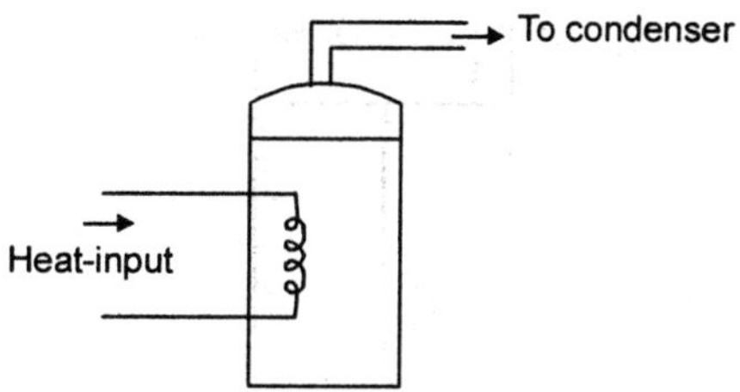

Fig. 1.54 Binary mixture

Solution :

For the distillation at constant pressure, the number of degrees of freedom

$$n = n_C - n_P + 1$$

The number of components $n_C = 2$

The number of phases $n_P = 2$

Therefore $n = 2 - 2 + 1 = 1$ (Temperature)

OR

For the process, the number of variables = 3 (F_O, T_O & E_i)

Number of system-defining equations = 2

(Conservation of energy and mass).

$\therefore$ $n = n_V - n_e = 3 - 2 = 1.$

Thus, no more than one automatic controller may be employed.

Problem 1.4

The flow rate through an exit pipe F_O in m^3 / sec is given by relation $F_O = 0.6\sqrt{h}$ where 'h' is the tank level in meter (Refer Fig. 1.8) . Find time constant τ_P for the steady state levels of 2 m and 5 m. Cross sectional area of the tank A' is 2 m^2.

Solution :

Refer section 1.7.10.1

$$\tau_P = \frac{2A\sqrt{hs}}{\beta} \qquad \text{for the relation } F_O = \beta\sqrt{h}$$

For $hs = 2$ m $\quad \tau_P = \dfrac{2 \times 2 \times \sqrt{2}}{0.6} = 9.43 \text{ sec}$

For $hs = 5$ m $\quad \tau_P = \dfrac{2 \times 2 \times \sqrt{5}}{0.6} = 14.9 \text{ sec}$

Problem 1.5

Derive the transfer function $H_{(s)}/Q_{(s)}$ for the liquid level system shown in Fig. 1.55. [H and Q are the deviation variables in 'h' and 'q' respectively.]

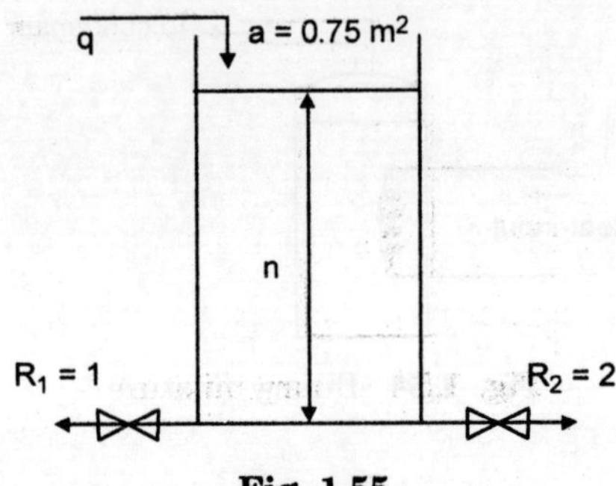

Fig. 1.55

Solution :

The material balance equation can be written as :

$$q - \frac{h}{R_1} - \frac{h}{R_2} = a\frac{dh}{dt}$$

at steady state :

$$q_S - \frac{h_S}{R_1} - \frac{h_S}{R_2} = 0$$

Substracting
$$(q - q_S) - \frac{(h - h_S)}{R_1} - \frac{(h - h_S)}{R_2} = a\frac{d(h - h_S)}{dt}$$

$$Q - \frac{H}{R_1} - \frac{H}{R_2} = a\frac{dH}{dt}$$

$$Q - H\left[\frac{1}{R_1} + \frac{1}{R_2}\right] = a\frac{dH}{dt}$$

$$Q - \frac{H}{R_{eq}} = a\frac{dH}{dt} \qquad \text{where} \quad R_{eq} = \frac{R_1 R_2}{R_1 + R_2} = \frac{2}{3}$$

Taking laplace transform

$$Q(s) - \frac{H(s)}{R_{eq}} = asH(s)$$

$$\frac{H(s)}{Q(s)} = \frac{R_{eq}}{aR_{eq}s + 1} = \frac{2/3}{0.5s + 1}$$

$$\frac{H(s)}{Q(s)} = \frac{0.67}{0.5s + 1}$$

Problem 1.6

A Tank system having a time constant of 0.5 min and a resistance of 0.25 min /m^2 is operating at steady state with an inlet flow of 2m^3/min. The flow is suddenly increased to 3m^3/min. Plot the response of the tank level (Assume area of cross section A = 2 m^2)

Solution :

From equation (1.23)

$$G(s) = \frac{\overline{h'(s)}}{F'_i(s)} = \frac{K_P}{\tau_P s + 1}$$

Where

K_P = R = 0.25 min/m^2

a = Cross sectional area of the tank = 2 m^2.

A = Change in inlet flow F_i from 2 m^3/min to 3 m^3/m (i.e., unit step)

τ_P = 0.5 min

$\therefore$

$$F_i(s) = 1/s$$

$$\frac{H(s)}{F_i(s)} = \frac{0.25}{0.5s + 1}$$

From equation 1.27, we get the response as

$$H_{(t)} = A.K_P(1 - e^{-t/\tau_P})$$

$$H_{(t)} = 1 \times 0.25(1 - e^{-t/0.5})$$

From the above equation the response of the tank level for the sudden increase of flowrate from 2 to 3 m^3/min can be tabulated and plotted as below : (Refer Table 1.2 and Fig 1.56).

Table 1.2 Response to step change

Time, min	H(t)
0	0
0.5	0.158
1.0	0.216
1.5	0.2376
2.0	0.245
2.5	0.2483
3	0.2493
4	0.25

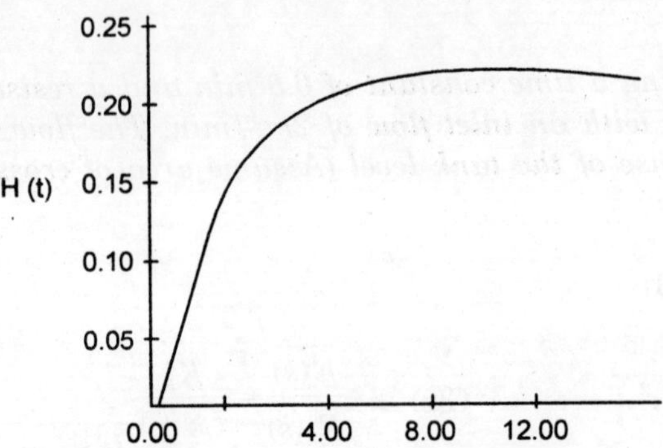

Fig. 1.56 Step response of the system

Problem 1.7 :

Derive the transfer function H(s)/Q(s) for the liquid level system shown in Fig. 1.57. When

(a) The tank operates about the steady state value of h_S *=0.3 m*

(b) The tank operates about the steady state value of h_S *=1 m*

The pump removes water at a constant rate of 0.3 m^3/min*, and is independent of head. The cross sectional area of the tank is 0.1* m^2 *and the resistance R is 11* m^2/min*.*

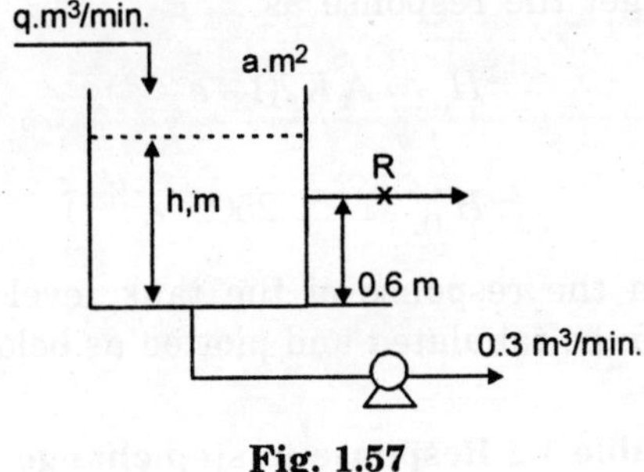

Fig. 1.57

Solution :

(*a*) The tank operates about the steady state value of 0.3 m. At this height the only outlet from the tank is through the pump.

Hence the mass balance equation can be written as

$$q - 0.3 = a\frac{dh}{dt}$$

At steady state

$$q_S - 0.3 = 0$$

substracting

$$(q - q_S) = a\frac{d(h - h_S)}{dt}$$

$$Q = a\frac{dH}{dt}$$

Taking laplace transform, we get

$$Q_{(S)} = a.s \ H_{(S)}$$

$$\therefore \qquad \frac{H(s)}{Q(s)} = \frac{1}{a.s} = \frac{1}{0.1s}$$

(*b*) The tank operates at a steady state value of h_S = 1 m. At this height the flow is through both the resistance and the pump.

The mass balance equation gives,

$$q - 0.3 - \left(\frac{h - 0.6}{R}\right) = a\frac{dh}{dt}$$

[Assuming the density of the liquid to be constant]

At steady state

$$q_S - 0.3 - \left(\frac{h_S - 0.6}{R}\right) = 0$$

Substracting

$$(q - q_S) - \frac{(h - h_S)}{R} = a\frac{d(h - h_S)}{dt}$$

$$Q - \frac{H_{(t)}}{R} = a\frac{dH}{dt}$$

Taking Laplace transform we get,

$$Q(s) - \frac{H(s)}{R} = a.s \ \text{H(s)}$$

$$RQ(s) - H(s) = R.a.s(H(s))$$

$$\frac{H(s)}{Q(s)} = \frac{R}{a.R.s + 1} = \frac{R}{\tau_p s + 1} = \frac{11}{11 \times 0.1 \times s + 1}$$

$$G(s) = \frac{11}{1.1s + 1}$$

Problem 1.8

A Thermometer having a time constant of 0.5 min is placed in a temperature bath and after thermometer comes to equilibrium with the bath, the temperature of the bath T_i is increased linearly at the rate of 1°C / min. What is the difference between the indicated and bath temperatures.

(a) 0.25 min after the change in temperature begins.

(b) 3 min after the change in temperature begins.

(c) *What is the maximum deviation between the indicated and bath temperature and when does it occur.*

(d) *How many minutes does the response lag after long enough time is elapsed?*

Solution :

Time constant, $\tau_P = 0.5$ min. Here $A = 1$

From equation (1.50)

$$y(t) = A(t - \tau_P) + A\tau_P e^{-1/\tau_P}$$

$$T = 1(t - 0.5) + 1 \times 0.5e^{-t/0.5} = (t - 0.5 + 0.5 + 0.5e^{-t/0.5})$$

Difference between indicated temperature, T, and bath temperature, T_i, is

$$\Delta = T - T_i = (t - 0.5 + 0.5e^{-t/0.5}) - t$$

$$= -0.5(1 - 0.5e^{-t/0.5})$$

(a) at $t = 0.25$ min, $\Delta = -0.5(1 - 0.5e^{-0.2/0.5}) = 0.1967°$

(b) at $t = 3.0$ min, $\Delta = -0.5(1 - 0.5e^{-3.0/0.5}) = 0.4987°$

(c) as $t \to \infty$, $\Delta \to -0.5$ which is the maximum value of the temperature deviation.

(d) as $t \to \infty$, $T_{(t)} = t - \tau_P$ and $T_i = t$ so the response lag the input by 0.5 minutes (τ_P).

Problem 1.9

Determine the transfer function $H(s)/Q\,i(s)$ for the liquid level system shown in Fig. 1.58. [Resistances R_1, R_2 and R_3 are linear. The flowrate from tank-4 is maintained constant at 'a' by means of a pump.]

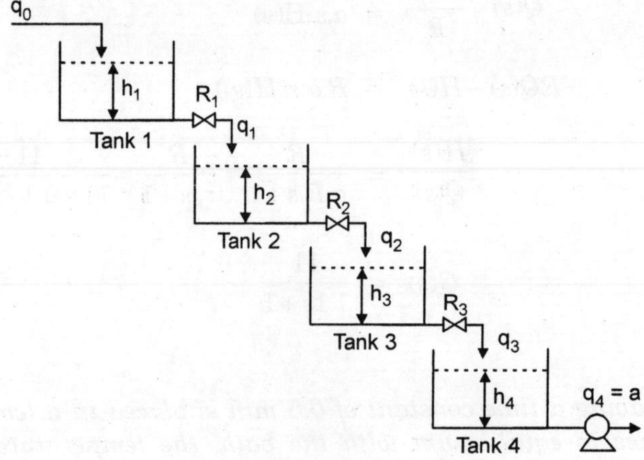

Fig 1.58

Solution :

Defining $Q_i = q_i - q_s$, $Q_1 = q_1 - q_{1s}$, $Q_2 = q_2 - q_{2s}$, $Q_3 = q_3 - q_{3s}$ and $H = h - hs$

The material balance for the tank-1 gives,

$$q_i - q_1 = a_1 R_1 \frac{dq_1}{dt}$$

$$Q_i(s) - Q_1(s) = a_1 R_1 Q_1(s)$$

Or

$$\frac{Q_1(s)}{Q_i(s)} = \frac{1}{a_1 R_1 s + 1}$$

Similarly for tank-2 and tank-3 we get

$$\frac{Q_2(s)}{Q_1(s)} = \frac{1}{a_2 R_2 s + 1}$$

$$\frac{Q_3(s)}{Q_2(s)} = \frac{1}{a_3 R_3 s + 1}$$

For tank-4, material balance gives,

$$q_3 - a = a_4 \frac{dh}{dt}$$

At steady state $q_{3s} - a = 0$

$$(q_3 - q_{3S}) = a_4 \frac{dH}{dt} \rightarrow Q_3(s) = a_{4S}\, H(s)$$

$$\frac{H(s)}{Q_3(s)} = \frac{1}{a_4 s}$$

$$\therefore \quad \frac{H(s)}{Q_i(s)} = \frac{Q_1(s)}{Q_i(s)} \times \frac{Q_2(s)}{Q_i(s)} \times \frac{Q_3(s)}{Q_2(s)} \times \frac{H(s)}{Q_3(s)}$$

$$= \frac{1}{a_1 R_1 s + 1} \times \frac{1}{a_2 R_2 s + 1} \times \frac{1}{a_3 R_3 s + 1} \times \frac{1}{a_4 s}$$

$$\frac{H(s)}{Q_i(s)} = \frac{1}{(a_1 R_1 s + 1)(a_2 R_2 s + 1)(a_3 R_3 s + 1) a_4 s}$$

Problem 1.10

There are N storage tanks of volume V arranged so that when water is fed into first tank, equal volume of liquid overflows from the first tank to the second tank and so on. Each tank initially contains zero concentration of component A and equipped with a perfect stirrer. At time zero, a stream of concentration 'Co' of component 'A' is fed into first tank at a volumetric flowrate 'q'. Find the resulting concentration in each tank as a function of time. [Refer Fig. 1.59]

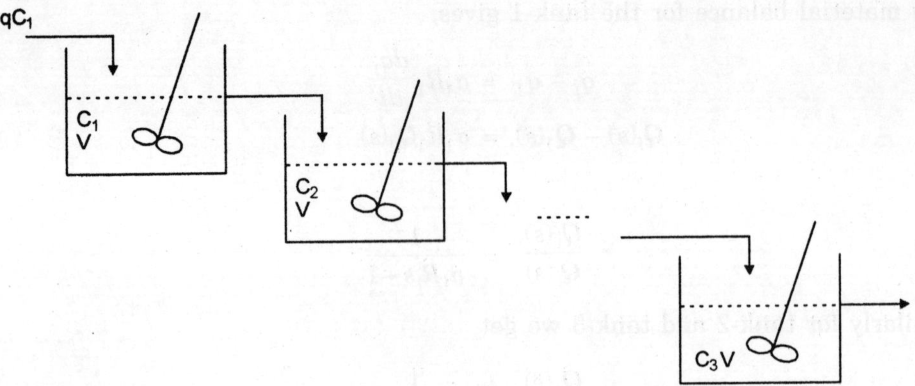

Fig. 1.59

Solution :

For tank-1

$$\frac{C_1}{C_i} = \frac{1}{\tau_P s + 1} \qquad \text{where } \tau_P = V/q$$

Since
$$C_i(s) = \frac{C_O}{s}, C_1(t) = C_O[1 - e^{-t/\tau p}]$$

In tank-2
$$\frac{C_2(s)}{C_i(s)} = \frac{1}{\tau_P s + 1} \quad or \quad C_2(s) = \frac{C_O}{s(\tau_P s + 1)^2}$$

$$\therefore \quad C_{2(t)} = C_O\left[1 - \left(1 + \frac{t}{\tau_P}\right)e^{-t/\tau p}\right]$$

Similarly,
$$\therefore \quad C_3(t) = C_O\left[1 - \left(1 + \frac{t}{\tau_P} + \frac{t^2}{2\tau_P{}^2}\right)e^{-t/\tau p}\right]$$

$$\therefore \quad C_n(t) = C_O\left[1 - \left(1 + \frac{t}{\tau_P} + \frac{t^2}{2!\tau_P{}^2} + \frac{t^3}{3!\tau_P{}^3} + ... + \frac{t^{n-1}}{n!\tau_P{}^{n-1}}\right)e^{-t/\tau p}\right]$$

Problem 1.11

Find the transfer function $H_2(s)/Q(s)$ and $H_3(s)/Q(s)$ for a three tank system in Fig 1.60. Where H_2, H_3 and Q are deviation variables. For a unit step change in Q, determine the initial and final heights in tank-3.

Solution :

Tanks 1 and 2 are interacting

$$\tau_{P1} = a_1R_1 = 2, \tau_{P2} = a_2R_2 = 2, \tau_{P3} = a_2R_3 = 1.$$

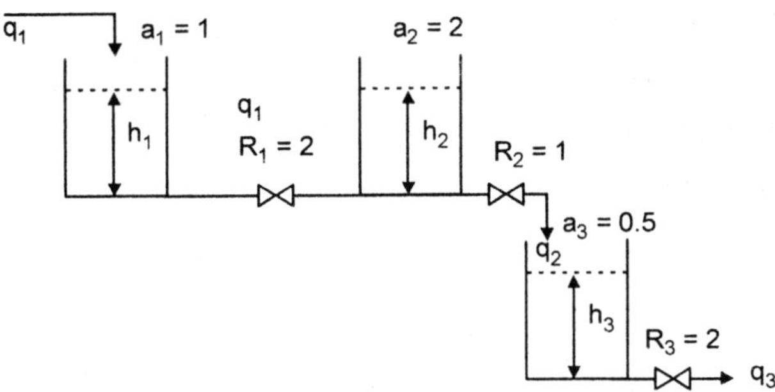

Fig. 1.60

Hence from equation (1.95)

$$\frac{H_2(s)}{Q_i(s)} = \frac{R_2}{\tau_{P1}\tau_{P2}s^2 + (\tau_{P1} + \tau_{P2} + a_1R_2)s + 1}$$

$$Q_2(s)/Q_i(s) = \frac{1}{4s^2 + 5s + 1} \qquad Q\ H_2(s) = \frac{Q_2(s)}{R_2}$$

Tank -3 is a first order process and hence from equation (1.23)

$$\frac{H_3(s)}{Q_2(s)} = \frac{K_{P3}}{\tau_{P3}s + 1} = \frac{R_3}{a_3R_3s + 1} = \frac{2}{s+1}$$

$$\therefore \qquad \frac{H_3(s)}{Q_i(s)} = \frac{H_3(s)}{Q_2(s)}\frac{Q_2(s)}{Q_i(s)} = \frac{2}{s+1} \times \frac{1}{4s^2 + 5s + 1}$$

For $Q_i(s)$ =1/s

$$H_3(s) = \frac{2}{s(s+1)(4s^2 + 5s + 1)}$$

(*i*) By initial value theorem

$$\underset{t\to 0}{Lt}\ [\mathrm{H}_3(t)] = \underset{S\to\infty}{Lt}\ [\mathrm{sH}_3(s)]$$

i.e., $H_3(o) = 0$

The initial height in tank 3 = 0

(*ii*) By final value Theorem

$$\underset{t\to\infty}{Lt}\ [\mathrm{H}_3(t)] = \underset{S\to 0}{Lt}\ [\mathrm{sH}_3(s)]$$

i.e., $H_3(\infty) = 2$

The final height in Tank 3 = 2.

Problem 1.12

A step change of magnitude 5 is introduced into a system having the transfer function $\frac{Y_{(S)}}{X_{(S)}} = \frac{8}{s^2 + 1.6s + 4}$. *Determine*

(a) Percent Overshoot

(b) Ultimate value of Y(t)

(c) Maximum value of Y(t)

(d) Period of Oscillation

(e) Rise time

Solution :

$$\frac{Y(s)}{X(s)} = \frac{8}{s^2 + 1.6s + 4} \qquad \text{and} \qquad X(s) = \frac{5}{s}$$

$$Y(s) = \frac{10}{s(0.25s^2 + 0.4s + 1)}$$

and $\tau^2 = 0.25,\ \tau = 0.5,\ 2\xi\tau = 0.4,\ \xi = 0.4$

(*a*) From equation (1.76)

$$\text{Overshoot} = \exp\left[\frac{-\pi\xi}{\sqrt{1-\xi^2}}\right]$$

$$\text{Percent overshoot} = 100 \exp\left[\frac{-\pi\xi}{\sqrt{1-\xi^2}}\right]$$

$$= 100\ e^{-\pi \times 0.4/\sqrt{1-0.4^2}} = 25.\ 4\ \%$$

(*b*) $Y_{\text{ultimate}} = Y(\infty)$, using final value Theorem

$$\underset{t\to\infty}{Lt}\ Y(t) = \underset{S\to 0}{Lt}\ sY(s) = \underset{S\to 0}{Lt} \frac{10}{0.25s^2 + 0.4s + 1}$$

Therefore, $Y_{\text{ultimate}} = 10$

(*c*) Overshoot $= \dfrac{Y_{\max} - Y_{(\infty)}}{Y_{(\infty)}}$

$$0.254 = \frac{Y_{\max} - 10}{10}$$

$$Y_{\max} = 2.54 + 10 = 12.54$$

(*d*) From equation (1.79)

$$\text{Period of oscillation T} = \frac{2\pi\tau}{\sqrt{1-\xi^2}}$$

$$T = \frac{2\pi \times 0.5}{\sqrt{1-0.42}} = 3.42$$

(*e*) From equation (1.75)

When $t = t_r$ i.e., risetime

$$\sin(\omega t_r + \phi) = 0$$

$$\therefore \quad \omega t_r + \phi = \pi$$

$$\therefore \quad t_r = \frac{\pi - \phi}{\omega} = \frac{\pi - \tan^{-1}\left[\frac{\sqrt{1-\xi^2}}{\xi}\right]}{\sqrt{1-\xi^2}/\tau} = \frac{\tau}{\sqrt{1-\xi^2}}\left[\pi - \tan^{-1}\left(\frac{\sqrt{1-\xi^2}}{\xi}\right)\right]$$

$$= \frac{0.5}{\sqrt{1-0.16}}\left[\pi - \tan^{-1}\left(\frac{\sqrt{1-0.16}}{0.4}\right)\right] = \mathbf{1.08\ min.}$$

1.14 PROBLEMS AND QUESTIONS

1. Name the variables of a liquid-heating tank heated by electric current in an electrical heater.
2. If the outflow at a vessel is proportional to the square root of head,
   a. What shape vessel results in a steady change of head ?
   b. What shape vessel results in a rate of change of head proportional to the head?
3. A system known to have a time-constant response requires 5 minutes to indicate 98 % of response. What is the time constant?
4. A tank operating at 3 m head, 5 lpm outflow through a valve and has a cross sectional area of 2 square metre. Calculate the time constant.
5. Define interacting system and give an example.
6. With an example, explain self-regulation of a process.
7. What is meant by non-self-regulation ?
8. Write any two characteristics of first-order process modeling.
9. What is a first order system and how do you derive the transfer functions of a first order lag or of a purely capacitive process.
10. What is the principle characteristic of the first-order processes, and what causes the appearance of a purely capacitive process.
11. Write the effect of time delay in process control system
12. What is the difference between interacting and non-interacting systems ?
13. Define the decay ratio of underdamped system.
14. Explain the term degree of freedom of a process.
15. In a stirred tank heater system shown in Fig. 1.61 below, the flow rate of the effluent system is proportional to the square root of the liquid level in the tank. Obtain the relation between change in outflow temperature for change in inflow temperature. Assume the inflow rate and steam temperature are constants.

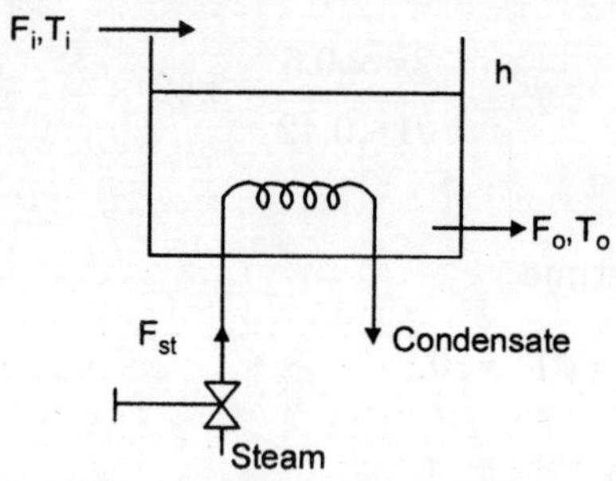

Fig 1.61

16. Consider a first order system with $\tau_P = 0.5$ and $K_P = 1$. Initially, the system is at steady state. Then the input changes linearly with time : m(t) = t.
    (*a*) Develop an expression that shows how the output changes with time in response to the input above.
    (*b*) What is the minimum and what is the maximum difference between the output $y(t)$ and input $m(t)$? When do these extreme points occur?
    (*c*) Plot the input m(t) and output y(t) in the same graph as functions of time.
17. Consider a second-order system with the following transfer function :

$$G(s) = \frac{\bar{y}(s)}{m(s)} = \frac{1}{s^2 + s + 1}$$

Introduce a step change of magnitude 10 into the system and find (*a*) Percent overshoot, (*b*) decay ratio, (*c*) maximum value of $y(t)$, (*d*) ultimate value of $y(t)$, (*e*) rise time, and (*f*) Period of oscillation.

18. Consider a conical water tank shown below. (Fig. 1.62). Write the dynamic material balance equation if the flow rate out of tank is a function of the square root of height of the water in the tank ($q_O = \beta\sqrt{h}$).

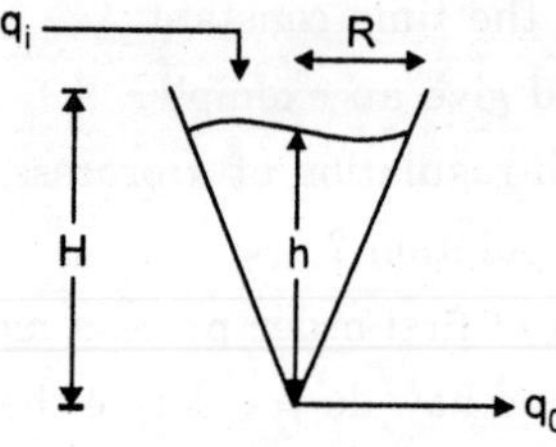

Fig 1.62

19. The two tank liquid system shown in Fig. 1.63 is operating at steady state when a unit step change is made in the flow rate to tank-1. The transient response is critically dampened and it takes 1 min. for the change in level of tank 2 to reach 50% of the total change. If the ratio of the cross-sectional area of the tanks $a_1/a_2 = 2$, calculate the ratio R_1/R_2. Calculate the time constant of each tank. How long does it take for the change in level of the first tank to reach 90 % of the total change ?

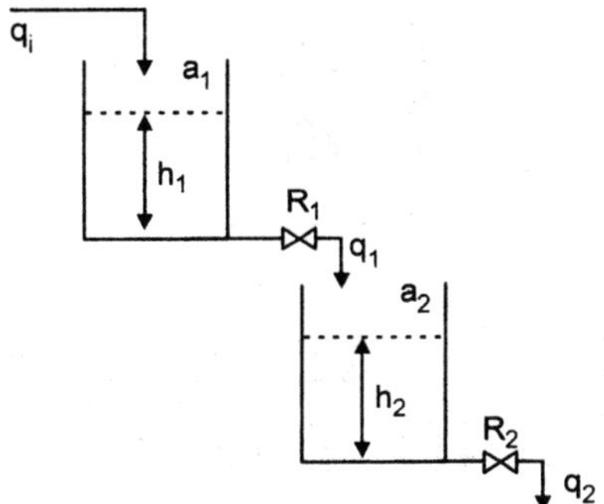

Fig 1.63

20. In the liquid level system shown in Fig. 1.64, the deviation in flow rate to the first tank is an impulse function of magnitude 5. The following data apply :

$a_1 = 0.3 \text{ m}^2$, $a_2 = a_3 = 0.6 \text{ m}^2$, $R_1 = 3.33 \text{ m/m}^3\text{/min}$ $R_2 = 5 \text{ m/m}^3\text{/min}$

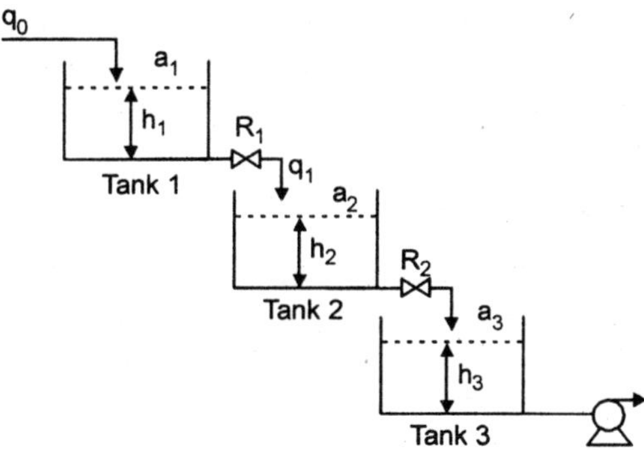

Fig 1.64

(*a*) Determine $H_1(s)$, $H_2(s)$ and $H_3(s)$ where H_1, H_2 and H_3 are the deviations in tank level.

(*b*) Sketch the response of H_1, H_2 and H_3.

(*c*) Determine H_1 $(t = 3.46 \text{ min})$, H_2 (3.46) and H_3 (3.46)

2

CONTROL CHARACTERISTICS AND TUNING

2.1 THE AUTOMATIC CONTROLLER

The automatic controller determines the value of the controlled variable, compares the actual value to the desired value, determines the deviation, and produces the counteraction necessary to maintain the smallest possible deviation between desired value and actual value. The method by which the automatic controller produces the counteraction is called the mode of control or control action.

2.1.1 Concept of Feedback control

Fig. 2.1 shows an automatic feedback control system. It consists of a process [with output 'y' (or 'c'), a potential load disturbance 'd' (or 'u'), and an available manipulated variable 'm'], measuring device, a controller and a final control element. The disturbance 'd' (also known as load or process load) changes in an unpredictable manner and our control objective is to keep the value of the output 'y' (or 'c') at desired level 'y_{sp}' (reference 'r'). A feedback control action takes the following steps.

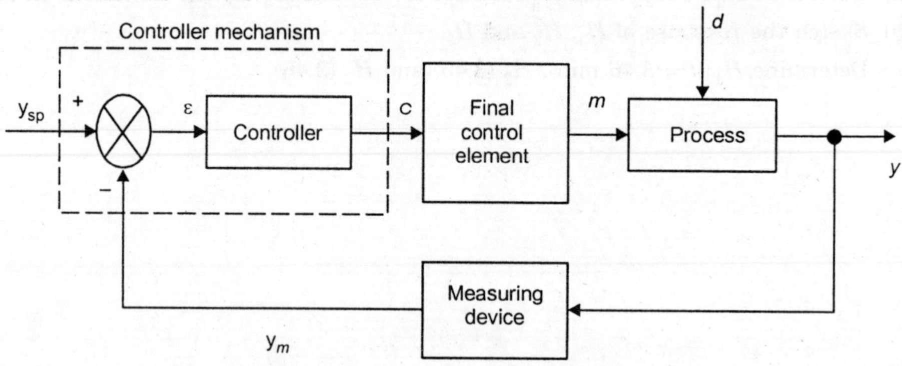

Fig. 2.1 Feedback control system

1. Measures the value of the output (flow, pressure, temperature, liquid level, composition etc.) using the appropriate measuring device. Let 'y_m' (or 'b') be the value indicated by the measuring sensor.

2. Compares the indicated value (actual value) to the desired value (set point or reference) of the output. Let the deviation be 'ε' or 'e' (error) which is $y_{sp}-y_m$ (or $r-b$)
3. The value of the deviation or error is supplied to the main controller. The controller in turn changes the value of the manipulated variable 'm' in such a way as to reduce the magnitude of the deviation 'ε' or error 'e' through the final control element (usually a control valve with actuator).

The system with process alone with 'm' and 'd' as inputs and 'y' as output is called an 'open loop'. The feedback-controlled system shown in Fig. (2.1) is called 'closed loop'. The Fig. (2.2) illustrates an automatic closed loop control system for a level control system.

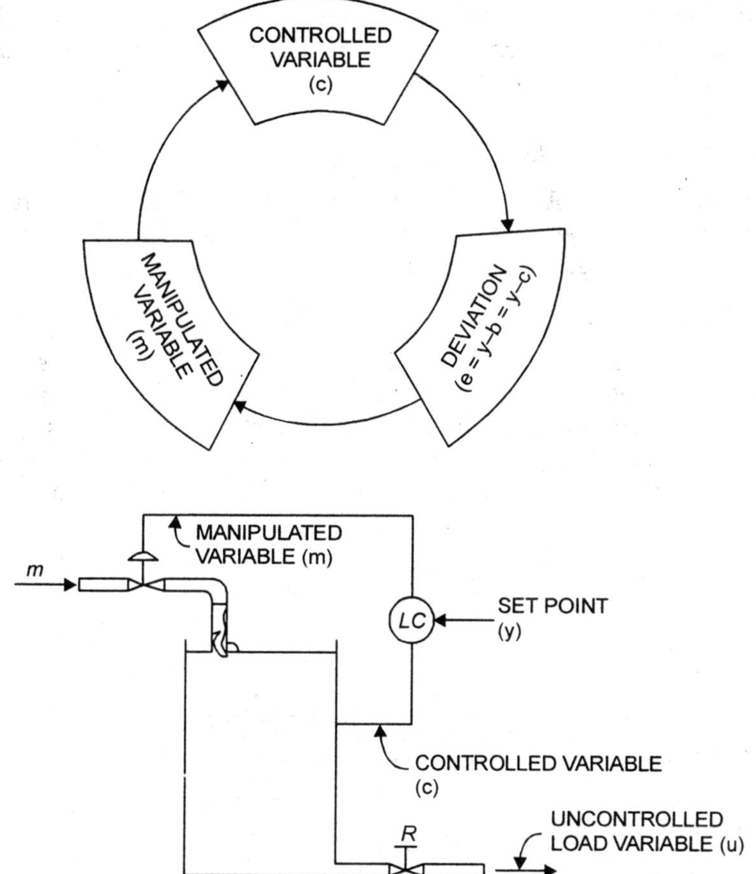

Fig. 2.2 The automatic closed loop control system

2.1.2 Basic Hardware Components

The basic hardware components for a feedback control loop are the following :

1. **Process :** The material equipment along with the physical or chemical operations which take place (tanks, heat exchangers, reactors, separators, etc.).
2. **Measuring system** : Sensors like Thermocouples, Bellows, Diaphragms, Orifice plates, Gas chromatographs, various analysers along with transducers.

3. **Controller** : Also includes the function of the comparator. This is the unit with logic that decides by how much to change the value of the manipulated variable to achieve the desired value (set point)
4. **Final control element** : Usually, a control valve or a variable-speed metering pump. This is the device that receives the control signal from the controller and implements it by physically adjusting the value of the manipulated variable.
5. **Transmission lines** : Used to carry the measurement signal from the sensor to the controller and the control signal from the controller to the final control element. These lines can be either electrical or pneumatic (compressed air) or hydraulic (liquid).

Each of the elements above should be viewed as a physical system with an input and an output. Consequently their behaviour can be described by a differential equation or equivalently by a transfer function.

2.2 PROCESS CHARACTERISTICS

The selection of what controller modes to use in a process is a function of the characteristics of the process. In this section we will define a few properties of processes that are important for selecting the proper modes.

2.2.1 Process Equation

A process-control loop regulates some 'dynamic variable' in a process. This 'controlled variable', a process parameter may depend on many other parameters in the process and thus suffer changes from many different sources. We select normally one of these other parameters as 'controlling parameter'. If a measurement of the controlled variable shows a deviation from the set point, then the controlling parameter is changed, which in turn changes the controlled variable.

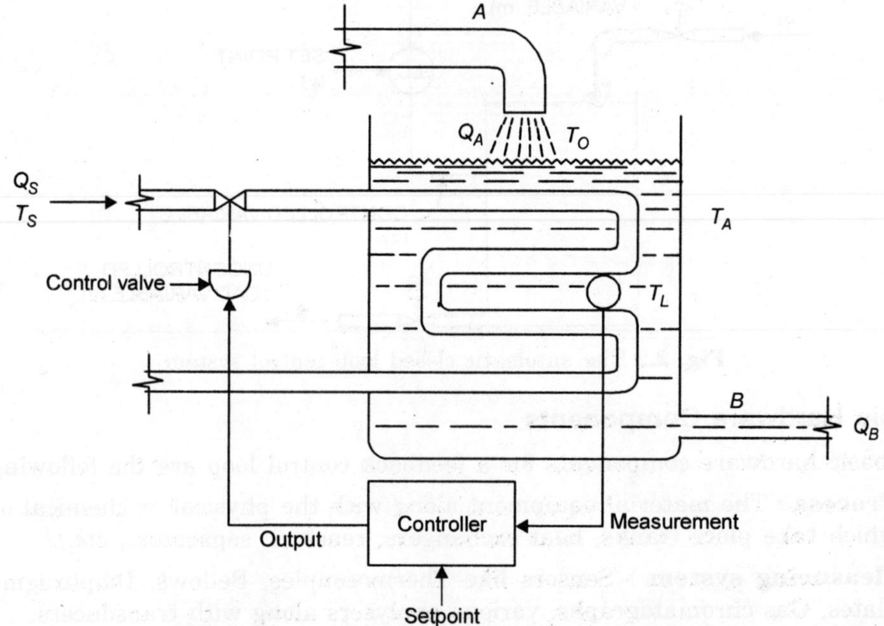

Fig. 2.3 Control of temperature by process control

As an example, consider the control of liquid temperature in a tank shown in Fig. 2.3. The 'Controlled Variable' is the liquid temperature T_L. This temperature depends on many parameters in the process such as the input flow rate via pipe A, the output flowrate via pipe B, the ambient temperature T_A, the steam temperature T_S, inlet temperature T_O, and the steam flow rate Q_S. In this case, the steam flow rate is the controlling parameter chosen to provide control over the controlled variable, liquid temperature T_L. If one of the other parameters changes a change in temperature results. To bring the temperature back to the set point, we only change the steam flow rate, that is heat input to the process. This process could be described by a 'Process Equation' where liquid temperature T_L is a function as

$$T_L = F\,(Q_A, Q_B, Q_S, T_A, T_S, T_O) \tag{2.1}$$

where Q_A, Q_B = Flow rates in pipes A and B

Q_S = Steam flow rate

T_A = Ambient temperature

T_O = Inlet fluid temperature

T_S = Steam temperature

To provide control via Q_S, we do not need to know the functional relationship exactly, nor do we require linearity of the function. The control loop adjusts Q_S and thereby regulates T_L, regardless of how the other parameters in equation (2.1) vary with each other. In many cases the relationship of equation is not even analytically known.

2.2.2 Process Load

It is possible to identify a set of values for the process parameters that results in the controlled variable having the set point value. 'Process Load' refers to this set of parameters excluding the controlled variable. When all parameters have their nominal values, we speak of 'nominal load' on the system. Whenever the parameters change from nominal values, we say that a process load change has occurred. The controlling variable is adjusted to compensate for this load change and its effect on the dynamic variable to bring it back to the set point.

Transient

Another type of change involves a temporary variation of one of the load parameters. After the excursion, the parameter returns to its nominal value. This variation is called a 'Transient'. A transient causes variations of the controlled variable and the control system must take equally transient changes of the controlling variable to keep error to a minimum. A transient is not a load change because it is not permanent.

2.2.3 Process Lag

A process load change or transient causes a change in the controlled variable. The process-control loop responds to assure, some finite time later, that the variable returns to set point value. Part of this time is consumed by the process itself and is called the 'Process lag'. Thus, referring to Fig. 2.3, assume the inlet flow is suddenly doubled. Such a large process load change radically changes (reduces) the liquid temperature. The control loop responds by opening the steam inlet valve to allow more steam and heat input to bring the liquid temperature back to the setpoint. The loop itself reacts faster than the process. In fact, the physical opening of the control valve is the slowest part of the loop. Once steam is flowing at the new rate, however, the body of liquid must be heated by the steam before the setpoint value is reached again. This time delay or 'process lag' in heating is a function of the process and not the control system. Clearly, there is no advantage in designing control systems many times faster than the process lag.

2.2.4 Self-Regulation

A significant characteristics of some processes is the tendency to adopt a specific value of the controlled variable for nominal load with no control operations. The control operations may be significantly affected by such 'self-regulation'. Refer section (1.11) for more details on self-regulation.

2.3 CONTROL SYSTEM PARAMETERS

The basic characteristics of the process that are related to control are seen in section 2.2. In this section we will examine the general properties of the controller (Refer Fig. 2.4). Inputs to the controller are measured indication of the 'controlled variable' and 'set point' representing the desired value of the variable, expressed in the same fashion as the measurement. The controller output is a signal representing action to be taken when the measured value of the controlled variable deviates from set point.

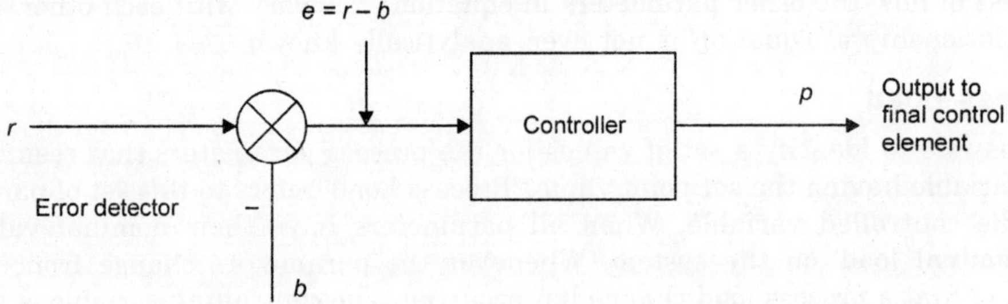

Fig. 2.4 The error detector and controller block diagram

The measured indication of a variable is denoted by 'b', while the actual variable is denoted by 'c'. Thus if a sensor measures temperature by conversion to resistance, the actual variable is temperature in degrees celsius, but the measured indication is resistance in ohms. Further conversion may be performed by transducers or transmitters to provide a current in 'mA', for example. In such a case, the current becomes the measured indication of the variable.

2.3.1 Error

The deviation or error of the controlled variable from the setpoint is given by

$$e = r - b \tag{2.2}$$

where e = error

b = measured indication of variable

r = setpoint variable (reference)

The equation (2.2) expresses error in an absolute sense, or in units of the measured analog of the control signal.

To describe controller operation in a general way, it is better to express the error as percent of the measured variable range (*i.e.*, the span). The measured value of a variable can be expressed as percent of span over a range of measurement by equation.

$$c_p = \frac{c - c_{\min}}{c_{\max} - c_{\min}} \times 100 \qquad (2.3)$$

where c_p = measured value as percent of measurement range.

c = actual measured value

$c_{\max}$ = maximum of measured value

$c_{\min}$ = minimum of measured value

To express error as percent of span, the measured indications of minimum and maximum can be used as below:

$$e_P = \frac{r - b}{b_{\max} - b_{\min}} \times 100 \qquad (2.4)$$

where e_p = error expressed as percent of span

2.3.2 Variable Range

Generally, the variable under control has a range of values within which control is maintained. This range can be expressed as the minimum and maximum value of the variable or the nominal value plus and minus the spread about this nominal. If a standard 4–20 mA signal transmission is employed, then 4 mA represents the minimum value of the variable and 20 mA the maximum.

2.3.3 Control Parameter Range

The controller output range is the translation of output to the range of possible values of the final control element. This range also is expressed as the 4–20 mA standard signal again with the minimum and maximum effects in terms of the minimum and maximum current.

2.3.4 Control Lag

The control system also has a lag associated with its operation that must be compared to the process lag discussed in section (2.2.3). When a controlled variable experiences a sudden change, the process-control loop reacts by outputting a command to the final control element to adopt a new value to compensate for the detected change. 'control lag' refers to the time for the process-control loop to make necessary adjustments to the final control element. Thus in Fig. (2.3), if a sudden change in liquid temperature occurs, it requires some finite time for the control system to physically actuate the steam control valve.

2.3.5 Dead Time

Another time variable associated with process control is both a function of the process-control system and the process. This is the elapsed time between the instant deviation (error) occurs and the correction action first occurs. An example of dead time occurs in the control of a chemical reaction by varying reactant flowrate through a very long pipe. When a deviation is detected, a control system quickly changes a valve setting to adjust flow rate. But if the pipe is quite long, there is a period of time during which no effect is felt in the reaction vessel. This is the time required for the new flowrate to move down the length of the pipe. Such dead times can have a very profound effect on the performance of control operations on a process.

2.3.6 Cycling

We frequently refer to the behaviour of the dynamic variable error under various modes of control. One of the most important modes is an oscillation of the error about zero. This means the variable is cycling above and below the set point value. Such cycling may continue indefinitely, in which case we have 'steady-state cycling'. Here we are interested in both the peak amplitude of the 'error' and the 'period of the oscillation'. If the cycling amplitude decays to zero, however, we have a cyclic transient error. Here we are interested in the 'initial error', the period of the cyclic oscillation, and 'decay time' for the error to reach zero.

2.3.7 Controller Modes

There are in general two modes of controller operations, namely 'discontinuous' and 'continuous'. In discontinuous mode, the controller command initiates a discontinuous change in the controller parameter. Two-position (ON/OFF) mode controller, multiplication mode controller, single speed and multiple speed floating controllers are the examples of discontinuous mode controllers. In continuous mode, smooth variation of the control parameter is possible. Proportional, integral, derivative and composite control modes are examples of continuous modes.

The choice of operating mode for any given process-control system is a complicated decision. It involves not only process characteristics but also cost-analysis, product rate, and other industrial factors. At the outset, the process-control technologist should have a good understanding of the operational mechanism of each mode and its advantages and disadvantages.

In each mode, the output of the controller is described by a factor 'p'. This is the 'Percent of controller output' relative to its total range as defined in the following equation.

$$p = \frac{u - u_{\min}}{u_{\max} - u_{\min}} \tag{2.5}$$

where
p = controller output as percentage of full scale
u = value of the output
$u_{\max}$ = maximum value of controlling parameter
$u_{\min}$ = minimum value of controlling parameter

2.3.8 Control Actions (Reverse and Direct)

The error that results from the measurement of the controlled variable may be positive or negative, because the value may be greater or less than the set-point. How this polarity of the error should change the controller output can be selected according to the nature of the process.

Direct action

A controller is said to be operating with 'direct action' when an increasing value of the controlled variable causes an increasing value of the controller output. An example is a level-control system which outputs a signal to an output valve. If the level rises (controlled variable increases) the controller output should increase to open the valve more to keep the level under control.

Reverse action

A controller is said to be operating with 'reverse action' when an increasing value of the controlled variable causes a decreasing value of the controller output. An example of this would be a simple temperature control of a furnace with fuel as heat energy. If the temperature increases, the controller output should decrease to close the valve for decreasing the fuel input to bring the temperature under control.

2.4 DISCONTINUOUS CONTROLLER MODES

There are various modes that show discontinuous changes in controller output as controlled variable error occurs. They are frequently used in process control.

2.4.1 Two-Position Mode (ON-OFF controller)

It is the simplest, the cheapest and undoubtedly the most widely used type of control for both industrial and domestic service. Two-position control is a position type of controller action in which the manipulated variable is quickly changed to either a maximum or minimum value depending upon whether the controlled variable is greater or less than the set point. The minimum value of the manipulated variable is usually zero (off) and the maximum value is the full amount possible (On). Although an analytic equation cannot be written, we can, in general, write

$$p = \begin{cases} 0\% & e_P < 0 \\ 100\% & e_P > 0 \end{cases} \tag{2.6}$$

This relation shows that when the measured value is less than the set point, full controller output results. When it is more than the set point, the controller output is zero.

2.4.1.1 Electrical Two-Position Control

A room heater or a water heater is a common example. If the temperature drops below a setpoint, the heater is turned on. A liquid level control using two-position controller is shown in Fig. 2.5. A float in the vessel operates an electric switch which controls power to a solenoid valve. When the liquid level rises, the switch contacts are closed, the solenoid valve closes, and the inflow is cut off. When the liquid level falls, the switch contacts are opened, the solenoid valve opens, and the inflow resumes. If the float level has no bearing friction and the electrical contacts draw no arc, the action is sharp or 'knife-edge'. In virtually any practical implementation of the two-position controller, there is an overlap as e_P increases through zero or decreases through zero. In this span, no change in controller output occurs. This phenomeno is graphically represented in the left hand side of Fig. 2.5. This is again best shown in Fig. 2.6, which plots p versus e_p for a two-position controller. We see that until an increasing error changes by Δe_P above zero, the controller output will not change state. In decreasing, it must fall Δe_P below zero before the controller changes to the 0% rating. The range $2\,\Delta e_P$, which is referred to as the 'neutral zone' or 'differential gap' is often purposely designed above a certain minimum quantity to prevent excessive cycling. The existence of such a neutral zone is an example of desirable hysteresis in a system.

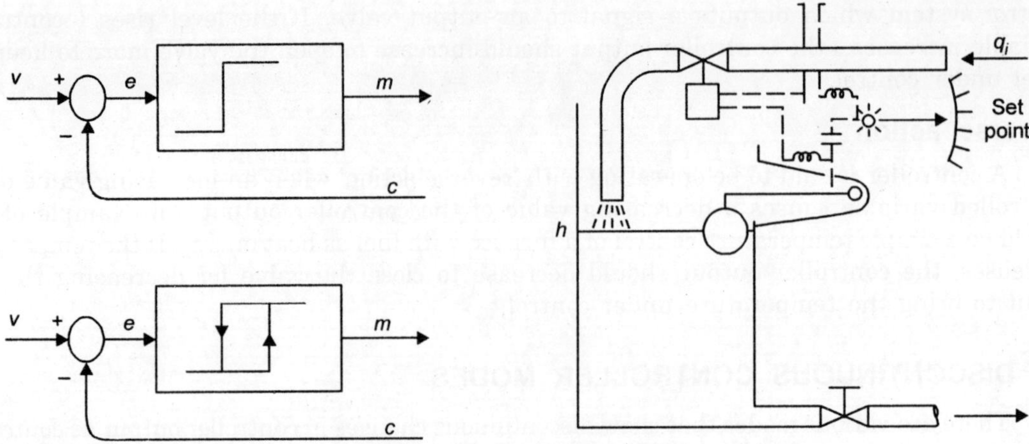

Fig. 2.5 Two-position control

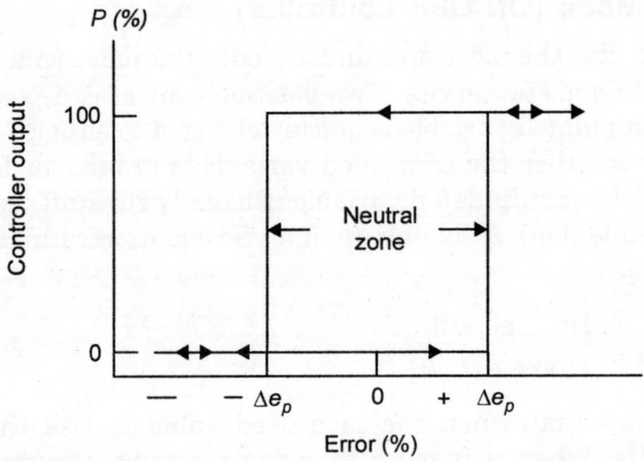

Fig. 2.6 Two-position controller action with neutral zone

The 'differential gap' in two-position controller causes the manipulated variable to maintain its previous value until the controlled variable has moved slightly beyond the set point. A differential gap is caused in the two-position controller of Fig. (2.5) if small static friction exists at the bearing on the float arm. The liquid level must then rise slightly above the desired value to create sufficient buoyant force to overcome friction when the level is rising. Also, the liquid level must fall slightly below the desired value when the level is falling so that the weight force may overcome the friction. This kind of differential gap may be caused by unintentional friction and lost motion.

Fig. 2.7 shows the typical output of an 'ON-OFF' controller to a sinusoidal variation of the inputs (controlled variable).

2.4.1.2 Pneumatic Two-position Controller

The device shown in Fig. (2.8) illustrates how physical elements are configured to give the 'on-off' control mode. It consists of a flapper-nozzle and an 'air relay' or power amplifier and is the central component of many pneumatic controllers. The back pressure (P_b) in the chamber of flapper-nozzle is controlled by the position of the flapper with respect to the nozzle. If the flapper is fully closed (X = 0), the back pressure will be equal to the supply pressure (P_s). If the flapper is fully open (X is very large), the back pressure will be approximately equal to the ambient pressure (P_a). The output of the air relay (P_o) is a direct function of the back pressure. The relay is essentially a power amplifier necessary to supply the air flow required to any pneumatic controller. The relay is termed reverse acting because for an increase in P_b there is a corresponding decrease in P_o.

$$- P_b = P_o - K X \tag{2.7}$$

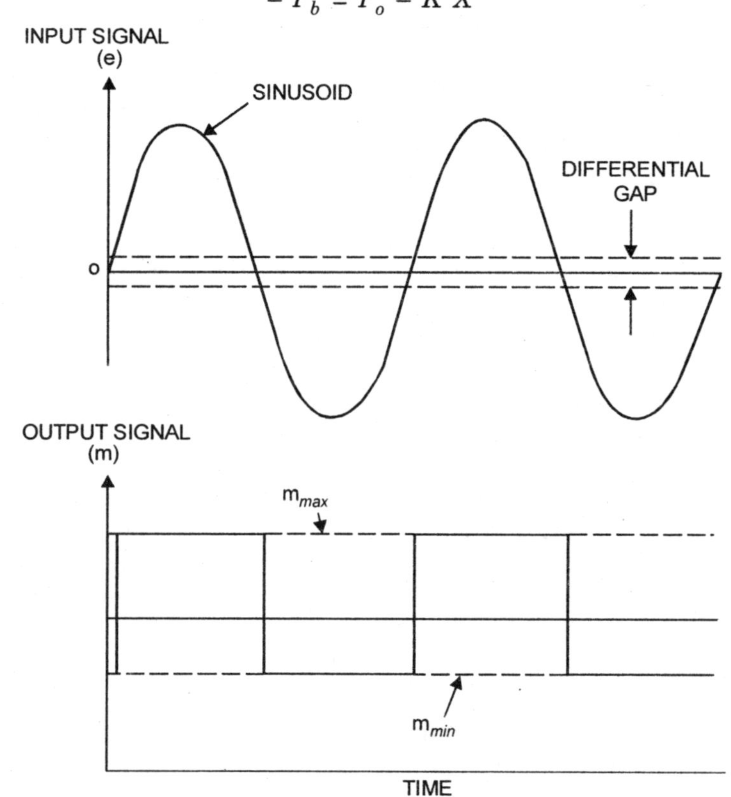

Fig. 2.7 Response of a two-position controller

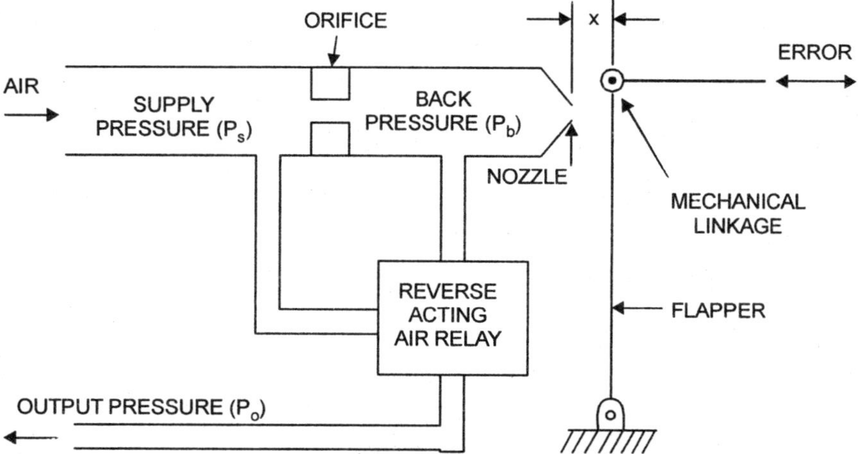

Fig. 2.8 Pneumatic two-position controller

In the equation (2.7) K is typically in the neighbourhood of 13 to 21 bars/mm (5000 to 800 PSI/Inch). This high sensitivity makes the device undesirable for proportional control, but it can be used as a 'two-position' or 'bang-bang' controller because any small positive value of X will result in the maximum value of P_o and any small negative value of X will result in the minimum value of P_o. The terms negative and positive are used relatively to the normal flapper position.

2.4.1.3 Applications

Generally, the two-position control mode is best adapted to large-scale systems with relatively slow process rates. Thus, in a room heating or air-conditioning system, the capacity of the system is very large in terms of air volume, and the overall effect of the heater or cooler is relatively slow. Other examples of two-position control applications are liquid bath temperature control, level control etc. The process under two-position control must allow continued oscillation in the controlled variable because, by its very nature, this mode of control always produces such oscillation. For large systems, these oscillations are of long duration, which is partly a function of the neutral-zone size.

2.4.2 Multiposition Mode

A logical extension of the previous two-position control mode is to provide 'several intermediate' rather than only two settings of the controller output. A three-position mode of control is one in which the manipulated variable takes one of three values: high, medium or low, depending upon whether the deviation is large positively, close to zero, or large negatively. Similarly, four-five- and multiposition control may be used. The multiposition control mode is used in an attempt to reduce the cycling behaviour and overshoot and undershoot inherent in the two-position mode. The multiposition mode is represented by the following equation.

$$p = p_i \qquad e_p > |e_i| \qquad i = 1, 2, \ldots\ldots n \tag{2.8}$$

As the error exceeds certain set limits $\pm e_i$, the controller output is adjusted to preset values p_i. The most common example is the three-position controller where

$$p = \begin{cases} 100 & e_p > e_2 \\ 50 & -e_1 < e_p < e_2 \\ 0 & e_p < -e_1 \end{cases} \tag{2.9}$$

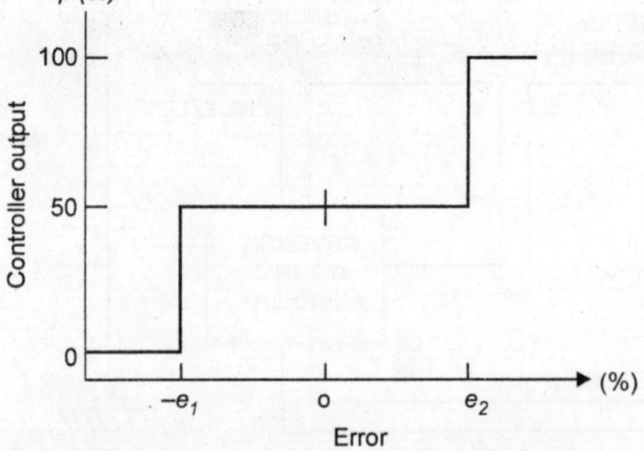

Fig. 2.9 Three-position controller action

As long as the error is between e_2 and e_1 of the setpoint, the controller stays at some nominal setting indicated by a controller output 50%. If the error exceeds the setpoint by e_2 or more, then the output is increased by 100%. If it is less than the setpoint by $-e_1$ or more, the controller output is reduced to zero.

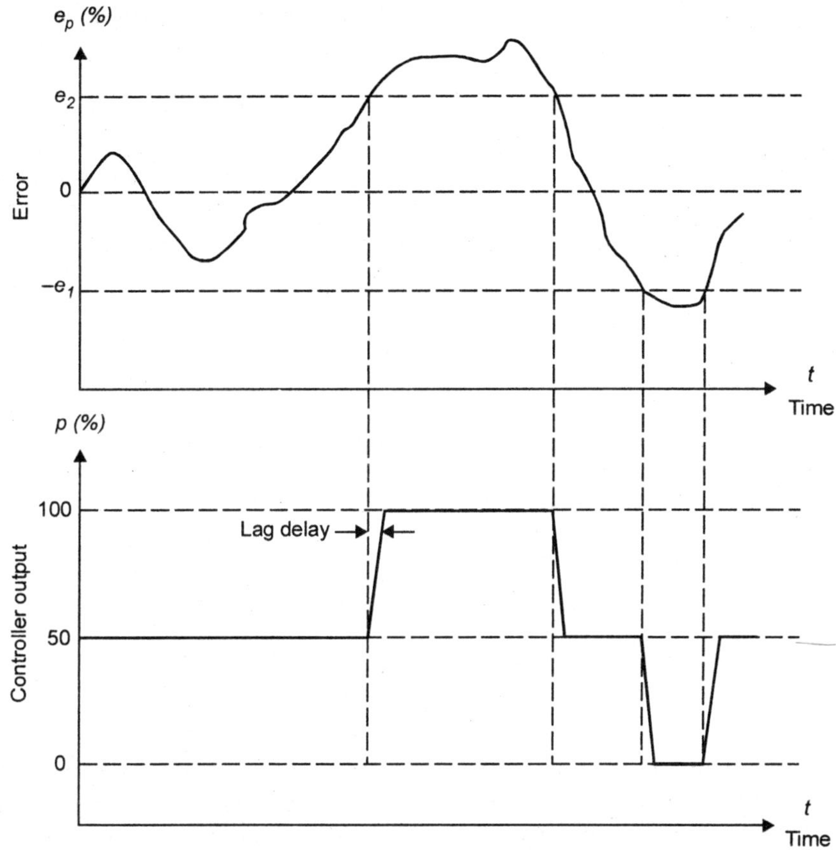

Fig. 2.10 Relationship between error and three-position controller action, including the effects of lag

Fig. 2.9 illustrates this mode graphically. Some small neutral zone usually exists about the change points, but not by design and thus they are not shown. This type of control mode usually requires a more complicated final control element because it must have more than two settings. Fig 2.10 shows a graph of dynamic variable and final control element setting versus time of a hypothetical case of three-position control. Note the change in control element setting as the variable changes about the two trip points. On this graph the finite time required for the final control element to change from one position to another position is also shown. Note the overshoot and undershoot of the error around the upper and lower setpoints. This is due to both the process lag time and the controller lag time indicated by the finite time required for the control element to reach a new setting.

2.4.3 Floating—Control Mode

In the two previous modes of controller action, the output was uniquely determined by the magnitude of the error input. If the error exceeded some preset limit, the output was changed to a new setting as quickly as possible. In floating control, the specific output of the controller is not uniquely determined by the error. If the error is zero, the output will not change but remains (floats) at whatever setting it was when the error went to zero. When the error moves off zero, the controller output again begins to change. Actually, as with the two-position mode, there is typically a neutral zone about zero error where no change in controller position occurs.

2.4.3.1 Single-Speed Floating Control

In the single-speed floating mode of control the manipulated variable changes at a constant rate in one direction when the deviation is positive and in the opposite direction at a constant rate when the deviation is negative. That means the output of the control element changes at a fixed rate when the error exceeds the neutral zone. The action can be represented by an equation:

$$\frac{dp}{dt} = \pm K_F \qquad |e_p| > \Delta e_p \tag{2.10}$$

where $\frac{dp}{dt}$ = rate of change of controller output with time.

K_F = rate constant (%/s)

Δe_p = half the neutral zone

If equation (2.10) is integrated for the actual controller output, we get

$$p = \pm K_F t + p_{(0)} \qquad |ep| > \Delta e_p \tag{2.11}$$

where $p_{(0)}$ = controller output at $t = 0$.

which shows that the present output depends on the time history of errors that have previously occurred. Because such a history is usually not known, the actual value of 'p' floats at an undetermined value. If the deviation persists, then equation (2.10) shows that the controller saturates at either 100% or 0% and remains there until an error drives it toward the opposite extreme. A graph of single-speed floating control, a graph showing controller output versus time and error versus time for a hypothetical case illustrating typical operation is shown in Fig. 2.11. In this example, we assume the controller is reverse acting, which means the controller output decreases when the error exceeds the neutral zone. This corresponds to a negative K_F in equation (2.10). Most controllers can be adjusted to act in either the reverse or direct mode. Here the controllers start at some output $p(0)$. At time 't_1', the error exceeds the neutral zone. The controller output decreases at a constant rate until 't_2' when the error again falls below the neutral zone limit. At 't_3', the error fall below the lower limit of the neutral zone, causing controller output to change until the error again moves within the allowable band.

Single-speed floating control is commonly associated with systems in which the final control element is a single-speed reversible motor. The output of the reversible motor is either forward, reverse, or off depending on the error condition. A typical response of a single-speed floating controller to a sinusoidal input is shown in Fig. (2.12).

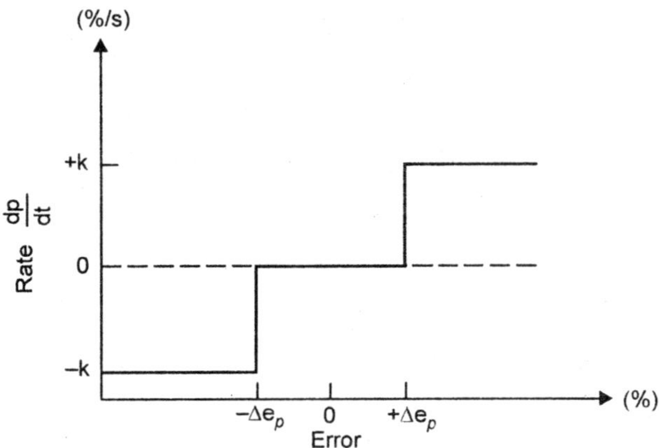

(*a*) Single-speed floating controller action. The ordinate is the rate of change of controller output with time

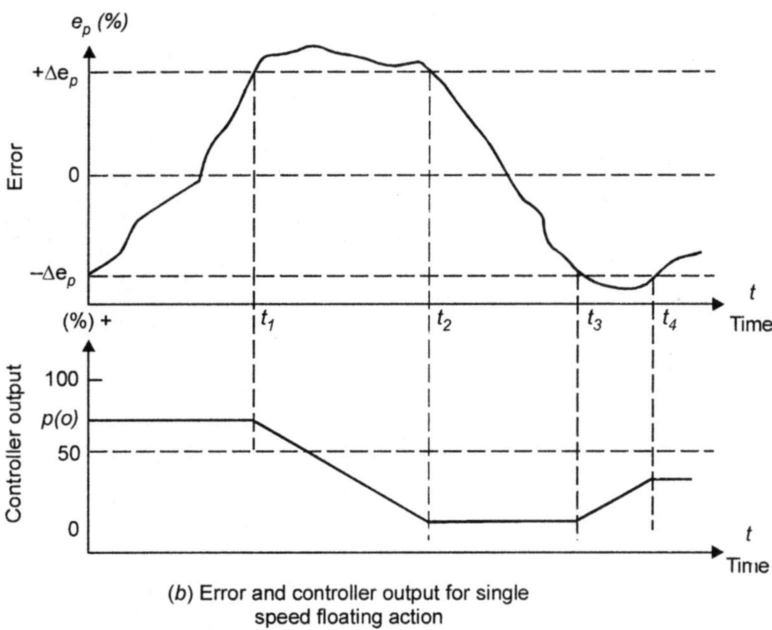

(*b*) Error and controller output for single speed floating action

Fig. 2.11 Single-speed floating controller

Single-speed floating control action may be visualized from Fig. (2.5) if the solenoid is replaced by a reversible motor with gear reducer to move the control valve system. When the level rises, the switch contact is made, and the motor-reducer slowly closes the control valve. As soon as the level falls, the switch contact is broken and the motor-reducer reverses its direction of rotation and opens the control valve. A double-throw electrical relay may be required. A neutral zone is used so that the motor remains stationary when the deviation is small.

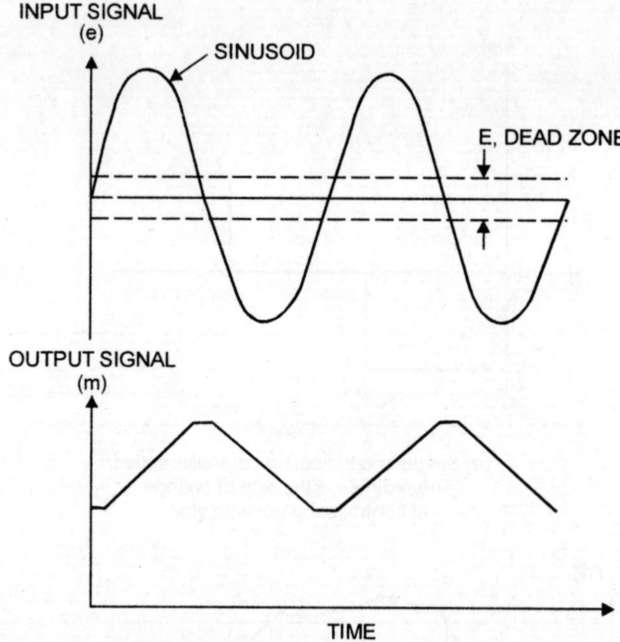

Fig. 2.12 Response of a single-speed floating controller

2.4.3.2 Multiple-Speed Floating Control

Multiple-speed floating control is used in which several rates of change of the manipulated variable correspond to several magnitudes of deviation. Usually, the rate increases as the deviation exceeds certain limits. Thus, if we have certain speed change points 'e_{pi}' depending on the error, then each has its corresponding output rate change K_i. We can then say

$$\frac{dp}{dt} = \pm K_{Fi} \qquad |e_p| > e_{Pi} \tag{2.12}$$

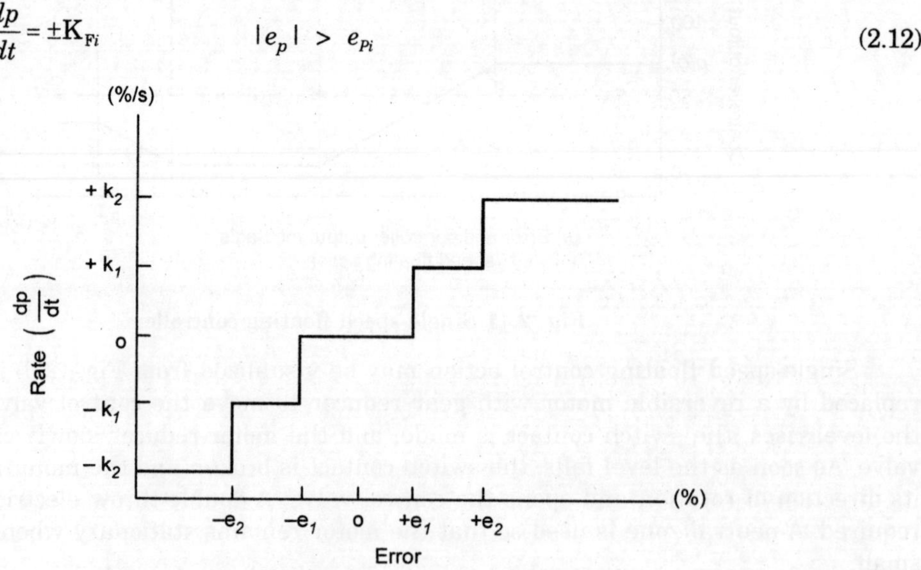

Fig. 2.13 Multiple-speed floating control action

If the error exceeds e_{pi}, then the speed is K_{F1}. If the error rises to exceed e_{p2}, the speed is increased to K_{F2}, and so on. Actually, this mode is a discontinous attempt to realise an 'Integral mode' to be discussed under 'continuous controller modes'. A graph of this mode is shown in Fig. 2.13.

2.4.3.3 Applications

Primary applications of the floating-control mode are for the single-speed controllers with a neutral zone. This mode has an inherent cycle nature much like the two-position, although this cycling can be minimized depending on the application. Generally, the method is well suited to self-regulation processes with very small lag or dead time, which implies small-capacity processes. When used with large-capacity systems, the inevitable cycling must be considered.

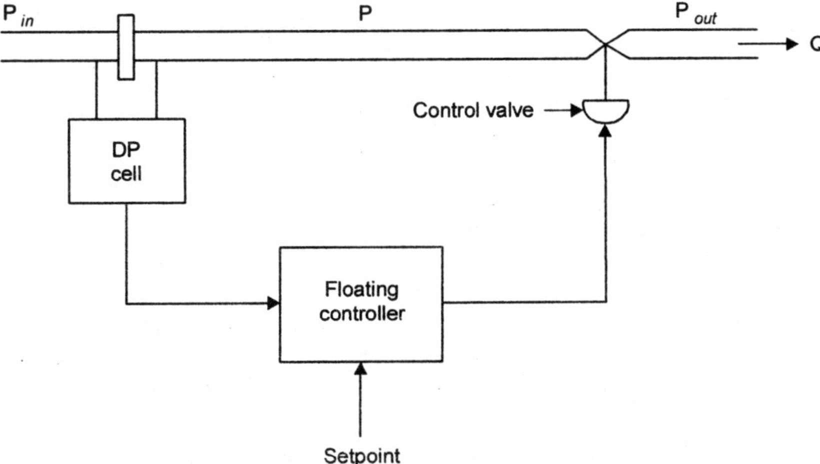

Fig. 2.14 Single-speed floating control action applied to a flow control system

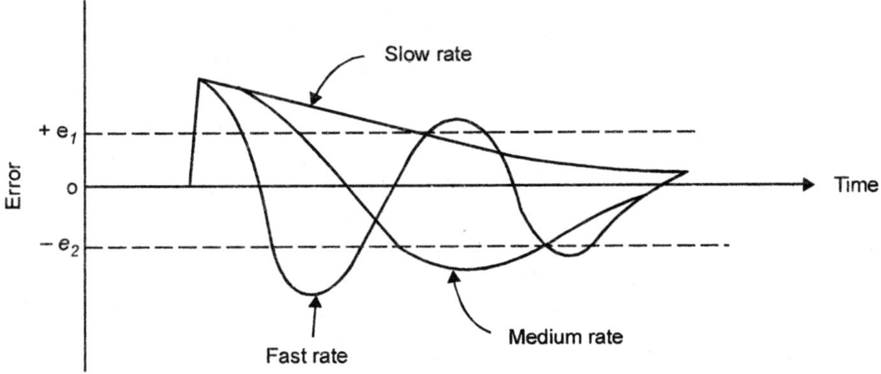

Fig. 2.15 The rate of controller output change has a strong effect on error recovery in a floating controller

An example of single-speed floating control is a liquid flow rate through a control valve (Refer Fig. 2.14). The load is determined by the inlet and outlet pressures P_{in} and P_{out}, and the flow is determined in part by the pressure P within the DP cell and control valve. This is an example of a system with self-regulation. We assume some valve opening has been found commensurate with the desired flow rate. If the load changes (either P_{in} and P_{out},), then an error occurs. If larger than the neutral zone, the valve begins to open or close at a constant rate until an opening is found that supports the flow rate at the new load conditions. Clearly the rate is very important because very fast process lags cause the valve to continue opening (or closing) beyond that optimum self-regulated position. This is shown in Fig. 2.15, where the response to a sudden deviation is shown for various floating rates.

2.5 CONTINUOUS CONTROLLER MODES

Continuous controller modes are an extension of the discontinuous types discussed in section 2.4. The most common controller action used in process control is one or a combination of continuous controller modes. In these modes, the output of the controller changes smoothly in response to the error or rate of change of error. Continuous controller modes include proportional (*P*), integral (*I*) and differential (*D*) ones.

2.5.1 Proportional Control Mode

The two-position mode had the controller output of either 100% or 0% depending on the error being greater or less than neutral zone. In multiple-step modes, more divisions of controller outputs versus error are developed. The natural extension of this concept is the 'Proportional Mode', where a smooth, linear relationship exists between the controller output and the error. In other words, proportional action is a mode of controller action in which there is a continuous linear relation between values of the deviation and manipulated variable. Thus the action of the controller variable is repeated and amplified in the action of the final control element. Several synonymous names in common use are proportional action, correspondence control, droop control, and modulating control. The adjustable parameter of the proportional mode K_p is called the 'proportional gain' or 'proportional sensitivity'. Over some range of errors about the set point, each value of error has unique value of controller output in one-to-one correspondence. The range of error to cover the 0% to 100% controller output is called the 'proportional band (*PB*)' because the one-to-one correspondence exists only for errors in this range. The proportional band is equivalent to the inverse of the proportional gain and is defined by the equation.

$$\text{PB} = \frac{100}{K_p} \tag{2.13}$$

The proportional band characterizes the range over which the error must change in order to drive the actuating signal of the controller over its full range. And the proportional mode can be expressed by the equation

$$p = K_p e_p + p_o \tag{2.14}$$

where K_p = Proportional gain between error and controller output (% per %)

p = Controller output (%)

e_p = Error (%)

p_o = Controller output with no error (%)

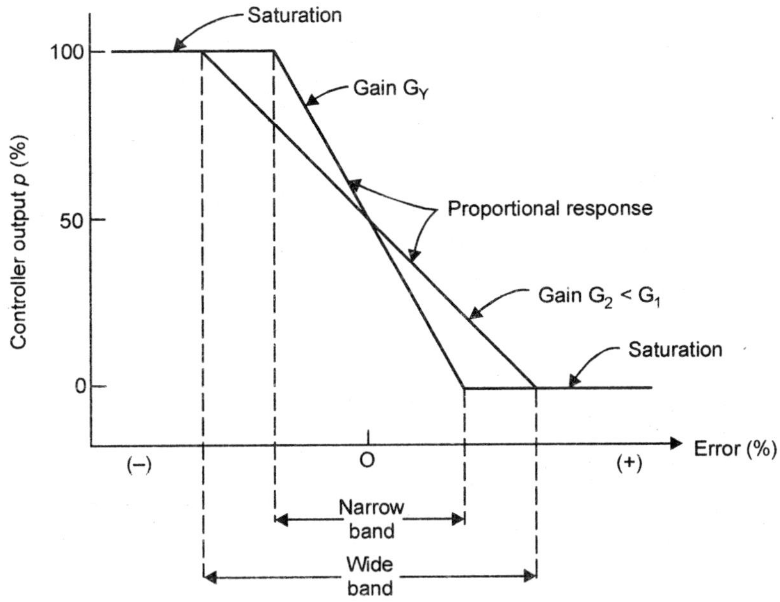

Fig. 2.16 The proportional band of a proportional controller depends on the gain, in an inverse fashion

A plot of the proportional mode output versus error for equation (2.14) is shown in Fig 2.16. In this case, p_O has been set to 50% and two different gains have been used. Note that the proportional band is dependent on the gain. A high gain means large response to an error but also a narrow error band within which the output is not saturated. That means a high percentage of *PB* (wide band) correspond to less sensitive controller settings. The characteristics of the proportional control mode can be summarized as below [Refer equation 2.14 and Fig. 2.16].

1. If the error is zero, the output is a constant equal to p_O.
2. If there is error, for every 1% of error a correction of K_P percent is added to or subtracted from p_O, depending on the reverse or direct action of the controller.
3. There is a band of error about zero of magnitude PB within which the output is not saturated at 0% or 100%.

The transfer function of the proportional controller is K_P.

Offset

Whenever a change in load occurs, the proportional control mode produces a permanent 'residual error' in the operating point of the controlled variable which is referred to as 'offset'. It can be minimized by a larger constant K_P, which also reduces the proportional band. Consider a system under nominal load with the controller at 50% and the error zero as shown in Fig. 2.17. If a transient error occurs, the system responds by changing controller output in correspondence with the transient to effect a return to zero error. Suppose, however, a load change occurs that requires a permanent change in controller output to produce the zero error state. Because a one-to-one correspondence exists between controller output and error, it is clear that a new zero error controller output can never be achieved. Instead, the system produces a small permanent offset in reaching a compromise position of controller output under new loads.

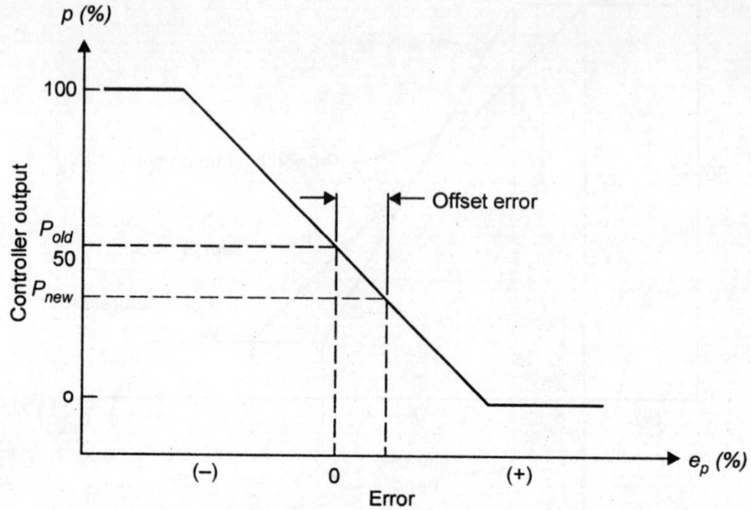

Fig. 2.17 An offset error must occur if a proportional controller requires a new nominal controller output following a load change.

Application

Proportional control generally is used in processes where large load changes are unlikely or with moderate to small process lag times.

2.5.2 Integral Control Mode (Reset Action Mode)

Integral action is a mode of control action in which the value of the manipulated variable is changed at a rate proportional to the deviation. Thus if the deviation is doubled over a previous value, the final control element is moved twice as fast. When the controlled variable is at the set point (zero deviation), the final control element remains stationary.

This mode represents a natural extension of the principle of floating control in the limit of infinitesimal changes in the rate of controller output with infinitesimal changes in error. Instead of single speed or even multispeeds, we have a continuous change in speeds depending on error. This mode is often referred to as 'reset action'. This mode can be expressed by the following equation.

$$\frac{dp}{dt} = K_I e_P \tag{2.15}$$

where $\frac{dp}{dt}$ = rate of controller output change (%/s)

K_I = constant relating the rate to the error (%/s)/%

In some cases the inverse of K_I, called the integral time $T_I = \frac{1}{K_I}$, expressed in seconds or minutes, is used to describe the integral mode. T_I is defined as the time of change of controlled variable caused by a unit change of deviation.

If we integrate equation (2.15), we can find the actual controller output at anytime as

$$p(t) = K_I \int_0^t e_P(t)dt + p_{(0)} \tag{2.16}$$

where $p_{(0)}$ = the controller output at $t = 0$

This equation shows that the present controller output p(t) depends on the history of errors from when observation started at t =0. We see from equation (2.15), for example, that if the error doubles, the rate of controller output change also doubles. The constant K_I expresses the scaling between error and controller output. Thus, a large value of K_I means that a small error produces a large rate of change of p and vice versa.

Fig. 2.18(*a*) graphically illustrates the relationship between the p rate of change and error for two different values of K_I. Fig. 2.18(*b*) shows how, for a final fixed error, the different K_I values produce different values of p as a function of time as predicted by equation (2.16). Thus, we see that the 'faster' rate provided by K_I causes a 'much greater' controller output at a particular time after the error is generated.

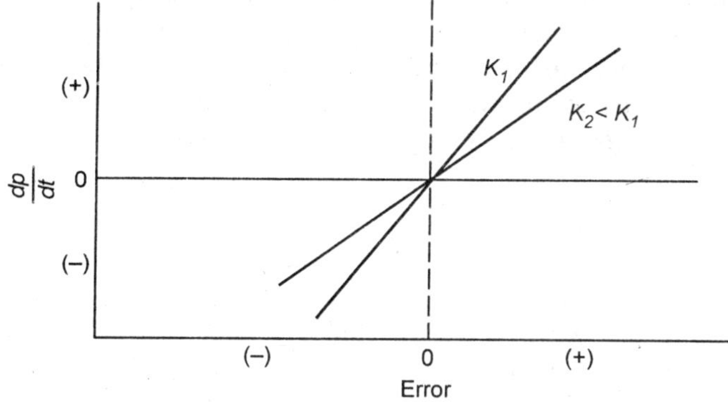

(*a*) The rate of output change depends on gain and error

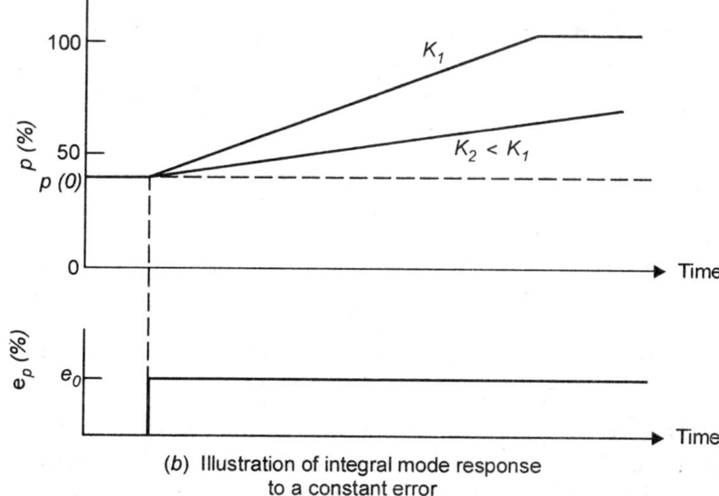

(*b*) Illustration of integral mode response to a constant error

Fig. 2.18 Integral controller mode action

The characteristics of the integral mode with reference to equation (2.16) may be summarized as below:

1. If the error is zero, the output stays fixed at a value to what it was when the error went to zero.
2. If the error is not zero, the output will begin to increase or decrease at a rate of K_I percent/second for every one percent of error.

(The integral controller constant K_I may be expressed in percentage change per minute per percentage error, whenever a typical process-control loop has characteristic response time in minutes rather than in seconds.)

The transfer function of the Integral control is given by $\frac{K_I}{s}$ or $\frac{1}{T_I s}$.

Applications

The integral mode eliminates the 'offset' which plain proportional control cannot remove. The reason proportional control must result in an offset is because it disregards the past history of error leaving the accumulated effect of past errors uncorrected. The integral mode, on the other hand, continuously looks at the total past history of the error by continuously integrating the area under the error curve and eliminates the offset by forcing the addition (or removal) of mass or energy, which should have been added (or removed) in the past.

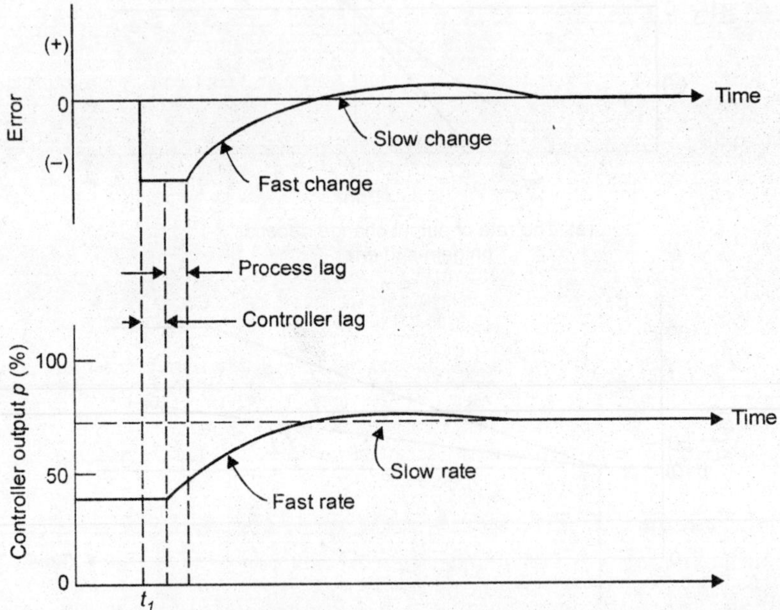

Fig. 2.19 Illustration of integral mode controller output and error showing the effect of process and control lag

Use of the integral mode is shown by the flow control system in Fig. (2.14), except that we now assume that the controller operates in the integral mode. Operation can be understood with the help of Fig. (2.19). A load change introduced error occurs at sometime t_1. The proper valve position under the new load to maintain the constant flow rate is shown as a dashed line in the p graph of the Fig. (2.19). In the integral mode, the value initially begins to change very rapidly, as predicted by equation (2.15). As the valve opens, the error decreases and slows the valve opening rate as shown. The ultimate effect is that the system drives the error to zero at a slowing controller rate. The effect of process and control system lag is shown as simple delays in the controller output change and in the error reduction when the controller action occurs. If the process lags are too large, the error can oscillate about zero or even be cyclic. Typically, the integral mode is not used alone but can be for systems with small process lags and correspondingly small capacities.

2.5.3 Derivative Control Mode

The derivative mode of controller operation provides that the controller output depends on the rate of change of error. The other terms for derivative response are 'rate response', 'lead component' or 'anticipatory response'. The mode cannot be used alone because when the error is zero or constant, the controller has no output. The analytic expression for derivative control mode is given by the following equation.

$$p = K_D \frac{de_P}{dt} \tag{2.17}$$

where K_D = derivative gain constant (%-s/%)

$\frac{de_P}{dt}$ = rate of change of error (%/s)

With the presence of derivative term ($\frac{de_P}{dt}$), the controller anticipates what the error will be in the immediate future and applies control action which is proportional to the current rate of change in the error (Anticipatory control). One major drawback with this mode is that for a noisy response with almost zero error it can compute large derivatives and thus yield large control action, although it is not needed.

The derivative gain constant K_D is also called as the rate or derivative time and is commonly expressed in minutes. Derivative time is defined as the amount of lead, expressed in units of time, that the control action is given. In other words, derivative time is the time interval by which the rate action advances the effect of the proportional control action. That is, if the derivative mode is set for a time 'T_d', it will generate a corrective action immediately when the error starts changing and the size of that correction will equal in size the correction which the proportional mode would have generated T_d time later. The longer the T_d setting, the further into the future the derivative mode predicts and the larger is its corrective contribution.

The proportional mode considers the present state of the process error, and the integral mode looks at its past history, while the derivative mode anticipates its future state and acts on that prediction. This third control mode became necessary as the size of processing equipment increased and, correspondingly, the mass and the thermal inertia of such equipment. For such large processes it is not good enough to respond to an error when it has already evolved, because the flywheel effect (the inertia or momentum) of those large processes makes it very difficult to stop or reverse a trend once it has evolved. The purpose of the derivative mode is to predict process errors before they have evolved and take corrective action in advance of that occurrence.

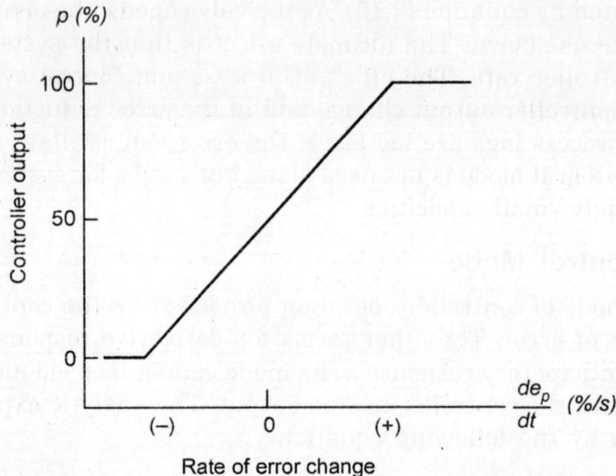

Fig. 2.20 Derivative mode of controller action where an output of 50 % has been assumed for the zero derivative state.

The characteristics of the derivative controller can be noted from the graph of Fig. (2.20) which shows controller output for the rate of change of error. This shows that, for a given rate of change of error, there is a unique value of controller output. The time plot of error and controller response further shows the behaviour of this mode, as shown in Fig. 2.21. The extent of controller output depends on the rate at which this error is changed and not on the value of the error.

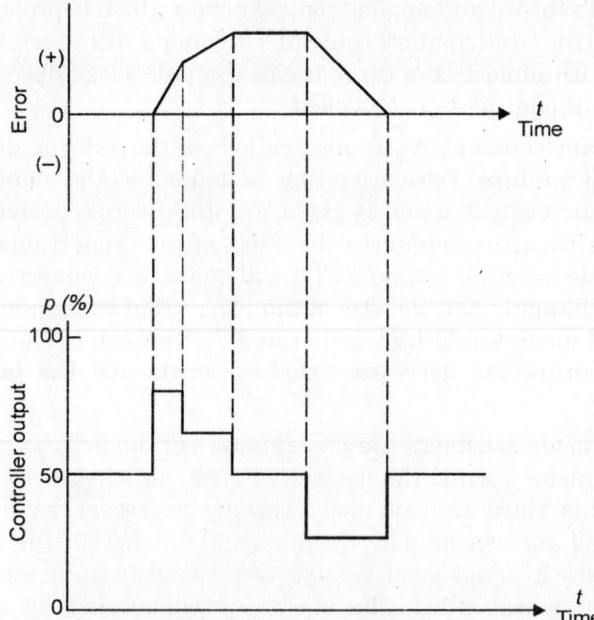

Fig. 2.21 Derivative mode of controller action for a sample error signal

The characteristics of the derivative mode with reference to equation (2.17) may be summarized as below :

1. If the error is zero, the mode provides no output.
2. If the error is constant in time, the mode provides no output.
3. If the error is changing in time, the mode contributes an output of K_D percent for every 1% per second rate of change of error.
4. For direct action, a positive rate of change of error produces a positive derivative mode output.

The transfer function of the derivative control is given by K_Ds or T_Ds.

2.6 COMPOSITE CONTROL MODES

It is very common in the complex of industrial processes to find control requirements that do not fit the application norms of any of the previously considered controller modes. It is possible to combine several basic modes, thereby gaining the advantages of each mode. In some cases, an added advantage is that the modes tend to eliminate some limitations they individually posses. Though Proportional control mode and Integral control mode are used for limited applications, normally the following combinations are commonly used.

1. Proportional-Integral Control (PI)
2. Proportional-Derivative Control (PD)
3. Proportional-Integral-Derivative Control (PID)

2.6.1 Proportional-Integral Control (PI)

This control mode results from a combination of the proportional mode and the integral mode. Certain advantages of both control actions can be obtained from this mode. This mode is also called as Proportional plus reset action controller (or in short PI Controller). Equations for proportional mode and integral mode are combined to have analytic expression for this mode and given below.

$$p = K_P e_P + K_P K_I \int_0^t e_P dt + p_{t(0)} \ldots\ldots \quad (2.18)$$

where $p_{t(0)}$ = Integral term value at t = 0 (initial value).

The main advantage of this composite control mode is that the one-to-one correspondence of the proportional mode is available and the integral mode eliminates the inherent offset. Notice that the proportional gain, by design, also changes the net integration mode gain, but that the integration gain, through K_I, can be independently adjusted. Recall that the proportional mode offset occurred when a load change required a new nominal controller output and could not be provided except by a fixed error from the set point. In the present mode, the integral function provides the required new controller output, thereby allowing the error to be zero after a load change. The integral feature effectively provides a 'reset' of the zero error output after a load change occurs. This is illustrated by the graphs of Fig. 2.22. At time t_1 a load change occurs that produces the error shown. Accommodation of the new load condition requires a new controller output. It can be seen that the controller output is provided through a sum of proportional plus integral action that finally leaves the error at zero. The proportional part is obviously just an image of the error.

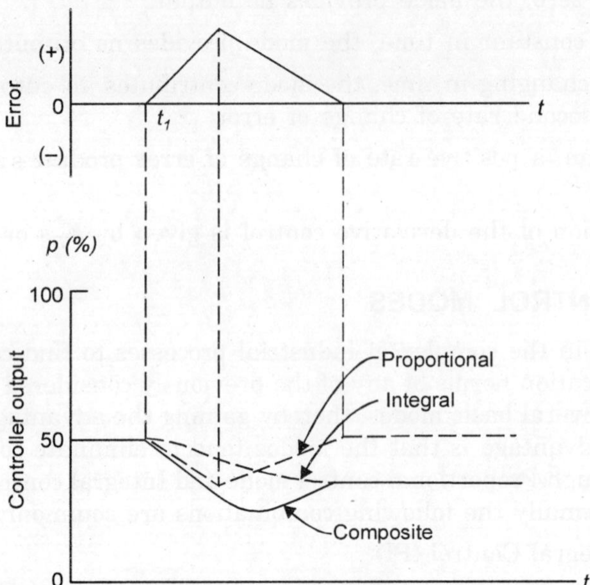

Fig. 2.22 Proportional-Integral (PI) action showing the reset action of the integral contribution (reverse action)

The characteristics of the PI mode with reference to equation (2.18) may be summarized as below:

1. When the error is zero, the controller output is fixed at the value that the integral term had when the error went to zero. This output is given by $p_{t(0)}$ in equation (2.18) simply because we chose to define the time at which observation starts as $t = 0$.
2. If the error is not zero, the proportional term contributes a correction and the integral term begins to increase or decrease the accumulated value [initial $p_{t(0)}$], depending on the sign of the error and the direct or reverse action. The integral term cannot become negative. Thus, it will saturate at zero if the error and action try to drive the area to a net negative value.

The transfer function of PI controller is given by $K_p\left(1+\dfrac{K_I}{s}\right)$.

Reset rate/repeats per minute

The integral action adjustment is the integral time $T_I\left(=\dfrac{1}{K_I}\right)$. For a step change of deviation 'e', the integral time or reset time is the time required to add an increment of response equal to the original step change of response of the proportional action as indicated in Fig. 2.23. 'Reset rate' is defined as the number of times per minute that the proportional part of the response is duplicated. Reset rate is therefore called 'repeats per minute' and is the inverse of integral time (= K_I).

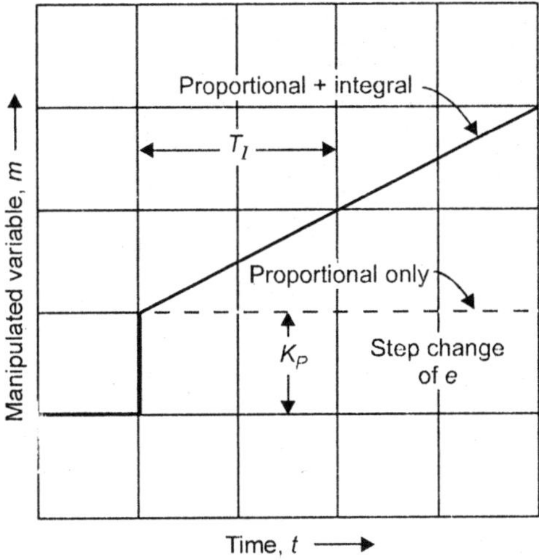

Fig. 2.23 Proportional-reset control action

Application

PI mode can be used in systems with frequent or large load changes. Because of the integration time, however, the process must have relatively slow changes in load to prevent oscillations induced by the integral overshoot. Another disadvantage of the system is that during startup of a batch process, the integral action causes a considerable overshoot of the error and output before settling to the operation point.

Integral windup

Often when error cannot be eliminated quickly, and given enough time this mode produces larger and larger values for integral term, which in turn keeps increasing the control action until it is saturated (valve fully open/close). This condition is called 'Integral Windup'. This occurs during changeover operations and shutdowns etc.

2.6.2 Proportional-Derivative Control Mode (PD)

A derivative control action may be added to proportional control action and the combination termed as a proportional-derivative control action. Equations for proportional mode and derivative mode are combined to have analytical expression for PD mode.

$$p = K_p e_p + K_p K_D \frac{de_p}{dt} + p_0 \quad \text{...(2.19)}$$

It is clear that this system cannot eliminate the offset of proportional controllers. It can, however, handle fast process load changes as long as the 'load change offset error' is acceptable. An example of the operation of this mode for a hypothetical load change is shown in Figure 2.24. Note the effect of derivative action in moving the controller output in relation to the error rate change.

The transfer function of PD control is given by $K_P(1 + K_D s)$.

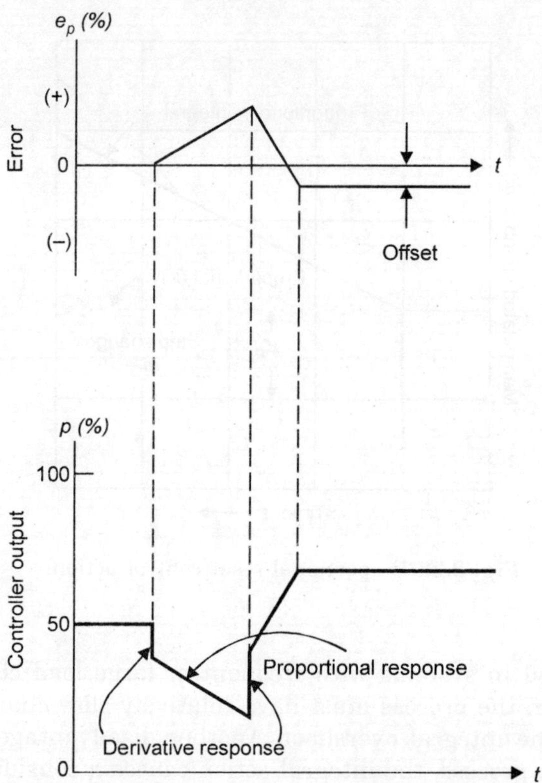

Fig. 2.24 Proportional-derivative (PD) action showing the offset error from the proportional mode (reverse action)

2.6.3 Proportional-Integral-Derivative Control Mode (PID)

The PID Controller is also called as three-mode controller. In the industrial practice it is commonly, known as 'Proportional-plus-reset-plus-rate controller'. The combination of proportional, integral, and derivative modes is one of the most powerful but complex controller mode operations. This system can be used for virtually any process condition. The equations of proportional mode, integral mode and derivative mode are combined to have analytical expression for PID mode.

$$p = K_P e_P + K_P K_I \int_0^t e_P dt + K_P K_D \frac{de_P}{dt} + p_{t(0)} \tag{2.20}$$

This mode eliminates the offset of the proportional mode and still provides fast response. The response of the three-mode system to an error is shown in Fig. 2.25. The three adjustment parameters here are proportional gain or PB, integral time and derivative time. PID controller is the most complex of the conventional control mode combinations. In theory, the PID controller can result in better control than the one-or two mode controllers. In practice, the control advantage can be difficult to achieve because of the difficulty of selecting the proper tuning parameters. The addition of the derivative mode to the PI controller is often specified to compensate not for process lags but for hardware lags that could be corrected at the source. PID controllers are used in the process industry to control slow variables such as temperature, pH, and other analytical variables.

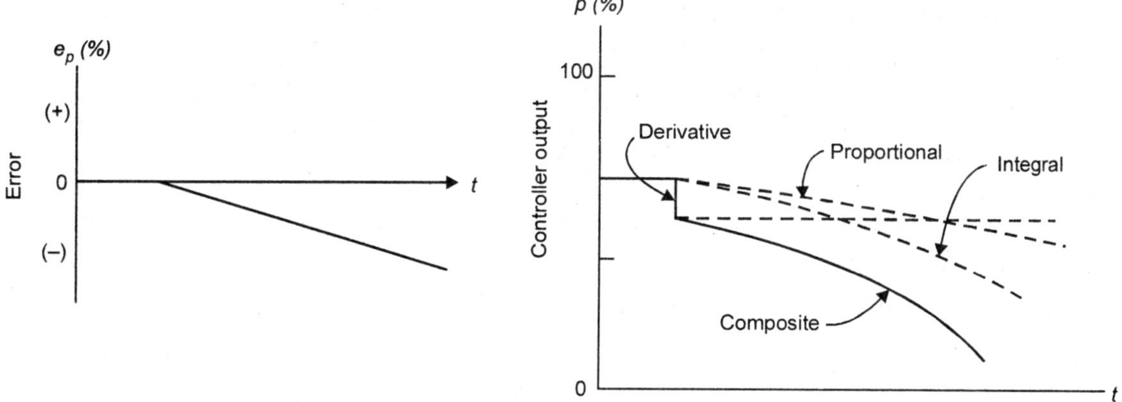

Fig. 2.25 The three-mode controller action exhibits proportional, integral, and derivative action

The transfer function of the PID control is given by $K_P\left(1+\frac{K_I}{s}+K_D s\right)$.

Table 2.1 illustrates transient response characteristics of continuous controllers. The transient response is a means of characterizing a system. It represents the trend versus time of the output (manipulated variable) for a step input (controlled variable).

Table 2.1 Transient response characteristics of continuous controllers

Variation of the input variable (controlled variable X)

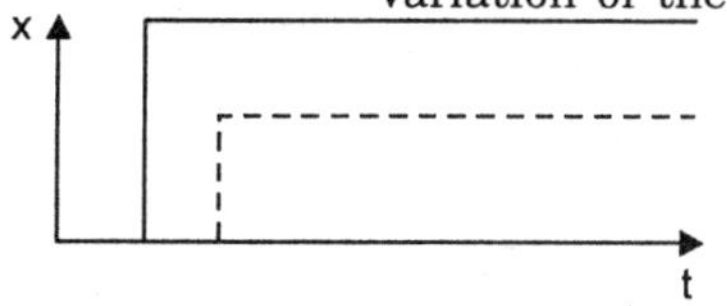

The input signal undergoes a step change.

(Two mutually independent processes are shown.)

Trend versus time of the output variable (manipulated variable Y.)

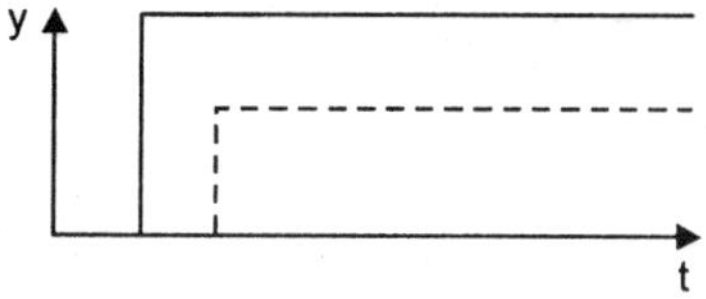

Proportional action controller.

The final control element is operated by an amount proportional to the deviation.

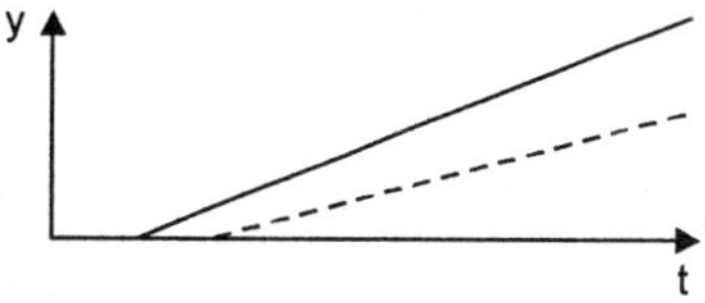

Integral action controller (floating action). The final control element is operated by an amount depending on the period of the deviation.

The rate of change of the manipulated variable is proportional to the deviation.

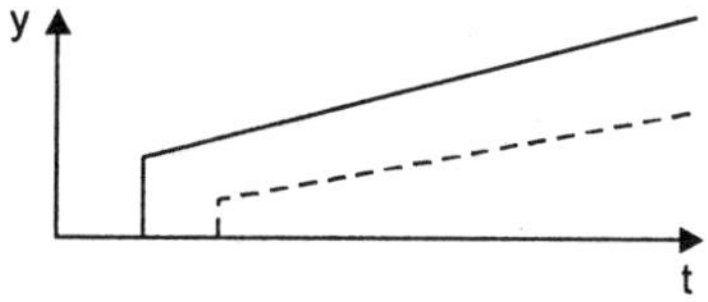

Proportional plus reset action controller.

The final control element first operated by an amount proportional to the deviation and then by an amount depending on its period.

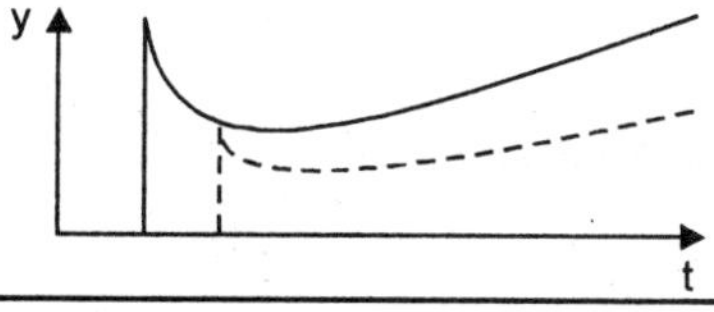

Proportional plus reset plus rate action controller.

The final control is first operated by an amount depending on the rate of change of the controlled variable and then as with proportinoal plus reset action control.

2.7 ELECTRONIC CONTROLLERS

We have so far seen the defining principles of various controller modes. Selection of the mode to use and appropriate gain depends on many factors involved in the process operation. This decision is made by engineers who are familiar with the process itself and who are guided by process-control technique experts who understand the characteristics of each mode. In this section and the next section we will study how modes of controller action are realized using analog techniques. In this section the emphasis is on electronic techniques, using op amps as the active element.

2.7.1 Error Detector: (Comparator)

The detection of an error signal is accomplished in electronic controllers by taking the difference between voltages. One voltage is generated by the process signal current passed through a resistor. The second voltage represents the set point which is usually generated by a voltage divider using a constant voltage as a source. A typical error detector is shown in Fig. 2.26. An error detector using a differential amplifier with a grounded reference transducer is shown in Fig. 2.27.

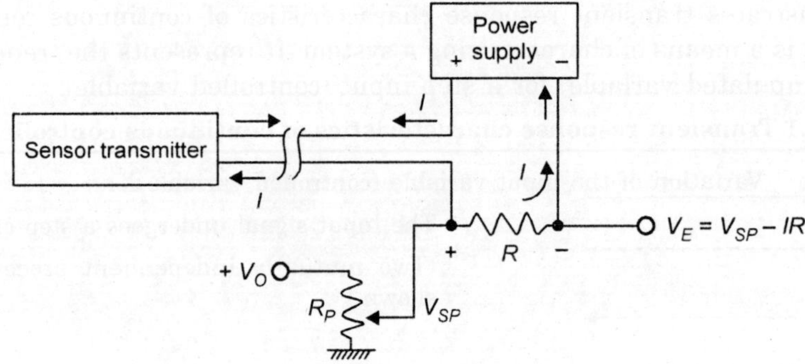

Fig. 2.26 Typical divider error detector

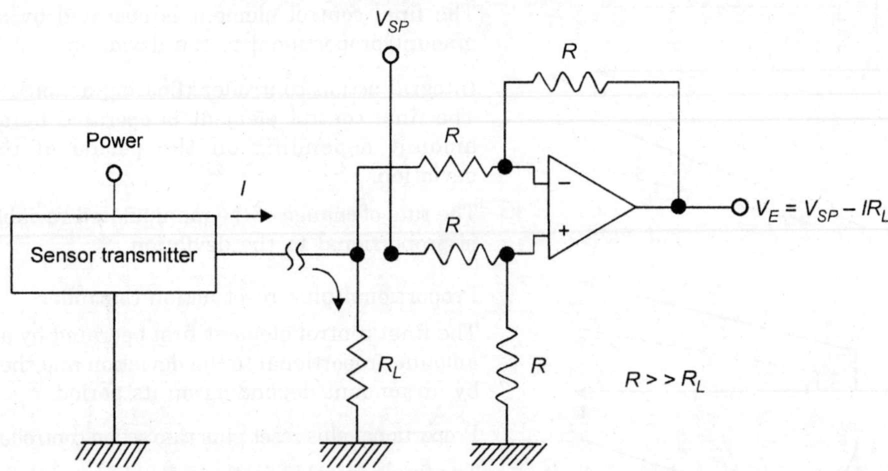

Fig. 2.27 An error detector using a differential amplifier.

2.7.2 Single Mode Controllers

The following methods illustrate the implementation of the pure modes of controller action with *op* amp circuits.

2.7.2.1 Two-Position

A two-position controller can be implemented by a great variety of electronic and electromechanical designs. Bimetal strip and mercury switch combination is employed for many two-position controller applications. One such switch operation is illustrated in Fig. 2.28. The bimetal strip bends, when temperature decreases, and reaches a point where the mercury slides down to close an electrical contact. The inertia of the mercury tends to keep the system in that position until the temperature increases to a value above the set point temperature. This provides the required neutral zone to present excessive cycling of the system.

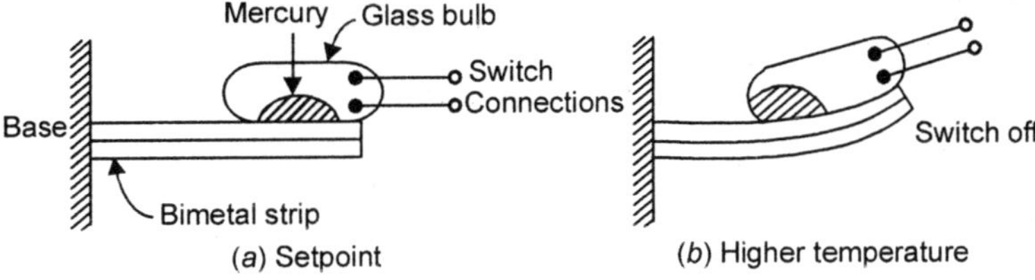

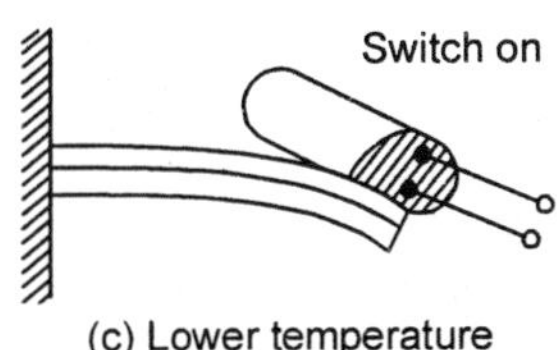

Fig. 2.28 A mercury switch on a bimetal strip is often used as a two-position temperature controller

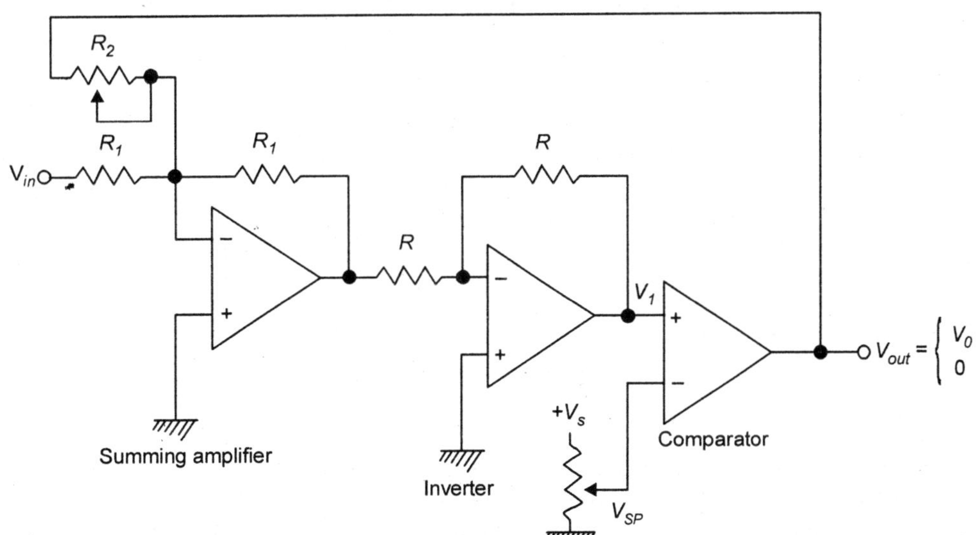

Fig 2.29 A two-position controller with neutral zone constructed from *op* amps and a comparator

A method using *op* amp implementation of ON-OFF control with adjustable neutral zone is given in Fig. 2.29. Here the controller input signal is assumed to be a voltage level with an 'ON' voltage of V_H and an 'OFF' voltage of V_L, and the output is comparator output of zero or V_0. The comparator output switches states when the voltage on its input V_1 is equal to the set point value V_{sp}. Analysis of the circuit shows that the high (ON) switch voltage is

$$V_H = V_{sp} \tag{2.21}$$

And the low (OFF) switching voltage is

$$V_L = V_{sp} - \frac{R_1}{R_3}V_0 \tag{2.22}$$

Fig. 2.30 shows the typical two-position relationship between input and output voltage for this circuit. The width of the neutral zone between V_L and V_H can be adjusted by variation of R_2. The relative location of the neutral zone is calculated from the difference between equations (2.21) and (2.22).

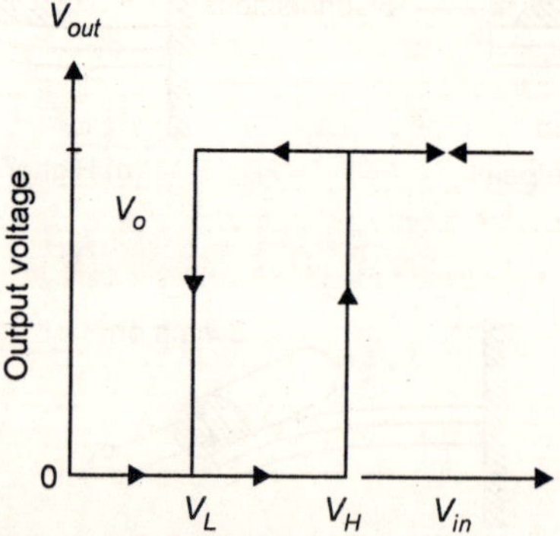

Fig. 2.30 Characteristics two-position response

The controller can be made reverse acting by placing an inverter in the feedback and reversing the comparator. Multiposition controllers can be devised by a similar process of *op* amp circuit development.

2.7.2.2 Floating-Type Controller

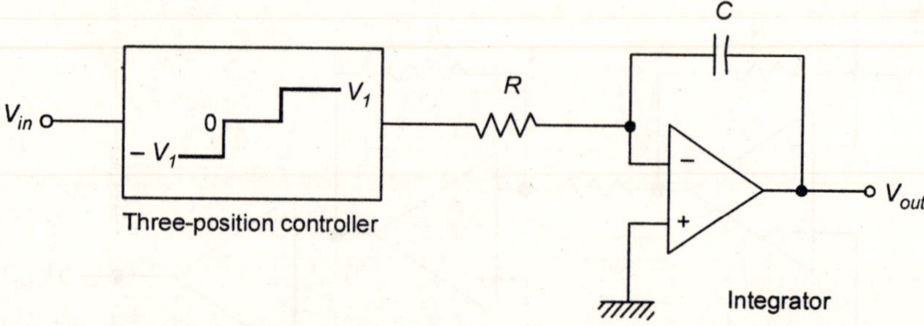

Fig. 2.31 Construction of a floating controller from a three-position controller and integrator

The floating type controller can be generated by connecting the output of a three-position controller into an integrator. Such a circuit is shown in Fig. 2.31. We assume the three-position controller was designed to provide outputs of V_1, Zero, or $-V_1$, depending on input. As an input to the integrator, this produces possible outputs of

$$V_{out} = \begin{cases} -\dfrac{V_1}{RC}(t - t_a) + V_a & V_{in} < V_{s1} \\ V_b & V_{s1} < V_{in} < V_{s2} \\ \dfrac{V_1}{RC}(t - t_C) + V_C & V_{in} > V_{s2} \end{cases} \tag{2.23}$$

where V_{s1} = lower setpoint voltage

V_{s2} = upper setpoint voltage

V_a, V_b, V_c = values of output when the input condition occurs.

t_a, t_b, t_c = times at which input reaches the setpoints.

The trip voltages can be set to provide the desired band of inputs producing no output, a positive rate, or a negative rate. The actual rate of output change depends on the values of resistor and capacitor in the integrator and the output level of the three-position circuit preceding the integrator. Remember that the output floats at whatever the latest value of output is when the input falls within the neutral zone.

2.7.2.3 Proportional mode

Implementation of this mode requires a circuit that has a response given by the equation.

$$p = K_P e_P + p_0 \tag{2.14}$$

where p = controller output 0–100%

K_P = proportional gain

e_P = error in percent of variable range

p_0 = controller output with no error.

If we consider both the controller output and error to be expressed in terms of voltage, the equation requires simply a summing amplifier. The *op* amp circuit in Fig. 2.32 shows such an electronic proportional controller. In this case, the analog electronic equation for the output voltage is

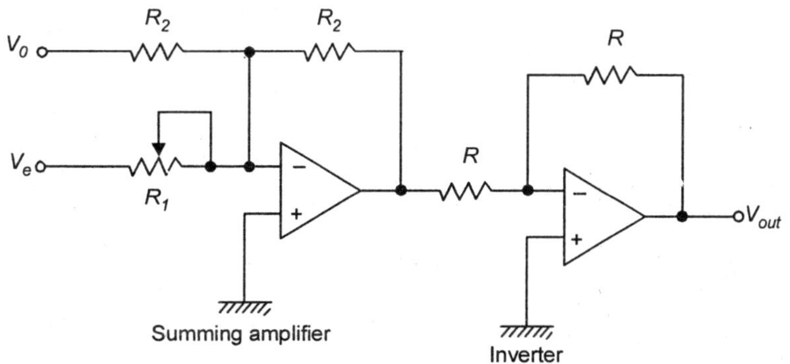

Fig. 2.32 An electronic proportional controller

$$V_{out} = G_p V_e + V_0 \tag{2.24}$$

where V_{out} = output voltage

$G_P = \frac{R_2}{R_1} = Gain$

V_e = error voltage

V_0 = output with zero error.

Gain G_P must be determined so that its effect in voltage is the same as that required by K_P. The output voltage variation must be matched as per the requirement of final control element. A zener diode can be used to fix the output swing to match. Generally, a voltage-to-current converter is used on the output to convert the output voltages to a current signals (4–20 mA nowadays) to drive the final control element.

2.7.2.4 Integral Mode

The integral mode is characterized by an equation of the form

$$p_{(t)} = K_I \int_0^t e_p(t)dt + p_I(0) \quad (2.16)$$

where $p_{(t)}$ = controller output in percent of full scale

K_I = Integration gain (s^{-1})

$e_p(t)$ = Deviations in percent of full-scale variable value

$p_I(0)$ = Controller output at $t = 0$

This function is easy to implement when op amps are used as the building blocks. A diagram of integral controller is shown in Fig. 2.33 . The corresponding voltage equation relating input to output is

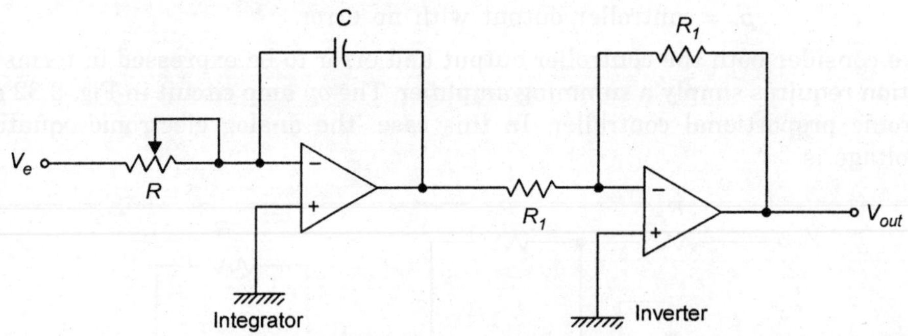

Fig. 2.33 Electronic integral-mode controller

$$V_{out} = G_I \int_0^t Ve\, dt + V_{out}(0) \quad (2.25)$$

where V_{out} = output voltage

G_I = 1/RC = Integral gain

V_e = error voltage

$V_{out}(0)$ = initial output voltage

The value of R and C can be adjusted to obtain the desired integration time. The initial controller output is the integrator output at $t = 0$. The integration time constant determines the rate at which the controller output increases when the error is constant.

If K_I is made too large, the output rises so fast that overshoots of the optimum setting occur and cycling is produced. The actual value of G_I, and therefore R and C, is determined from K_I and the input and output voltage ranges.

2.7.2.5 Derivative Mode

The derivative mode is never used alone because it cannot provide a controller output when the error is zero. Nevertheless, we show here how it is implemented with op amps for combination with other modes in the next section. The control mode equation is given earlier as

$$p = K_D \frac{de_P}{dt} \tag{2.17}$$

where p = controller output in percent of full output

K_D = derivative time constant(s)

e_P = error in percent of full scale range.

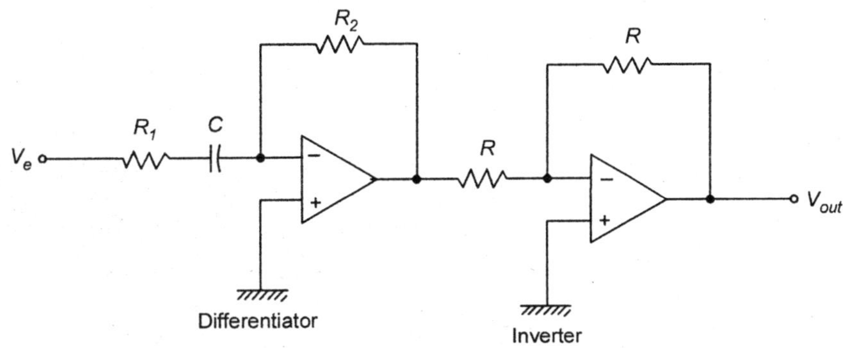

Fig 2.34 Electronic derivative-mode controller circuit

This function is implemented by op amps in the configuration shown in Fig. 2.34. Resistance R_1 is added for stability of the circuit against rapidly changing signals. The response of this circuit for slowly varying inputs is

$$V_{out} = G_D \frac{dV_e}{dt} \tag{2.26}$$

where V_{out} = Output voltage

$G_D = R_2C$ = Derivative time in seconds

V_e = error voltage

Without R_1 the circuit has a gain that increases with increasing frequency. Therefore, the circuit will be unstable and tend toward spontaneous oscillation when energized. Adding R_1 causes the circuit gain to revert to that of an inverting amplifier with a gain of R_2/R_1 at higher frequencies. The value of $G_D = R_2C$ is determined from K_D and knowledge of the measurement and output voltage ranges.

2.7.3 Composite Controller Modes

The circuits of the previous section (2.7.2) show that the pure modes of controller operation are easily constructed from op amps. But a pure mode is seldom used in process control because of the advantages of composite modes in providing good control. In this section, implementation of composite modes using op amps is considered .

The combination of several control modes combine the advantages of each mode and, in some cases, eliminate disadvantages. Composite modes are also implemented easily using op amp techniques. Basically, this consists of simply combining the mode circuits introduced in the previous section.

2.7.3.1 Proportional-Integral

A simple combination of the proportional and integral circuits provides the proportional-integral mode of controller action. The resulting circuit is shown in Fig. 2.35. For this case the relation between input and output is most easily found by applying op amp circuit analysis. Including the inverter we get,

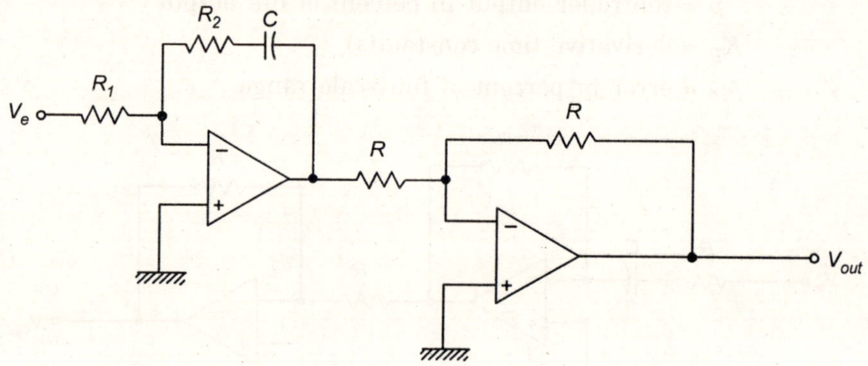

Fig. 2.35 Electronic proportional-integral (PI) controller

$$V_{out} = \left(\frac{R_2}{R_1}\right)V_e + \frac{1}{R_1C}\int_0^t V_e dt \tag{2.27}$$

The definition of the proportional-integral controller includes the proportional gain in the integral term, so we write

$$V_{out} = \left(\frac{R_2}{R_1}\right)V_e + \left(\frac{R_2}{R_1}\right)\frac{1}{R_2C}\int_0^t V_e dt + V_{out}(0) \tag{2.28}$$

Equation (2.28) has the same form as equation (2.18) for this mode. The adjustments of this controller are the proportional band through $G_P = R_2/R_1$, and the integral gain through $G_I = \dfrac{1}{R_2C}$.

2.7.3.2 Proportional-Derivative

A powerful combination of controllers mode is the proportional and derivative modes. This combination is implemented using a circuit similar to that shown in Fig. 2.36. Analysis shows that this circuit responds according to the equation.

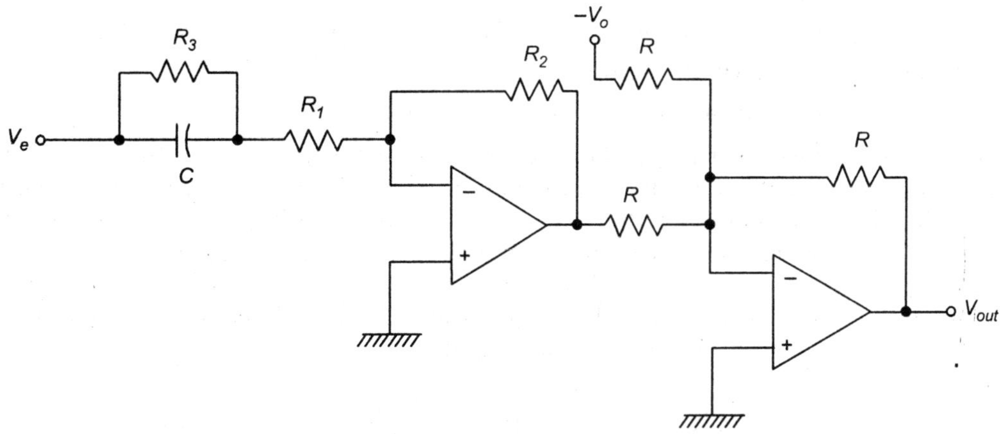

Fig. 2.36 An electronic proportional-derivative (PD) controller

$$V_{out} + \left(\frac{R_1}{R_1 + R_3}\right) R_3 C \frac{dV_{out}}{dt} = \left(\frac{R_2}{R_1 + R_3}\right) V_e + \left(\frac{R_2}{R_1 + R_3}\right) R_3 C \frac{dV_e}{dt} + V_0$$

where the quantities are defined in the figure and the output inverter has been included. We make the derivative coefficient on the left small to eliminate instability.

One choice is

$$\left(\frac{R_1}{R_1 + R_3}\right) R_3 C = \frac{0.1}{2\pi} T$$

where T is the fastest variable time change to be expected in the process. Then, the equation for the proportional derivative becomes

$$V_{out} = \left(\frac{R_1}{R_1 + R_3}\right) V_e + \left(\frac{R_2}{R_1 + R_3}\right) R_3 C \frac{dV_e}{dt} + V_0 \tag{2.29}$$

where the proportional gain $G_P = R_2/R_1 + R_3$ and the derivative gain is $G_D = R_3C$.

$$V_{out} = G_P\, V_e + G_P\, G_D \frac{dV_e}{dt} + V_0 \tag{2.30}$$

The equation (2.30) now corresponds to the form given in equation (2.19) for the proportional-derivative controller. Ofcourse, this mode still has the offset error of a proportional controller because the derivative term cannot provide reset action.

2.7.3.3 Proportional-Integral-Derivative

The ultimate process controller is the one that exhibits proportional, integral and derivative response to the process error input. In the previous section we saw that this mode was characterized by the equation :

$$p = K_P e_P + K_P K_I \int_0^t e_P dt + K_P K_D \frac{de_P}{dt} + p_I(0) \tag{2.20}$$

where p = controller output in percent of full scale

e_p = process error in percent of the maximum

K_P = proportional gain

K_I = integral gain

K_D = derivative gain

$p_I(0)$ = initial controller integral output.

The zero error term of the proportional mode is not necessary because the integral automatically accommodates for offset and nominal setting. This mode can be provided by a straight application of op amp circuits resulting in the circuit of Fig. 2.37. It must be noted, however, that it is possible to reduce the complexity of the circuitry and still realize the three-mode action, but in these cases an interaction results between derivative and integral gains. This circuit under consideration (Fig. 2.37) illustrates the principles of implementing this mode. Analysis of the circuit shows that the output is

$$-V_{\text{out}} = \left(\frac{R_2}{R_1}\right)V_e + \left(\frac{R_2}{R_1}\right)\frac{1}{R_1C_1}\int_0^t V_e dt + \left(\frac{R_2}{R_1}\right)R_D C_D \frac{dV_e}{dt} + V_{out}(0) \qquad (2.31)$$

where R_3 has been chosen from $2\pi R_3 C_D << T$ for stability. Comparison with equation (2.20) shown that this implements the three-mode controller if

$$G_P = \frac{R_2}{R_1}, \qquad G_D = R_D\ C_D \quad \text{and} \quad G_I = \frac{1}{R_1C_1}$$

$$-V_{out} = G_P V_e + G_P G_I \int V_e dt + G_P G_D \frac{dV_e}{dt} + V_{out}(0) \qquad (2.32)$$

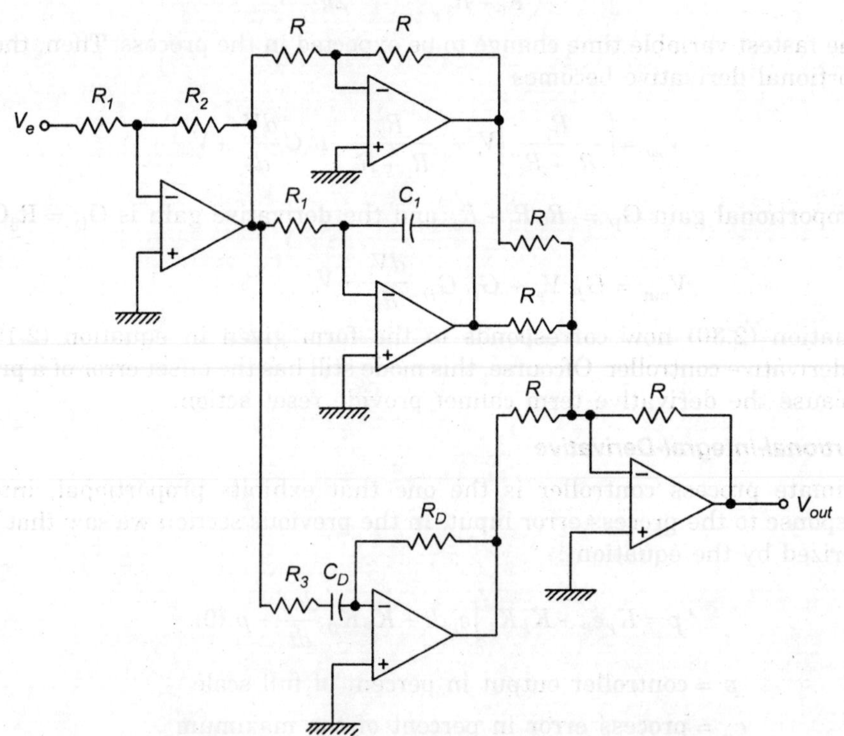

Fig. 2.37 An electronic three-mode (PID) controller.

The equation (2.32) is now corresponds to the form given in equation (2.20) for the PID controller.

These circuits have shown that the direct implementation of controller modes can be provided by standard op amp circuits. It is necessary, of course, to scale the measurement as a voltage with in the range of operation selected by the circuit. Furthermore, the outputs of the circuits shown have been voltages that may be converted to currents for use in an actual process-control loop. These circuits are only examples of basic circuits that implement the controller modes. Many modifications are employed to provide the controller action with different set of components.

2.8 PNEUMATIC CONTROLLERS

Historically, the reason for using pneumatics in process control was probably that electronic methods were not yet competitive in cost or reliability. Safety was and still is a factor where the danger of explosion from electrical malfunctions exists. It is also true that the final control element is often pneumatically or hydraulically operated, which suggests that an all-pneumatic process-control loop might be advantageous. It appears that analog or digital electronic methods will eventually replace most pneumatic installations. But we will still have pneumatic equipment for many years until these are depreciated in industry. A good understanding of process control principles can be applied to either electronic or pneumatic techniques, but it is necessary to consider some special features of pneumatic technology. This section provides a brief description of operations by which controller modes are pneumatically implemented.

2.8.1 Single Mode Controllers

The following methods illustrate the implementation of the pure modes of controller action with Flapper-nozzle arrangements as the basic mechanism of operation, much the same as the op amps is used in Electronics.

2.8.1.1 Two-Position

A two position controller with flapper-nozzle arrangements has already been explained in section 2.4.1.2 with reference to Fig. 2.8.

2.8.1.2 Proportional

The design shown in Fig. 2.38 provides proportional control from the flapper-nozzle arrangement. A feedback bellows and a spring have been added to the bottom of the flapper in Fig. 2.8. The flapper has also been extended so that the input signal E, which positions the flapper, is mechnically linked to the flapper above the nozzle. The signal E is the deviation of actual from set value *i.e.*, error 'e'. The relative position X of the flapper to the nozzle is determined by both the position Y of the feedback bellows and the position 'E' of the input signal. The output signal can be described by the equation:

$$P_O = \frac{L_2 K_f}{L_1 A_f} e \qquad (2.33)$$

The addition of the internal feedback mechanism effectively reduces the sensitivity of the device to an acceptable range for proportional control. Note that the proportional gain in the above equation can be easily adjusted by the ratio L_1/L_2, which can be changed by varying the position on the flapper at which the error signal is applied.

2.8.2 Composite Controller Modes

In this section implementation of composite modes using flapper-nozzle mechanism is considered.

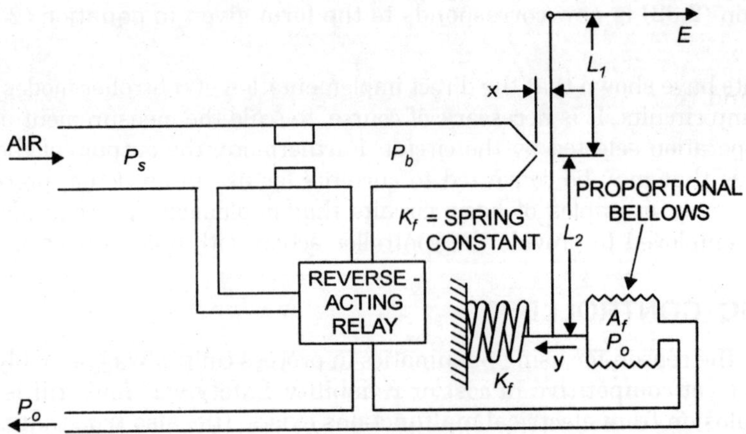

Fig. 2.38 Pneumatic proportional controller

2.8.2.1 Proportional-Integral

PI control mode is implemented using pneumatics by the system shown in Fig. 2.39. The additional bellows included in the feed back loop is mounted to act opposite the proportional bellows. Suppose there is sudden change in error signal. This drives the flapper toward the nozzle, increasing output pressure until the proportional bellows balances the error. The integral bellows is still at the original output pressure because the restriction prevents pressure changes from being transmitted immediately. As the increased pressure on the output bleeds through the restriction, the integral bellows slowly moves the flapper close to the nozzle, thereby causing a steady increase in output pressure (as dictated by the integral mode). The variable restriction allows for variation of the leakage rate and hence the integral time.

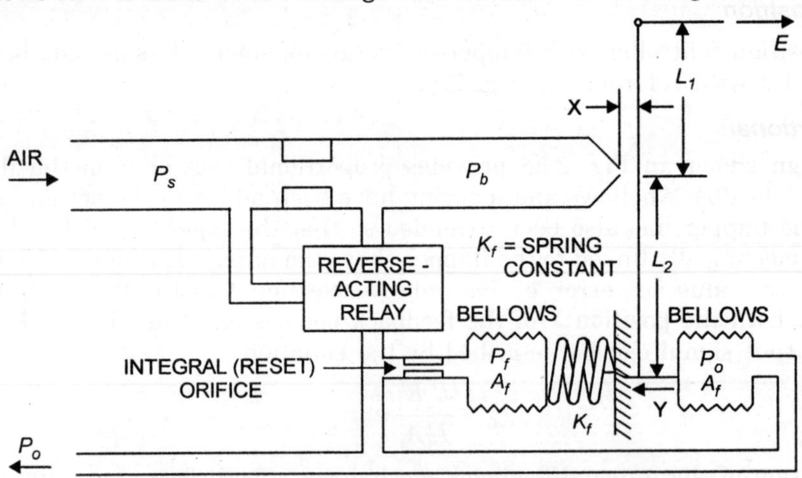

Fig. 2.39 Pneumatic proportional-plus-integral controller

2.8.2.2 Proportional-Derivative

This controller action can be accomplished pneumatically by the method shown in Fig. 2.40. The addition of a variable restriction in the line leading to the feedback bellows offers resistance to flow into the bellows and therefore creates a time lag effect. The proportional gain can be adjusted by varying the resistance of the restriction in the airline to the bellows.

As the error varies, the flapper is moved toward the nozzle with no resistance because the restrictions prevent an immediate response of the balance bellows. Thus, the output pressure rises very fast and then, as the increased pressure leaks into the balance bellows, decreases as the balance bellows moves the flapper back away from the nozzle.

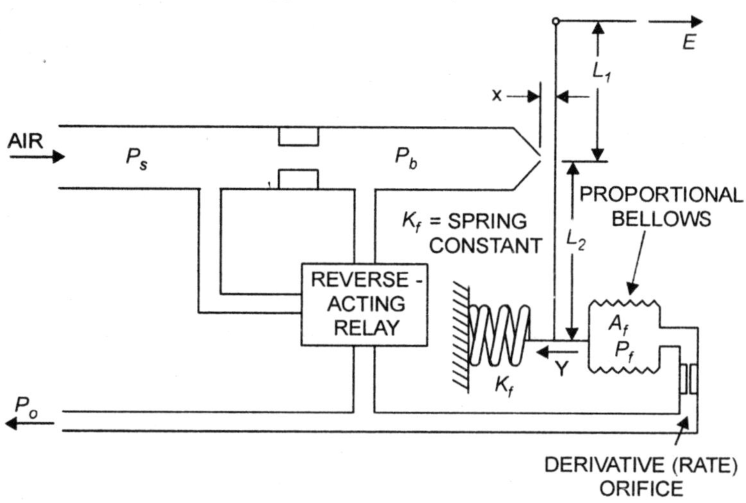

Fig. 2.40 Pneumatic proportional-plus-derivative controller

2.8.2.3 Proportional-Integral-Derivative : (Three mode)

The three mode (PID) controller is actually the most common type produced because it can be used to accomplish any of the previous modes by setting of restrictions. This device is shown in Fig. 2.41, and, as can be seen, is simply a combination of the three systems (P, PI and PD) presented. By opening or closing restrictions the three-mode controller can be used to implement the other composite modes. Proportional gain, reset time and rate are set by adjustment of bellows separation and restriction size. Because their three settings are interconnected, if one setting changed, it will affect all the others. Hence the actual working settings are to be done in practice.

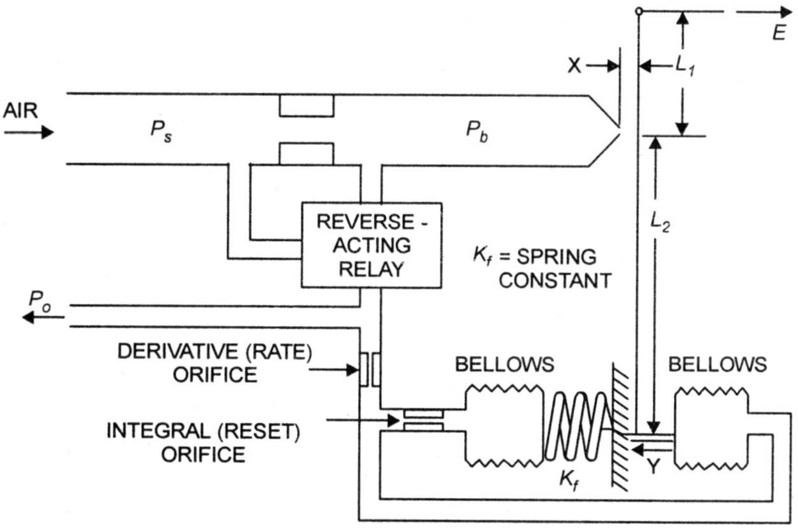

Fig. 2.41 Pneumatic proportional-plus-integral-plus-derivative controller

2.9 EVALUATION CRITERIA

The first problem encountered in selection and tuning of controllers is to define what is 'good' control. This unfortunately differs from process to process. In this section we will discuss how to select the type of feedback controller (*i.e.*, *P*, *PI*, *PD* or *PID*) and how to adjust the parameters of the selected controller (*i.e.*, K_P, K_I, K_D) for a particular process in order to achieve an 'optimum' response for the controlled process.

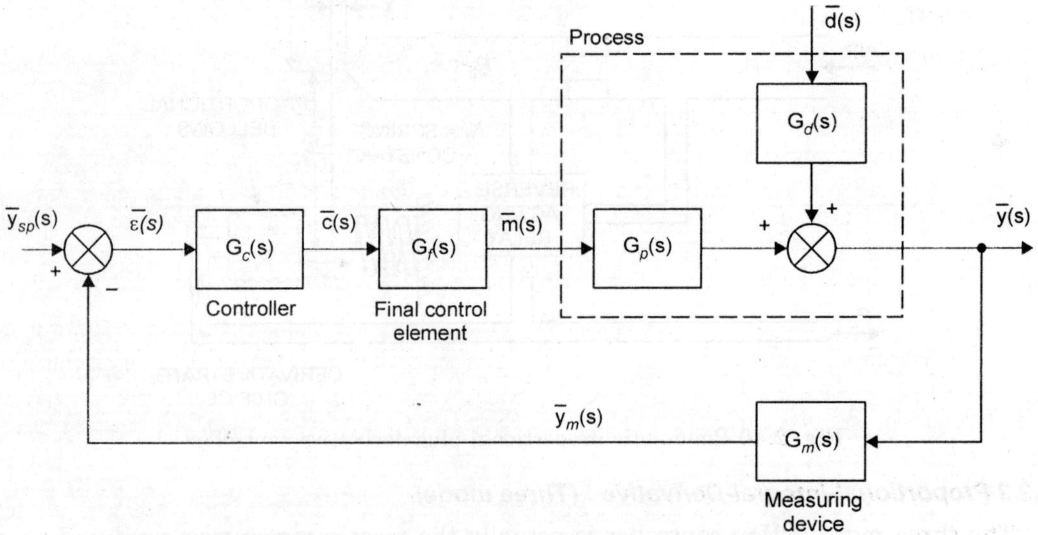

Fig. 2.42 Block diagram of generalized closed-loop system

Consider the block diagram of a general closed-loop system shown in Fig. 2.42. When the load or the set point changes, the response of the process deviates and the controller tries to bring the output again close to the desired set point. Fig. 2.43 shows the response of a controlled process to a unit step change in the load when different types of controllers have been used. We notice that different controllers have different effects on the response of the controlled process. It clearly demands that we should use some performance criterion for the selection and the tuning of the controller.

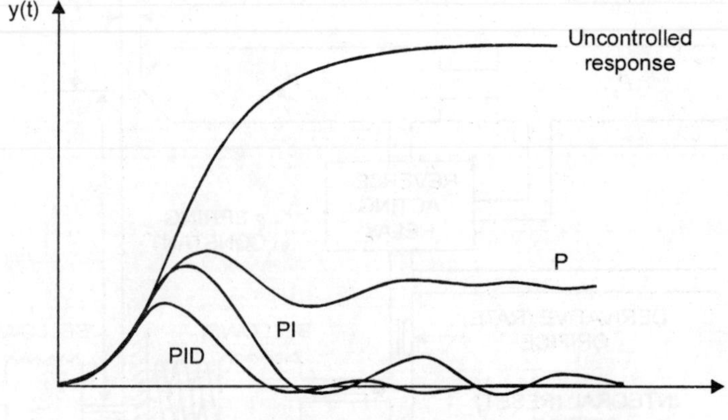

Fig. 2.43 Response of a system to unit step change in load with no control, And various types of feedback controllers

There are a variety of performance criteria we could use, such as :

1. Keep the maximum deviation (error) as small as possible.
2. Achieve short settling times.
3. Minimize the integral of the errors until the process has settled to its desired set point, and so on.

2.9.1 Performance Criteria

Consider two different feedback control systems producing the two closed-loop responses shown in Fig. 2.44. Response '*A*' has reached the desired level of operation faster than response '*B*'. if our criterion for the selection of controller had been 'return to the desired level of operation as soon as possible', then, clearly, we would select the controller which gives the closed-loop response of type '*A*'. But, if our criterion had been 'Keep the maximum deviation as small as possible' or 'return to the desired level of operation and stay close to it in the shortest time, we would have selected the other controller, yielding the closed-loop response of type '*B*'. Similar dilemmas will be encountered quite often during the selection of a controller.

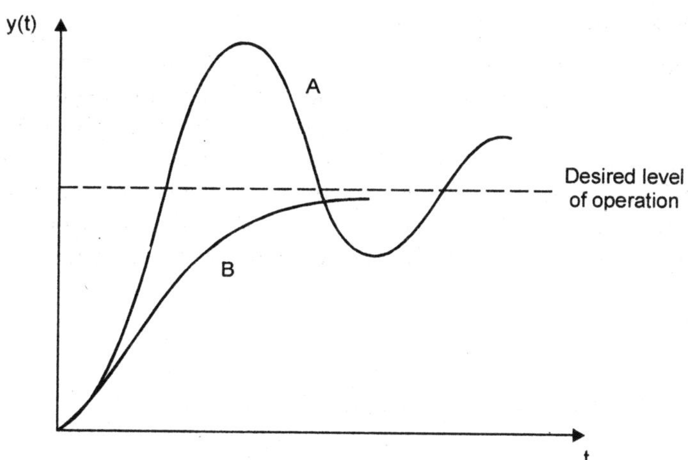

Fig. 2.44 Alternative closed-loop responses

For every process control application, we can distinguish

1. Steady-State performance criteria and
2. Dynamic response performance criteria.

Steady state performance criteria

The principal steady-state performance criterion usually is 'zero error at steady state'. We have seen already that in most situations the proportional controller cannot achieve zero steady-state error, while a PI controller can. Also, we know that for proportional control the steady state error (offset) tends to zero as $K_P \to \infty$. No further discussion is needed on this.

Dynamic response performance criteria

The evaluation of the dynamic performance of a closed-loop system is based on two types of commonly used criteria :

1. Simple performance criteria : criteria that use only a few points of the response They are simpler, but only approximate.
2. Time Integral Performance Criteria : Criteria that use the entire closed-loop response from time t = 0 until t = very large. These are more precise but also more cumbersome to use.

2.9.2 Simple Performance Criteria

The simple performance criteria are based on some characteristic features of the closed loop response of a system. The most often quoted are (Refer Fig. 1.25) : overshoot, risetime, settling time, decay ratio and frequency of oscillation of the transient. From all these performance criteria, the decay ratio has been the most popular by the practicing engineers. Specifically, experience has shown that a decay ratio $\left(\frac{C}{A}=\frac{1}{4}\right)$ is a reasonable trade-off between a fast rise time and a reasonable settling time. This criterion is usually known as the 'one-quarter decay ratio criterion'.

One-quarter Decay Ratio Criterion

The decay ratio criterion has the advantage of being readily measurable, as it is based on only two points on the step response. The general response curve is shown in Fig. 2.45. This measure of decay ratio is found by adjusting the control loop until the deviation from the disturbance is such that each deviation peak is down by one quarter from the preceding peak, as shown in Fig. 2.46. In this case, the actual magnitude of the deviation is not included in the measure, nor is the time between each peak. In this sense, neither duration nor magnitude of the deviation is directly involved in quarter-amplitude criterion.

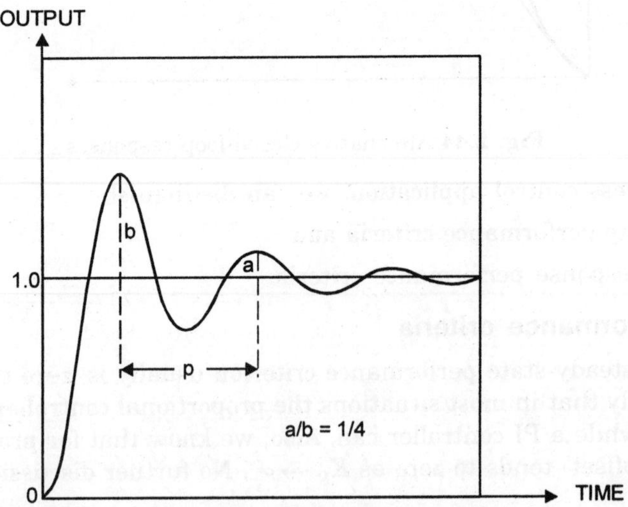

Fig 2.45 Response curve for ¼ decay ratio

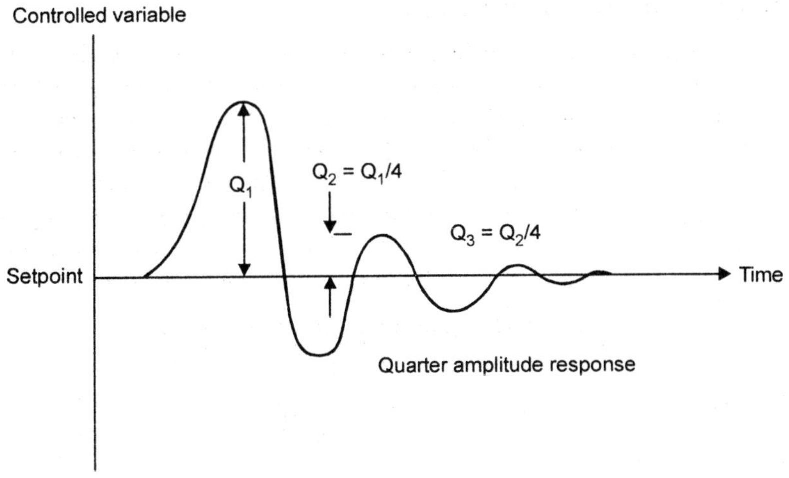

Fig. 2.46 In one type of cyclic response, the system is adjusted to make each Peak down by one quarter of the previous peak

Note : In cases of cyclic underdamped response, the most critical element is sometimes a combination of duration and deviation, which must be minimized. Thus, if minimum deviation occurs at one loop setting and minimum duration at another, then neither is optimum. One type of optimum measure of quality in these cases is to minimise the net area of the deviation as a function of time. In Fig. 2.47 this is shown as the sum of the shaded areas. Analytically, this can be expressed as

$$A = \int |r - b|\, dt \tag{2.34}$$

where A = area of deviation

b = measured value

r = set point value

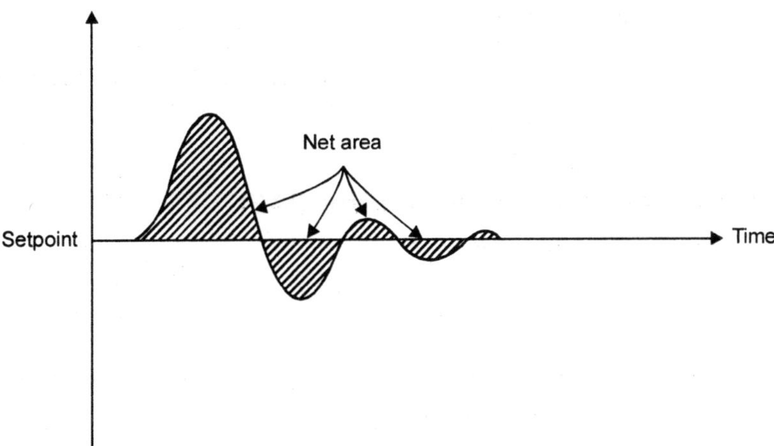

Fig. 2.47 The minimum area criterion for cyclic response adjusts the process-control loop until the net area is minimum

2.9.3 Time Integral Performance Criteria

The shape of the complete closed-loop response from time $t = 0$ until steady state has been reached, could be used for the formation of a dynamic performance criterion. Unlike the simple criteria that use only isolated characteristics of the dynamic response. (*e.g.*, decay ratio, settling time), the criteria of this category are based on the entire response of the process. The integral criteria have the advantage of being more precise, that is, more than one combination of controller settings will usually give a ¼ decay ratio, but only one combination will minimize the respective integral criteria. Table 2.2 gives the three most often used integral criteria in addition to the earlier explained ¼ decay criteria.

Table 2.2 Criteria for controller tuning

1. Specified Decay Ratio, Usually 1/4	$\text{Decay ratio} = \dfrac{\text{second peak overshoot}}{\text{first peak overshoot}}$
2. Minimum Integral of Square Error (ISE)	$\text{ISE} = \int_0^\infty \lvert e(t)\rvert^2 dt$ where $e(t)$ = (setpoint process output)
3. Minimum Integral of Absolute Error (IAE)	$\text{IAE} = \int_0^\infty \lvert e(t)\rvert dt$
4. Minimum Integral of Time and Absolute Error (ITAE)	$\text{ITAE} = \int_0^\infty \lvert e(t)\rvert t dt$

1. Integral of the Square Error (ISE)

The mathematical equation for this criteria is

$$\text{ISE} = \int_0^\infty e^2(t)\, dt \tag{2.35}$$

Where $e(t) = y_{SP}(t) - y(t)$ is the deviation (error) of the response from the desired setpoint.

ISE criteria gives more weight to larger deviations.

2. Integral of the Absolute value of the Error (IAE)

IAE is expressed by the following equation.

$$\text{IAE} = \int_0^\infty \lvert e(t)\rvert\, dt \tag{2.36}$$

IAE seems the best criterion for process control, since the penalty for control is generally a linear function of the error.

3. Integral of the Time-weighted Absolute Error (ITAE)

ITAE is given by the equation

$$\text{ITAE} = \int_0^\infty t\lvert e(t)\rvert\, dt \tag{2.37}$$

ITAE criteria weights deviations more heavily as time increases.

Which one of the three criteria above we will use depends on the characteristics of the system we want to control and some additional requirements we impose on the controlled response of the process. The following are some general guidelines.

1. If we want to strongly suppress large errors, ISE is better than IAE because the errors are squared and thus contribute more to the value of the integral.
2. For the suppression of small errors, IAE is better than ISE because when we square small numbers (smaller than one) they become even smaller.
3. To suppress errors that persist for longer times, the ITAE criterion will tune the controllers better because the presence of large 't' amplifies the effect of even small errors in the value of the integral.

Fig. 2.48 demonstrates, in a qualitative manner, the shape of the expected closed-loop responses. When we tune the controller parameters using ISE, IAE, or ITAE performance criteria, we should remember the following two points :

1. Different criteria lead to different controller designs.
2. For the same time integral criterion, different input changes lead to different designs.

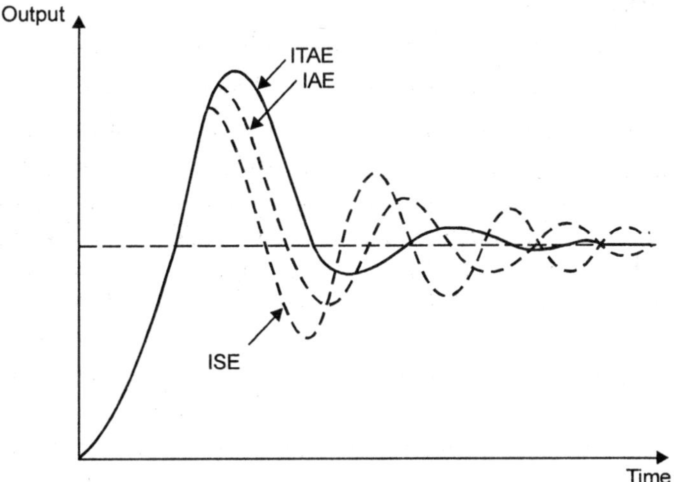

Fig. 2.48 Closed-loop responses using various time integral criteria

2.9.4 Selection of Feedback Controller

The systematic way of selecting one of the popular feedback controllers for a given process can be stated in three steps :

1. Define an appropriate performance criterion (*e.g.*, ISE, IAE or ITAE)
2. Compute the value of the performance criterion using a P or PI or PID controller with the best settings for the adjusted parameters K_P, K_I and K_D.
3. Select the controller which gives the 'best' value for the performance criterion.

The above procedure has got so many drawbacks. It is very tedious. It relies on models (transfer functions) for the process, sensor, and final control element which may not be known exactly. It incorporates certain ambiguities as to which is the most appropriate criterion and what input changes to consider. Instead, we can select the most appropriate type of feedback controller using only general qualitative considerations of the effect of the proportional, integral, and derivative control modes on the response of a system. They can be summarized as follows :

1. Proportional control

(*a*) Accelerates the response of a controlled process.

(*b*) Produces an offset (*i.e.*, non zero steady-state error) for all processes except those with terms 1/s (integrators) in their transfer function, such as the liquid level in a tank or the gas pressure in a vessel.

2. Integral control

(*a*) Eliminates any offset.

(*b*) The elimination of offset usually comes at the expense of higher maximum deviations.

(*c*) Produces sluggish, long oscillating responses.

(*d*) If we increase the gain K_P to produce faster response, the system becomes more oscillatory and may be led to instability.

3. Derivative control

(*a*) Anticipates future errors and introduces appropriate action.

(*b*) Introduces a stabilizing effect on the closed-loop response of a process.

Fig. 2.43 Reflects in a very simple way all the characteristics noted above.

It is clear from the above that a three-mode PID controller should be the best. This is true in the sense that it offers the highest flexibility to achieve the desired control response by having three adjustable parameters. At the same time, it introduces a more complex tuning problem because we have to adjust three parameters . To balance the quality of the desired response against the tuning difficulty we can adopt the following rules in selecting the most appropriate controller.

1. 'If possible, use simple proportional controller'. Simple proportional controller can be used if

(*a*) We can achieve acceptable offset with moderate values of K_P or

(*b*) the process has an interacting action (*i.e.*, a term 1/s in its transfer function) for which the proportional control does not exhibit offset. Therefore, for gas pressure or liquid-level control we can use only P controller.

2. 'If a simple P controller is unacceptable, use a PI'. A 'PI' controller should be used when proportional control alone cannot provide sufficiently small steady-state errors (offsets). Therefore, PI will seldom be used in liquid-level, or gas pressure control systems, but very often (almost always) for flow control . The response of a flow control system is rather fast. Consequently, the speed of the closed-loop system remains satisfactory despite the slow down caused by the integral control mode.

3. 'Use a PID controller to increase the speed of the closed-loop response and retain robustness'. The PI eliminates the offset but reduces the speed of the closed-loop response. For a multicapacity process whose response is very sluggish, the addition of a PI controller makes it even more sluggish. In such cases the addition of the derivative control action with its stabilizing effect allows the use of higher gains which produce faster responses without

excessive oscillations. Therefore derivative action is recommended for Temperature and composition control where we have sluggish, multicapacity processes.

2.10 CONTROLLER TUNING

After the type of feedback controller has been selected, as discussed in the previous section, we still have the problem of deciding what values to use for its adjusted parameters. This is known as 'Controller Tuning' problem. There are three general approaches we can use for tuning a controller.

1. Use simple criteria such as the one-quarter decay ratio, minimum settling time, minimum largest error, and so on. Such an approach is simple and easily implementable on an actual process. Usually, it provides multiple solutions. Additional specifications on the closed-loop performance will then be needed to break the multiplicity and select a single set of values for the adjusted parameters.
2. Use time integral performance criteria such as ISE, IAE or ITAE. This approach is rather cumbersome and relies heavily on the mathematical model (transfer function) of the process. Applied experimentally on an actual process, it is time consuming.
3. Use semiemprical rules which have been proven in practice.

In this section we will be discussing some of the empirical tuning methods.

2.10.1 Process Reaction Curve Method : (Cohen and Coon Method)

The basic approach is to open the process-control loop so that no control action (feed back) occurs. This is usually done by disconnecting the controller output from the final control element. All of the process parameters are held at their nominal values. This method can be used only for systems with self-regulation. This is also called as Open Loop Transient response method.

Consider the control system in Fig. 2.49 which has been 'opened' by disconnecting the controller from the final control element. Introduce a step change of magnitude A in the variable '*c*' which actuates the final control element. In the case of a valve, '*c*' is the stem position. Record the value of the output with respect to time. The curve $y_m(t)$ thus obtained is called the 'Process Reaction Curve' . Between y_m and *c*, we have the following transfer function.

$$G_{PRC}(s) = \frac{\overline{y_m}(s)}{\bar{c}(s)} = G_f(s).G_P(s).G_m(s) \tag{2.38}$$

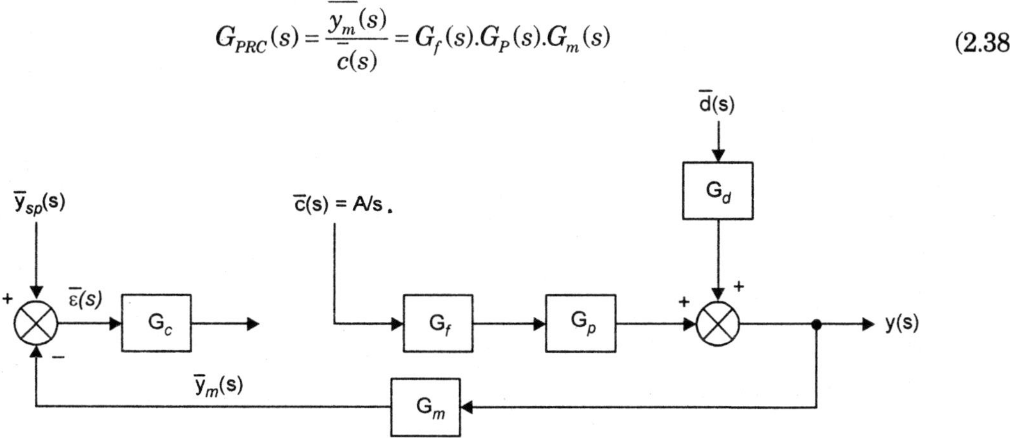

Fig. 2.49 Opened control loop

The equation shows that the process reaction curve affected not only by the dynamics of the main process but also by the dynamics of the measuring sensor and final control element.

Cohen and Coon observed that the response of most processing units to an input change, such as the above, had a sigmoidal shape as shown in Fig. 2.50 (*a*) which can be adequately approximated by the response of a first order system with dead time [Dashed curve in Fig. 2.50(*b*)].

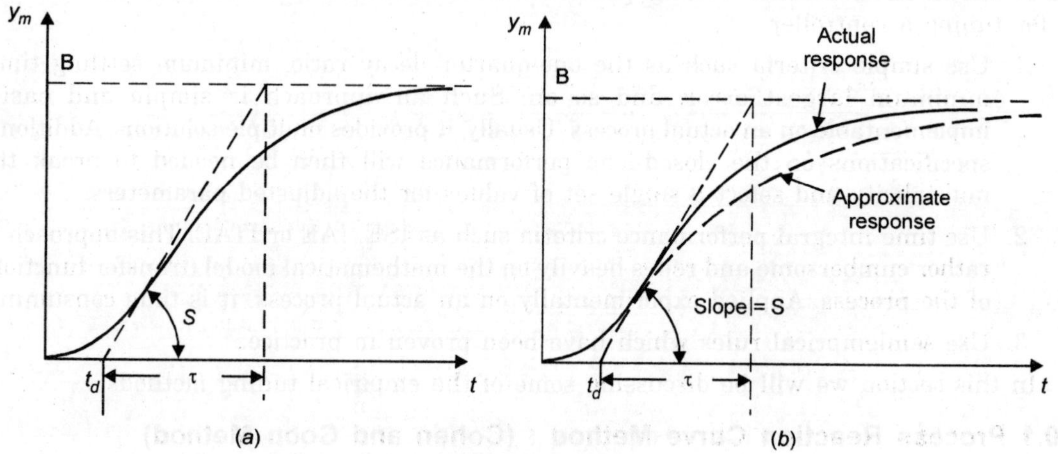

Fig 2.50 (*a*) Process reaction curve
(*b*) Its approximation with a first order plus Dead-time system

$$G_{PRC}(s) = \frac{\overline{y_m}(s)}{\overline{c}(s)} \cong \frac{K\,e^{-t_d s}}{\tau\,s+1} \tag{2.39}$$

which has three parameters : static gain 'K', dead time 't_d' and time constant 'τ'. From the approximate response it is easy to estimate the values of the three parameters. Thus,

$$K = \frac{B}{A} = \frac{\text{Output (at steady state)}}{\text{Input (at steady state)}}$$

$\tau = \dfrac{B}{S}$, where 'S' is the slope of the sigmoidal response at the point of inflection.

t_d = time elapsed until the system responded.

Cohen and Coon used the approximate model of equation (2.39) and estimated the values of the parameters K, τ and t_d as indicated above. Then they derived expressions for the 'best' controller settings using load changes, and various performance criteria, such as one-quarter decay ratio, minimum offset and minimum integral square error (ISE).

The results of their analyse are summarized below :

1. Proportional controller

For the proportional mode, the proportional gain setting K_P is found from

$$K_P = \frac{\tau}{K\,t_d} \tag{2.40}$$

Corrections to the value of K_P are sometimes used to obtain the quarter-amplitude criterion of response. The one given by Cohen and Coon is shown bracketed.

$$K_P = \frac{1}{K} \cdot \frac{\tau}{t_d} \cdot \left[1 + \frac{t_d}{3\tau}\right] \tag{2.41}$$

2. Proportional-integral controllers

For P-I mode, we are to use

$$K_P = \frac{1}{K} \cdot \frac{\tau}{t_d} \cdot \left[0.9 + \frac{t_d}{12\tau}\right] \tag{2.42}$$

$$T_I = t_d \left(\frac{30 + 3\dfrac{t_d}{\tau}}{9 + 20\dfrac{t_d}{\tau}} \right) \tag{2.43}$$

3. Proportional-integral-derivative controllers

For P-I-D mode, we use

$$K_P = \frac{1}{K} \cdot \frac{\tau}{t_d} \cdot \left[\frac{4}{3} + \frac{t_d}{4\tau}\right] \tag{2.44}$$

$$T_I = t_d \left(\frac{32 + 6\dfrac{t_d}{\tau}}{13 + 8\dfrac{t_d}{\tau}} \right) \tag{2.45}$$

$$T_D = t_d \left(\frac{4}{11 + 2\dfrac{t_d}{\tau}} \right) \tag{2.46}$$

The controller settings given above are based on the assumption that the first-order plus dead-time system is a good approximation for the sigmoidal response of the local loop real process. It is possible, though, that the approximation may be poor. In such a case, the Cohen and Coon settings should be viewed only as first guesses needing certain online correction.

It is noticed that almost all physical processes encountered in a chemical plant are simple first order or multicapacity processes whose response has the general overdamped shape. The oscillatory underdamped behaviour is produced mainly by the presence of feedback controllers. Therefore when we open the loop and thus disconnect the controller, the response takes the sigmoidal shape of an overdamped system.

We notice the following when we get the value of the proportional gain for the three controllers.

(*a*) The gain of the PI controller is lower than that of the 'P' controller. This is due to the fact that the integral control mode makes the system more sensitive (may even lead to instability) and thus the gain value needs to be more conservative.

(*b*) The stabilizing effect of the derivative control mode allows the use of higher gains in the PID controller. (higher than the gain for P or PI controllers).

2.10.2 Ziegler-Nichols Method : (Ultimate Cycle Method)

The 'Process reaction curve' was originally developed by Ziegler-Nichols. The corrections were developed by Cohen and Coon (when the quarter-amplitude response criterion is indicated) which has become popular later on.

Ziegler and Nichols have developed another method of controller setting assignment that has come to be associated with their name. This technique, also called the 'Ultimate Cycling Method', is based on adjusting a closed loop until steady oscillations occur. Controller settings are then based on the conditions that generate the cycling. This method is based on frequency response analysis.

Unlike the process reaction curve method which uses data from the open-loop response of a system, the Ziegler-Nichols tuning technique is a closed-loop procedure . It goes through the following steps :

1. Bring the system to the desired operational level (Design Condition).
2. Reduce any integral and derivative actions to their minimum effect.
3. Using proportional control only and with the feedback loop closed, introduce a set point change and vary proportional gain until the system oscillates continuously. The frequency of continuous oscillation is the cross over frequency, 'ω_0'. Let '*M*' be the amplitude ratio of the system's response at the cross over frequency.
4. Compute the following two quantities :

$$\text{'Ultimate gain'} = K_U = \frac{1}{M}$$

$$\text{'Ultimate period of sustained cycling'} = P_U = \frac{2\pi}{\omega_{CO}} \text{ min/cycle.}$$

5. Using the values of K_U and P_U, Ziegler and Nichols recommended the following settings for feedback controllers.

Mode	K_P	T_I (Min)	T_d (Min)
Proportional	$K_U/2$	–	–
Proportional-Integral	$K_U/2.2$	$P_U/1.2$	–
Proportional-Integral-Derivative	$K_U/1.7$	$P_U/2$	$P_U/8$

The settings above reveal the rationale of the Ziegler–Nichols methodology.

1. For proportional control alone, use a gain margin equal to 2.
2. For PI control use a lower proportional gain because the pressure of the integral control mode introduces additional phase lag in all frequencies with destabilizing effects on the system. Therefore, lower K_P, maintains approximately the same gain margin.

3. The presence of the derivative control mode introduces phase lead with strong stabilizing effects in the closed-loop response. Consequently, the proportional gain K_P for a PID controller can be increased without threatening the stability of the system.

2.10.3 Damped-Oscillation Method

In many plants, sustained oscillations for testing purposes are not allowable, and the ultimate frequency method as discussed under Ziegler-Nichols method cannot be used to secure the optimum settings. The following modification is easy to follow and perhaps more accurate than the ultimate method. By using only proportional action and starting with a low gain, the gain is adjusted until the transient response of the closed loop shows a decay ratio of ¼. The reset time and derivative time are based on the period of oscillation, P, which is always greater than the ultimate period P_U.

For Proportional-Integral-Derivative control,

$$T_D = \frac{P}{6}, \qquad T_I = \frac{P}{1.5}$$

With the derivative and reset times at the above values, the gain for ¼ decay ratio is again established by transient-response tests.

2.10.4 Frequency Response Methods of Tuning

Stability of operation of the process and automatic control system is very important in proess control industries. The study of stability is greatly enhanced by the use of the Laplace transform. There are number of ways by which the stability of a control system can be analysed. Some of the important ones are :

1. Routh-Hurwitz condition for stability
2. Direct substitution analysis
3. Nyquist Stability Criterion
4. Bode Magnitude-Phase method of Analysis.
5. Root Locus Technique.

No. 1 and 2 are Algebraic method, No. 3 and 4 are classified under frequency response method and No. 5, the Root Locus Technique is a graphical method. As the topic 'stability' is very important one under 'Control Systems', we consider this topic to be outside the preview of this book. Interested readers may study this topic in any of the standard 'Control Systems' book. We will be just touching the Bode stability criterion for the purpose of tuning the controllers.

The frequency response method of process controller tuning involves use of Bode plots for the process and control loops. The method is based on an application of the Bode plot stability criteria and the effects that proportional gain, integral time, and derivative time have on the Bode plot.

2.10.4.1 Gain and Phase Margin : (Refer Fig. 2.51)

The stability criteria represent limits of stability. For example, if the gain is slightly less than one when the phase lag is 180°, the system is stable. But if the gain is slightly greater than one at 180°, the system is unstable. It would be good to design a system with a margin of safety from such limits to allow for variations in components and other unknown factors. This consideration leads to the revised stability criteria, or more properly, a margin of safety

provided to each condition. The exact terminology is in terms of a 'gain margin' and 'phase margin' from the limiting values quoted. Although no standards exist, a common condition is:

1. If the phase lag is less than 140° at the unity gain frequency, the system is stable. This, then, is a 40° 'phase margin' from the limiting value of 180°.
2. If the gain is 5 dB below unity (or a gain of about 0.56) when the phase lag is 180°, the system is stable. This is a 5 dB 'gain margin'.

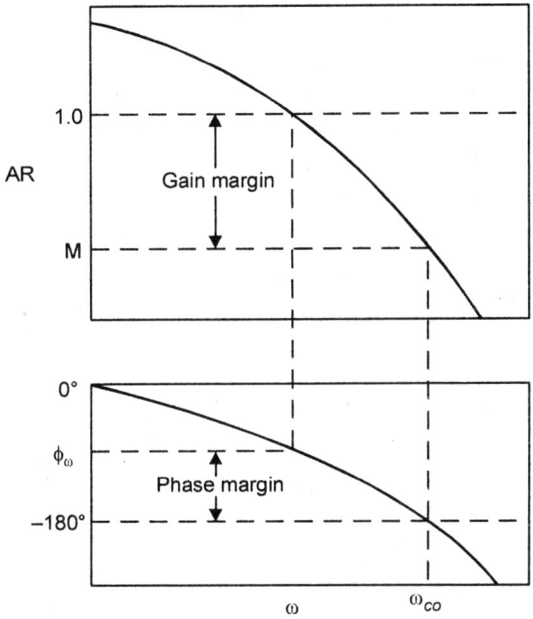

Fig. 2.51 Definition of gain and phase margins

2.10.4.2 Tuning

The operations of tuning using the frequency response method involve adjustments of the controller parameters until the stability is proved by the appropriate phase and gain margins. If the process and control elements' transfer functions are known, the correct settings can be determined analytically. If not, the Bode plot can be determined experimentally by opening the loop and providing a variable frequency disturbance of the controlling variable. If the measurements of phase and gain are made, then the Bode plot can be constructed. The significance of the unity gain cross over in frequency is that the system can correct any disturbances of frequency less than that of unity gain frequency. Any disturbance of higher frequency has no effect on the control system.

The tuning operation is based on the fact that the gains of each mode have a particular effects on the system Bode plot. By adjusting these gains, we can alter the bode plot until it satisfies the gain and phase margins of safety for a stable system. Remember that on the Bode plot, the gains appear as products and phases of the modes simply and algebraically.

2.10.4.3 Proportional Action

This mode of the controller simply multiplies the gain curve by a constant independent of frequency and has no phase effect at all. Thus, if a system gain curve is found for a proportional gain of 2 and we increase this to 4, then the entire gain curve is multiplied by 2.

This term can be used to move the intact gain curve up (increased gain) or down (decreased gain). Generally, moving the curve up extends the range of frequency that the system can control, provided the stability margins are maintained.

2.10.4.4 Integral Action

In its pure form, the integral mode contributes

$$\text{Integral Gain} = \frac{K_I}{\omega}$$

$$\text{Integral Phase} = -90° \text{ (lag)}$$

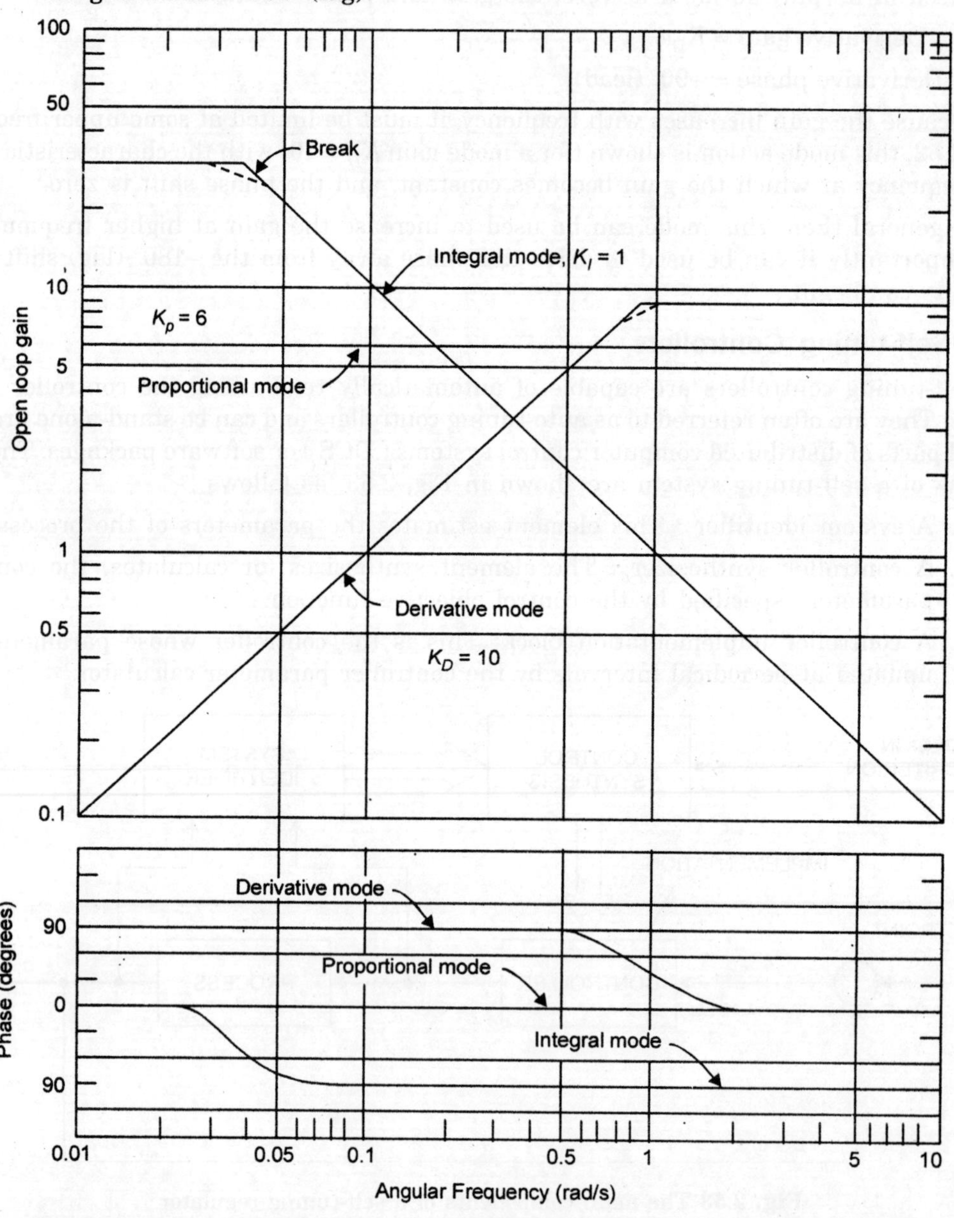

Fig. 2.52 Bode-plots of proportional, integral, and derivative controller modes

Thus we see that the integral mode contributes more gain at lower frequency and less gain at higher frequency because of the inverse dependence on the frequency ω. The phase shift is made more lagging by 90°. The integral mode gain K_I can increase the gain at lower frequencies. In practical form, the integral mode has some lower frequency (called the 'break point') at which its effects cease, and the gain curve becomes flat while the phase lag goes to zero. This is shown in Fig. 2.52, where $K_I = 1$ has been assumed.

2.10.4.5 Derivative Action

Because of problems of instability at higher frequencies, the derivative mode is almost never used in its pure form. If it were, the gain and phase would be

$$\text{Derivative gain} = K_D\omega.$$

$$\text{Derivative phase} = +90° \text{ (lead)}$$

Because the gain increases with frequency, it must be limited at some upper frequency. In Fig 2.52, this mode action is shown (for a mode gain $K_D = 10$) with the characteristic upper-limit frequency at which the gain becomes constant, and the phase shift is zero.

In general then, this mode can be used to increase the gain at higher frequency, but more importantly it can be used to drive the phase away from the –180° (lag) shift where instability can begin.

2.10.5 Self-tuning Controllers

Self-tuning controllers are capable of automatically readjusting the controller tuning settings. They are often referred to as auto-tuning controllers and can be stand-alone products, integral parts of distributed computer control systems (DCS) or software packages. The main elements of a self-tuning system are shown in Fig. 2.53, as follows :

1. A system identifier : This element estimates the parameters of the process.
2. A controller synthesizer : The element synthesizes (or calculates) the controller parameters specified by the control objective function.
3. A controller implementation block: This is the controller whose parameters are updated at periodical intervals by the controller parameter calculator.

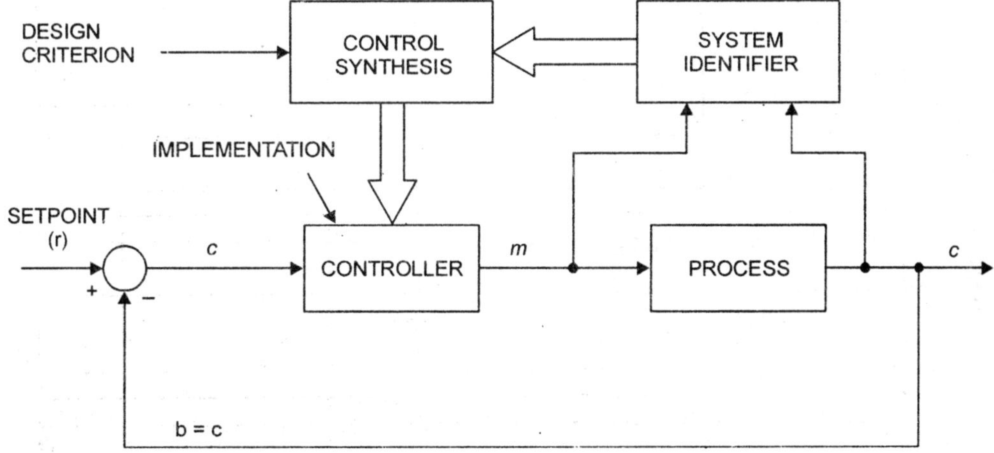

Fig. 2.53 The main components of a self-tuning regulator

The process identifier determines the response of the controlled variable (c) to a change in the manipulated variable (*m*). The model-based self-tuning algorithms use methods based on recursive estimators. Many commonly used industrial products, however, use pattern recognition or expert/fuzzy logic methods to extract key dynamic response features from a transient excursion in the system dynamics. The transient excursion may be deliberately introduced by the controller ; preferably, it is in the start up transient or another normally occurring transient of the process. The desired tuning settings for PID, optimal or other types of control are determined by the control synthesis block, which is also called the controller parameter calculator. The control synthesizer can be simple or sophisticated depending on the rules used.

Most self-tuning controllers will self-tune during start up and will retune on request, retune on naturally occurring upsets, or continuously tune using excitation introduced into the process by the controller. This excitation can be deterministic set point steps or doublets (used in pattern recognition methods) are persistent set point variations (often pseudo noise, and usually of the order of 5% of set point). These self-tuning controllers give control performance that is superior to manually tuned, static PID controllers when the process gain and the process dead time are both low. As the process gain and/or the dead time to time constant ratio rises their performance deteriorates.

2.11 PROBLEMS AND SOLUTIONS

Problem 2.1

The temperature range of a temperature controller is 250°C to 550°C. The set point is kept at 400°C. Find the percent of span error when the temperature is

(a) 395° C (b) 400° C and (c) 410° C.

Solution :

The percent error is given by the equation (2.4)

$$e_P = \frac{r-b}{b_{max}-b_{min}} \times 100,$$

Given : r = 400°C, b_{max} = 550°C and b_{min} = 250°C

(*a*) b = 395°C

$$\therefore \quad e_P = \frac{400-395}{550-250} \times 100 = \frac{5}{300} \times 100 = +1.67\%$$

(*b*) b = 400°C

$$\therefore \quad e_P = \frac{400-400}{550-250} \times 100 = \pm 0\%.$$

(*c*) b = 410°C

$$\therefore \quad e_P = \frac{400-410}{550-250} \times 100 = \frac{-10}{300} \times 100 = -3.33\%$$

Note : A positive error indicates that the temperature is below set point, a negative error indicates that the temperature is above set point and zero error indicates that the actual temperature and the set point temperature match one another.

Problem 2.2

For the situations given in Problem 2.1, calculate the measured value as percentage of measurement range.

Solution :

Refer to the equation (2.3)

$$c_P = \frac{c - c_{min}}{c_{max} - c_{min}} \times 100$$

Given :

$$c_{max} = 550°C \quad \text{and} \quad c_{min} = 250°C$$

(*a*) $c = 395°C$

$$\therefore \quad c_P = \frac{395-250}{550-250} \times 100 = \frac{145}{300} \times 100 = 48.3\%$$

(*b*) $c = 400°C$

$$\therefore \quad c_P = \frac{400-250}{550-250} \times 100 = \frac{150}{300} \times 100 = 50.0\%$$

(*c*) $c = 410°C$

$$\therefore \quad c_P = \frac{410-250}{550-250} \times 100 = \frac{160}{300} \times 100 = 53.3\%$$

Problem 2.3

The standard measured indication range of a transducer is 4–20 mA. If we have a set point value of 11 mA and a measurement of 11.5 mA, calculate the error expressed as percent of span.

Solution :

Where

$$r = \text{set point value} = 11 \text{ mA}$$

$$b = \text{Actual value} = 11.5 \text{ mA}$$

$$b_{max} = 20 \text{ mA and } b_{min} = 4 \text{ mA}$$

$$\therefore \quad e_P = \frac{11-11.5}{20-4} \times 100 = \frac{-0.5}{16} \times 100 = -3.125\ \%.$$

Note : Without even knowing what is being measured error can be calculated only by knowing current values.

Problem 2.4

A liquid-level control system linearly converts a displacement of 5 to 10 meters into a 4–20 mA control signal. A relay serves as the two-position controller to open or close an inlet valve. The relay closes at 12 mA and opens at 10 mA. Find (a) the relation between displacement level and current, and (b) the neural zone and displacement gap in metres.

Solution :

(*a*) The relation between level and current is a linear equation such as

$$H = KI + H_O.$$

We find K and H_O by writing two equations

$$5 \text{ m} = K \text{ (4 mA)} + H_O$$

$$10 \text{ m} = K \text{ (20 mA)} + H_O$$

Solving these simultaneous equations yields

$$16\, K = 5,\ K = {}^{5}\!/_{16} = 0.3125 \text{ m/mA}$$

$$H_O = 10 - 0.3125 \times 20 = 3.75 \text{ m}$$

∴ The relation between displacement and current is given by :

$$\text{H} = 0.3125 \text{ I} + 3.75$$

(*b*) The relay closes at 12 mA corresponding to high level limit H_H (say).

$$H_H = (0.3125) \times 12 + 3.75 = 7.5 \text{ m.}$$

The low level limit (H_L) occurs at 10 mA,

$$H_L = (0.3125) \times 10 + 3.75$$

$$= 3.125 + 3.750 = 6.875 \text{ m.}$$

Thus, the neutral zone is $H_H - H_L = 7.5 - 6.875$ m.

$$= 0.625 \text{ m.}$$

Problem 2.5

As a water tank loses heat, the temperature drops by 2 K per minute. When a heater is on, the system gains temperature at 4 K per minute. A two-position controller has a 0.5 minute control lag and a neutral zone of ± 4 % of the set point about a set point of 323 K. Plot the heater temperature versus time. Find the oscillation period.

Solution :

Let us assume that we start at the set point value ; then the temperature will drop linearly at,

$$T_1(t) = T(t_s) - 2(t - t_s) \tag{2.47}$$

Where t_s = time at which we start observation.

The heater will start at a temperature of 310 K (4% below set point), after which the temperature will rise accordingly,

$$T_2\ (t) = T\ (t_h) + 4\ (t - t_h) \tag{2.48}$$

where t_h = time at which heater goes on.

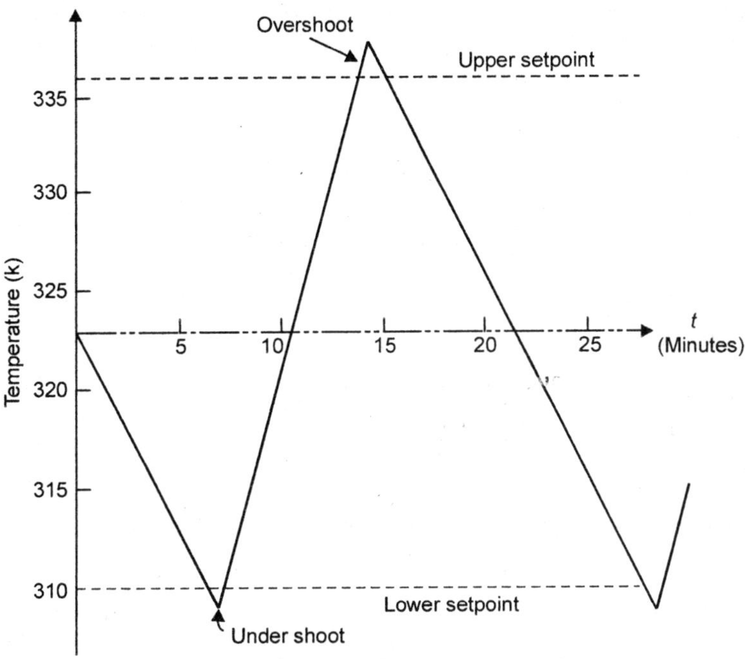

Fig. 2.54

When the temperature reaches 336°*K*, the heater goes off and the system temperature drops by 2 *K*/m until 310°*K* is reached. The system response is then plotted as in Fig. 2.54, using equations (2.47) and (2.48). Notice the period is 21.5 minutes. There is also a 1*K* undershoot and +2*K* overshoot because of the lag.

Problem 2.6

Suppose, in a single speed floating-control mode, a process error lie' ''within the neutral zone with controller output (p) = 25 %. At t = 0, the error falls below the neutral zone. If the rate constant K_F = ± 2 % per second, find the time when the output saturates.

Solution :

The relation between controller output and time is given in the equation (2.11)

$$p = K_F t + p_{(0)}$$

Given : K_F = 2%/sec and $p_{(0)}$ = 25 %.

The output saturates when p = 00 %.

$$\therefore \quad 100 = 2 \times t + 25\ \%$$

$$t = 75/2 = 37.5\ s$$

The output saturates after the lapse of 37.5 seconds from the time the error falls below the neutral zone.

Problem 2.7

Consider the proportional mode level-control system of Fig. 2.55. Valve 'A' is linear with a flow scale factor of 10 m^3/hr per percent controller output. The controller output is nominally 50 % with a proportional gain K_P = 10%. A load change occurs when flow through valve 'B' changes from 500 m^3/hr to 600 m^3/hr. Calculate the new controller output and offset error.

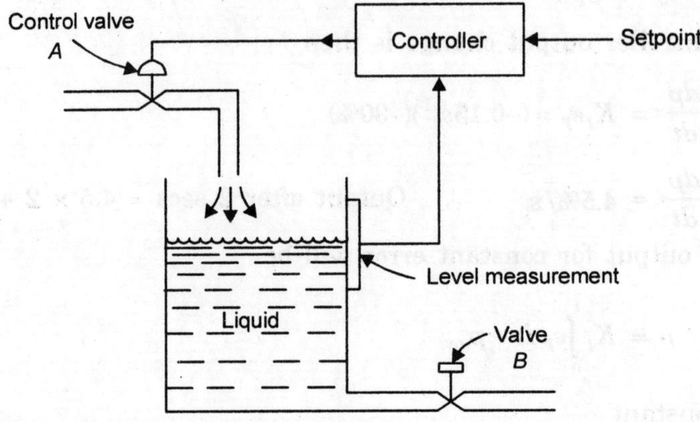

Fig. 2.55 Level control system

Solution :

Certainly, valve 'A' must move to a new position for 600 m³/hr flow or otherwise the tank will get emptied.

This can be accomplished by a 60% 'new controller output' because

$$QA = \left(10\frac{\text{m}^3/\text{hr}}{\%}\right)(60\%) = 600 \text{ m}^3/\text{hr as required.}$$

For proportional controller, we have

$$p = K_P\, e_P + p_O \tag{2.14}$$

Given : $p_O = 50\ \%,\ K_P = 10\%$

$p = 60\%.$

$$\therefore \quad e_P = \frac{p - p_O}{K_P} = \frac{60-50}{10} = 1\%$$

$\therefore$ Offset error = 1 %.

Problem 2.8

An integral controller is used for speed control with a set point of 12 rpm within a range of 10 to 15 rpm. The controller output is 22% initially. The constant K_I = –0.15% controller output per second per percentage error. If the speed jumps to 13.5 rpm, calculate the controller output after two seconds for a constant e_P.

Solution :

e_P can be found out using equation (2.4)

$$e_P = \frac{r-b}{b_{max}-b_{min}} \times 100 \qquad (2.4)$$

$$e_P = \frac{12-13.5}{15-10} \times 100$$

$$e_P = -30\%$$

The rate of controller output change is then

$$\frac{dp}{dt} = K_I e_P = (-0.15s^{-1})(-30\%)$$

$$\frac{dp}{dt} = 4.5\%/s \qquad \therefore \quad \text{Output after 2 secs} = 4.5 \times 2 + 22 = 31\%.$$

The controller output for constant error will be

$$p = K_I \int_0^t e_P dt + p_{(0)} \qquad (2.16)$$

but because e_p is constant

$$p = K_I e_P t + p_{(0)}$$

After 2 seconds we have

$$p = (-0.15)(-30)(2) + 22 = 31\%.$$

Problem 2.9

Given the error of Figure 2.56 (top), plot a graph of a proportional -integral- controller output as a function of time. Assume $K_P = 5$, $K_I = 1.0s^{-1}$ and $p_{I(0)} = 20\%$.

Solution :

We find the solution by an application of

$$P = K_P e_P + K_P K_I \int_0^t e_P dt + p_{I(0)} \qquad (2.18)$$

To find the controller output, we solve equation (2.18) in time. The error can be expressed in three time regions.

(1) $0 \le t \le 1$ (t between 0 and 1 second) . The error rises from 0 % to 1% in 1 second. Thus, it is given by $e_P = t$.

$1 \le t \le 3$ For this time span the error is constant and equal to 1%; therefore it is given by $e_P = 1$.

$t \ge 3$ For this time the error is zero, $e_P = 0$.

We now write out and solve equation (2.18) for each of these time spans.

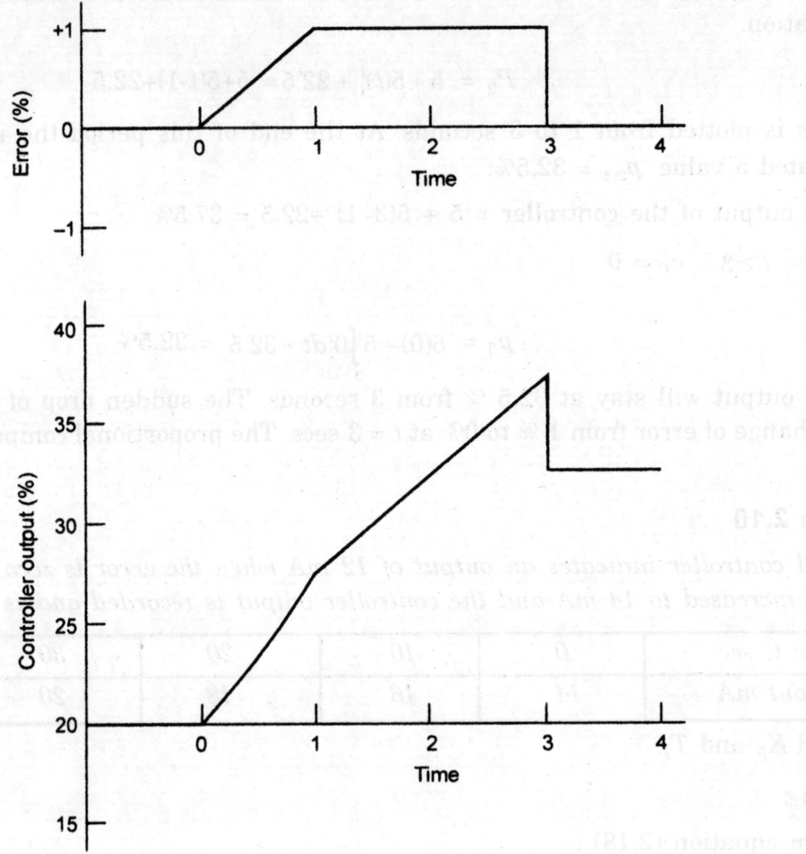

Fig. 2.56

(1) $0 \leq t \leq 1$, $e_P = t$

$$p_1 = 5t + 5\int_0^t t\,dt + 20 = 5t + 5\left[\frac{t^2}{2}\right]_0^t + 20$$

$$p_1 = 5t + 2.5t^2 + 20$$

This is plotted in Fig. 2.56 (bottom) from 0 to 1 second. Notice the curvature because of the squared term. At the end of 1 second, the integral term has accumulated a value of $p_{I(1)} = 22.5\%$. The output of the controller $p_{I(1)} = 5 \times 1 + 22.5 = 27.5\%$.

(2) $1 \leq t \leq 3$, $e_P = 1$

$$p_2 = 5 + 5\int_1^t 1\,dt + 22.5$$

The integral term accumulation from 0 to 1 second forms the initial condition for this new equation.

$$P_2 = 5 + 5(t)_1^t + 22.5 = 5+5(t-1)+22.5$$

This is plotted from 1 to 3 seconds. At the end of this period the integral term has accumulated a value $p_{2(3)} = 32.5\%$.

The output of the controller = 5 + 5(3–1) +22.5 = 37.5%.

(3) $t \geq 3$, $e_P = 0$

$$p_3 = 5(0) + 5\int_3^t 0\,dt + 32.5 = 32.5\%$$

The output will stay at 32.5 % from 3 seconds. The sudden drop of 5% is due to the sudden change of error from 1 % to 0% at t = 3 secs. The proportional component has become zero now.

Problem 2.10

A PI controller indicates an output of 12 mA when the error is zero. The set point is suddenly increased to 14 mA and the controller output is recorded and is given below.

Time t, sec	*0*	*10*	*20*	*30*
Output mA	*14*	*16*	*18*	*20*

Find K_P and T_I.

Solution :

From equation (2.18) :

$$p = K_P e_P + K_P K_I \int_0^t e_P dt + p_{I(0)},$$

$p_{I(0)}$ = 12 mA (∵ controller output is 12 mA when error is zero.

Set point = Output; then the set point is increased to 14 mA causing a step change of 2 mA)

Error created at t = 0 = 14 – 12 = 2 mA.

∴ e_P = 2 mA.

From the table we can plot the curve as shown in Fig. 2.57.

The proportional component $K_P\ e_P$ = 14 – 12 = 2.

$$\therefore \qquad K_P = \frac{K_P e_P}{e_P} = \frac{2}{2} = 1 \text{ mA/mA}$$

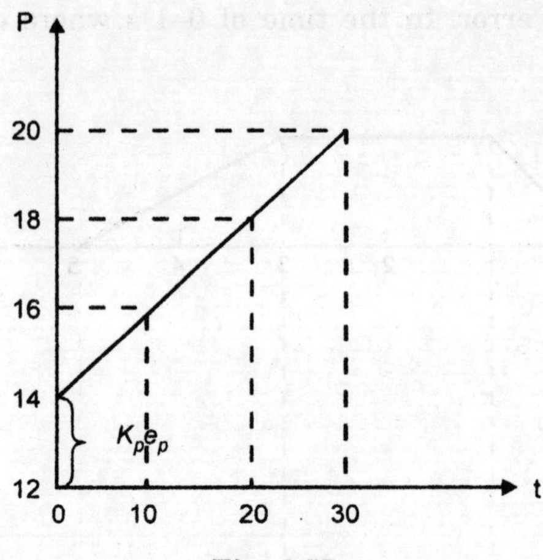

Fig. 2.57

The Integral component

$$K_P K_I e_P [t]_0^t + p_I(0) = K_P K_I t + p_{I(0)}$$

$$\therefore \qquad p_{(t)} = K_P e_P + K_P K_I e_P t + p_{I(0)}$$

Substituting the values from table for any one point, (say 18 mA, 20 Secs).

$$18 = 1 \times 2 + 1 \times K_I \times 2 \times 20 + 12$$

$$K_I = \frac{18-14}{40} = \frac{4}{40} = 0.1$$

$$\therefore \qquad T_I = \frac{1}{K_I} = \frac{1}{0.1} = 10 \text{secs}$$

May be verified taking one another point (say 16 mA, 10 secs)

$$16 = 1 \times 2 + 1 \times K_I \times 2 \times 10 + 12.$$

$$K_I = \frac{16-14}{20} = 0.1$$

$$\therefore \qquad T_I = 10 \text{ secs}$$

Problem 2.11

Suppose the error shown in Fig. 2.58(a) is applied to a proportional - derivative controller with $K_P = 5$, $K_D = 0.5$ s and $p_0 = 20\%$. Draw a graph of the resulting controller output.

Solution :

In this case we evaluate

$$p = K_P e_P + K_D K_P \frac{de_P}{dt} + p_{(0)} \tag{2.19}$$

over the three spans of the error. In the time of 0–1 s where $e_P = a\ t$,

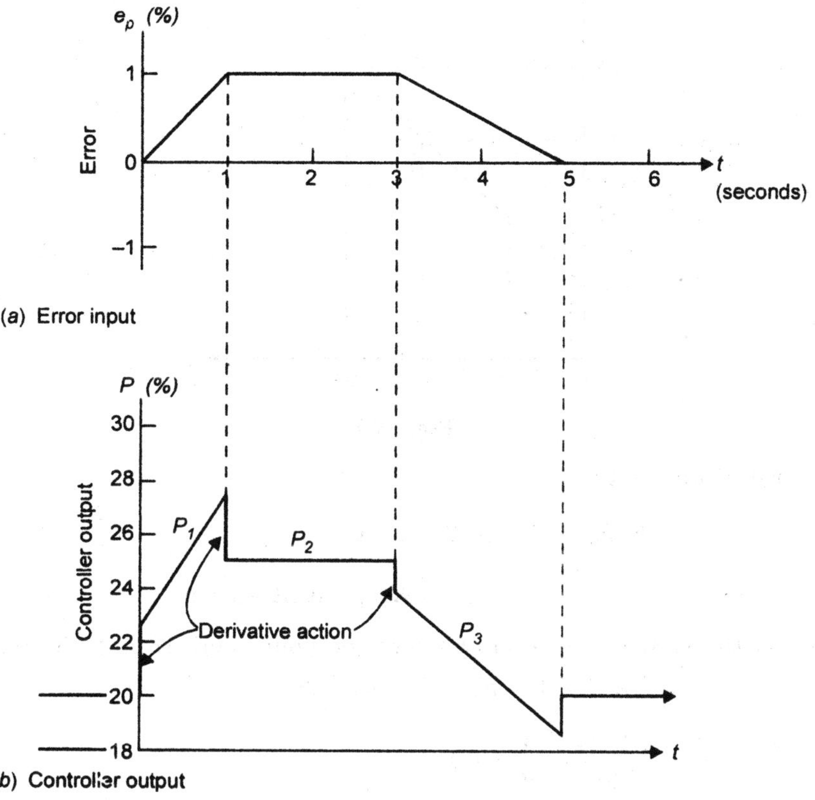

Fig. 2.58

We have $$p_1 = K_P a t + K_D K_P a + p_O$$

$$p_1 = 5t + 2.5 + 20 \qquad (\because \quad a = 1\%/\text{s})$$

Note the instantaneous change of 2.5% produced by this error. In the span from 1–3 s, we have

$$p_2 = 5 + 20 = 25$$

The span from 3-5s has an error of $e_P = -0.5t + 2.5$ so that we get for 3–5s

$$p_3 = -2.5t + 12.5 - 1.25 + 20 \qquad \text{or}$$

$$p_3 = -2.5t + 31.25$$

This controlled output is plotted in Fig. 2.58(*b*).

Problem 2.12

Draw a plot of the three-mode (PID) controller output for the error shown in Figure 2.58(a). Assume $K_P = 5$, $K_I = 0.7s^{-1}$, $K_D = 0.5s$ and $p_I(0) = 20\%$.

Solution :

From Fig. 2.58(*a*), the error can be expressed as follows :

$$0 - 1\text{s},\ e_P = t\%.$$

1 – 3s, $e_P = 1\%$

3 – 5s, $e_P = -\frac{1}{2}t + 2.5\%$

We must apply each of these spans to the three-mode equation for controller output.

$$p = K_P e_P + K_P K_I \int_0^t e_P dt + K_P K_D \frac{de_P}{dt} + p_{I(o)} \qquad (2.20)$$

or

$$p = 5e_P + 3.5\int_0^t e_P dt + 2.5\frac{de_P}{dt} + 20$$

From 0–1s we have,

$$p_1 = 5t + 3.5\int_0^t t dt + 2.5 + 20 \quad \text{or} \quad p_1 = 5t + 1.75t^2 + 22.5$$

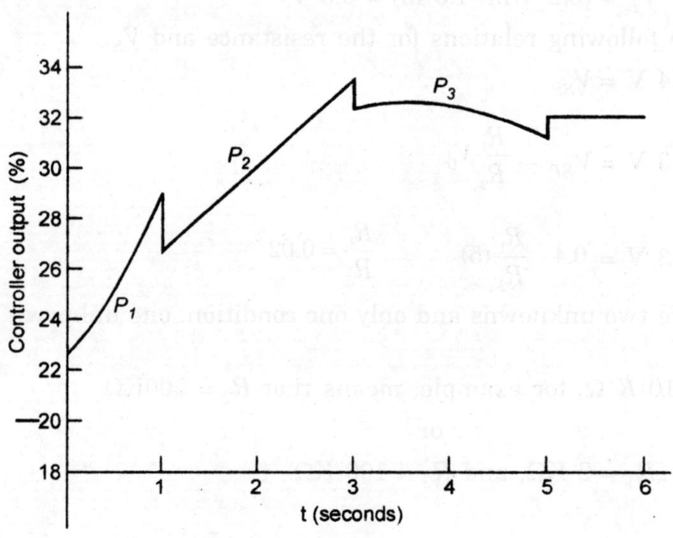

Fig. 2.59

This is plotted in Fig. 2.59 in the span of 0–1s. At the end of 1s, the integral term has accumulated to $p_{I(1)} = 21.75\%$. Now from 1–3s, we have

$$p_2 = 5 + 3.5\int_1^t (1)dt + 21.75 \quad \text{or} \quad p_2 = 3.5(t-1) + 26.75$$

This controller variation is shown in the figure from 1–3s. At the end of 3s, the integral term has accumulated a value of $p_{I(3)} = 28.75\%$.

Finally, from 3–5s, we have

$$p_3 = 5\left(-\frac{1}{2}t + 2.5\right) + 3.5\int_3^t \left(-\frac{1}{2}t + 2.5\right) dt - \frac{2.5}{2} + 28.75$$

or

$$p_3 = -0.875t^2 + 6.25\ t + 21.625$$

This is plotted in the figure from 3–5s. After 5 seconds, the error is zero. Therefore, the output will be simply be the accumulated integral response providing a constant output of p_t = 32.25%.

Problem 2.13

Level measurement in a sump tank is provided by a transducer scaled as 0.2 V/m. A pump is to be turned on by application of +5 V when the sump level exceeds 2.0 m. The pump is to be turned back off when the sump level drops to 1.5 m. Develop an electronic two-position controller.

Solution :

Let us use the circuit of Fig. 2.29. The high and low trip voltages will be determined by the conditions of the problem. From these the values of the resistances can be determined.

$$V_H = (0.2 \text{ V/m})(2.0 \text{ m}) = 0.4 \text{ V}$$

and

$$V_L = (0.2 \text{ V/m})(1.5 \text{ m}) = 0.3 \text{ V}$$

This gives the following relations for the resistance and V_{SP}.

$$0.4 \text{ V} = V_{SP}$$

$$0.3 \text{ V} = V_{SP} - \frac{R_1}{R_2} V_O$$

$$\therefore \quad 0.3 \text{ V} = 0.4 - \frac{R_1}{R_2}(5) \qquad \frac{R_1}{R_2} = 0.02$$

Since there are two unknowns and only one condition, one unknown can be reasonably selected.

Picking R_1 = 10 K Ω, for example, means that R_2 = 500KΩ.

or

R_1 = 2 KΩ, and R_2 = 100 KΩ.

Problem 2.14

A control signal varies from 0–5 Volts. A floating controller, such as that in Fig. 2.31, has trip voltages of 2 Volts and 4 Volts, and a three position controller has outputs of 0 and ± 2 volts. The integrator consists of a 1 MΩ resistor and a 1 μF capacitor. Plot the controller output in response to the input of Fig. 2.60 (a).

Solution :

We can find the output by applying the relations of equation(2.33) to the input voltage. The result is shown in Fig. 2.60(b). This result is arrived at as follows :

1. From 0–1 second, the output is zero (an assumed starting point) and remains so because the input is within the neutral zone.
2. From 1–3 seconds, the lower set point has been reached and the output is given by

$$V_{out} = -\frac{V_1}{RC}(t-1) = -2(t-1)\text{V}.$$

At t = 3s, the output is –4V.

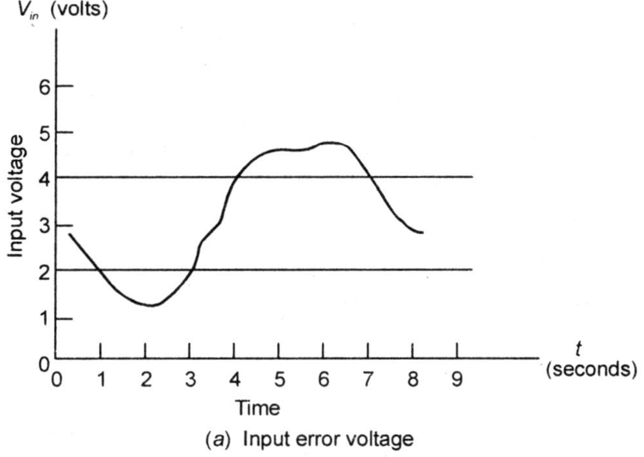

(a) Input error voltage

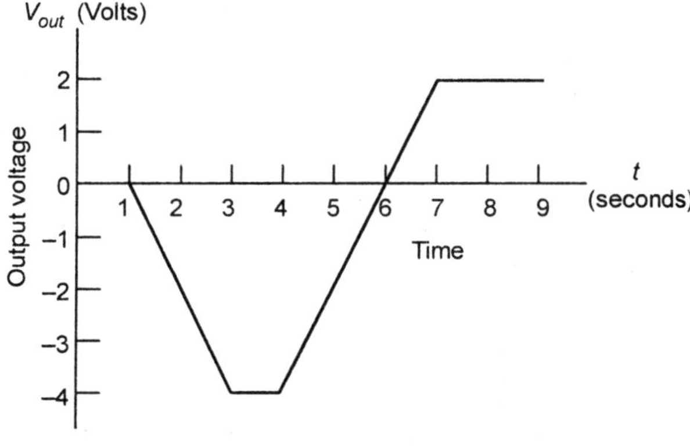

(b) Output voltage

Fig. 2.60 Input and Output voltage for the floating controller of problem 2.14

3. From 3–4 seconds, the output remains at –4 V because the input is in the neutral zone.
4. From 4–7 seconds, the input reaches the upper set point and the output becomes

$$V_{out} = +\frac{V_1}{RC}(t-4) - 4\text{V}.$$
$$= +\ 2(t - 4) - 4\text{V}.$$

At $t = 7$ seconds, the input again falls within the neutral zone and the output becomes

$$V_{out} = +\ 2(7–4) - 4\text{V} = +2\ \text{V}$$

5. The output will remain at +2 V until the input again hits a set point value.

Problem 2.15

A controller shown in Figure 2.32 have scaling so that 0–10 volts corresponds to a 0–100% output. Assume R_2 = 10 KΩ and full-scale error range is 10 volts.

(a) Find the values of V_O and R_1 to support a 20% proportional band and a 50% zero error controller output.

(b) *If the load changes such that a new controller output of 40% is required, find the corresponding offset error.*

Solution :

(a) The value of V_O is simply 50% of 10 volts *i.e.*, 5 volts, to provide the zero error controller output.

$$\therefore \quad V_O = 5 \text{ V}$$

20% proportional band means that a change of error of 20% must cause the controller output to vary 100%.

Thus, from

$$V_{out} = G_P V_e + V_O$$

we note that when the error has changed 20% of 10 volts or 2 volts, we must have full controller output change. Thus

$$G_P = \frac{\Delta V_{out}}{\Delta V_e} = \frac{10}{2} = 5$$

So that if R_2 = 10 KΩ, then $$R_1 = \frac{R_2}{G_P} = \frac{10}{5} = 2\Omega$$

(b) In this case we need a negative error so that the output is 40% of 10 volts = 4 volts.

$$V_{out} = G_P V_e + V_O$$

$$4 = 5 V_e + 5$$

V_e = –1/5 volts and because the full-scale signal in 10 volts, we have an error of

$$\frac{-0.2}{10} \times 100 = -2\%$$

Problem 2.16

An integral control system will have a measurement range of 0.4 to 2.0 V and an output range of 0 to 6.8 V. Design an Opamp integral controller to implement a gain of K_I = 4%/(%-min). Specify the values of G_I, R and C.

Solution : (Refer Fig. 2.33)

The input range is 2.0 –0.4 = 1.6 V and the output range is 6.8 V. We must convert K_I into units of seconds.

[4%/(%-min)] [1 min/60 s] = 0.0667%/(%-s)

1% of the input for 1 s = (0.01)(1.6 V)(1s) = 0.016 V –s

0.0667% of the output = (0.000667) (6.8V) = 0.00454 V.

Thus, the gain is

$$G_I = (0.00454 \text{ V}/0.016 \text{ V-s}) = 0.283 \text{ s}^{-1}$$

Because G_I = 1/*RC*, we have

$$RC = \frac{1}{G_I} = \frac{1}{0.283} = 3.53s$$

Assuming a reasonable value of C =100 μF, R can be calculated.

$$R = \frac{3.53}{100x10^{-6}} = 35.3 \text{ K}\Omega$$

Problem 2.17

Design a proportional – integral controller with a proportional band of 30% and an integration gain of 0.1 %/(%-s). The 4–20 mA input converts to a 0.4 –2 volt signal, and the output is to be 0–10 volts. Calculate the values of G_P, G_I, R_2, R_1 and C respectively with reference to the Fig. 2.35.

Solution :

A proportional band of 30% means that when the input changes by 30% of range or 0.48 volts, the output must change by 100% or 10 volts. This gives a gain of

$$G_P = \frac{10}{0.48} = 20.83 \left(= \frac{R_2}{R_1} \right)$$

A K_I of 0.1%/(%-s) says that a 1% error for 1 second should produce an output change of 0.1%. One percent of 1.6 V is 0.016 V and 0.1% of 10V is 0.01V, so

$$G_I = \frac{0.01}{0.016} = 0.625s^{-1} \left(= \frac{1}{R_2 C} \right)$$

We have now two equations to solve for three variables

$$G_P = \frac{R_2}{R_1} = 20.83 \qquad \text{and} \quad G_I = \frac{1}{R_2 C} = 0.625s^{-1}$$

Assuming a reasonable value of C=10 µF we can calculate,

$$R_2 = \frac{1}{0.625 \times 10 \times 10^{-6}} = 160\text{K}\Omega$$

$$R_1 = \frac{R_2}{20.83} = \frac{160}{20.83} = 7.68\text{K}\Omega$$

Problem 2.18

A proportional-derivative controller has a 0.4–2.0V input measurement range, 0 to 5 V output, K_P = 5 %/%, and K_D = 0.08 % per (% /min). The period of the fastest expected signal change is 1.5 seconds. Implement this controller with an opamp circuit.

Solution :

A circuit shown in Fig. 2.36 may be considered. G_P and G_D are to be found out.

A K_P = 5%/% means a 20 % P_B. So we can write

$$G_P = \frac{(100\%)(5\text{V})}{(20\%)(1.6\text{V})} = \frac{5\text{V}}{0.32\text{V}} = 15.625 \left(= \frac{R_2}{R_1 + R_3} \right)$$

To find G_D we must first change K_D to seconds

$$K_D = [0.08\%/(\%\text{-min}) \times 60\text{s/min}] = 4.8\%/(\%/\text{s})$$

Now we can write

$$G_D = \frac{(4.8\%)(5\text{V})}{(1\%)(1.6\text{V})} = \frac{5\text{V}}{1.6\text{V}} = 15s \ \ (= R_3C)$$

The period limitation allows us to write

$$\left(\frac{R_1}{R_1 + R_3}\right)R_3C = \frac{0.1}{2\pi}(1.5s) = 0.024s$$

Now we have three relations and four unknowns

$$\frac{R_2}{R_1 + R_3} = 15.625, \quad R_3C = 15 \text{ and } \frac{R_1}{R_1 + R_3}R_3C = 0.024s$$

Assuming a reasonable value of C = 100 μF we get

$$R_3 = \frac{15}{C} = \frac{15}{100 \times 10^{-6}} = 150\text{K}\Omega$$

$$\frac{R_1}{R_1 + R_3} = \frac{0.024}{R_3C} = \frac{0.024}{150 \times 10^3 \times 100 \times 10^{-6}}$$

Solving for R_1 we get R_1 = 240 Ω

$$\frac{R_2}{R_1 + R_3} = 15.625, \quad R_2 = 15.625(240 + 150 \times 10^3) = 2.35 \text{ M}\Omega$$

Problem 2.19

A temperature – control system inputs the controlled variable as a range from 0 to 4 V. The output is to be given to a heater requiring 0 to 8 V. A PID controller is to be used with K_P = 2.4%/%, K_I = 9%/(% -min) and K_D = 0.7% /(%/min). The period of the fastest expected change is estimated to be 8 seconds. Develop the PID circuit with opamps.

Solution : (Refer Fig. 2.37)

The input range is 4V and the output range is 8 V.

Let us first find out the circuit gains.

G_P = For the proportional mode, a 1% error means a voltage change of (0.01) (4 V) = 0.04 V.This should cause an output change of 2.4% or (0.024)(8 V) = 0.192 V.

Thus $$G_P = \frac{0.192\text{V}}{0.04\text{V}} = 4.8$$

For the integral term, an error of 1% should cause the output to change by 9%/min, which is $\left(\frac{9}{60}\right) = 0.15\%/s$.

Thus $G_I = (0.0015 s^{-1})$ (8 V) (0.04 V) = 0.3 s^{-1}.

For the derivative term, an error change of 1% per min or (0.04 V/60) = 6.67×10^{-4} V/s should cause an output change of 0.7% or (0.007)(8 V) = 0.056 V. Thus

$$G_D = (0.056 \text{ V}/ 6.67 \times 10^{-4} \text{ V/s}) = 84 \text{ s}$$

These results provide the following relations.

$$G_P = \frac{R_2}{R_1} = 4.8, \qquad G_I = \frac{1}{R_I C_I} = 0.3 \text{ s}^{-1}, \qquad G_D = R_D C_D = 84\text{s}.$$

From the fastest period specification, using a factor of 100 for the inequality, we form the relationship

$$2\pi R_3 C_D = (0.01)(8s) = 0.08s$$

which gives seven unknowns and four equations. We can pickup three quantities and try for the rest four.

Assuming R_1 = 10 KΩ, $C_1 = C_D$ = 10 μF, we get

R_2 = 4.8 R_1 = 48 KΩ.

$$R_I = \frac{1}{0.3 \times 10 \times 10^{-6}} = 333\text{K}\Omega$$

$$R_D = \frac{84}{C_D} = \frac{84}{10 \times 10^{-6}} = 8.4 \text{ M}\Omega$$

$$R_3 = \frac{0.08}{2\pi C_D} = \frac{0.08}{2\pi \times 10 \times 10^{-6}} = 1.27\text{K}\Omega$$

As 8.4 MΩ for R_D seems to be very high for practical considerations, let us assume C_D = 100 μF. We get

R_D = 840 KΩ and R_3 = 127Ω which seems more reasonable.

Problem 2.20

Suppose the deviation following a step-function disturbance is a 4.7% error in the controlled variable. If a quarter-amplitude criterion is used to evaluate the response, find the error of the second and third peak.

Solution : Refer Fig. No: 2.46

The amplitude of each peak must be one quarter of the previous peak.

Given the amplitude of the first peek as 4.7%.

∴ a_1 = 4.7%

Amplitude of the second peak $a_2 = \frac{a_1}{4}$

$$a_2 = \frac{4.7}{4} = 1.18\%$$

Amplitude of the third peak $a_3 = \frac{a_2}{4} = \frac{a_1}{16}$

$$a_3 = \frac{1.18}{4} = 0.3\%$$

Problem 2.21

Given the two response curves of Fig. 2.61 for deviation versus time, following an initial disturbance, find the response preferred using the minimum area criterion.

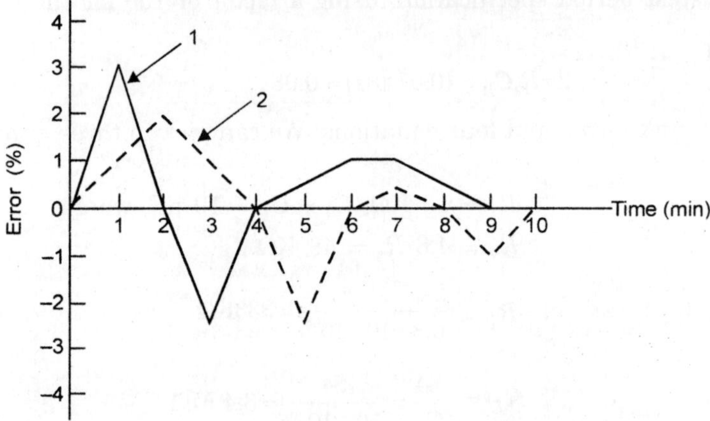

Fig. 2.61 Error vs Time

Solution :

We find the area by application of

$$A_P = \int |e_P| dt$$

In these cases we find that the areas geometrically by finding the net area of each curve.

For curve1, we have A_1 = 13%–minute and

For curve2, we get A_2 = 100%–minute.

Thus it is clear that curve 2 is preferred under the minimum area criterion.

Problem 2.22

Consider a servo control problem of a first-order process with PI controller. Select the best values of K_P and T_I for one-quarter decay ratio criterion.

Solution :

It can be easily shown that the closed-loop response is given by the following equation, when $G_m = G_f = 1$ (*i.e.*, Transfer function of measuring unit $G_m = 1$ and transfer function of final control element $G_f = 1$).

$$\bar{y}(s) = \frac{T_I s + 1}{\tau^2 s^2 + 2\varsigma\tau s + 1} \bar{y}_{SP(S)}$$

where $$\tau = \sqrt{\frac{\tau_P T_I}{K_{P1} K_P}}$$

K_{P1} = Static process gain

K_P = Proportional gain of the controller

T_I = Integral time

τ_P = Process time constant

and $$\varsigma = \frac{1}{2}\sqrt{\frac{T_I}{\tau_P K_{P1} K_P}}(1 + K_{P1} K_P)$$

We notice that the closed-loop response is second-order.

Selecting the one-quarter decay ratio criterion, we can write from equation (1.77)

$$\text{Decay ratio} = \exp\left(\frac{-2\pi\varsigma}{\sqrt{1-\varsigma^2}}\right)$$

Therefore, for our problem we have.

$$\exp\left[\frac{-2\pi \times \frac{1}{2}\sqrt{\frac{T_I}{\tau_P K_{P1} K_P}}(1 + K_{P1} K_P)}{\sqrt{1 - \frac{1}{4}\frac{T_I}{\tau_P K_{P1} K_P}(1 + K_{P1} K_P)^2}}\right] = \frac{1}{4}$$

After algebraic simplifications we take,

$$-2\pi\sqrt{\frac{T_I}{4\tau_P K_{P1} K_P - T_I(1 + K_{P1} K_P)^2}}(1 + K_{P1} K_P) = \log\left(\frac{1}{4}\right)$$

The above equation has two unknowns K_P and T_I. Therefore, we will have several controller settings which satisfy the one-quarter decay ratio criterion. Let K_{P1} =0.1 and τ_P = 10. Then, we find the following solutions.

K_P	10	30	50	100
T_I	0.464	0.348	0.258	0.153

and so on.

The question is which one to select. Usually, we select first the proportional gain K_P, so that the controller has the necessary strength to push the response back to the desired set point and then we choose the corresponding T_I value so that the one-quarter decay ratio is satisfied.

Problem 2.23

For the feedback system shown in Fig. 2.62, select the best values for K_P and T_I using Time-Integral criteria.

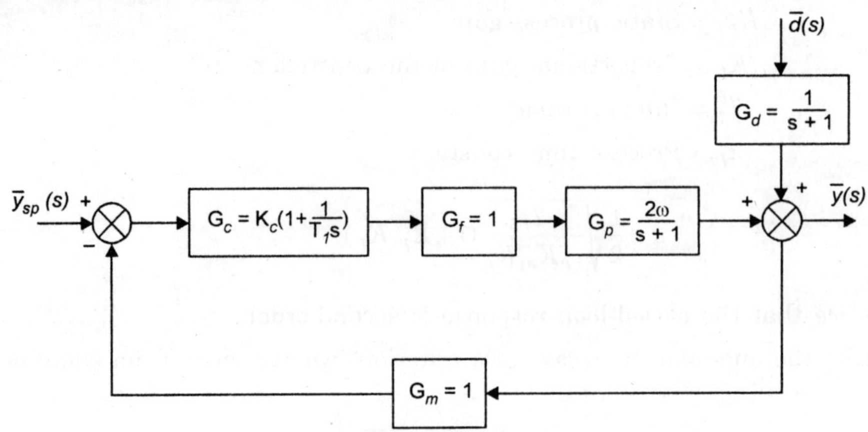

Fig. 2.62 Closed-loop system

Solution :

The closed loop response is given by

$$\bar{y}(s) = \frac{T_I s + 1}{\frac{T_I}{20K_c} s^2 + T_I\left(1 + \frac{1}{20K_c}\right)s + 1}\,\bar{y}_{SP(S)} + \frac{\left(\frac{T_I}{20K_c}\right)s}{\frac{T_I}{20K_c} s^2 + T_I\left(1 + \frac{1}{20K_c}\right)s}\,\bar{d}'(s) + 1 \tag{2.49}$$

or

$$\bar{y}(s) = \frac{T_I s + 1}{\tau^2 s^2 + 2\varsigma\tau s + 1}\,\bar{y}_{SP(S)} + \frac{\left(\frac{T_1}{20K_P}\right)s}{\tau^2 s^2 + 2\varsigma\tau s + 1}\,\bar{d}'(s)$$

where
$$\tau = \sqrt{\frac{T_I}{20K_c}} \tag{2.50}$$

and
$$\varsigma = \frac{1}{2}\sqrt{\frac{T_I}{20K_c}}(1 + 20K_c) \tag{2.51}$$

To select the best values for K_P and T_I we can use one of the three criteria ISE, IAE or ITAE. Further more, we can consider changes either in the load or set point. Finally even if we select set point changes, we still need to decide what kind of changes we will consider (*i.e.*, step, impulse, sinusoidal etc). Let us say that we select ISE as the criterion and unit step changes in the set point. From equation (2.49) we have,

$$\bar{y}(s) = \frac{T_I s + 1}{\tau^2 s^2 + 2\varsigma\tau s + 1} \times \frac{1}{s}$$

inverting the equation we find (if $\varsigma < 1$)

$$y(t) = 1 + \frac{e^{-\varsigma t/\tau}}{\sqrt{1-\varsigma^2}}\left[\frac{T_I}{\tau}\sin(\sqrt{1-\varsigma^2})\frac{t}{\tau} - \sin\left(\sqrt{1-\varsigma^2}\,\frac{t}{\tau} + \tan^{-1}\frac{\sqrt{1-\varsigma^2}}{\varsigma}\right)\right] \tag{2.52}$$

Then solve the following optimization problem

$$\text{Minimise} \qquad \text{ISE} = \int_0^\infty [y_{SP} - y(t)]^2\, dt$$

by selecting the values of τ and ζ, where $y(t)$ is given by equation (2.52).

The optimal values of τ and ζ are given by the solution of the following equations (conditions for optimality) :

$$\frac{\partial(\text{ISE})}{\partial\tau} = \frac{\partial(\text{ISE})}{\partial\varsigma} = 0$$

Let τ^* and ζ^* be the optimal values. Then from equations (2.50) and (2.51), we can find the corresponding optimal values for the controller parameters T_I and K_P.

If the criterion was the ITAE, we would have to solve the following :

$$\text{Minimise ITAE} = \int_0^\infty t\,|y_{SP} - y(t)|\,dt$$

by selecting the values of τ and ζ where $y(t)$ is given by equation (2.52).

The solution τ^* and ζ^* is given by the equations

$$\frac{\partial(\text{ITAE})}{\partial\tau} = \frac{\partial(\text{ITAE})}{\partial\varsigma} = 0$$

and inturn, from equations (2.50) and (2.51) we can find the optimal K_P and T_I.

It is clear that the solutions of the foregoing two problems with different criteria will be, in general, different.

Let us consider now unit step changes in the load. Equation (2.49) yields

$$\bar{y}(s) = \frac{(\tau_I/20K_c)s}{\tau^2 s^2 + 2\varsigma\tau s + 1} \times \frac{1}{s} \qquad \text{and after inversion we get}$$

$$y(t) = \frac{(\tau_I/20K_c)e^{-\varsigma t/\tau}}{\tau\sqrt{1-\varsigma^2}}\sin\left(\sqrt{1-\varsigma^2}\,\frac{t}{\tau}\right) \tag{2.53}$$

We can find the optimal values of K_c and τ_I, following a similar procedure as previously. Since the response $y(t)$ is now different than it was for a unit step change in the set point [Compare equations (2.52) and (2.53)], we expect that the optimal settings of K_c and τ_I will be different even if we use the same criterion (ISE or ITAE).

Problem 2.24

A transient disturbance test is run on a process loop. The results of a 9% controlling variable change give a 'process reaction graph' as shown in Fig. 2.63.

(a) Find settings for three-mode (PID) action.

(b) Find the three mode settings for a quarter-amplitude response with corrections developed by Cohen and Coon.

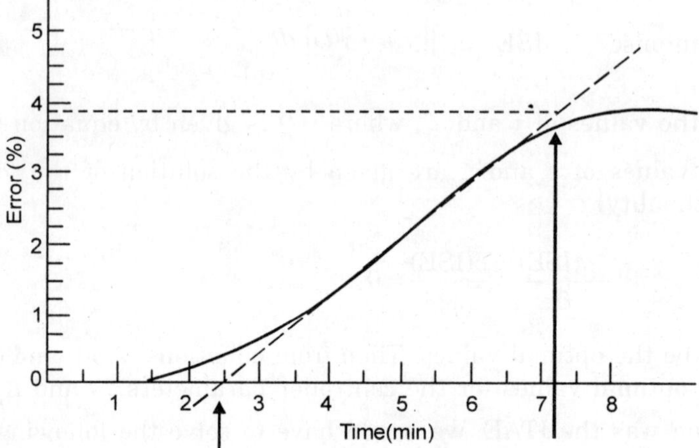

Fig. 2.63 Process reaction graph

Solution :

(*a*) By drawing the inflection point tangent on the graph, we find a time lag t_d = 2.4 minutes, and a τ = Process reaction time of (7.2 – 2.4) = 4.8 minutes.

$$\therefore \text{ The reaction rate (time constant) } S = \frac{B}{\tau}$$

$$S = \frac{3.9\%}{4.8\,\text{min}} = 0.8125\%/\text{min}.$$

The controller settings are found from equations (2.40) to (2.46) after omitting the correction factors of Cohen and Coon, but adding Ziegler and Nichols factor.

$$\text{Proportional gain } K_P = \frac{1.2}{K} \cdot \frac{\tau}{t_d}$$

where
$$K = \frac{\text{output (B)}}{\text{input (A)}} = \frac{3.9}{9} = 0.433$$

$$\therefore \quad K_P = \frac{1.2}{0.433} \times \frac{4.8}{2.4} = 5.5$$

or proportional band
$$= \frac{100}{K_p} = \frac{100}{5.5} = 18\%$$

Integral time $T_I = 2 \times t_d = 4.8$ minutes.

Derivative time $T_D = \dfrac{t_d}{2} = \dfrac{2.4}{2} = 1.2$ mts

[The factors 1.2 for K_P, 2 for T_I and 0.5 for T_D were assumed by Ziegler and Nichols before the corrections developed by Cohen and Coon.]

(*b*) With Cohen and Coon corrections :

$$K_P = \frac{1}{K} \cdot \frac{\tau}{t_d}\left(\frac{4}{3} + \frac{t_d}{4\tau}\right)$$

$$= \frac{1}{0.433} \times \frac{4.8}{2.4}\left(\frac{4}{3} + \frac{2.4}{4x4.8}\right) = 4.62 \times 1.455$$

$$= 6.72 \text{ (or PB = 14.9\%)}$$

$$T_I = t_d \left[\frac{32 + 6\dfrac{t_d}{\tau}}{13 + 8\dfrac{t_d}{\tau}}\right] = 2.4 \left[\frac{32 + 6 \times 0.5}{13 + 8 \times 0.5}\right] = 4.94 \text{ minutes}$$

$$T_D = t_d \left[\frac{4}{11 + 2\dfrac{t_d}{\tau}}\right] = 2.4 \left[\frac{4}{11 + 2 \times 0.5}\right] = 0.8 \text{ minutes.}$$

Problem 2.25

In an application of the Ziegler-Nichols method, a process begins oscillation with a 30% proportional band in an 11.5-min period. Find (a) The nominal three-mode controller settings and (b) Settings to give quarter-amplitude response.

Solution :

(*a*) 30 % PB means, $K_u = \dfrac{100}{PB} = \dfrac{100}{30} = 3.33$

$$P_u = 11.5 \text{ min.}$$

Proportional gain : $K_P = \dfrac{K_u}{1.7} = \dfrac{3.33}{1.7} = 2$

$$T_I = \frac{P_u}{2} = \frac{11.5}{2} = 5.75 \text{ min}$$

$$T_D = \frac{P_u}{8} = \frac{11.5}{8} = 1.44 \text{ min}$$

(*b*) $K_P = 2$ (same as above)

$$T_I = \frac{P_u}{1.5} = \frac{11.5}{1.5} = 7.67 \text{ min}$$

$$T_D = \frac{P_u}{6} = \frac{11.5}{6} = 1.92 \text{ min}$$

Problem 2.26

Determine if the system of Fig. 2.64 is stable according to the rules given below :

Rule 1: A system is stable if the phase lag is less than 180° at the frequency for which the gain is unity.

Rule 2: A system is stable if the gain is less than one at the frequency for which the phase lag is 180°.

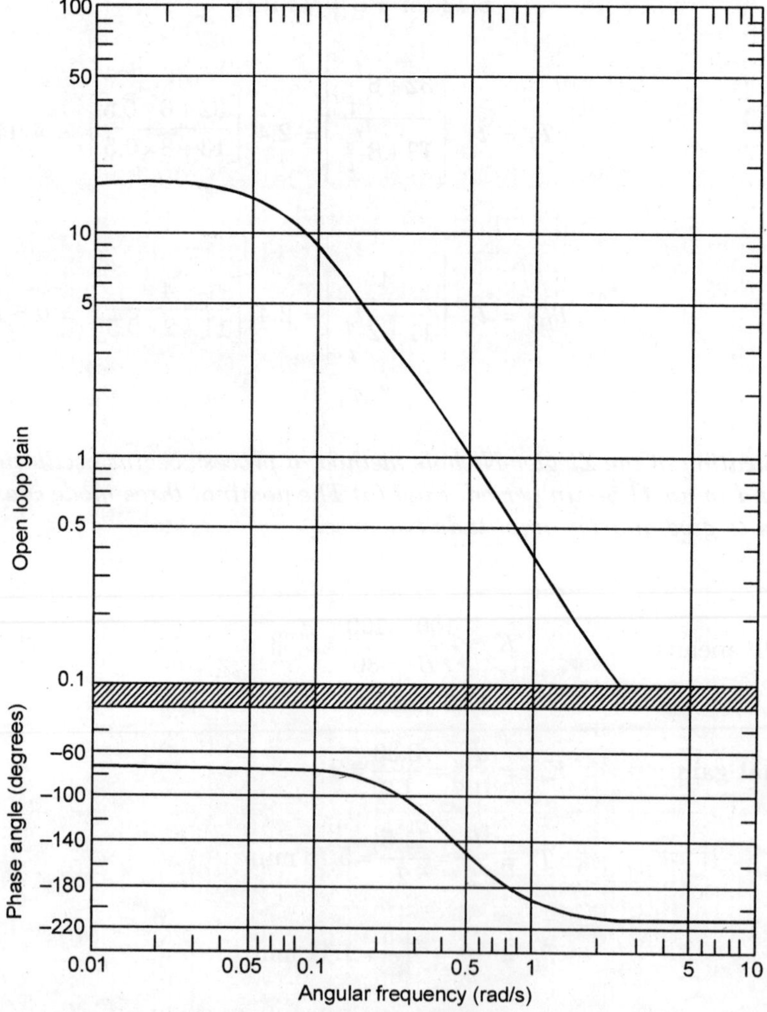

Fig. 2.64 Bode plot with an example of process-control loop frequency response

Solution :

Consider Rule 1:

We find the frequency for which the gain is unity as 0.5 rad/s (or 0.5/20 = 0.08 Hz.] For this frequency, the phase lag is 140° which is less than 180°. Hence the system is stable.

Consider Rule 2:

Frequency for phase lag 180° = 0.8 rad/sec.

The gain for this frequency = 0.5 which is less than one. Hence the system is stable.

Problem 2.27

Determine if the system with a Bode plot of Fig. 2.65 satisfies the phase margin of 40° and gain margin of 5 dB (0.56 gain). If not, what is to be done ?

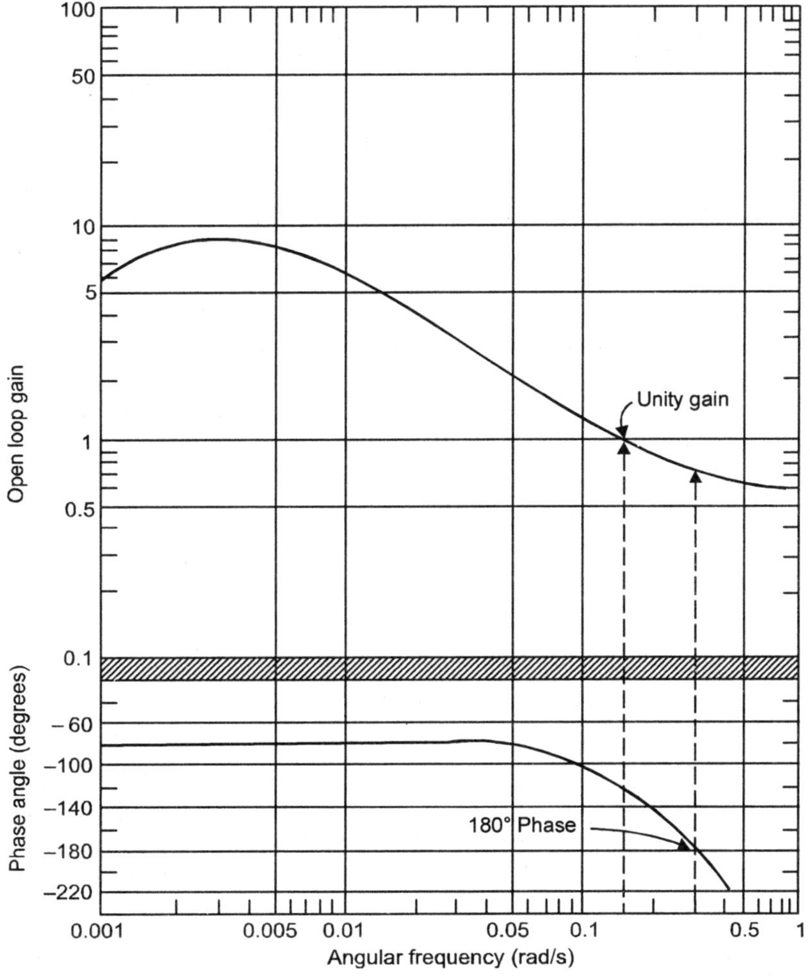

Fig. 2.65 Bode plot for the problem 2.27

Solution :

Phase margin : Unity gain occurs at an angular frequency of 0.15 rad/s, and the corresponding phase is –120°C. The phase margin is 180 – 120° = 60° and hence the condition is satisfied.

Gain margin : 180° phase lag occurs at 0.3 rad/s and the corresponding gain is 0.7, which is too high. Gain should be less than 0.56 to satisfy the gain margin of 5 dB.

To satisfy this condition, the controller gain will have to be reduced slightly and brought below 0.56 instead of 0.7.

2.12 PROBLEMS AND QUESTIONS

1. Define process load, process lag and self-regulation.
2. A velocity control system has a range of 220 to 460 mm/s. If the set point is at 320 mm/s and the measured value is 315 mm/s. Calculate the error as percentage of span.
3. A controlling variable is a motor speed that varies from 1000 to 1500 rpm. If the speed is controlled by a 30 to 50 V dc signal, calculate (*a*) the speed produced by an input of 39 volts and (*b*) the speed calculated as a percent of span.
4. What is meant by 'Dead time' ?
5. Differentiate discontinuous and continuous controller modes.
6. Define reverse and direct actions of a controller
7. What is called 'Neutral zone' ?
8. In what way single speed floating control is different from 'ON-OFF' control. ?
9. What is the difference between single speed and multispeed floating controls ?
10. A 10-m diameter cylindrical tank is emptied by a constant outflow of 2 m^3/min. A two-position controller is used to open and close a full valve with an open flow of 4 m^3/min. For level control, the neutral zone is 1m and the set point is 12m . Calculate the cycling period and plot the level versus time.
11. Where do the floating control mode find applications ?
12. Name the three different continuous control modes.
13. What are called composite control modes? Name atleast three such composite modes.
14. What is called an 'offset' in a proportional controller ?
15. How is offset eliminated with the help of integral control mode ?
16. Draw the response of outputs of different controller modes for a step error.
17. Define: (*a*) Proportional band (*b*) Proportional gain (*c*) Repeats per minute (*d*) Integral time (*e*) Rate gain and (*f*) Differential time.
18. For a proportional controller, the controlled variable is a process temperature with a range of 50 to 250°C and a set point of 150°C. Under normal conditions, the set point is maintained with an output of 50%. Find the proportional offset that results from a load change which requires a 55% output if the proportional gain is (*a*) 0.7 and (*b*) 2.0
19. Why derivative control mode cannot be used alone ?
20. A PI controller is reverse acting with PB = 20 and repeats per minute = 12. Find (*a*) the proportional gain, (*b*) the integral gain, and (*c*) the time that the controller output will reach 0% after a constant error of 1.5% starts. The controller output when the error occurred was 75%.

21. A PID controller is reverse acting with PB = 25, repeats per minute = 12. and rate gain of 0.2 minutes. The controller output when the error occurred was 70%. Find (*a*) the derivative gain, and the time at which the controller output reach 0% if the input error is $e_p = 0.9\ t^2$.
22. A closed loop system has time constants of 1 min and 10 min and a proportional controller. Obtain the response to a ramp change in set point at a controller gain that gives a damping coefficient of 0.3. Would a higher or lower gain be advantageous ?
23. Under what conditions pneumatic controllers are preferred over electronic controllers?
24. In a certain integral controller the deviation changes sinusoidally with time. Show that the phase of the manipulated variable is always 90 degrees behind the deviation. ($e = \sin \omega t$)
25. Write the transfer function of P, PI and PID control.
26. What are the major components of process control system?
27. Define phase cross over frequency in the Bode stability criterion.
28. Draw the block diagram of a control loop for measurement of the process reaction curve.
29. Bring out the relative advantages and disadvantages of proportional, integral, and derivative control actions.
30. State how the disadvantages of P, I and D control actions are overcome in composite mode operations.
31. Explain their characteristics effects of PI controller on the closed loop response of a first order process for unit step change in (*a*) set point and (*b*) load.
32. Discuss about the factors to be considered while selecting the type of controller for various processes.
33. Describe the evaluation criteria of Integrals of the square error (ISE), absolute value of the error (IAE) and time-weighted absolute error (ITAE) of the controller settings.
34. Give the optimum controller settings for the model $G(s) = \dfrac{e^{-0.5s}}{4s+1}$ obtained by reaction curve method.
35. Discuss about the controller settings using Ziegler–Nichols continuous cycling method and write its limitations.
36. Explain the function of a controller.
37. Mention any two drawbacks of the derivative control action.
38. What are the parameters required to design a best controller ?
39. Write any two practical significance of the (*a*) Gain Margin and (*b*) Phase margin.
40. Compute the response of PD controller to a ramp change in the error '*e*'. Sketch the contributions of the proportional and derivative actions separately. Discuss the anticipatory nature of the derivative control term.
41. Explain briefly about the various control actions of (*a*) Electronic controllers and (*b*) Pneumatic controllers.
42. Why the Ziegler-Nichols tuning procedure is often called the continuous cyclic tuning method?
43. How would you select the most appropriate out of the three time-integral criteria ISE, IAE and ITAE for a particular application ?
44. What are the principal questions that arise during the design of a feedback controllers ? Discuss them on the basis of a physical example.
45. Write any two limitations of single-speed floating control.

46. Sketch pneumatic PID controller.
47. The differential gap (neutral zone) in an ON-OFF control is 1% of set point value. When set point is 300°C, duration of valve open is equal to 2 minutes and duration of valve closed is 2 minutes. Find the period of oscillation of controlled variable when set point is 250°C.
48. Design an electronic PID controller with the following specifications :

    Proportional Band (PB) = 25%

    Derivative Gain (K_D) = 20

    Integral Gain (K_I) = 2.5
49. Explain any one time response method of tuning controller.
50. Write brief note on ¼ decay ratio.
51. Why both gain margin and phase margin should be considered for tuning of a controller ?
52. Design a 40% PB controller for motor speed control. The motor speed varies from 50 to 150 rpm for an input control voltage of 0 to 5 volts. A speed sensor linearly changes from 2.0 to 5.0 KΩ over the speed range. A set point of 100 rpm is desired for which the motor control circuit input is 2.5 volts. Suppose the set point is changed to 125 rpm with no other adjustments. What offset error will occur ?
53. An integral controller has an input range of 1 to 10 Volts and an output range of 0 to 15 volts. If K_I = 12% (%-min), find G_I, R and C.
54. A liquid level system converts a 4–10 m level into a 4–20 mA current. Design a three mode controller that outputs 0–5 V with a PB = 50%, reset time = 0.03 min and derivative time 0.05 min. Fastest expected change time is 0.8 minutes.
55. An open loop transient test provides the process reaction graph as shown in Figure 2.66 for a 7.5% disturbance.

    (*a*) Find the standard Proportional-Integral gain settings.

    (*b*) Find the three-mode quarter-amplitude settings.

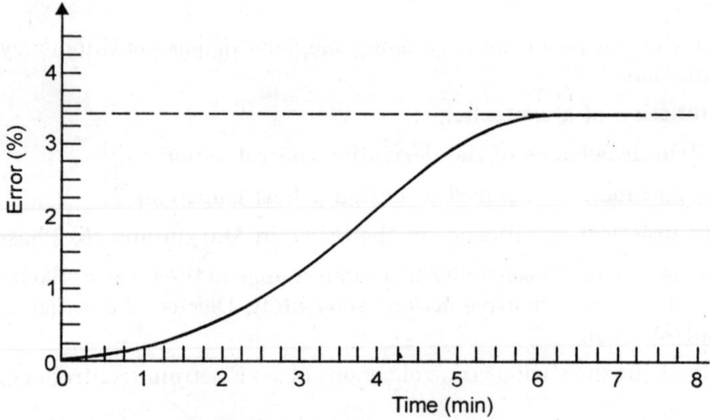

Fig. 2.66 Process reaction graph

56. In the Ziegler-Nichols method, the critical gain was found to be 4.0 and the critical period was 2.0 minutes. Find the standard settings for (*a*) proportional control (*b*) PI control and c) PID control.

57. Specify the gain and phase margins for Fig. 2.67. Is the system stable ?

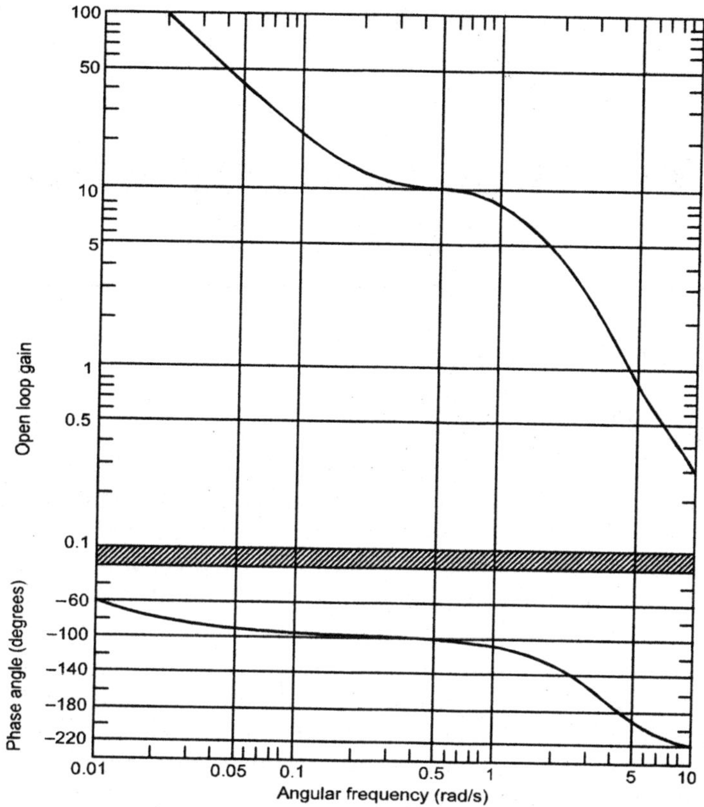

Fig. 2.67 Bode plot

58. If the nominal proportional gain for the process in Fig. 2.67 was 11.5, determine the gain that will just satisfy a gain margin of 0.56 and phase margin of 40°

59. State the two rules for specifying a stable system with reference to gain and phase shift.

60. Define (*a*) Gain Margin and (*b*) Phase margin.

3

CONTROL SYSTEMS WITH MULTIPLE LOOPS

3.1 ADVANCED CONTROL SYSTEMS

Feedback control is the type of control encountered most commonly in industrial processes and particularly in chemical processes. But it is not the only one used in industries. There exists a situation where feedback control action is insufficient to produce the desired response of a given process. In such cases other control configurations such as feed forward, ratio, cascade, override, adaptive, split range, auctioneering, inferential and multivariable controls are used. An introduction to the analysis and design of such control systems is dealt here. The discussion is limited to a simple qualitative presentation of these control systems, since a more rigorous presentation goes beyond the scope of this book.

The feedback control configuration involves one output (measurement) and one manipulated variable in a single loop. The other control configurations mentioned above may use more than one measurement and one manipulation or one measurement and more than one manipulated variables. In such cases control systems with multiple loops may arise. These control systems involve loops that are not separate but share either the single manipulated variable or the only measurement.

The systems with a single manipulated input and a single controlled output are called 'single-input and single output' systems (SISO in short). Chemical processes usually have two or more controlled outputs, requiring two or more manipulated variables. Such control systems are called 'multiple input and multiple output' systems (MIMO in short).

3.2 FEED FORWARD CONTROL

Feedback control system measures the controlled variable, compares that measurement with the set point or reference, and if there is a difference between the two, changes its output signal to the manipulated variable in order to eliminate the error. This means that feedback control cannot anticipate and prevent errors, it can only initiate its corrective action after an error has already developed. Thus we can conclude that feedback control loops can never achieve perfect control of a process, that is, keep the output of the process continuously at the desired set point value in the presence of load or set point changes.

Unlike the feedback systems, the feed forward control configuration react to variations in disturbance variables (or set point), predict the disturbance's effects and take corrective action to eliminate its impact on the process output. Therefore, the feed forward controllers have the theoretical potential for perfect control. But, as it is difficult to measure all possible disturbance variables and to predict their effect quantitatively, feedforward control is generally used along with feedback control. In most cases, a combination of feedforward and feedback techniques can correct process deviations in the shortest time. It may also be said that feed forward loops are usually corrected by feedback trimming. With feedforward, the feedback controller must only change its output by an amount equal to what the feedforward system fails to correct.

The feedforward system is more costly and requires more engineering effort than a feedback system. Hence, prior to design and installation, the control improvement it brings must be determined to be worthwhile. Most feedforward systems have been applied to processes that are very sensitive to disturbances and slow to respond to corrective action.

3.2.1 Structure of Feedforward Control Scheme

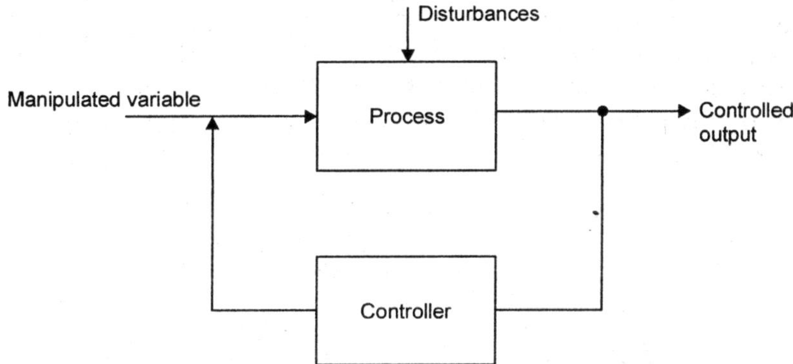

Fig. 3.1 Structure of feedback control scheme

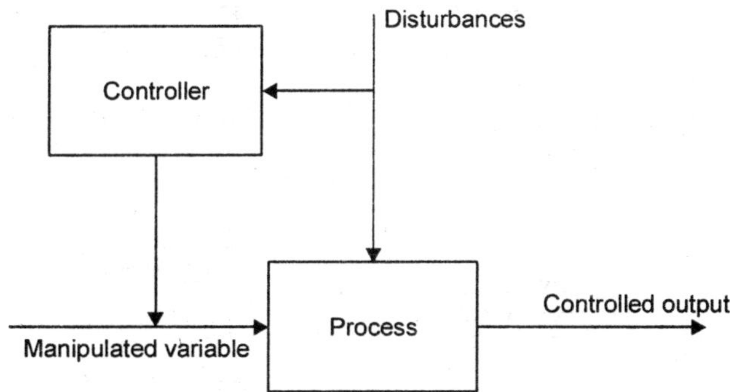

Fig. 3.2 Structure of feedforward control scheme

Fig. 3.1 shows a typical schematic of a feedback control system. In comparison to that we can see the general form of a feedforward control system in Fig. 3.2. It measures the disturbance directly and then it anticipates the effect that it will have on the process output. Subsequently, it changes the manipulated variable by such an amount as to eliminate completely the impact of the disturbance on the process output (controlled variable). Control action starts immediately after a change in the disturbance has been detected. It is clear from the Figs. 3.1 and 3.2 that feedback acts 'after the fact', in a compensatory manner, whereas feedforward acts 'beforehand' in an anticipatory manner.

As a practical example, consider the stirred tank heater shown in Fig. 3.3. The control objective is to keep the temperature of the liquid in the tank at a desired value (set point) despite any changes in the temperature of the inlet stream. Fig. 3.4 shows the conventional feedback loop, which measures the temperature in the tank and after comparing it with the desired value, increases or decreases the steam pressure, thus providing more or less heat into the liquid. A feedforward control system uses a different approach as shown in Fig. 3.5. It measures the temperature of the inlet stream (disturbance) and adjusts appropriately the steam pressure (manipulated variable). Thus it increases the steam pressure if the inlet temperature decreases and decreases the steam pressure when the inlet temperature increases.

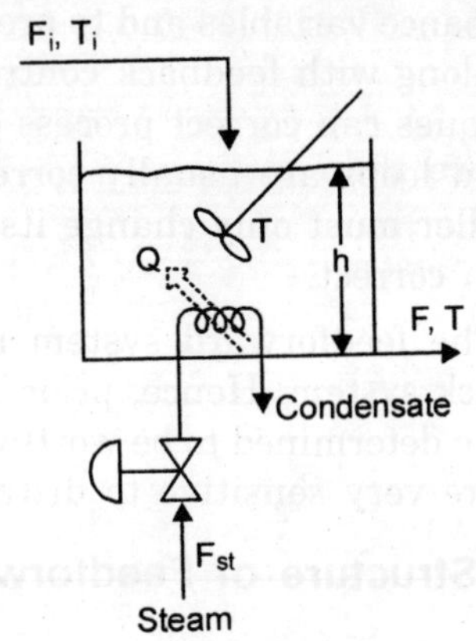

Fig. 3.3 Stirred tank heater

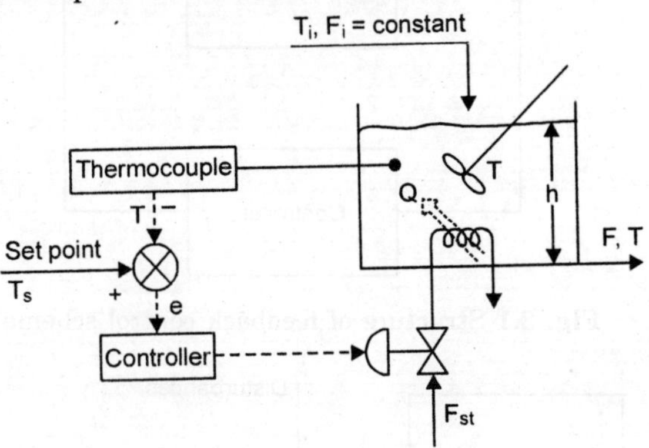

Fig. 3.4 Feedback temperature control for a tank heater

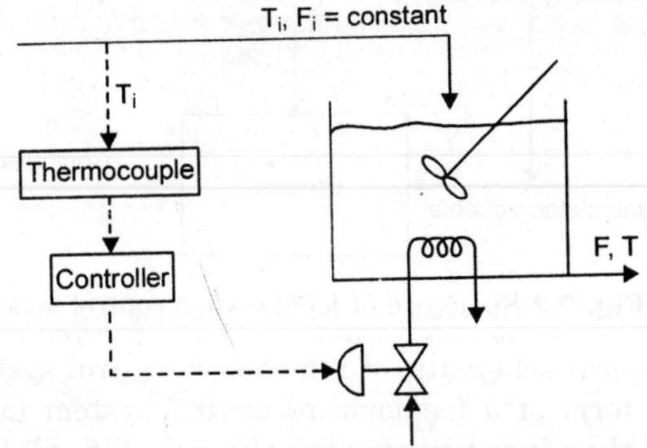

Fig. 3.5 Feedforward temperature control for stirred tank heater

3.2.2 Feedforward Control of Various Processing Units

Heat exchanger : (Refer Fig. 3.6)

The objective is to keep the exit temperature of the liquid constant by manipulating the steam pressure. There are two principal disturbances (loads) that are measured for feedforward control : liquid flow rate and liquid inlet temperature. Feedforward control can be developed for more than one disturbance also. The controller acts according to which disturbance changed value. Fig. 3.7 represents the general case of feedforward control with several loads (disturbances) and a single controlled variable. The major components of load are entered into a model to calculate the value of the manipulated variable required to maintain control at the set point.

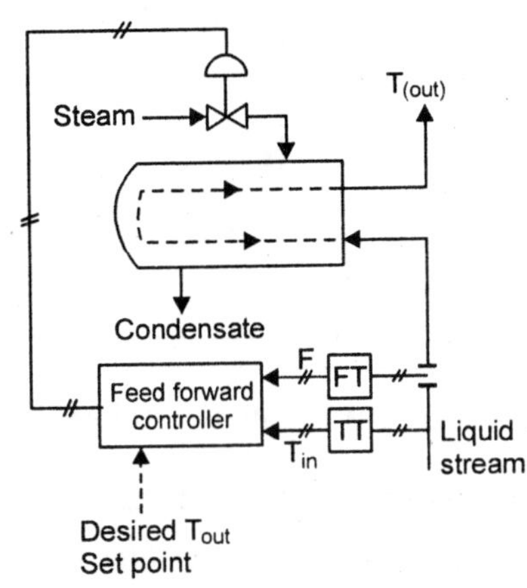

Fig. 3.6 Heat exchanger

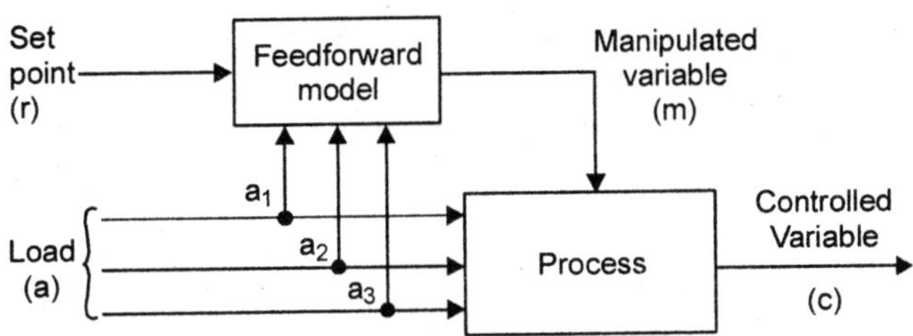

Fig. 3.7 The feedforward control loop

Boiler drum level control : (Refer Fig. 3.8)

Here the objective is to keep the liquid level in the drum constant. The two disturbances are the steam flow from the boiler, which is dictated by varying demand elsewhere in the plant, and the flow of the feedwater which is also the principal manipulated variable.

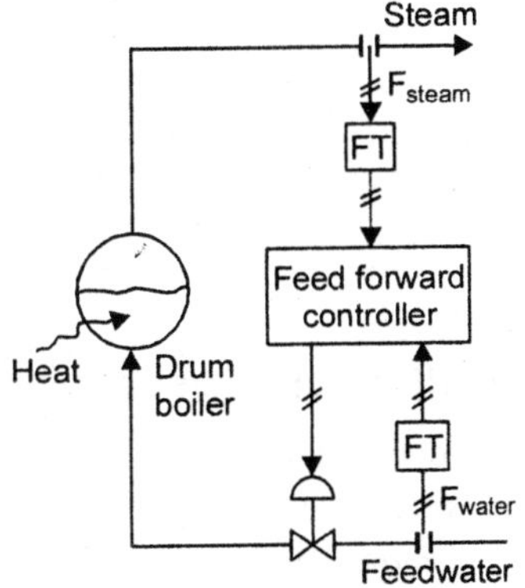

Fig. 3.8 Drum boiler

Distillation Column : (Refer Fig. 3.9)

The two disturbances here are the feed flow rate F and the composition C. The available manipulated variables are the steam pressure in the reboiler and the reflux ratio. The composition of overhead or bottom product is the control objective. Feedforward control is particularly useful for a distillation column, because its response time can be measured in hours leading to large amount of off-specification products.

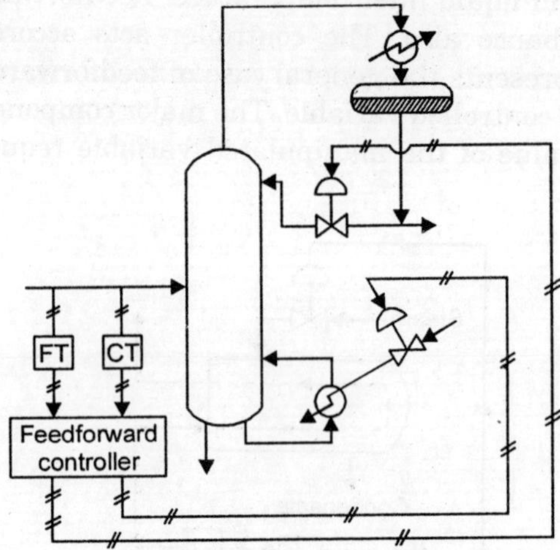

Fig. 3.9 Distillation column

Continuous stirred tank reactor (CSTR) : (Refer Fig. 3.10)

Inlet concentration and temperature are the two disturbances, and the product withdrawal flow rate and the coolant flow rate are the two manipulations. There are two objectives : to maintain constant temperature and composition within the CSTR. The feed forward control of a CSTR indicates that the extension to systems with multiple controlled variables should be rather straight forward.

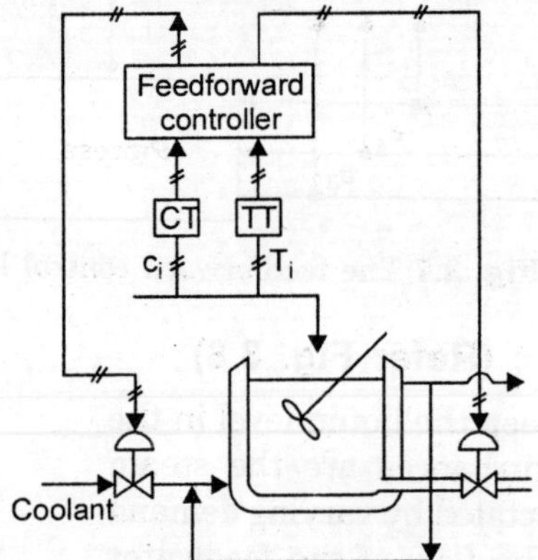

Fig. 3.10 Continuous stirred tank reactor

3.2.3 Feedforward-Feedback Control

Feedforward control has the potential for perfect control, but it also suffers from several inherent weaknesses as listed below:

1. It requires the identification of all possible disturbances and their direct measurements, something that may not be possible for many processes.
2. Any changes in the parameters of a process cannot be compensated by a feedforward controller because their impact cannot be detected.
3. Feedforward control requires a very good model for the process, which is not possible for many systems in industry.

On the other hand, feedback control is rather insensitive to all three of these drawbacks, but has poor performance for a number of systems and raises questions of closed -loop stability.

Table 3.1 summarises the relative advantages and disadvantages of the control systems.

Table 3.1 : Relative advantages and disadvantages of feedforward and feedback control

Advantages	Disadvantages
Feedforward	
1. Acts before the effect of a disturbance has been felt by the system	1. Requires identification of all possible disturbances and their direct measurement
2. Is good for slow systems (multicapacity) or with significant dead time	2. Cannot cope with unmeasured disturbances
3. It does not introduce instability in the closed-loop response	3. Sensitive to process parameter variations
Feedback	
1. It does not require identification and measurement of any disturbance	1. It waits until the effect of the disturbances has been felt by the system before control action is taken.
2. It is insensitive to modeling errors.	2. It is unsatisfactory for slow processes or with significant dead time
3. It is insensitive to parameter changes.	3. It may create instability in the closed-loop response.

We would expect that a combined feedforward-feedback control system will retain the superior performance of the first and the insensitivity of the second to uncertainities and inaccuracies. Indeed, any deviations caused by the various weaknesses of the feedforward control will be corrected by the feedback controller. This is possible because a feedback control loop directly monitors the behaviour of the controlled process (measures process output). Fig. 3.11 shows the configuration of a typical combined feedforward -feedback control system.

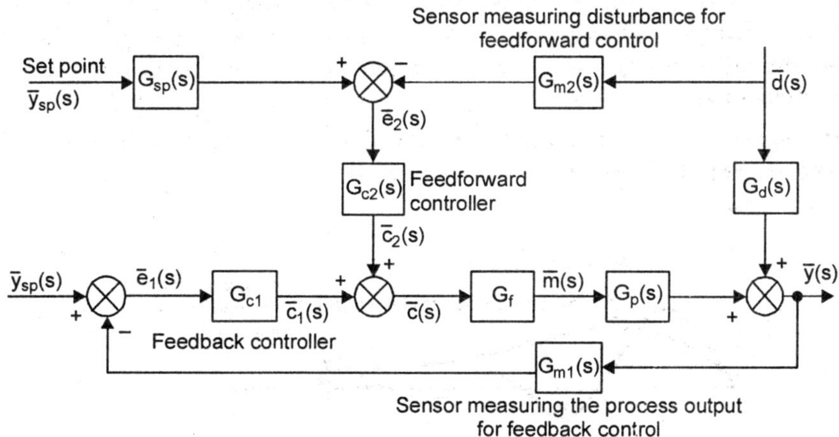

Fig. 3.11 Generated block diagram for feedforward control

Three element control for controlling boiler drum level can be referred as one of the practical applications of feedforward -feedback control system. A simplified block diagram of such an application is shown in Fig. 3.12. Feedforward controller receives load signals of Feedwater flow and steam flow and generates an output signal corresponding to the difference between the two (It is expected that at steady state feed water flow is equal to the steam flow and the level remains at set point). The feedback controller receives the controlled variable signal (level) to balance against the set point. Change in feed water flow or steam flow due to various plant conditions is recognized in anticipation and action taken much before the disturbance causes the level change. The offset which is likely to occur is taken care of by the feedback controller as it takes actual level also into account.

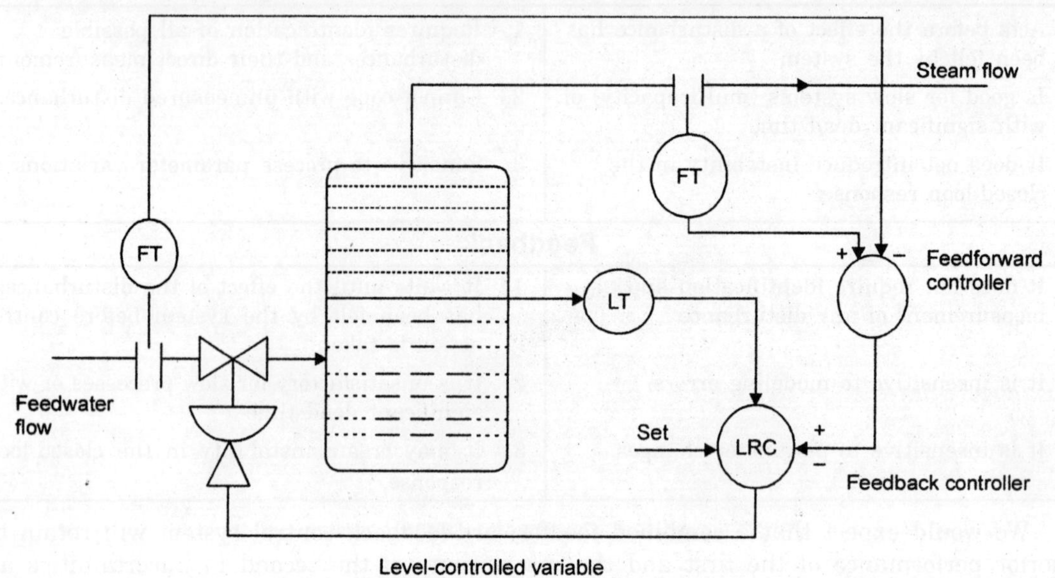

Fig. 3.12 Example of feedforward-feedback control–Drum Level Control

The use of feedback in a feedforward system does not detract from the performance improvement which was gained by feedforward control. Without feedforward, the feedback control loop was required to change its output follow all changes in load. With feedforward, the feed back controller must only change its output by an amount equal to what the feedforward system fails to correct.

A typical response of a feedforward control only and a combined feedforward- feedback control is shown in Fig. 3.13 which is self explanatory.

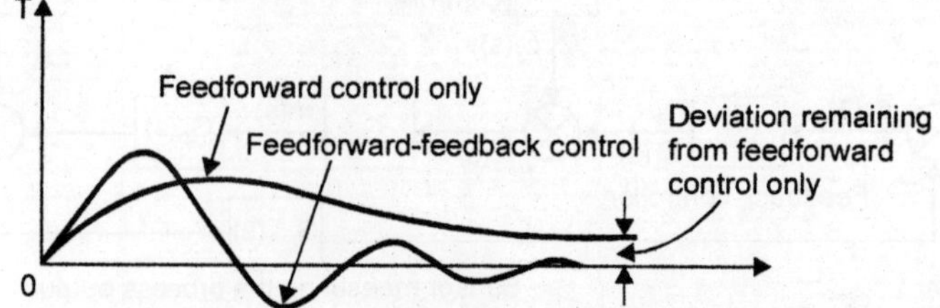

Fig. 3.13 Temperature response of a stirred tank heater under feedforward control alone and with feedback trimming

3.3 CASCADE CONTROL

Many processes are controlled by regulating the flow of a heating medium such as steam, gas, oil or fuel for supplying heat to a process. Variations in flow not dictated by the controller are caused by changes in pressure differential at the valve, which in turn, result from changes in pressure of supply, changes in downstream pressure, and so on. These changes are difficult to counteract since they must carry through the process before they are detected in the controller. Supply changes sometimes occur suddenly or over a wide range, and deviation may become excessive before a new balance of conditions can be established. Such conditions are normally overcome by a scheme called 'cascade control'.

In a cascade control configuration we have one manipulated variable and more than one measurement. In the scheme there will be two controllers namely Primary controller and Secondary controller. The output of the primary controller is used to adjust the set point of a secondary controller, which in turn sends a signal to the final control element (may be control valve). The process output is fed back to the primary controller, and a signal from an intermediate stage of the process is fed back to the secondary controller. The block diagram of such a cascade control system is shown in Fig. 3.14. Two measurements are taken from the system and each used in its own control loop. The outer loop (primary controller) controller output is the set point of the inner loop (secondary controller). Thus, if the outer loop controlled variable changes, the error signal that is input to the controller effects a change in set point of the inner loop. Eventhough the measured value of the inner loop has not changed, the inner loop experiences an error signal and thus new output by virtue of its setpoint change. Cascade control generally provides better control of the outer loop variable than is accomplished through a single variable system. The schematic representation of a cascade control is shown in Fig. 3.15. which clearly demonstrates that the disturbances arising within the secondary loop are corrected by the secondary controller before they can affect the value of the primary controller output. This important benefit has led to the extensive use of cascade control in industrial (especially in chemical) processes. In chemical processes, flow rate control loops are almost always cascaded with other control loops.

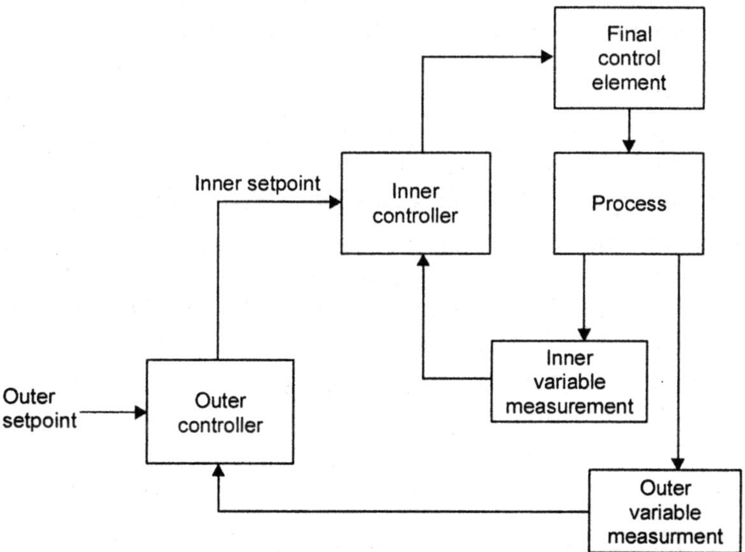

Fig. 3.14 General features of a cascade process-control system

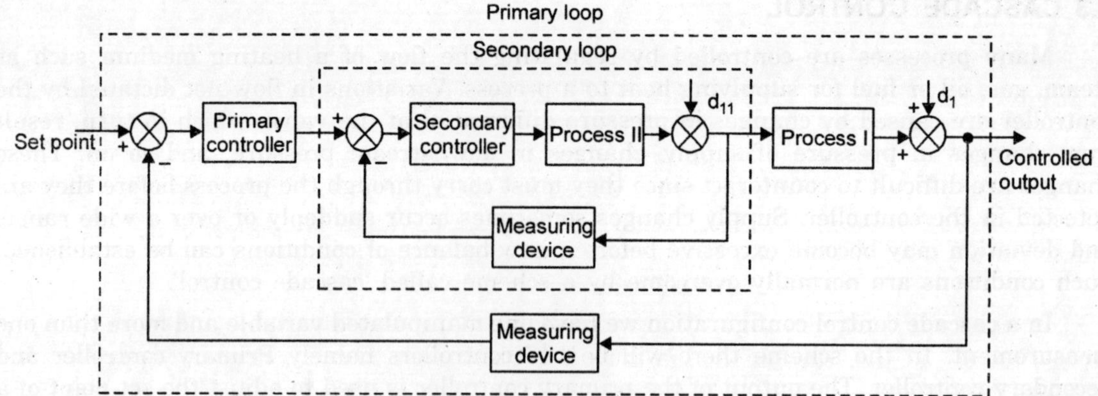

Fig. 3.15 Cascade control

The two distinct advantages gained with cascade control are :

1. Disturbance affecting the secondary variable can be corrected by the secondary controller before a pronounced influence is felt by the primary variable.
2. Closing the control loop around the secondary part of the process reduces the phase lag seen by the primary controller, resulting in increased speed of response.

3.3.1 Selection and Tuning of Cascade Control

The two controllers of a cascade control system are standard feedback controllers (*i.e.*,P,PI and PID). Generally, a proportional controller is used for the secondary loop, although a PI controller with small integral action is not unusual. Any offset caused by P control in the secondary loop is not important since we are not interested in controlling the output of the secondary process.

The dynamics of the secondary loop are much faster than those of the primary loop. Consequently, the phase lag of the closed secondary loop will be less than that of the primary loop. This feature leads to the following important result, which constitutes the rationale behind the use of cascade control : The cross over frequency for the secondary loop is higher than that for the primary loop. This allows us to use higher gains in the secondary controller in order to regulate more effectively the effect of a disturbance occurring in the secondary loop without endangering the stability of the system.

The tuning of the two controllers of a cascade system proceeds in two steps :

1. First, we determine the settings for the secondary controller using one of the methods that we studied in Chapter-2 (Cohen Coon, Ziegler-Nichols or others employing time integral criteria or phase and gain margin considerations).
2. Second, from the Bode plots of the overall system we determine the crossover frequency using the settings for the secondary loop we found above. Then, using the frequency response techniques, we choose the settings for the primary controller (using Bode plots).

3.3.2 Applications

1. Jacketed CSTR : Consider the CSTR shown in Fig. 3.16. The reaction is exothermic and the heat generated is removed by the coolant, which flows in the jacket around the tank.

The control objective is to keep the temperature of the reacting mixture, T, constant at a desired value. Possible disturbances to the reactor include the feed temperature Ti and the coolant temperature Tc. The only manipulated variable is the coolant flow rate Fc.

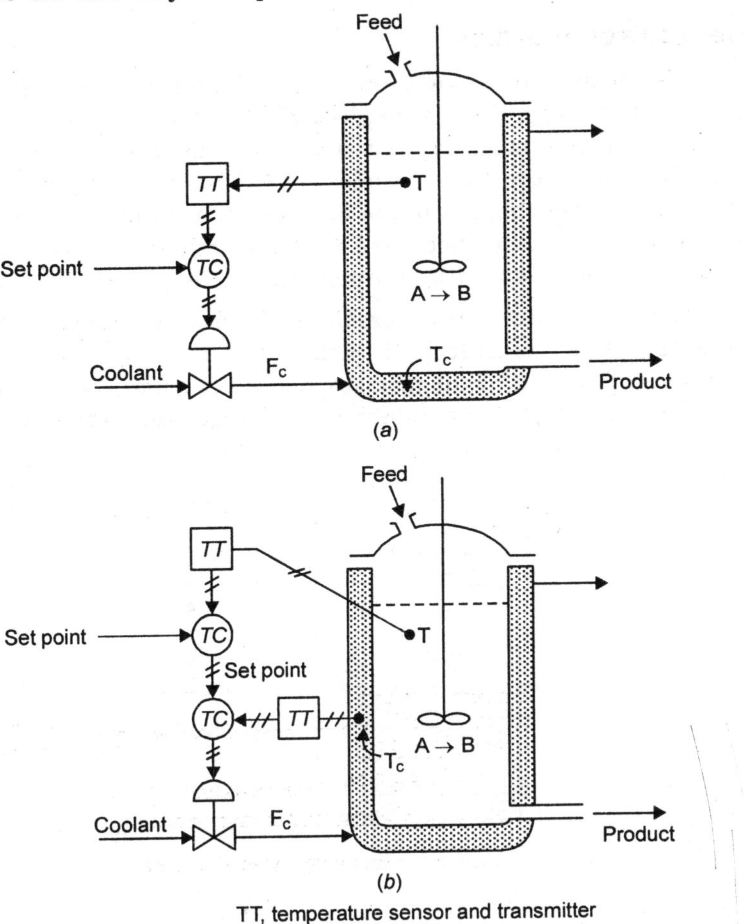

Fig. 3.16 Temperature control of jacketed CSTR
(*a*) Conventional feedback (*b*) Cascade

Simple feedback control : The control configuration shown in Fig. 3.16(*a*) is for simple feedback control. It is clear that T will respond much faster to changes in Ti than to changes in Tc. Therefore, the simple feedback control will be very effective in compensating for changes in Ti and less effective in compensating for changes in Tc.

Cascade Control : We can improve the response of the simple feedback control to changes in the coolant temperature by measuring Tc and taking control action before its effect has been felt by the reacting mixture. Thus, if Tc goes up, increase the flow rate of the coolant to remove the same amount of heat. Decrease the coolant flow rate when Tc decreases.

We notice, therefore, that we can have two control loops using two different measurements, T and Tc, but sharing common manipulated variable, Fc. How these loops are related is shown in Fig. 3.16 (*b*). The loop that measures T (controlled variable) is the dominant, or primary, or master control loop and uses a set point supplied by the operator. Whereas the loop that measures Tc, uses the output of the primary controller as its set point and is called the secondary or slave loop.

2. Liquid level control in a tank

Consider the problem of controlling the level of liquid in a tank through regulation of the input flow rate. Refer Fig. 3.17. A single-variable system to accomplish the control is shown in Fig. 3.17(*a*). A level measurement is used to adjust a flow-control valve as a final control element. The set point to the controller establishes the desired level. In this system, upstream load changes cause changes in the flow rate that result in level changes. The level change is, however, a second stage effect here. Consequently, the system cannot respond until the level has actually been changed by the flow rate change.

Fig. 3.17(*b*) shows the same control problem solved by a cascade system. The flow loop is a single-variable system as described earlier, but the set point is determined by measurement of level. Upstream load changes are never seen in the level of liquid in the tank because the flow control system regulates such changes before they appear as substantial changes in level.

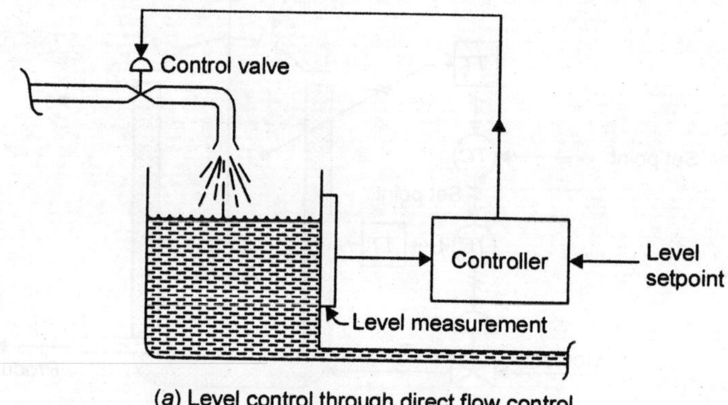

(*a*) Level control through direct flow control

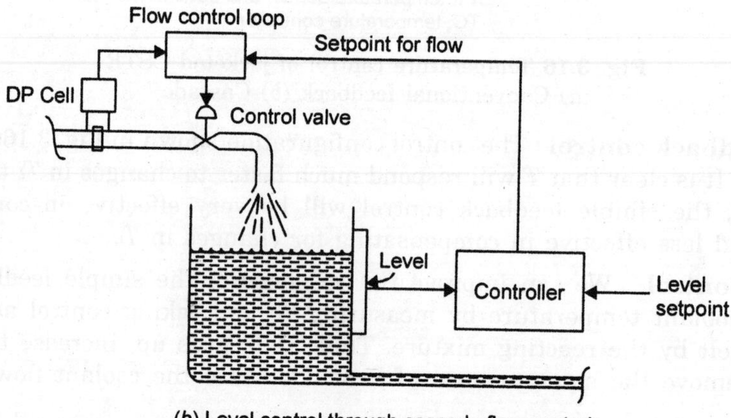

(*b*) Level control through cascade flow control

Fig. 3.17 Cascade control often provides better control than direct methods

3. Control system for a stripping column

Cascade control is especially effective if the inner loop is much faster than the outer loop and if the main disturbances affect the inner loop first. When a flow rate is the manipulated variable in a single loop control system, using a secondary loop to control the flow almost eliminates the effects of pressure change, which would otherwise cause unwanted changes in flow rate.

Fig. 3.18 shows a stripping column where the temperature near the bottom is used to adjust the set point of a steam flow controller. When the feed volatility decreases, the temperature in the column increases and the primary controller calls for a lower steam rate. The response of the system to changes in feed composition is only slightly improved by using cascade control, since the lags in the column are much larger than those in the flow control system. The response to changes in steam pressure or column pressure is very much better with cascade control; changes in flow rate are correctly very rapidly by the fast flow control loop.

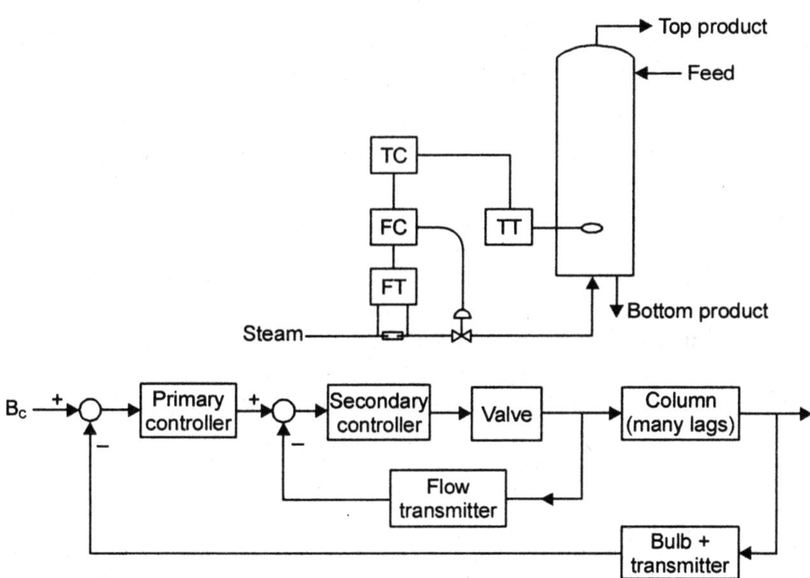

Fig. 3.18 Cascade control system for a stripping column

4. Control system for a furnace temperature

The furnace under discussion is heated by the fuel (gas, oil etc) whose flow rate is measured by a flow transmitter. The temperature of the furnace is measured with the help of a temperature transmitter. The temperature is the controlled variable and is a part of the primary controller loop. The temperature controller (primary controller) output is the set point to the flow controller (secondary controller). Any variations in the flow due to variations in line pressure, chokage etc will be taken care of by the secondary controller before the change affects the flow rate and hence the temperature. The block diagram of such a cascade control loop is shown in Fig. 3.19 which is self explanatory.

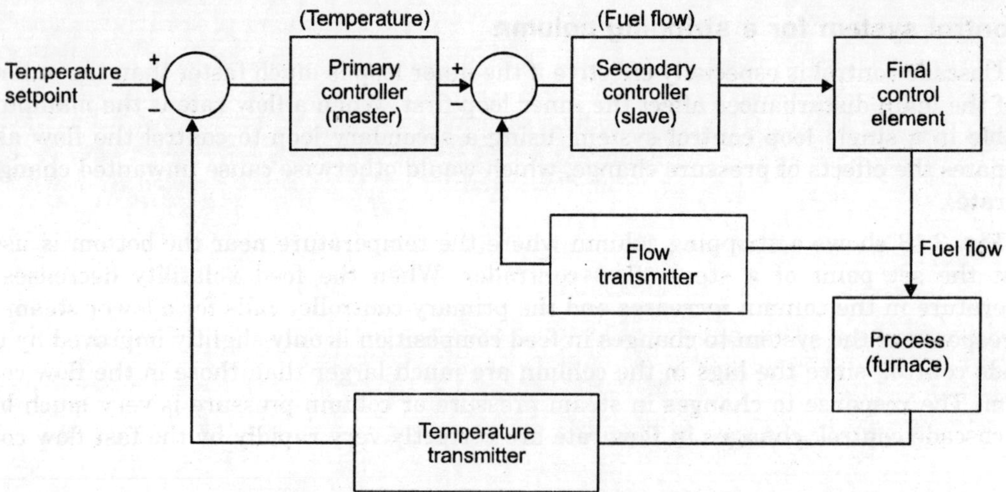

Fig. 3.19 Cascade control of furnace temperature

3.3.3 Selection of Secondary Variable

When there are several lags ahead of the main process lag, the correct choice of secondary variable may not be obvious. We may consider a system consisting of a column and reboiler (Refer Fig. 3.20) for our further discussions. The controlled variable is the temperature a few plates above the reboiler. The secondary variable could be steam flow, pressure on the steam side of the exchanger, or vapour flow from the reboiler. Controlling the steam flow could eliminate only the effects of changes in steam supply pressure or condensing pressure. The time constants for the steam flow system are an order of magnitude smaller than the pressure lags ; so making the flow response still faster by a closed-loop system has a negligible effect on the response of the column reboiler system. Using a secondary flow controller is rarely worthwhile in this case, probably because the major disturbances enter further on in the system. If control of the steam flow seems desirable, the pressure in steam header should be regulated.

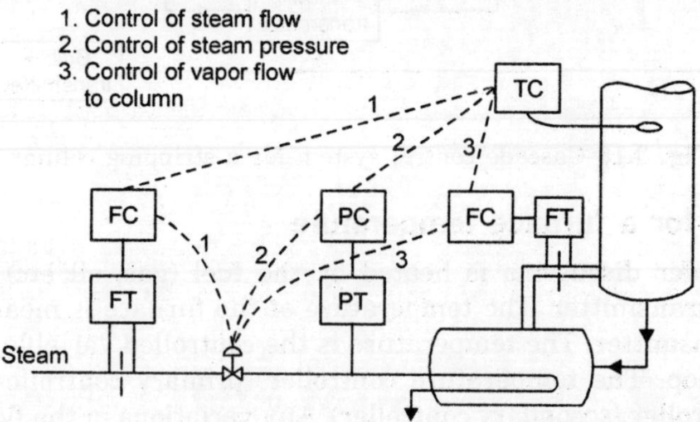

Fig. 3.20 Cascade control of column temperature

The steam pressure in the exchanger is a measure of the condensing temperature and to some extent a measure of the wall temperature. The time constant for the metal walls is usually several seconds, and by including this in the inner loop the critical frequency for the outer loop is significantly increased. However, keeping the steam pressure constant does not ensure a constant vapour rate, which is really the important variable. By using the vapour flow as a secondary variable, the inner loop would include the lag on the liquid side of the reboiler, which can be represented by a several-second time constant. With the enlarged inner loop, disturbances such as changes in reboiler level, reboiler temperature, or heat-transfer coefficient would enter the inner loop and be more quickly corrected than with the other two schemes. Of course the disturbances in steam supply pressure would not be corrected as rapidly as with the first scheme.

How many elements should be included in the inner loop depends on the frequency and magnitude of the load changes at various points. A general rule is that the response of the inner loop should be much faster than the response of the outerloop; there is little benefit from cascade control if the sum of the time constants in the inner loop exceeds that for the outer loop.

3.4 RATIO CONTROL

Ratio control is a special type of feedforward control where two disturbances (loads) are measured and held in a constant ratio to each other. Many industries require feed in specific ratio, examples being air-fuel ratio in burners, reactants ratio to blending unit and reactors. It is mostly used to control the ratio of flowrates of two streams. Both flow rates are measured but only one can be controlled. The stream whose flow rate is not under control is usually referred to as 'wild stream'.

Fig. 3.21 shows two different ratio control configurations for two streams. Stream *A* is the wild stream. In configuration 1 (Fig. 3.21*a*) we measure both flow rates and take their ratio. This ratio is compared to the desired ratio (set point) and the deviation (error) between the measured and desired ratios constitutes the actuating signal for the ratio controller.

In configuration 2 (Fig. 3.21*b*) we measure the flow rate of the wild stream *A* and multiply it by the desired ratio. The result is the flow rate that the stream *B* should have and constitutes the set point value which is compared to the measured flow rate of stream *B*. The deviation constitutes the actuating signal for the controller, which adjusts appropriately the flow of stream *B*. As the magnitude of the wild stream flow changes, the set point of the controller is automatically moved to new value by the ratio setter (sometimes called as 'ratio station') so that an exact ratio is maintained between flow rates of stream *A* and stream *B*.

3.4.1 Applications

Ratio control is used extensively in chemical processes.

1. Ratio of two reactants

A most common ratio control is to control the ratio of two reactants entering a reactor at a desired value. In this case, one of the flow rates is measured but allowed to float, that is, not regulated, and the other is both measured and controlled to provide the specified constant ratio. An example of this system is shown in Fig. 3.22. The flow rate of reactant A is measured and added, with appropriate scaling, to the measurement of flow rate *B*. The controller reacts to the resulting input signal by adjustment of the control valve in the reactant *B* input line. This configuration is similar to the one discussed as configuration 2 (Refer 3.21*b*).

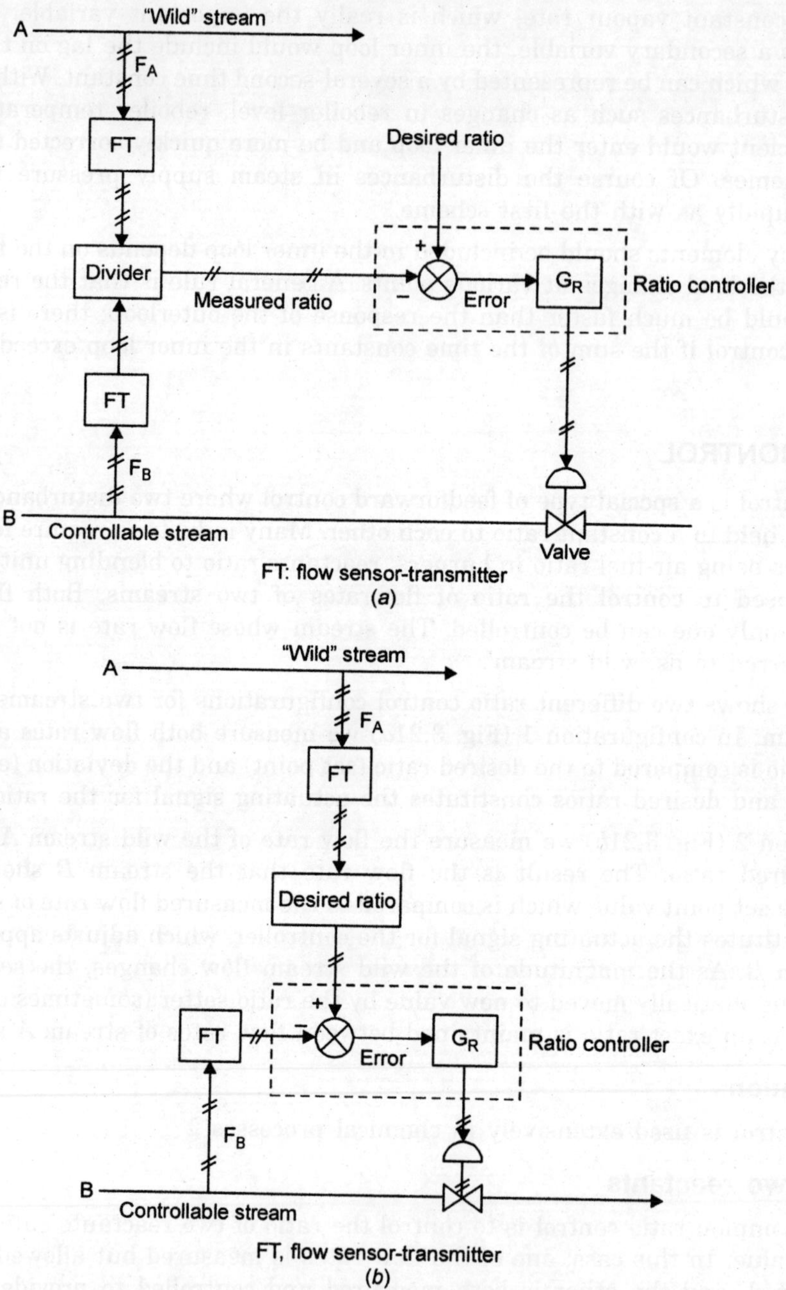

Fig. 3.21 Alternative configurations of ratio control systems

2. Fuel-air ratio control

This ratio control is used to keep the ratio of fuel/air in a burner (in the furnace) at its optimum value. This is to make sure the proper combustion of the fuel with the just required amount of air. To control the temperature of a furnace, the fuel demand is controlled by a cascade controller as discussed in section 3.3.2.4. This ratio controller may be used in series with temperature controller. The fuel flow rate measured as secondary variable can be used here as wild stream flow rate and given to ratio setter. The output of the ratio setter is the set point for the ratio controller (air controller) which in turn changes the valve position of the control valve in the air line to keep the desired fuel/air ratio. (Fig. 3.23).

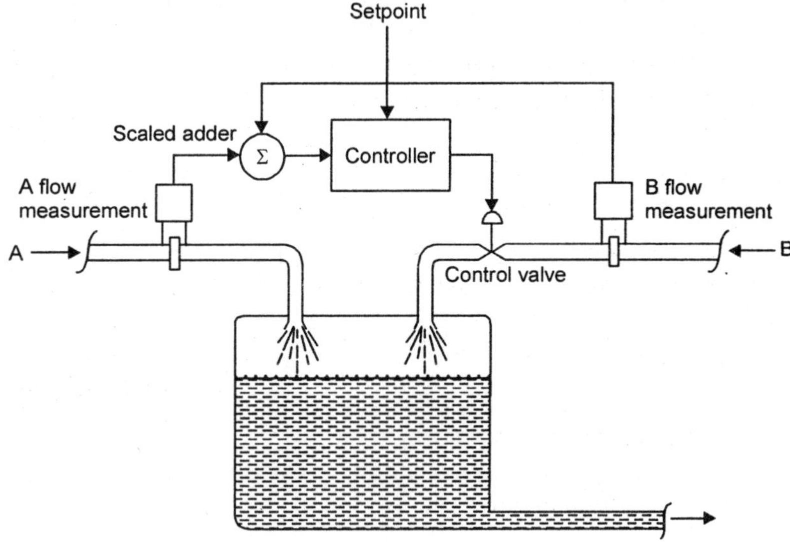

Fig. 3.22 A compound system for which the ratio of two flow rates is controlled

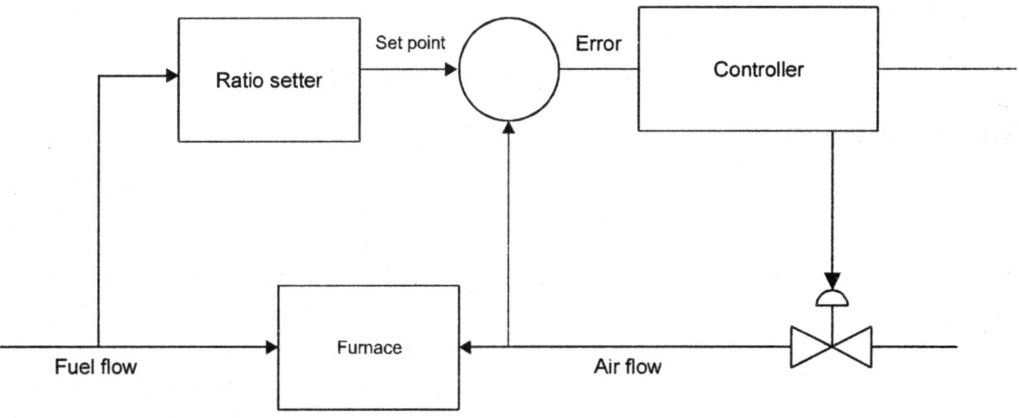

Fig. 3.23 Fuel/Air ratio control

The secondary variable (Air flow) is controlled by making its magnitude follow a primary variable (Fuel flow) at a set ratio.

3. Fuel-air Ratio control for boilers

The Fuel-Air ratio control discussed in the earlier section is known as series control. This means the change in air flow rate occurs as per ratio set only after the change has occurred in fuel flow rate. But in boilers used in power plants, fuel and air should be controlled in parallel rather than in series for safety reasons. This is necessary because a lag of only one or two seconds in measurement or transmission will seriously upset combustion conditions in a series system. This can result in alternating periods of excess and deficient combustion air. One such parallel fuel-air ratio control used in boilers is shown in Fig. 3.24. Firing rate demand (fuel demand) from the master control of the boiler is given as set point value to the fuel controller. Fuel flow rate changes as per the requirement of the master control. (Firing rate demand changes according to the power plant's output power demand). The same firing rate demand signal is simultaneously given to the air controller through a ratio setter so that the air flow rate is also controlled along with fuel flowrate maintaining the proper ratio required for efficient combustion.

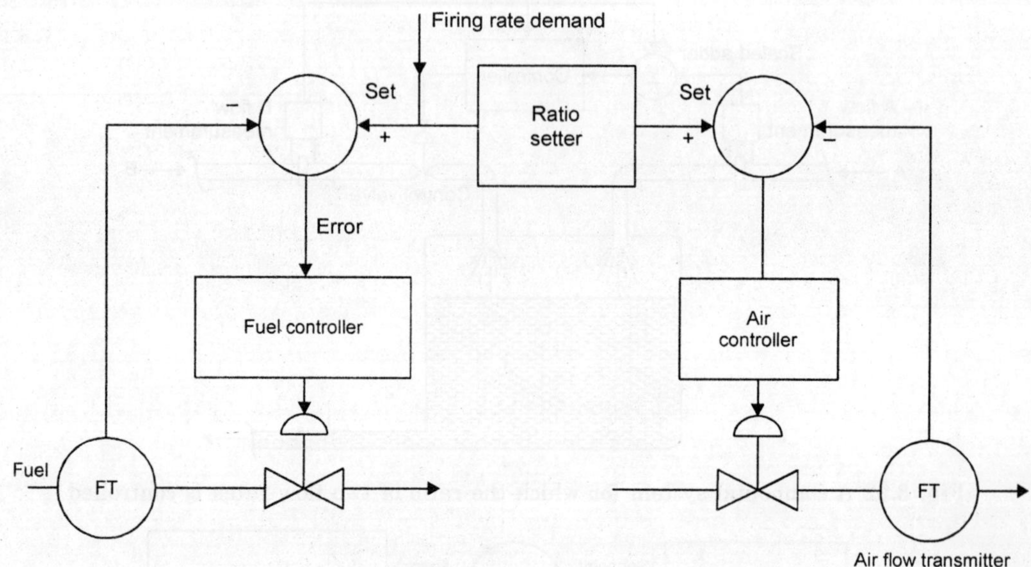

Fig. 3.24 Parallel fuel-air ratio control

4. Blending systems

The ratio of two blended streams is to be kept constant in order to maintain the composition of the blend at the desired value. Blending systems are applied to a variety of materials in a number of industries : solvents, paints, reactor feeds, fertilizers etc in chemical industry; gasoline, lube and fuel oils, distillates etc in petroleum industry ; cement, asbestos etc. in the building industry.

These applications provide the processor with economic advantages by controlling the consumption of materials (costly components and additives can be blended more precisely) and by reducing investment in floor space and batching tanks (costly blend tanks are eliminated). Through the use of continuous systems the time lags resulting from batch methods are eliminated, productivity is increased, man power needs are reduced, and inventory can be in the form of component base stocks rather than as partially blended or finished products. Blending systems provide technical advantages by accurately controlling the quality of the product and by providing the flexibility to blend a variety of finished products with minimum time required to change from one product to another.

These applications provide continuous control of the flow of each component with fixed ratios between components using ratio controls so that when the streams are combined to form the finished blend at a fixed through-put rate, the composition of the finished product is within specifications.

5. Other applications

Other commonly encountered applications include :

1. To keep a constant ratio between the feed flow rate and the steam in the reboiler of a distillation column.
2. To keep constant the reflux ratio in a distillation column.
3. To maintain the ratio of the liquid flow rate to vapour flow rate in an absorber constant, in order to achieve the desired composition in the exit vapour stream.

3.5 SELECTIVE CONTROL SYSTEMS

For each controlled variable in a control system, there must be atleast one manipulated variable. In many systems, however, the controlled variables out number the manipulated variables. When this happens, the system must decide how to share the manipulated variables. Switching between variables can be easily and smoothly accomplished using selective devices called signal selectors. The selective control systems with the help of signal selectors transfer control action from one controlled output to another according to need.

Selective control is the name given to the application of signal selectors in a control strategy. Signal selectors are devices that choose the lowest, highest, or median signal from among two or more signals. They are available both in analog and digital control hardware. In most applications selective control is a form of multivariable control where the selectors facilitate the on-line modification of control strategies as a function of changing operating conditions. The selectors allow the strategies to be changed smoothly without disturbing the process. Selective controls have a variety of uses as listed below :

1. Guarding against exceeding equipment or operating constraints (overrides)
2. Automatic start-up and shutdown.
3. Protection against instrument failures.
4. Selection of extreme values.

There are several types of selective control systems. We will be discussing two of such selective controls here namely

1. Override Control and 2. Auctioneering Control.

3.5.1 Override Control

During the normal operation of a plant or during its start up or shutdown it is possible that dangerous situations may arise which may lead to destruction of equipment and operating personnel. In such cases it is necessary to change from the normal control action and attempt to prevent a process variable from exceeding an allowable upper or lower limit. This can be achieved through the use of special types of switches. The 'high selector switch (HSS)' is used whenever a variable should not exceed an upper limit, and the 'low selector switch(LSS)' is employed to prevent a process variable from exceeding a lower limit.

The definition of override control can be stated as follows :

'Controllers that remain inactive until a constraint is about to be exceeded at which point they take over control of the manipulated variable from the normal controller through a selector and thereby prevent the exceeding of that constraint'.

Some common applications of overrides include the prevention of :

1. Flooding in a distillation column by throttling boil up or feed flow rates.
2. Exceeding level ranges.
3. High pressure or temperature in a reactor by reducing heat input.
4. The development of low oxygen levels in furnace of gas streams by reducing fuel flow.
5. The development of high steam header pressures by diverting some of the steam to a low-pressure header or condenser.

3.5.1.1 Applications of Override Control

1. Protection of a boiler system

Usually the steam pressure in a boiler is controlled through the use of a pressure control loop on the discharge line (Loop-1 in Fig. 3.25). At the same time the water level in the boiler should not fall below a lower limit which is necessary to keep the heating coil immersed in water and thus prevent its burning out. Fig. 3.25 shows the override control system using a low selector switch (LSS). According to this system, whenever the liquid level falls below the allowable limit, the LSS switches the control action from pressure control to level control (loop-2) and closes the valve on the discharge line.

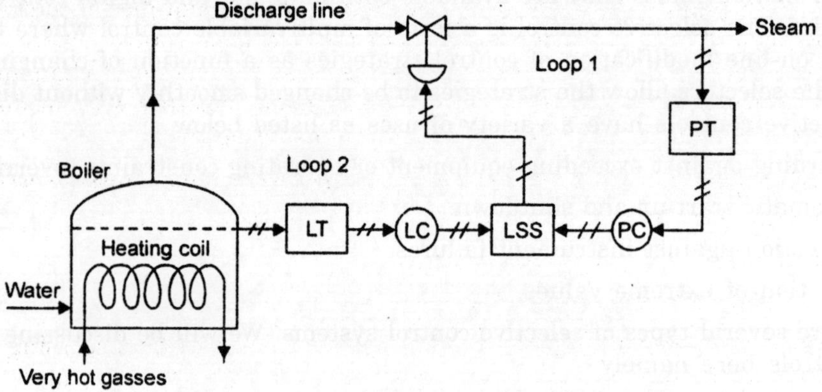

Fig. 3.25 Override control to protect a boiler system

2. Protection of a compressor system

The discharge of a compressor is controlled with a flow control system (loop-1 in Fig. 3.26). To prevent the discharge pressure from exceeding an upper limit, an override control with a high selector switch (HSS) is introduced. It transfers control action from the flow control to the pressure control loop (loop-2) whenever the discharge pressure exceeds the upper limit. Notice that flow control or pressure control is actually cascaded to the speed control of the compressor's motor.

3. Protection of a Steam distribution system

In any chemical process there is a network distributing steam, at various pressure levels, to the processing units. High pressure steam is 'letdown' to lower pressure levels at the let-down stations. The amount of steam let down at such stations is controlled by the demand on the low pressure steam line (loop-1 in Fig. 3.27) .To protect the high-pressure line from excessive pressures, we can install an override control system with a HSS, which transfers control action from loop-1 to loop-2 when the pressure in the high-pressure lines exceeds an upper limit.

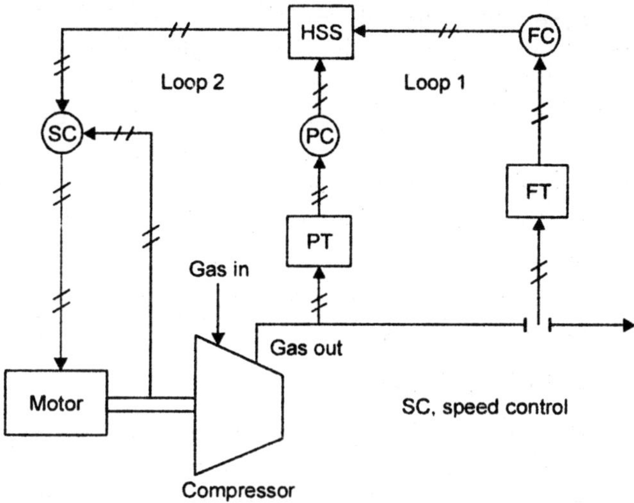

Fig. 3.26 Override control to protect a compressor

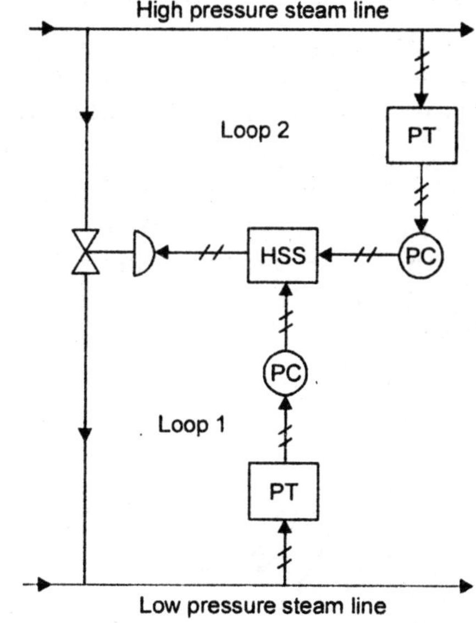

Fig. 3.27 Override control for steam distribution system

4. Start-up and shutdown controls

Some process start-ups are too complex to be accomplished by the operator without the aid of the some automatic start-up controls. Automatic start-up controls typically include a ramping signal to open valves and override controls to prevent exceeding of constraints. Signal selectors decide which control signal should manipulate each valve.

5. Manipulated variable limitations

In some control applications there can be one controlled variable and a choice of manipulated variables. A typical example is the firing of a process heater with either of two fuels. The choice between the fuels is usually made on an availability basis and on cost basis. The less costlier fuel-*A* may be burned to the limit of its availability and then it is supplemented with costlier fuel-*B*. This limit may be set by an overriding controller responding to the storage capacity of fuel *A*. A successful control system incorporating this limitation must have these two features:

1. Capability of manipulating the limited variable while staying below its limit.
2. Smooth transition from one manipulated variable to the other without adversely affecting the controlled variable.

Accommodating these features requires coordinating the manipulated variables and properly weighing their effects on the process.

3.5.2 Auctioneering Control Systems

Auctioneering control configurations select among several similar measurements the one with the highest (or lowest) value and feed it to the controller. Thus it is a selective controller, which possesses several measured outputs and one manipulated input.

3.5.2.1 Applications

1. Catalytic tubular reactors with highly exothermic reactions

Several highly exothermic reactions take place in tubular reactors filled with a catalyst bed. Typical examples are hydrocarbon oxidation reactions such as the oxidation of oxylene or naphthalene to produce phthalic anhyride. Fig. 3.28 shows the temperature profile along the length of the tubular reactor. The highest temperature is called the hot spot. The location of the hot spot moves along the length of the reactor depending on the feed conditions (temperature, concentration, flow rate) and the catalyst activity. In the Fig. 3.28, three such profiles are shown. The value of the hot-spot temperature also depends on the factors listed above and the temperature and flow rate of the coolant. The control of such systems is a real challenge for the chemical engineer.

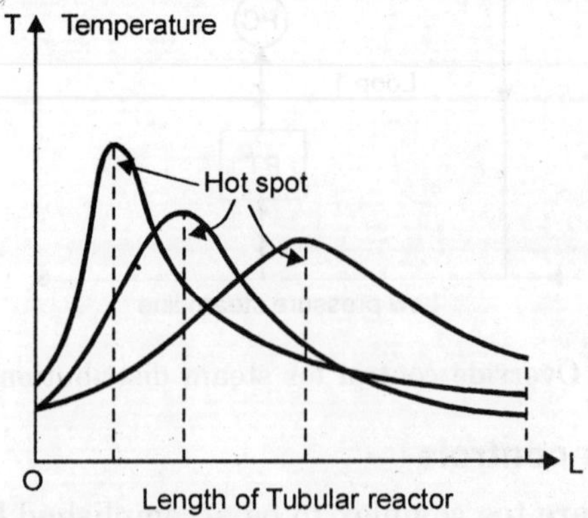

Fig. 3.28 Temperature profiles in a tubular catalytic reactor

The primary control objective is to keep the hot-spot temperature below an upper limit. Therefore, we need a control system that can identify the location of the hot spot and provide the proper control action. This can be achieved through placement of several thermocouples along the length of the reactor and use of an auctioneering system to select the highest temperature which will be used to control the flow rate of the coolant. (Fig. 3.29).

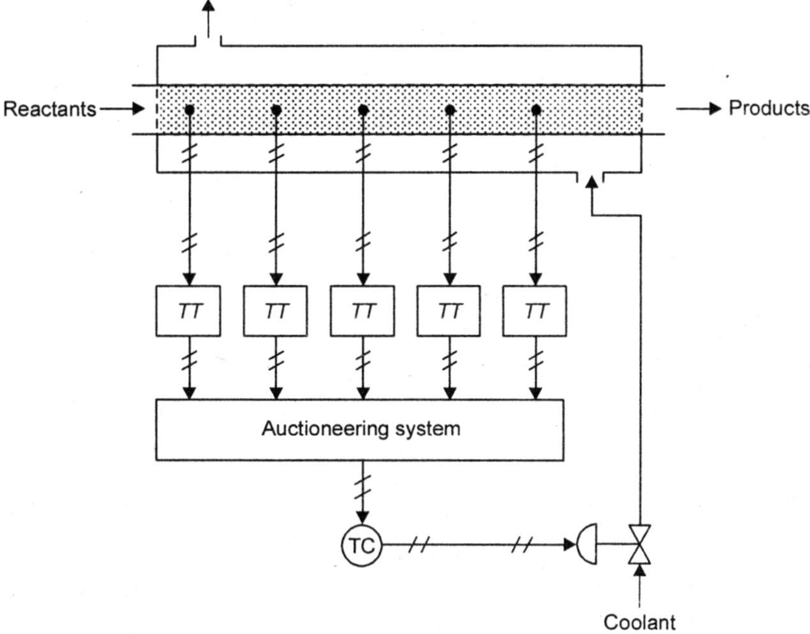

Fig. 3.29 Auctioneering control system for a tubular catalytic reactor

2. Regeneration of catalytic reactors

The catalyst in catalytic reactors undergoes deactivation as the reaction proceeds, due to carbonaceous deposits on it. It can be regenerated by burning off these deposits with air or oxygen. To avoid destruction of the catalyst, due to excessive temperature during the combustion of the deposits, we can use an auctioneering system which takes the temperature measurements from various thermocouples along the length of the reactors, selects the highest temperature that corresponds to the combustion front as it moves through the bed and controls appropriately the amount of air.

3.6 SPLIT-RANGE CONTROL

The split-range-control configuration has one measurement only (one controlled output) and more than one manipulated variable. Since there is only one controlled output, we have only one control signal, which is thus split into several parts, each affecting one of the available manipulations. In other words, we can control a single process output by coordinating the actions of several manipulated variables, all of which have the same effect on the controlled output. Such systems are not very common in chemical processes but provide added safety and operational optimality whenever necessary.

3.6.1 Applications

1. Split-range control of a chemical reactor

Consider the reactor shown in Fig. 3.30, where a gas-phase reaction takes place. Two control valves manipulate the flows of the feed and the reaction product. It is clear that in order to control the pressure in the reactor, the two valves cannot act independently, but should be coordinated. Fig. 3.30(*b*) indicates the coordination of the two valves' actions as a function of the controller's output signal. Table 3.1 below may also be referred.

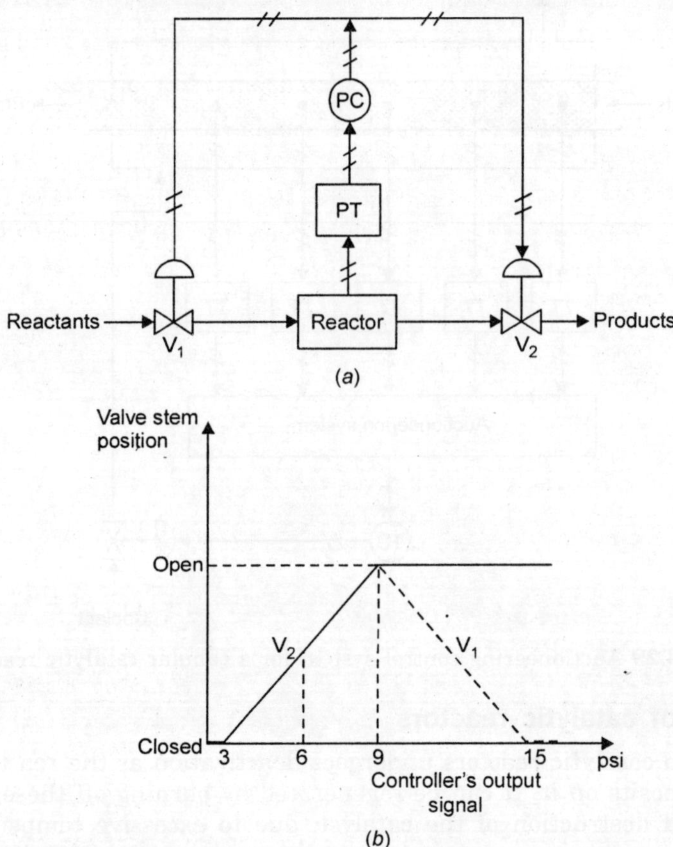

Fig. 3.30 (*a*) Reactor system with split-range control (*b*) Action of two valves

Table 3.1 Output signal and valve positions

Controller's output signal		Valve V_1	Valve V_2
Pneumatic	**Electronic**	**Stem position**	**Stem position**
3 psi	4 mA	open	closed
9 psi	12 mA	open	open
15 psi	20 mA	closed	open

Let the controller's output signal corresponding to the desired operation of the reactor be 6 psi (8 mA). From Fig. 3.30(*b*) we see that valve V_2 is partly open while valve V_1 is completely open. When for various reasons the pressure in the reactor increases, the controller's output signal also increases. Then it is split into two parts, affects the two valves simultaneously, and the following actions take place :

As the controller output increases from 6 psi (8 mA) to 9 psi (12 mA), the value V_2 open continuously, while valve V_1 remains completely open. Both actions lead to a reduction in the pressure.

For large increases in the reactor's pressure, the control output may exceed 9 psi (12 mA). In such a case, as we can see from Fig. 3.30(*b*), valve V_2 is completely open while V_1 starts closing. Both actions again lead to a reduction in pressure until the reactor has returned to the desired operation.

2. Split-range control of the pressure in a steam header

In this application, several parallel boilers discharge steam in a common steam header and from there to the process needs (Refer Fig. 3.31). The control objective is to maintain constant pressure in the steam header when the steam demand at the various processing units changes. There are several manipulated variables (steam flow from every boiler) which can be used simultaneously. The Fig. shows the structure of the resulting control system. It should be noted that instead of controlling the steam flow from each boiler, we could control the firing rate and thus the steam production rate at each boiler. It is also possible in both the cases that the load sharing percentage for the boilers can be fixed depending on the present capability of the boilers. Similar structures can be developed for the pressure control of a common discharge or suction header for *N* parallel compressors.

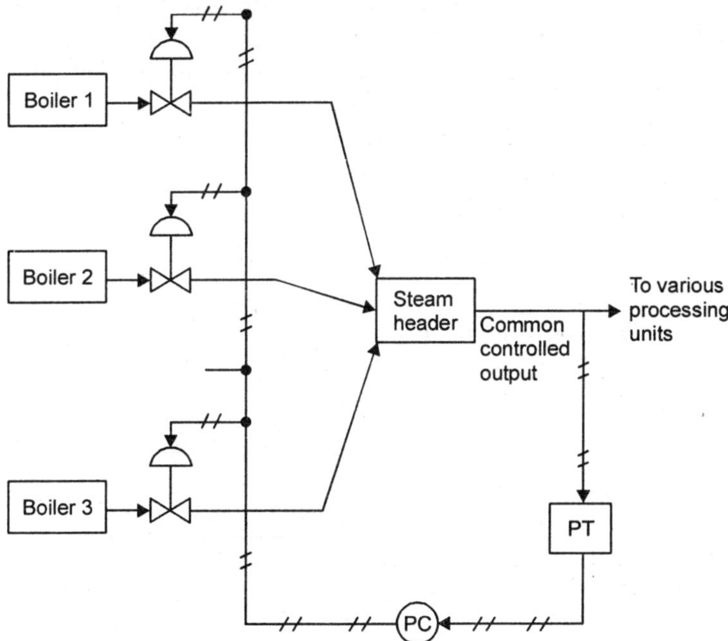

Fig. 3.31 Steam header with split-range control

3. Temperature control of a reactor vessel

Consider an application where temperature of a reactor vessel has to be controlled with the help of both cooling water and heating steam. The two manipulated variables are cooling water flow rate and steam flow rate. Initially when the reactor starts from cold condition, the steam value is fully open to provide heat energy to heat the reactor and the cooling water valve is fully closed for obvious reasons. When the reactor starts the temperature rises due to energy from steam as well as that of the self generated heat out of reaction. Both the valves remain as stated above till the temperature reaches the set point (the controller output remains at 0%). When the temperature exceeds set value, the steam valve starts closing as the controller output increases and closes fully when the controller output reaches 50%. Till such time the cooling water valve remained closed. It starts opening at 50% of controller output and opens fully at 100% of controller output. In practice the valve operates anywhere between full close (50% controller output) to full open (100% controller output) to control the temperature. Due to some reasons if the temperature falls below set value and the controller output goes below 50%, the cooling water valve comes to full close position and remains whereas the steam valve starts opening.

4. Distillation column pressure control with inerts present

In some separation processes the problem of pressure control is complicated by the presence of large percentages of inert gases. The non condensables must be removed, or they will accumulate and blanket off the condensing surface, thereby causing loss of column pressure control. Fig. 3.32 shows a schematic diagram to overcome the above problem using split control technique. As the inert or non condensables build up in the condenser, the pressure controller (PRC-1) will tend to open the control valve (PCV-1) to maintain the proper rate of condensation. The controller output signal could also be used to operate a purge control valve (PCV-2), as the opening of PCV-1 passes a certain operating point. This could be done either by means of a calibrated valve positioner or by use of second pressure controller (PIC-2).

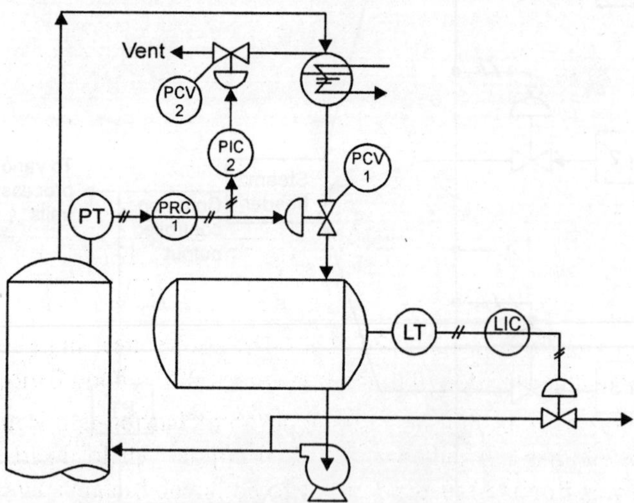

Fig. 3.32 Column pressure control with inerts present

5. Furnace recuperator temperature control

Recuperator in a furnace is an energy saving equipment. The flue gas coming out of a furnace after combustion normally carries lot of heat energy. Recuperator makes use of this energy to preheat the combustion air required for combustion in the furnace with the help of heat exchanger. Combustion air flows through a bundle of pipes in the recuperator where the flue gas passes through. The flue gas loses its heat to combustion air in the process and comes out to the atmosphere through induced draft fan (ID fan) and chimney as waste gas. The flue gas temperature has to be controlled (by spraying water) before reaching recuperator to avoid burning of recuperator tubes because of high flue gas temperature.

The flue gas temperature control used for the above purpose will have a controller and two control valves. That is, the controller output will operate two control valves in a split control mode. Here the application is little different from the cases already discussed. Normally the cooling water pipe line where the first valve or small valve is fixed will be able to deliver 10% of the cooling water flow rate with 100% open to that of the second valve or large valve in a bigger pipe line with 100% open. When the controller output is 0% (Flue gas temperature is less than set value) both the valves will remain closed. When the temperature of flue gas rises, the controller output increases, and the smaller valve starts opening. It will reach 100% opening for 50% of controller output. Till such time the large valve will remain closed. When the controller output increases about the 50%, the larger valve starts opening with the smaller one remaining at 100% open. When the output reaches 100%, the larger valves open 100%. That means, when the temperature difference is less, the control takes place only with the help of small valve. If the error grows high, the large valve comes into operation as the small one has exhausted its capability (Refer Fig. 3.34).

3.6.2 Control Valve Positioning and Sequencing

It was common to use positioners to accomplish 'split-ranging'. That is, with two valves controlled by one controller, each positioner moved its valve for only a part of the controller output range. One typical application is for temperature control with the cooling valve full open at 0% controller signal and closed at 50% signal and the heating valve beginning to open at 50% controller signal and fully open at 100%. Current preferred practice is to use a fixed-gain-plus-adjustable-bias relay to convert the controller output signal. With this approach, maintenance is simplified because standard calibrations are used and full positioner accuracy is retained. Fig. 3.33 and Table 3.2 may be referred for the above said temperature control.

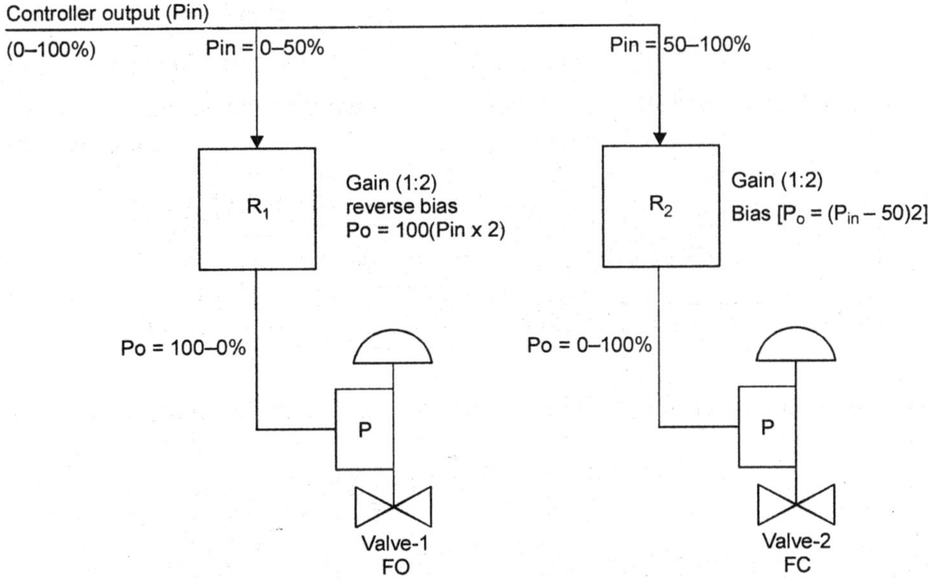

Fig. 3.33 Split-ranging two control valves with signal relays

Table 3.2 Controller output and valve positions in %

Controller output	Output relay-1	Output relay -2	Stem Position valve -1	Stem position valve-2
0	100	0	100	0
25	50	0	50	0
50	0	0	0	0
75	0	50	0	50
100	0	100	0	100

When the rangeability requirements of the process exceed the capabilities of a single valve, the control valve sequencing becomes necessary. In most cases where more than one valve is required to increase rangeability, the needed characteristic is equal-percentage. In order to have the two valves act as one without disturbing the smooth equal-percentage characteristics at transition, only one valve must open at any one time.

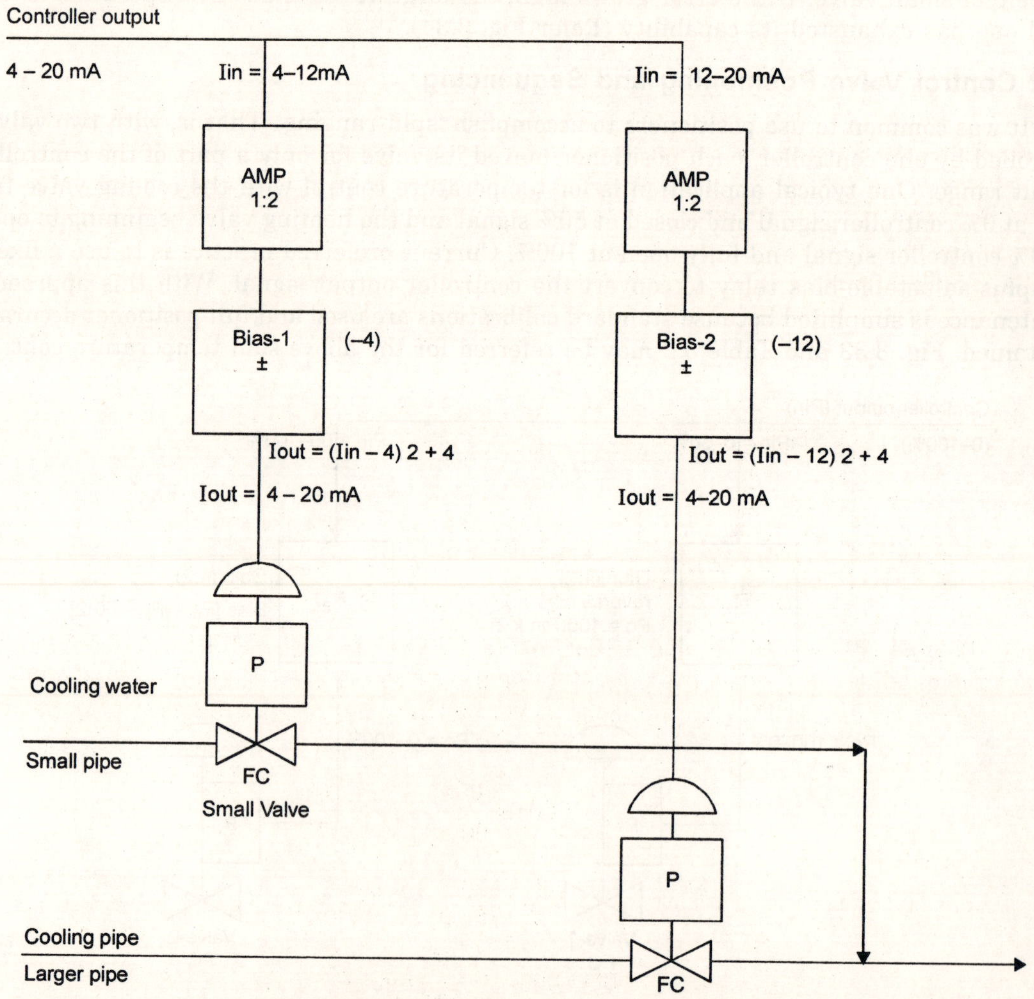

Fig. 3.34 Split-ranging loop-valve sequencing

Table 3.3 Controller output in mA and valve positions in %.

Controller output (mA)	Output bias-1	Output bias -2	Stem position valve -1	Stem position valve-2
4	4	(–12) 4 } suppressed	0	0
8	12	(–4) 4 }	50	0
12	20	4	100	0
16	20 } Saturated	12	100	50
20	20 }	20	100	100

The valve sequencing method for a particular application (application 5 of section 3.6.1-furnace recuperator temperature control) is illustrated in Fig. 3.34 and Table 3.3. The 'split-ranging' loop shown contains gain-plus-bias circuits. The controller under consideration is an electronic one with standard 4–20 mA output. The amplifiers and biases used are also matching ones to 4–20 mA standard. The control valve positioner also operates the control valve as per standard (0% for 4 mA and 100% for 20 mA). As explained earlier the small valve operates when the error is less and the cooling water flow rate supplied by the small pipe lines is sufficient to keep the temperature under control. If the error magnifies, the cooling water supplied by the small line with 100% opening of the valve may not be sufficient. Then with small valve remaining at 100%, the large valve starts opening to control the temperature.

As the purpose of valve sequencing is to increase the rangeability of the loop without upsetting its stability, the existence of more than one valve should not be noticeable by the controller. In other words, the controller would operate as if its final control element were a single valve, having the desired gain characteristics and a very wide rangeability. In order to keep the gain characteristics of the valve-pair correct, there should be no 'bumps' when the large valve is opened. This requires that only one valve be operated at a time and the other be either full closed or fully opened.

3.7 ADAPTIVE CONTROLS

Adaptive control system is one which can adjust its parameters automatically in such a way as to compensate for variations in the characteristics of the process it controls. The various types of adaptive control systems differ only in the way the parameters of the controller are adjusted. Although the basic objectives and functions can be easily described in a qualitative manner, their practical implementation is rather complicated, and involves extensive computations. In this section we will be discussing the basic logic giving examples of its practical applications.

Most processes, especially chemical processes, are non linear in some aspects. The process gain may vary with load (for example, the gain of heat transfer processes will drop with rising load) or can change with time, that is non stationary (for example, because of dirt or coating build up). The dead time and with it the period of oscillation can also vary (for example, the transportation lag or dead time represented by a reactor jacket increases as the water flow rate through the jacket drops). All the above aspects require the settings (P, PI or PID settings) of the controller revised. Adaptive control involves automatically detecting the changes that occur in the gain or dead time (period) of the process and readjusting the PID control mode settings, thereby adapting the tuning of the loop to the changing conditions.

Adaption can be based on the inputs (loads) entering the process (called as 'programmed' or 'feedforward' adaption) or can be based on the behaviour of the controlled variable (called as 'self-adaption' or 'feedback' adaption). The criteria for adaption specified in the steady state is called 'steady-state adaptive (or optimizing) control'. When the criteria for modifying the tuning constants is the damping of the controlled variable after an upset, the method of adaption is called 'dynamic adaption'. Now we may redefine the adaptive control system as below :

'An adaptive control system is one whose parameters are automatically adjusted to meet corresponding variations in the parameters of the process being controlled in order to optimize the response of the control loop'.

If the process nonlinearities are compensated by controller function (*i.e.*, changing controller gain, integral or derivative tuning parameters), the controller is a non linear controller. Non linear control will be considered here as a subset of adaptive control.

3.7.1 Programmed or Scheduled Adaptive Control

Where a measurable process variable produces a predictable effect on the gain of the control loop, compensation for its effect can be programmed into the control system. Suppose that the process is well known and that an adequate mathematical model for it is available. If there is an auxiliary process variable which correlates well with the changes in process dynamics, we can relate ahead of time the 'best' values of the controller parameters to the value of the auxiliary process variable. Consequently, by measuring the value of the auxiliary variable we can 'schedule or program' the adaption of the controller parameters. Fig. 3.35 shows the block diagram of a programmed adaptive control system. We notice that it is composed of two loops. The inner loop is an ordinary feedback control loop. The outerloop includes the parameter adjustment (adaptation) mechanism and it is comparable to feedforward compensation.

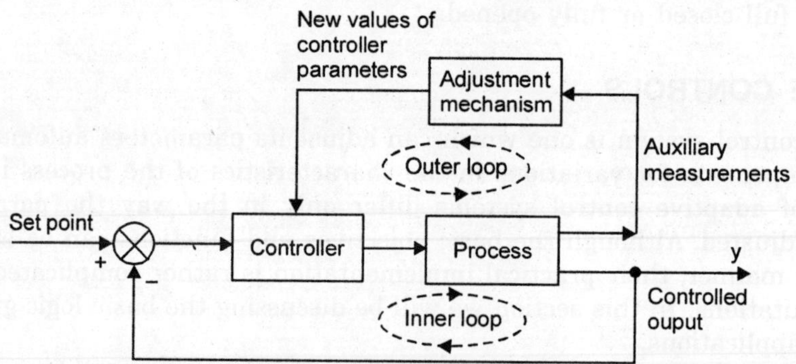

Fig. 3.35 Programmed adaptive control system

3.7.1.1 Applications

1. Gain scheduling adaptive controller

In a normal feedback control loop shown in Fig. 3.36, the control valve or another of its components may exhibit a non linear character. In such a case the gain of the non linear component will depend on the current steady state. Suppose that we want to keep the total gain of the overall system constant. From Fig. 3.36 we find easily that the open-loop gain is given by

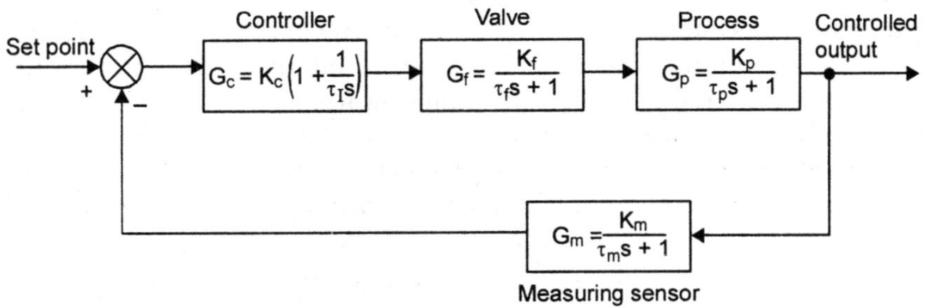

Fig. 3.36 Feedback control loop

$$K_{\text{overall}} = K_P\, K_m\, K_C\, K_f = \text{constant}$$

It is clear then that as the gain K_f of the nonlinear valve changes the gain of the controller, K_C, should change as follows :

$$K_C = \frac{\text{constant}}{K_P K_m K_f} \tag{3.1}$$

We assume that the gains K_P and K_m are known exactly. Furthermore, if the characteristics of the control valve are known well, then its gain, K_f can be calculated from the stem position. Therefore, by measuring the stem position (auxiliary measurement) we can compute K_f. Then equation (3.1) yields the adaptation mechanism of this simple gain scheduling adaptive controller. Fig. 3.37 shows the resulting control structure. Notice that the gain scheduling is comparable to feedforward compensation. There is no feedback to compensate for incorrect adaptation.

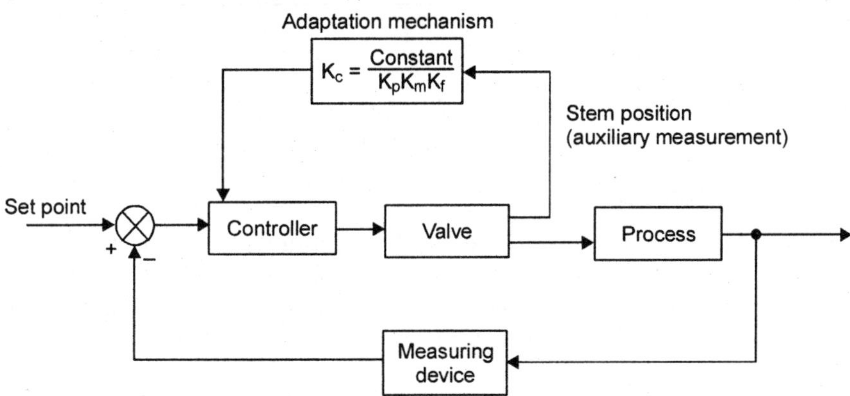

Fig. 3.37 Gain scheduling adaptive control

2. Programmed adaptive control of a combustion system

Consider a burner where the fuel/air ratio is kept at its optimal value to achieve the highest efficiency of combustion. Excess fuel or air will reduce the efficiency. The optimal fuel /air ratio is maintained through a ratio control mechanism as shown in Fig. 3.38(*a*). The optimal value of the fuel /air ratio, which maximizes the combustion efficiency, depends on the conditions prevailing within the process (for example, the temperature of the air). Consequently, as the temperature of the air changes, so does the optimal value of the fuel/air ratio.

From previous experimental data we know how the optimal fuel /air ratio changes with air temperature for maximum efficiency. Therefore to maintain the ratio continuously at its optimal value despite any changes in the air temperature, we can use a programmed adaptive control system. Such a system is shown in Fig. 3.38(*b*). It measures the temperature of the air (auxiliary measurement) and adjusts the value of the fuel /air ratio. Notice again the ratio adjustment mechanism is like feedforward compensation.

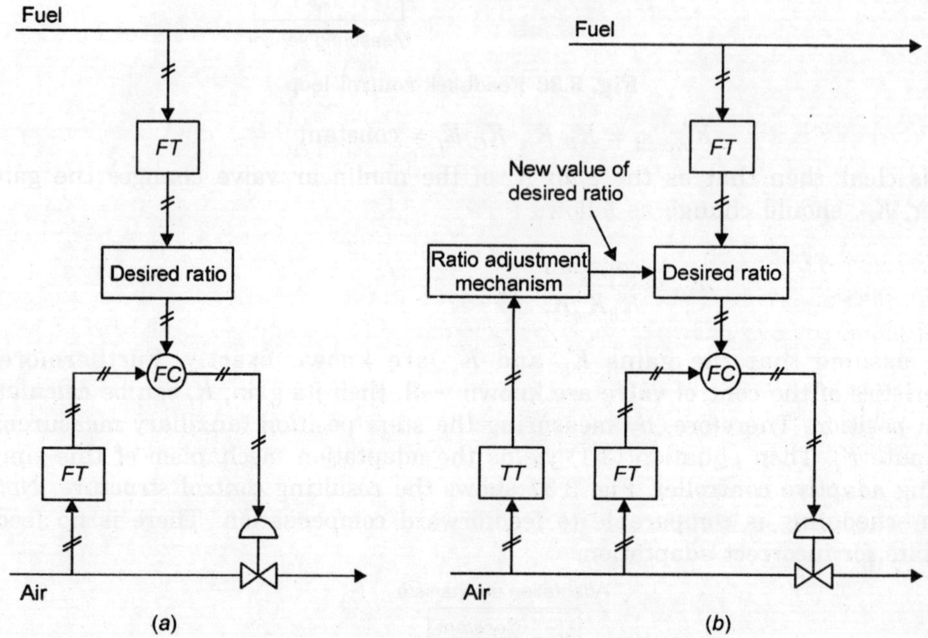

Fig. 3.38 (*a*) Ratio controller for a combustion system
(*b*) Corresponding ratio adjustment mechanism

3.7.2 Self-Adaptive Control

If the process is not known well, we need to evaluate the objective function on-line (while the process is operating) using the values of the controlled output. Then adaption mechanism will change the controller parameters in such way as to optimize (maximize or minimize) the value of the objective function (criterion).

3.7.2.1 Applications

1. Model-reference adaptive controller (MRAC)

Fig. 3.39 illustrates a different way to adjust the parameters of the controller. We postulate a 'reference model' which tells us how the controlled process output ideally should respond to the command signal (set point). The model output is compared to the actual process output. The difference (error–e_m) between the two outputs is used through a computer to adjust the parameters of the controller in such a way as to minimize the integral square error :

$$\text{Minimised ISE} = \int_0^t [e_m(t)]^2 \, dt \qquad (3.2)$$

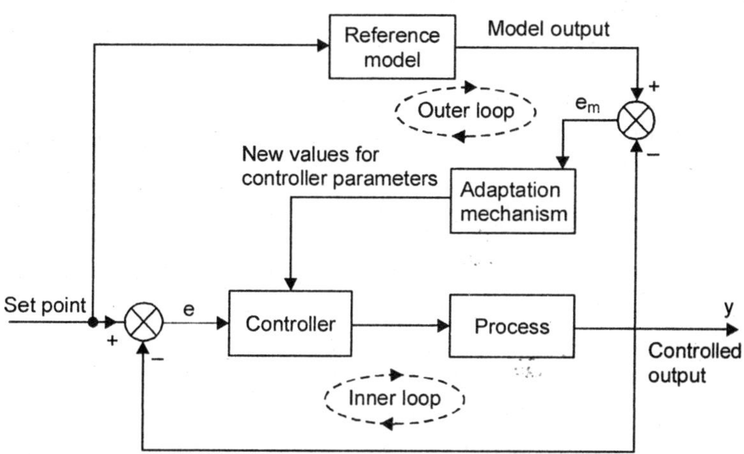

Fig. 3.39 Model-reference adaptive control

The model chosen by the control designer for reference purposes is to a certain extend arbitary. Most often a rather simple linear model is used.

We notice that the model-reference adaptive control is composed of two loops. The inner loop is an ordinary feedback control loop. The outer loop includes the adaptation mechanism and also looks like a feedback loop. The model output plays the role of the setpoint while the process output is the actual measurement. There is a comparator whose output is the input to the adjustment mechanism. The key problem is to design the adaptation mechanism in such a way as to provide a stable system (That is to bring the error–e_m to zero).

2. Self-tuning regulator (STR)

Consider the block diagram of Fig. 3.40. It represents the structure of a self-tuning regulator, which constitutes another way of adjusting the parameters of a controller. The STR is composed, again, of two loops. The inner loop consists of the process and an ordinary linear feedback controller. The outer loop is used to adjust the parameters of the feedback controller and is composed of a recursive parameter estimator and an adjustment mechanism for the controller parameters.

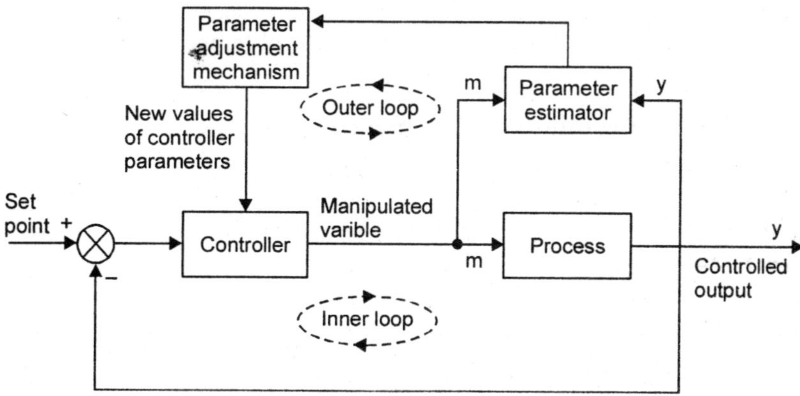

Fig. 3.40 Self-tuning regulator

The parameter estimator assumes a simple linear model for the process.

$$\frac{\overline{y}(s)}{\overline{m}(s)} = \frac{K_P e^{-t_d s}}{\tau s + 1} \tag{3.3}$$

Then, using measured values for the manipulated variable 'm' and the controlled output y, it estimates the values of the parameters K_P, τ and t_d, employing a least-squares estimation technique. Once the values of the process parameters K_P, τ and t_d, are known, the adjustment mechanism can find the 'best' values for the controller parameters using various design criteria, such as phase or gain margins, integral of the squared error, etc.

Both the parameter estimator and the adjustment mechanism require involved computations. For this reason the STR can be implemented only through the use of digital computers.

The range of adaptive control systems' applicability has expanded with the introduction of digital computers for process control. Most of the adaptive control systems require extensive computations for parameter estimation and optimal adjustment of controller parameters, which can be performed on-line only by digital computers.

3.8 INFERENTIAL CONTROL

Inferential control is one in which the desired parameter is controlled not by directly measuring but inferring from another linked parameter.

The reasons for giving in for such a control may be due to :

1. Controlled variable cannot be measured directly.
2. It is too difficult to measure economically.
3. No reliable measurement is possible.

One has to select a secondary parameter which represents the desired parameter and try to control the secondary parameter with available control system. It is expected that in turn the desired parameter is controlled. If the disturbances that create the control problems can be measured and an adequate process model is available, we could use feedforward control to keep the unmeasured output at its desired value. Inferential control comes to our rescue if the disturbances also cannot be measured. None of the control configurations studied so far can be used to control an unmeasured process output in the presence of unmeasured disturbances. This is the type of control problems where inferential control is the only solution.

3.8.1 Structure of an Inferential Control System

Consider the block diagram of the process shown in Fig. 3.41, with one unmeasured controlled output, y, and one secondary measured output, z. The manipulated variable m and the disturbance d affect both outputs. The disturbance is considered to be unmeasured. The transfer functions in the block diagram indicate the relationships between the various inputs and outputs, and they are assumed to be perfectly known. We can easily derive the following input-output relationships :

$$\overline{y} = Gp_1 \overline{m} + Gd_1 \overline{d} \tag{3.4}$$

$$\overline{z} = Gp_2 \overline{m} + Gd_2 \overline{d} \tag{3.5}$$

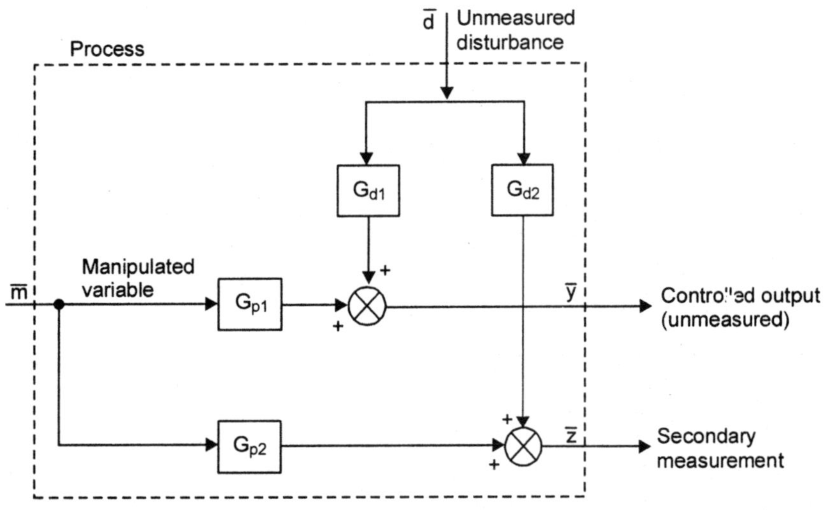

Fig. 3.41 Process with need for inferential control

From an equation (3.5) we can solve with respect to $\bar{d}$ and find the following estimate of the unmeasured disturbance :

$$\bar{d} = \frac{1}{Gd_2}\bar{z} - \frac{Gp_2}{Gd_2}\bar{m} \tag{3.6}$$

substitute the estimate above into equation (3.4) and find the following relationship :

$$\bar{y} = \left[Gp_1 - \frac{Gd_1}{Gd_2}Gp_2 \right]\bar{m} + \frac{Gd_1}{Gd_2}\bar{z} \tag{3.7}$$

The equation (3.7) provides the needed estimator which relates the unmeasured controlled output to measured quantities like m and z. Fig. 3.42 shows the structure of the resulting inferential control system. Notice that the estimated value of the unmeasured output plays the same role as a regular measured output ; that is, it is compared to the desired set point and the difference is the actuating signal for the controller.

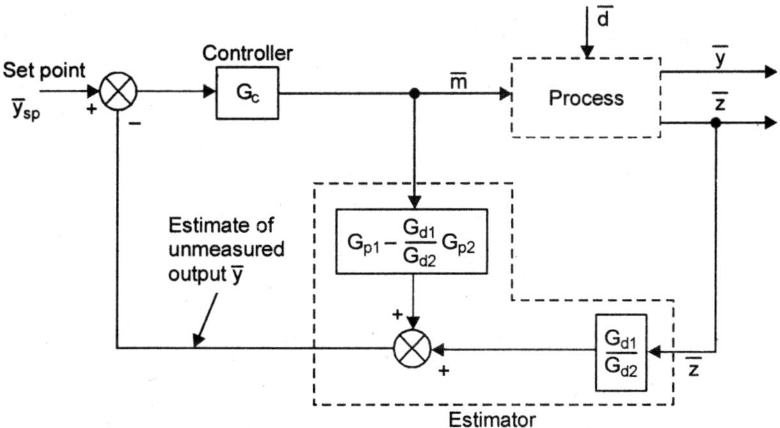

Fig. 3.42 Corresponding inferential control system

The inferential control will perfectly function provided :

1. The process transfer functions Gp_1, Gp_2, Gd_1 and Gd_2 are perfectly known.
2. The perfect estimator is available.

In chemical process control the variable that is most commonly inferred from secondary measurements is composition. This is due to the lack of reliable, rapid, and economical measuring devices for a wide spectrum of chemical systems. Thus inferential control may be used for the control of chemical reactors, distillation columns, and other mass transfer operations such as driers and absorbers. Temperature is the most common secondary measurement used to infer the unmeasured composition.

3.8.2 Application : Distillation Column's Overhead Composition Control

When the process transfer functions are only approximately known (which is usually the case), the inferential scheme provides control of varying quality, depending on how well the process is known. Let us consider for our discussions the control of overhead composition in a distillation column. The feed composition is the unmeasured disturbance and the control objective is to maintain the overhead composition. The reflux ratio is the manipulated variable. Since the feed and overhead compositions are considered unmeasured, we can only use inferential control. The secondary measurement employed to infer the overhead composition is the temperature at the top tray. Top tray temperature is normally selected because it will reflect better the condition of the overhead product.

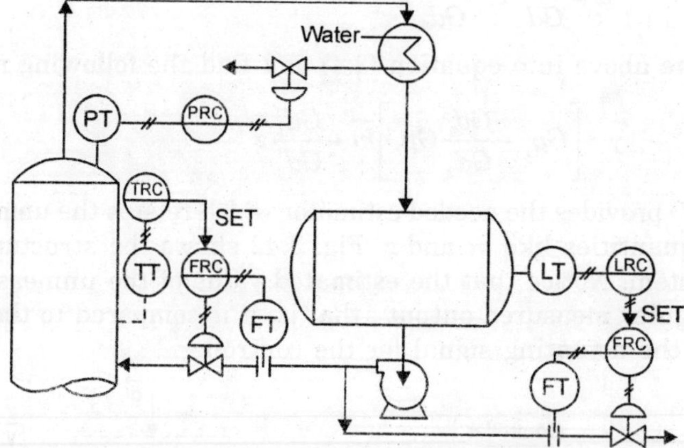

Fig. 3.43 Temperature cascaded reflux flow for improved overhead compositon control

The composition of the distillation is controlled indirectly by controlling the temperature, that is, by maintaining the temperature constant . But, the temperature is an indication of composition only when column pressure remains constant. (or if the temperature measurement is pressure compensated). Hence the point of column pressure control is near the temperature control point. Since distillation separates materials according to their difference in vapour pressures and since vapour pressure is a temperature-controlled function, temperature measurement has historically been used to indicate composition. If the feed composition is relatively constant, we can use a control system shown in Fig. 3.43 which is a temperature cascaded reflux flow control. The column pressure is controlled independently for the reasons discussed above.

3.9 CONTROL OF SYSTEMS WITH 'LARGE DEAD TIME'

Section 1.9.2 may also be referred where dynamic systems with deadtime has been discussed fundamentally. If you consider a general feedback control system, all the dynamic components of the loop may exhibit significant time delays in their response. The reasons for such delays are :

1. The main process may involve transportation of fluids over long distances or include phenomena with long incubation periods.
2. The measuring device may require long periods of time for completing the sampling and the analysis of the measured output (a gas chromatograph is such a device).
3. The final control element may need some time to develop the actuating signal.

In such situations, a conventional feedback controller would provide quite unsatisfactory closed-loop response, for the following reasons :

1. A disturbance entering the process will not be detected until after a significant period of time.
2. The control action that will be taken on the basis of the last measurement will be inadequate because it attempts to regulate a situation (eliminate an error) that originated a while back in time.
3. The control action will also take sometime to make its effect felt by the process.
4. As a result of all the factors noted above, significant dead time is a significant source of instability for closed-loop responses.

For example, let us consider the open-loop transfer function

$$G_{OL} = \frac{K_p e^{-t_d s}}{0.5s+1} \tag{3.8}$$

The effect of deadtime on cross over frequency and ultimate gain is plotted as shown in Fig. 3.44 . We see that as the dead time of an open-loop transfer function increases, the following two undesirable effects take place :

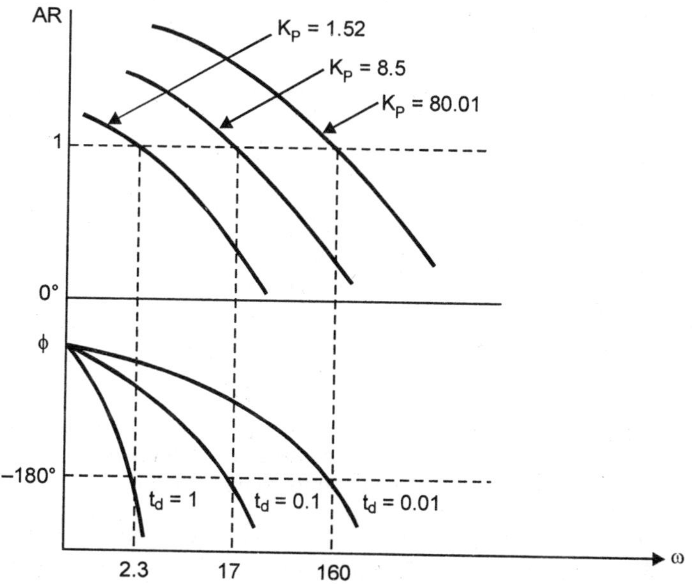

Fig. 3.44 Effect of dead time on crossover frequency

1. The crossover frequency decreases as dead time t_d increases. This implies that the closed-loop response will be sensitive even to lower-frequency periodic disturbances entering the system.
2. The ultimate gain also decreases as the dead time t_d increases. Therefore, to avoid the instability of the closed-loop response, we must reduce the value of the proportional gain K_P, which leads to sluggish response.

It is clear that a control system different from the conventional feedback is needed to compensate for dead time effects.

3.9.1 Dead-Time Compensation

Smith proposed a modification of the classical feedback control system for the compensation of dead-time effects. It is known as the 'Smith predictor' or 'the dead time compensator' and is considered for our further discussions.

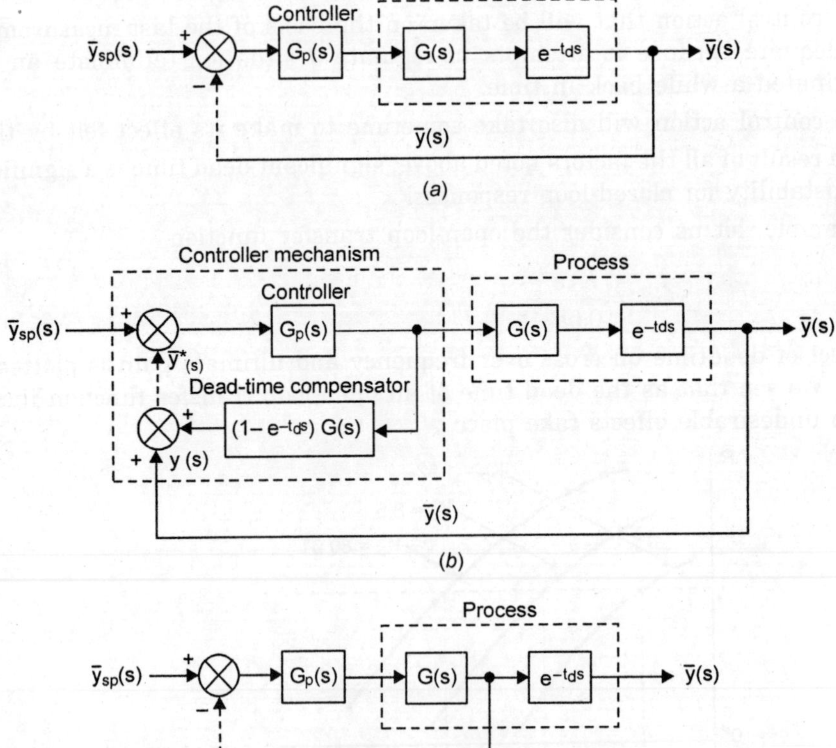

Fig. 3.45 (*a*) Feedback system with process dead time
(*b*) Feedback with complete dead-time compensation
(*c*) Net result of dead-time compensation

Consider the simple feedback loop with set point changes only as shown in Fig. 3.45(*a*). Assuming all the deadtime is caused by the process :

$$Gp(s) = G(s)e^{-t_d s}$$

and that for simplicity, $Gm(s) = Gf(s) = 1$, the open loop response to a change in the point-point can be written as :

$$\bar{y}(s) = G_P(s)\left[G(s)e^{-t_d s}\right]\bar{y}_{SP}(s) \tag{3.9}$$

In order to eliminate the undesired effects, we would like to have an open-loop feedback signal that carries current and not delayed information, such as

$$\overline{y^*}(s) = G_P(s)G(s)\bar{y}_{SP}(s) \tag{3.10}$$

This is possible if in the open-loop response $\bar{y}(s)$ we add the quantity

$$\bar{y}'(s) = \left(1 - e^{-t_d s}\right)G_P(s)G(s)\bar{y}_{SP}(s) \tag{3.11}$$

It is easy to verify that

$$\bar{y}'(s) + \bar{y}(s) = \overline{y^*}(s)$$

The implication of adding $\bar{y}'(s)$ to the signal $\bar{y}(s)$ is shown in Fig. 3.45(*b*). There we notice that the signal $\bar{y}'(s)$ can be taken by a simple local loop around the controller, which is called the 'dead-time compensator' or 'smith predictor'. The simplified loop of Fig. 3.45(*c*) is completely equivalent to that of Fig. 3.45(*b*) and indicates the real effect of the dead-time compensator. It moves the effect of dead time outside the loop.

The block diagram of Fig. 3.45 (*c*) is meant to give only a schematic representation of what is the effect of the dead-time compensator, not to depict physical reality. The dead-time compensator predicts the delayed effect that the manipulated variable will have on the process output. This prediction led to the term 'Smith predictor' and it is possible only if we have a model for the dynamics of the process (transfer function, dead time).

The dead-time in a chemical process is usually caused by material flows. Since the flow rate is not normally constant but shows variations during the operation of a plant, the value of the dead time changes. Therefore, if the dead-time compensator is designed for a certain value of the dead time, then when t_d takes on a new value the compensation will not be as effective.

3.10 INTERACTING CONTROL SYSTEMS

When two or more related variables are regulated by separate feedback controllers, the control systems interact and the interaction affects the stability of both systems. Interaction arise when a controller is used to regulate the pressure in a tank and other controllers act to change the liquid level or the flow of gas into the system. An increase in liquid level compresses the gas in the tank; so the block diagram for the pressure control system must include level changes as a load variable. Increases in pressure tend to increase the flow of liquid from the tank, and so pressure changes are load changes for the level control loop.

There is some degree of self-regulation in this example, since the higher pressure resulting from a sudden increase in level tends to increase the discharge flows of both liquid and gas. However, there are more lags to be considered in the combined system than in either system alone, and controller settings that are satisfactory for the individual loops may lead to cycling when both controllers are operating. The effect of the interaction is greatest when both loops have about the same critical frequency. Interaction exists when the pressure at the top of a distillation column and the reflux drum level are automatically controlled, but the pressure control loop is usually much faster than the level control system, and so the interaction are not serious.

An interesting example of pressure-level interaction can be noticed in a vacuum evaporator. A change of about 2mm of mercury vacuum can make a change of about 20cm of change in level. Hence slight changes in vacuum may cause saturation in the level control system. Interactions can arise when the relative humidity of air is controlled by spraying water into a chamber and the temperature is controlled by circulating the air over a heater. Interactions also occur when both top and bottom products of an absorber, distillation column, or other counter current transfer device are independently controlled. A thorough analysis of such systems requires an analog controller, though an analytical solution is possible if the system is simplified to two or three time constants.

A more serious type of interaction occurs when an attempt is made to control too many variables. The density and boiling point of the product from an evaporator should not be separately controlled, since they both depend primarily on concentration. The feed rates and flow rates of all streams leaving a process obviously cannot be controlled, since a slight error in adjusting the set points leads to an accumulation or depletion of material. These and similar situations can usually be avoided by using common sense and the phase rule.

3.11 MULTI VARIABLE CONTROL : (MVC)

As a matter of fact any reasonably complex industrial process is multivariable because many variables exist in the process and must be regulated. In general, however many of these are either non interacting or the interaction is not a serious problem in maintaining the desired control functions. In such cases, either single-variable or cascade loops suffice to effect satisfactory control of the overall process. The use of the word 'multivariable' refers to those processes wherein many strongly interacting variables are involved. Multivariable system can have such a complex interaction pattern that the adjustment of a single set point causes a profound influence on many other control loops in the process. In some cases, instabilities, cycling, or even runaway result from the indiscriminate adjustment of a few setpoints.

The control configurations we have examined so far were confined to processes with a single controlled output, requiring a single manipulated input. Such single-input, single-output (SISO) systems are very simple. Chemical processes usually have two or more controlled outputs, requiring two or more manipulated variables. The design of control systems for such multiple-input, multiple-output (MIMO) is complex and requires fairly good knowledge of the process. A MIMO controller can recognize and compensate for process interaction much more effectively than can standard SISO controllers on individual loops.

For our discussions, let us consider a general process with several inputs and outputs as shown in Fig. 3.46 which summarises all the classes of variables that one can have around an industrial process especially chemical process. Major aspects to be considered when addressing the multivariable control (MVC) problem are :

1. It is important to select the proper variable pairing for interactive feedback control.
2. Application of feedforward compensation is desirable to help minimize the required effort of interactive feedback loops reacting to process disturbance.
3. The use of selective or override controls where possible will minimize the need for interactive feedback loops.
4. Using interaction compensation will frequently help to provide a stabilising effect on interactive feedback loops and allow tighter tuning.

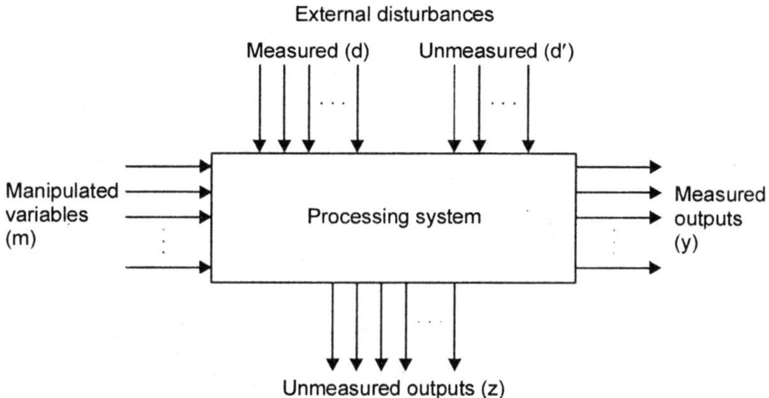

Fig. 3.46 Input and output variables around a chemical process

3.11.1 Design Questions of MVC

One must be able to find out suitable answers for the following questions before attempting to design a control system for MIMO processes.

1. How many and which of all possible variables should be controlled at desired values? (The answer identifies the objectives of the control)
2. What outputs should be measured ? Once control objectives have been identified one need to select the measurements necessary to monitor the operation of the process. Measured outputs can be classified into two categories :
   (*a*) **Primary measurements :** These are the controlled outputs through which we can determine directly if the control objectives are satisfied.
   (*b*) **Secondary measurements :** These are not used to monitor directly the control objectives but are auxiliary measurements employed for cascade, adaptive, or inferential control.
3. What inputs can be measured ? All the manipulated variables are assumed to be measurable. With respect to the disturbances only a few can be measured easily and reliably.
4. What manipulated variables should be used ? A MIMO system possesses several manipulated variables which can be used for the design of a control system. The selection of the most appropriate manipulations is a very critical problem and should be approached with care. Some manipulations have a direct, fast, and strong effects on the controlled inputs; others do not. Furthermore, some variables are easy to

manipulate in real life (like flowrates of liquids, air and fuel); others are not. (like the flow of solids, slurries etc).

5. What is the configuration of the control loops?

   Once all the possible measurements and manipulations have been identified, one need to decide how they are going to be interconnected through the control loops. In other words, what measurement will actuate a given manipulated variable or what manipulation will be used to regulate a given controlled output at its desired value?

For MIMO systems there is a large number of alternative control configurations. The selection of the most-appropriate configuration for a particular process requires the thorough knowledge of the process. We will be limiting our discussion on MVC here as the detailed design of such MIMO systems are beyond the scope of this book. Some applications of MIMO systems will be covered in chapter 5.

3.12 PROBLEMS AND SOLUTIONS

Problem 3.1

In a compound control system, the ratio between two variables is to be maintained at 4 to 1. If each has been converted to a 0–5 Volt range signal, devise a signal conditioning system that will output a zero signal to the controller when the ratio is correct.

Solution :

We may use a standard summing amplifier using op amps. The output is related to the input by

$$V_{out} = -\frac{R_f}{R_1}V_1 - \frac{R_f}{R_2}V_2$$

where R_f = Feedback resistance

R_1, R_2 = Input resistances for V_1 and V_2 respectively.

For $V_{out} = 0$,

$$0 = -\frac{R_f}{R_1}V_1 - \frac{R_f}{R_2}V_2, \quad -\frac{R_f V_1}{R_1} = \frac{R_f V_2}{R_2}$$

$$\therefore \frac{V_1}{V_2} = -\frac{R_1}{R_2}$$

One input voltage must be negative as we cannot use negative resistance. The ratio of the resistances should be 4 to 1 as required.

$$R_1 = 4\,R_2.$$

Because the gain is unspecified, we can use $R_1 = R_f$, for example. Accomplishing all the above, we can have the summing amplifier circuit as shown in Fig. 3.47.

$$Vout = -\left(\frac{4K\Omega}{4K\Omega}\right)(-V_1) - \left(\frac{4K\Omega}{1K\Omega}\right)V_2$$

or $$Vout = V_1 - 4V_2$$

Thus whenever $V_1 = 4V_2$, the output is zero.

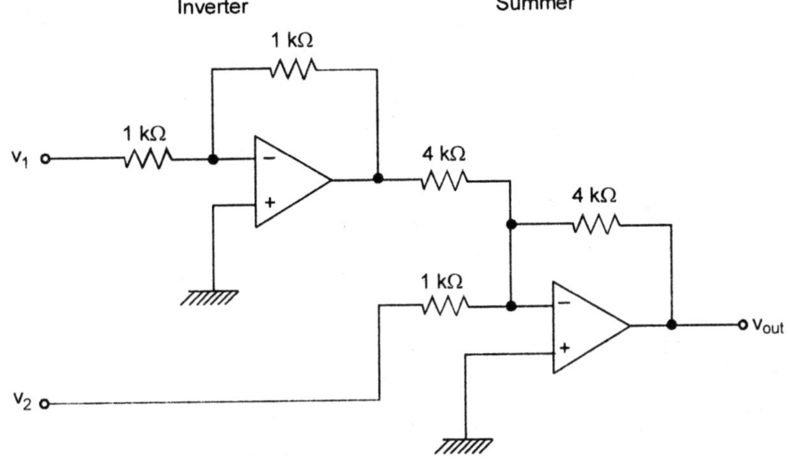

Fig. 3.47 Cicuit for example

Problem 3.2

A gas fired furnace has to be supplied with air at the ratio of 1 to 5 ± 1 (Gas:Air:1:5 ±1). If standard transmitters are to be used and the controller is also to follow the standard signal of either 4-20 mA or 1 to 5V, what should be the calibration technique to be used if the ratio is to be maintained almost through out the operating range ? (Assume orifices are the head producers for both gas and air).

Solution :

The furnace is designed for operating with Gas fuel having such a calorific value that requires about 5 times of air quantity for proper combustion. To take care of variations in calorific value time to time, a provision is to be kept for changing the ratio in the control system. In the problem under consideration the ratio is to be changed from 1:4 to 1:6. Hence while designing we should see that the head producer and transmitter should take care of the maximum ranges of air flow and gas flow. With this assumption let us try to solve this problem.

Since standard current signals are to be used in the control system, we have two solutions to offer.

Case 1: Identical Transmitters for both GAS and AIR

Here the input (differential head) to the transmitter is same for both the transmitters. In other words, the maximum differential pressure (h_{max}) applied to both the transmitters to get 20 mA (I_{max}) is same. Hence orifices design has to be done in such a manner that 'h_{max}' produced by maximum Gas flow [$G_{F(max)}$] is same as that of 'h_{max}' produced by maximum Air Flow. [$A_{F(max)}$].

$$\therefore \quad \frac{\text{Designed maximum air flow}}{\text{Designed maximum gas flow}} = 6$$

[1: 6 ratio turns out to be 1:1 current ratio so that electrically the control is possible]

Case 2 : Different transmitters : Due to practical reasons if we find it difficult to design two orifice head produces which can produce same h_{max} for flowrates of 6 times different from one another, we can follow different strategy. Orifices can be designed to have two different 'h_{max}' values, one for Gas and another for Air. The transmitters here are to be different so that they produce standard current output for different 'h_{max}' values.

[Gas transmitter (h_{max}) gas → 20 mA output, Air transmitter (h_{max}) air → 20 mA output.]

For example, if the maximum gas requirement for the furnace is 1000 Nm^3 / hr and the orifice develops 100 mm WG differential pressure at 1000 Nm^3/hr (h_{max} = 100 mm WG), the transmitter has to be calibrated for h_{max} = 100 mmWG (20 mA).

The maximum Air requirement is 6000 Nm^3/hr (as we assumed) Air line orifice develops 250 mm WG for this maximum, Air transmitter has to be calibrated for 250 mmWG Max (20 mA). When practically 1:6 ratio has to be achieved, the ratio in currents has to be maintained at 1:1. Here the transmitters are not interchangeable, whereas in case-1 it is.

Problem 3.3

If the temperature of a furnace with fuel as heat energy has to be controlled economically by maintaining proper combustion, suggest a control system utilizing cascade and ratio controls.

Solution :

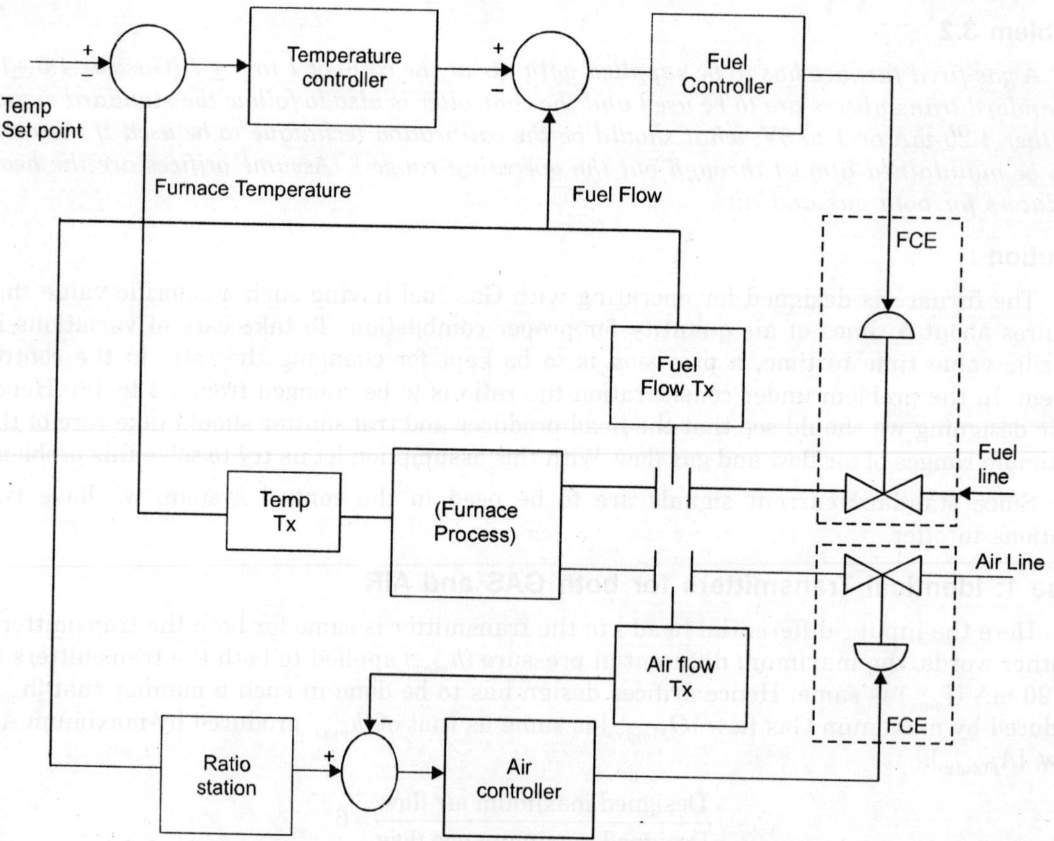

Fig. 3.48 Furnace temperature control (Refer problem 3.3)

Solution :

Let us first understand the process as follows. Furnace is to be heated up and maintained at desired temperature. If there is any deviation, the heat energy flow has to be controlled by manipulating the fuel flow rate. Fuel line pressure variations, conditions of fuel line etc are disturbances in the system which will affect the control system if it is a simple feedback system (temperature is controlled variable and fuel flow rate is manipulated variable). Hence we can have in primary loop the temperature as the controlled variable. The output of the primary controller can be given as set point to secondary controller whose manipulated variable is Fuel flow rate. Thus we can have a 'cascade control system' for temperature control. As the secondary controller takes care of unmeasured disturbances also to some extend, 'feedforward action' is also incorporated to some extent.

To take care of proper combustion in the furnace, we may have to adjust the amount of air in the furnace with required ratio in respect to fuel entering in. There has to be another controller to perform this function. This can be a Ratio controller with Fuel flow multiplied by the set ratio as set point and air flow as manipulated variable. Fuel flow signal can be used for both the controllers. A schematic diagram taking care of the above discussed points is shown in Fig. 3.48.

Problem 3.4

A split range control has been designed for a cooling operation with two control valves (one small and another large valve). The controller output varies from 4–20 mA. The small valve positioner has been designed to open from 0% to 100% when the controller output varies from 4 to 12 mA and remain open afterwards. The large valve positioner has been designed to operate for 10 mA to 20 mA from 0% to 100% (from 4 mA to 10 mA range, the valve will remain closed). Assuming both the valves are linear and maximum flow through small valve is 10% of the maximum flow through large valve, draw controller output current Vs stem position of both valves as well as that of total flow if maximum flow through large valve is 2000 m^3/hr.

Solution :

We may tabulate first the stem positions and corresponding flow rates for different controller outputs as below and then plot stem positions and total flow rate vs controller output as shown in Fig. 3.49.

Controller output	Stem position small valve %	Stem position large valve %	Flow through small valve	Flow through large valve	Total flow m^3/hr
4 mA	0	0	0	0	0
6 mA	25	0	50	0	50
8 mA	50	0	100	0	100
10 mA	75	0	150	0	150
12 mA	100	20	200	400	600
14 mA	100	40	200	800	1000
16 mA	100	60	200	1200	1400
18 mA	100	80	200	1600	1800
20 mA	100	100	200	2000	2200

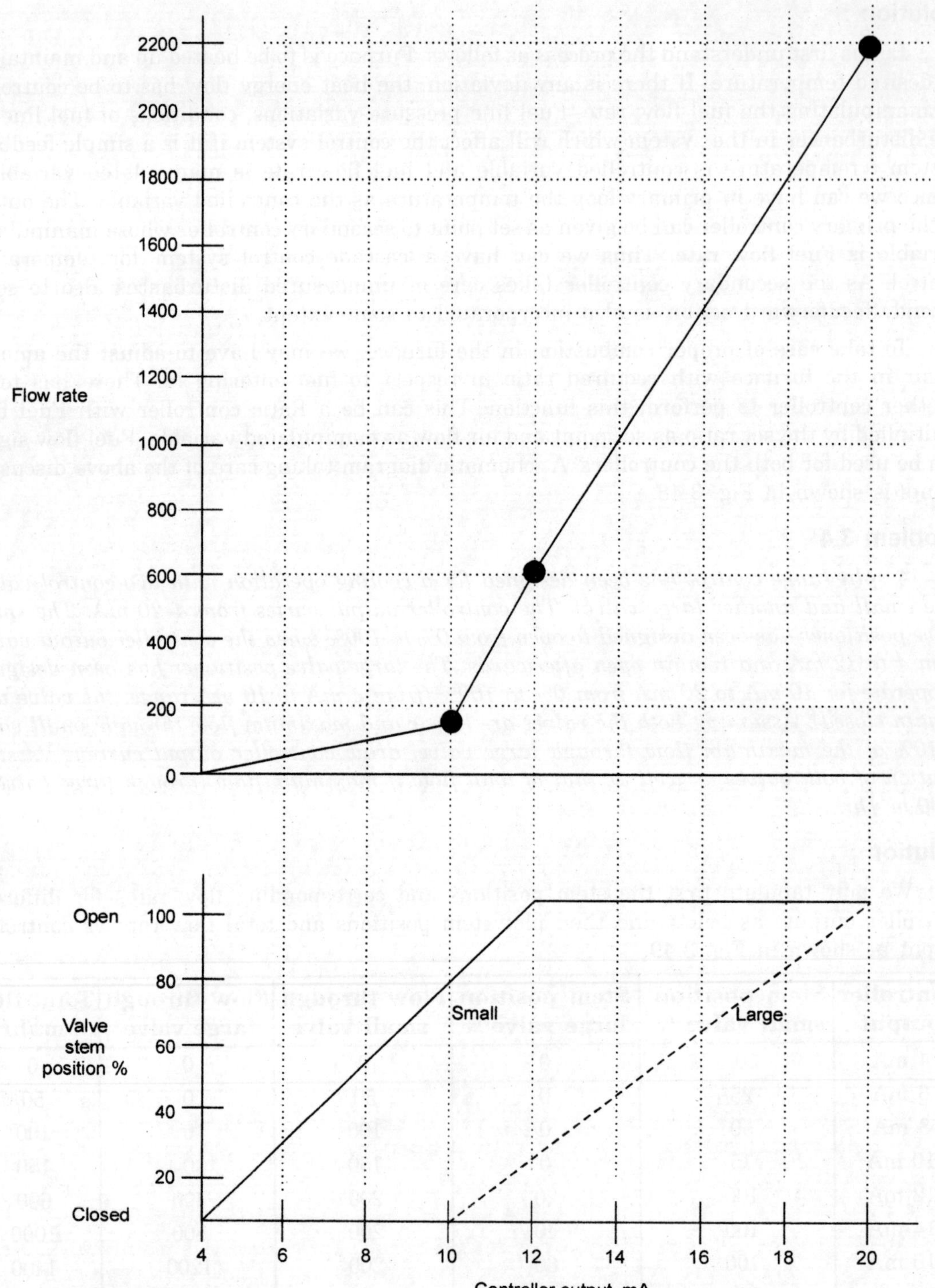

Fig. 3.49 Controller output vs stem position and flowrate (Problem 4)

3.13 PROBLEMS AND QUESTIONS

1. What is meant by feedforward control?
2. Why feedforward control is preferred along with feedback control?
3. Explain feedforward control strategy for a typical process.
4. Explain the purpose of cascade control for heat exchanger.
5. What is meant by auctioneering control?
6. Discuss the rationale of a cascade control system and demonstrate why it provides better response than simple feedback?
7. Compare the inner loop and outer loop in a cascade control system with reference to loop gain and speed of response.
8. What are the main advantages and disadvantages of cascade control? For what kind of processes can you employ cascade control?
9. What is meant by selective control systems?
10. How many types of selective control systems are available? Discuss their characteristics.
11. What is split range control? Describe a situation where you could use split range control?
12. When cascade control will give improved performance than conventional feedback control?
13. Explain feedforward control with an example from distillation column.
14. List the benefits we can expect out of feedforward control when compared with feedback controller.
15. What is ratio control? Explain with an example.
16. What for 'ratio station' is required in a ratio control loop?
17. Expand 'MVC', 'MIMO' and 'SISO'.
18. Explain the control of MIMO systems with an example.
19. A ratio control is used to keep two flow rates '*A*' and '*B*' at 1:1. The flow transmitters used are standard 4-20 mA output with square root extractors inbuilt. If transmitter for measuring '*A*' is calibrated for a maximum differential pressure (h_{max}) of 144 mmWG for maximum flow of 15,000 m^3 /hr, to what 'h_{max}' the '*B*' transmitter should be calibrated for its maximum flow 30,000 m^3/hr to avoid ratio station in the control system. (Assume identical pipe size and orifice for both '*A*' and '*B*') (Ans : 576 mmWG)
20. What are reasons for 'deadtime' in a process control?
21. How are 'large dead time' delays are compensated?
22. Why a conventional feedback control system is not suitable for processes with large deadtime? Give reasons.
23. Explain about 'Smith Predictor'.
24. Design a 'Smith predictor' (Deadtime compensator) for the process with transfer function:

$$G(s) = \frac{5e^{-0.55}}{4s+1}.$$

25. Describe Inferential Control . Under what circumstances is it recommended?
26. A split control is used to control a reactor temperature. The two manipulated variables are steam flow when temperature is below set point and coolant flow when temperature goes above set point. Draw a graph between controller output and stem position in percentage of both the control valves assuming the controller is 50% when set point matches with the temperature measured.

27. Differentiate between programmed adaptive control and self-adaptive control.
28. Give some examples of interacting control systems. How to eliminate interaction effects?
29. Gas A is sparged into a stirred tank to react continuously with B, which is dissolved in a aqueous salt solution. A temperature bulb in the tank is used to control the flow of cooling water to the jacket. A differential pressure-type level controller connected to the base of the tank adjusts the flow of liquid from the bottom of this tank. The flow of gas *A* is automatically controlled to maintain 50% conversion of B. The tank pressure is controlled at 2Kg/cm^2 by throttling the gas exit stream, which contains unreacted A plus water vapour (the normal temperature is 110°C). Draw a complete block diagram for this system, showing all interactions.
30. Draw a block diagram, and give the main transfer functions for cascade control of a continuous-flow chemical reactor. The reactor temperature is used to adjust the set point of the jacket temperature controller, which regulates the input temperature of the cooling fluid. Assume that the tank and the jacket are perfectly mixed. Show how the interaction between the jacket and kettle influence the time constants by making up a typical numerical example.

4

FINAL CONTROL ELEMENTS

4.1 INTRODUCTION

The final control element is the mechanism which alters the value of manipulated variable in response to the output signal from the automatic control device. The position of the final control element in the automatic control loop is shown in Fig. 4.1 as G_2. The final control element system often consists of three parts :

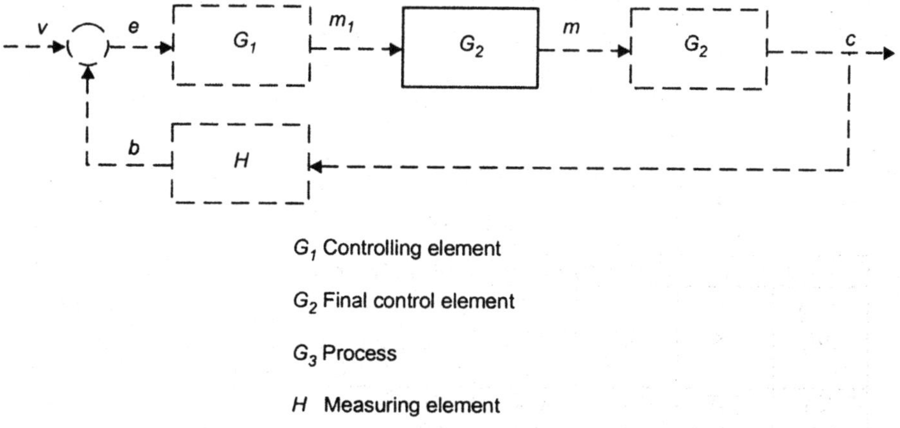

Fig. 4.1 The final control element

1. Signal converter for signal conversions (Interface)
2. Actuator : To convert the signal received from signal converter to a rotary or linear motion.
3. A device to adjust the value of the manipulated variable (control valve/control element).

For a typical process-control application the conversion of a process-controller signal to a control function can be represented by the steps shown in Fig. 4.2. The input control signal may take many forms, including an electric current (say 4–20 mA), digital signal, or pneumatic pressure. The signal converter modifies the control signal to properly interface with the next stage of control, that is, the actuator. Thus, if a valve control element is to be operated by an electric motor actuator, then a 4–20 mA dc control signal must be modified to operate the motor.

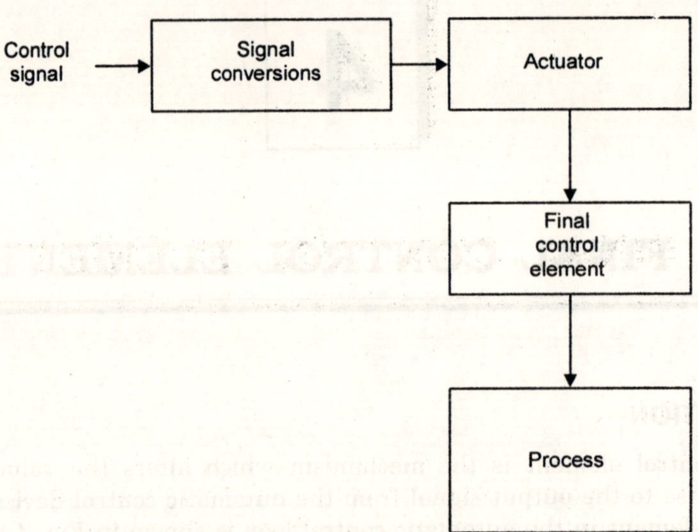

Fig. 4.2 Elements of the final control operation

The results of signal conversions provide an amplified and/or converted signal that is designed to operate (actuate) a mechanism that changes a manipulated variable and thus changing the controlled variable. The direct effect is usually implemented by something in the process such as a valve or heater that must be operated by some device. The 'actuator' is a translation of the converted control signal into action on the control element. Thus, if a valve is to be operated, then the actuator is a device that converts the control signal into physical action of opening or closing the valve.

Then comes the final control element itself. This device, normally a control valve, has direct influence on the process dynamic variable and is designed as an integral part of the process. Thus, if flow is to be controlled, then the control element, a valve, must be built directly into the flow system. Similarly, if temperature is to be controlled, then some mechanism or control element that has a direct influence on temperature must be involved in the process. This could be a heater/cooler combination that are electrically actuated by relays or a pneumatic valve to control influx of reactants.

In short, the above three parts, namely the signal converter, actuator and control element, are the hardware components of the control loops that implement the control action. They receive the output of a controller (actuating signal) and adjust accordingly the value of the manipulated variable.

4.2 SIGNAL CONVERSIONS : (SIGNAL CONVERTERS)

The principal objective of signal conversion is to convert the low-energy control signal to a high-energy signal to drive the actuator. Controller output signals are typically in one of three forms : (1) electrical current, usually 4-20 mA, (2) pneumatic pressure, usually 3–15 psi (or 0.2 to 1 Kg/cm^2), and (3) digital signals, usually *TTL* level voltages in serial or parallel format. There are many different schemes for conversion of these signals to other forms depending on the desired final form and evolving technology.

4.2.1 Analog Electrical Signals

Relays : A common conversion is to use the controller signal to activate a relay when simple on/off or two position is sufficient. In some cases where the low current signal is insufficient to drive a heavy industrial relay, an amplifier is used to boost the control signal to a level sufficient to do the job.

When motor actuator is used, the forward and reverse rotations are again controlled by relays. The controller output current is compared with the position of the valve (valve position transmitters are used to sense the stem position of the valve in terms of standard resistance or current signal)in a comparator whose 'more' or 'less' outputs are used to activate the corresponding 'more' or 'less' relays. When 'more' relay activates, the motor is driven in forward direction so that the valve opens and when 'less' relay activates, the motor is driven in reverse direction so that the valve closes. The power connections to the motor is accordingly made through those relays. Such an arrangement of comparator cum relays is called 'interface' units. (act as interface between controller and actuator).

Amplifiers : High-Power ac or dc amplifiers often can provide the necessary conversion of the low-energy control signal to a high-energy form. Such amplifiers may serve for motor control, heat control, light level control, and a host of other industrial needs.

Motor control : Many motor control circuits are designed as packaged units that accept a low-level dc signal directly to control motor speed. Such circuits are built using amplifiers and SCRs or TRIACs to perform this control.

4.2.2 Digital Electrical Signals

Mostly used along with Digital controllers of distributed control systems (DCS) and with computerized control applications. Interfaces are required to convert the digital signals to forms required by final control operations.

ON-OFF control

There are many cases in process control where the control algorithm is accomplished by simple commands to outside equipment to change speed, turn on (or off), move up, and so on. In such cases, the computer can simply load a latched output line with a '1' or '0' as appropriate. Then it is a simple matter to use this signal to close a relay or activate some other outside circuit.

Digital to analog converter : (DAC)

When the digital output must provide a smooth control, as it does in valve positioning, the computer must provide an input to a DAC that then determines an appropriate analog output. When a computer must provide outputs to many final control elements, a data output module can be employed. These integrated modules contain channel addressing, DAC, and other required elements of a self-contained output interface system.

Direct action

As the use of digital and computer techniques in process control becomes more widespread, new methods of final control have been developed that can be actuated directly by the computer. Thus, a stepping motor interfaces very easily to the digital signals that a computer outputs. In another development, special integrated circuits are made that reside within the final control element (embedded systems) and allow the digital signal to be connected directly.

4.2.3 Pneumatic Signals

In a pneumatic system, information is carried by the pressure of gas (normally air) in a pipe. If we have a pipe of any length and raise the pressure of air in one end, this increase in pressure will propagate down the pipe until the pressure throughout is raised to the new value. The pressure signal travels down the pipe at a speed in the range of the speed of sound in air, which is about 330 m/s. Thus if a transducer varies air pressure at one end of a 330 metre pipe in response to some controlled variable, then that same pressure occurs at the other end of the pipe after a delay of approximately one second. For many process-control installations, this time delay is of no consequence, although it is very slow compared to an electrical signal. This type of signal propagation was used for many years in process control before electrical/electronics technology advanced to a level of reliability and safety to enable its use with confidence. Pneumatic is till employed in many installations either because of danger to electrical equipment or as carryover from previous years, where conversion to electrical methods would not be cost effective. In general, pneumatic signals are carried with dry and clean air (called as instrument air) where signal information has been adjusted to lie within the range of 0.2 to 1 kg/cm^2 (3 to 15 psi). There are basically three types of signal conversion possible namely 1) Amplification, 2) Flapper/Nozzle system and 3) Current-to-pressure converter or simply *I/P* converter.

4.2.3.1 Amplification

A pneumatic amplifier, also called a booster or relay, raises the pressure and /or air flow volume by some linearly proportional amount from the input signal. Thus, if the booster has a pressure gain of 10, the output would be 2 to 10 kg/cm^2. This is accomplished via a regulator that is activated by the control signal. A schematic diagram of one fundamental type of pressure booster is shown in Fig. 4.3. As the signal pressure varies, the diaphragm motion will move the plug in the body block of the booster. If motion is down, the air leak is reduced and the pressure in the output line is increased. The device shown is reverse acting because a high-signal pressure will cause output pressure to decrease. Many other designs are available.

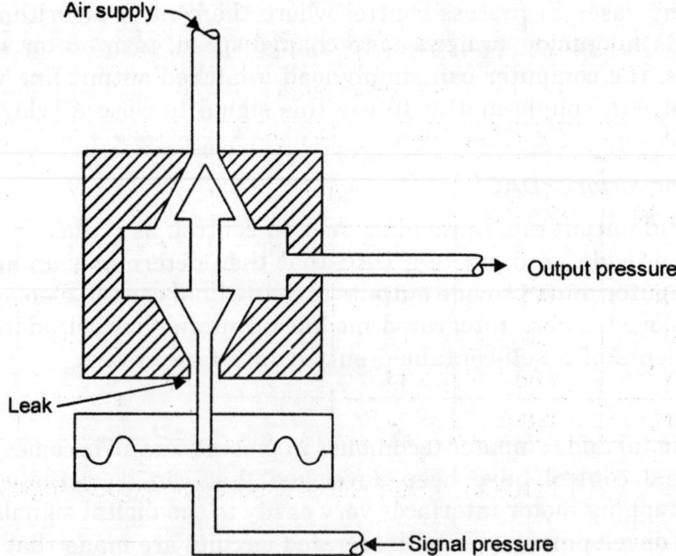

Fig. 4.3 A pneumatic amplifier or booster converts the signal pressure to a higher pressure or the same pressure but with greater air volume

4.2.3.2 Flapper/Nozzle System : (Nozzle/Baffle System)

A very important signal conversion is from pressure to mechanical motion and vice versa. This conversion can be provided by a flapper/nozzle system (sometimes called a nozzle/baffle system). A simplified diagram of this device is shown in Fig. 4.4(*a*). A regulated supply of pressure, usually over 1.5 kg/cm^2, provides a source of air through the restriction. The nozzle is open at the end where the gap exists between nozzle and flapper, and air escapes in this region. If the flapper moves down and closes off the nozzle opening so that no air leaks, the signal pressure will rise to the supply pressure. As the flapper moves away, the signal pressure will drop because of the leaking air. Finally, when the flapper is far away, the pressure will stabilize at some value determined by the maximum leakage through the nozzle. Fig. 4.4(*b*) shows the relationship between signal pressure and gap distance. Note that the great sensitivity is in the central region where the slope of the line is the greatest. In this region, the response will be such that a very small motion of the flapper can change the pressure by an order of greater magnitude. We will be coming across this flapper-nozzle system very often in this chapter.

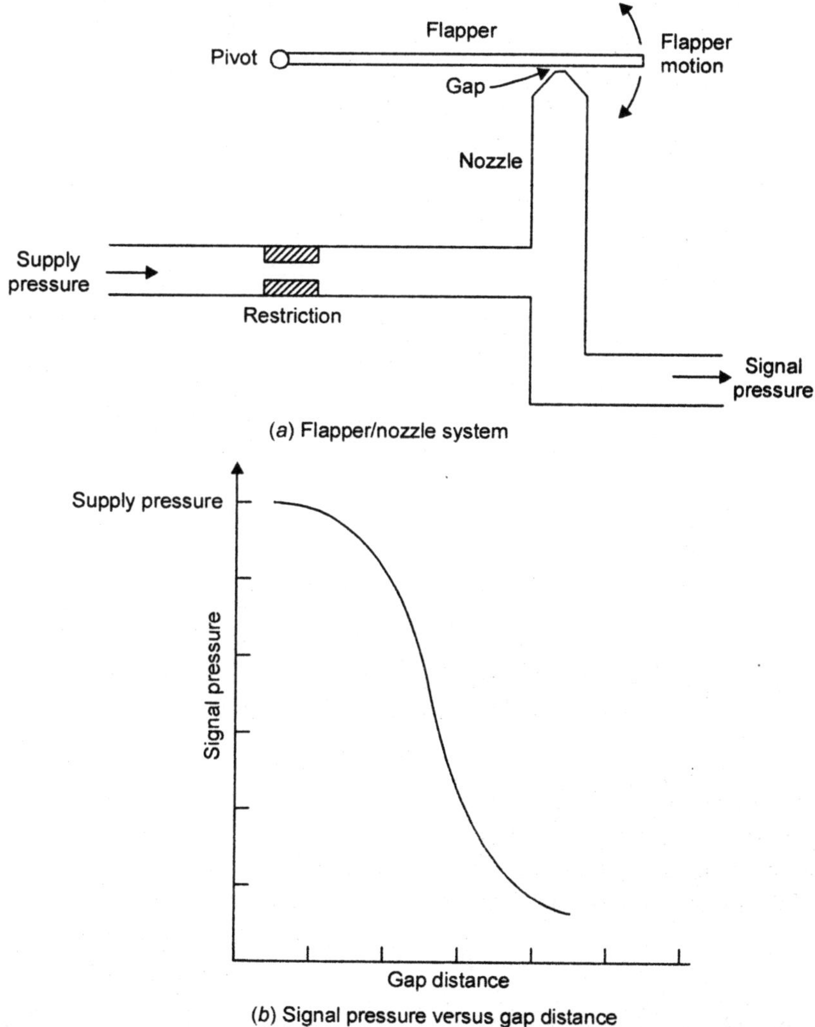

Fig. 4.4 Principles of the flapper/nozzle system

4.2.3.3 Current-to-Pressure Converters : (I/P Converter)

The current-to-pressure converter, or simply *I/P* converter, is a very important element in process control. Often, when we want to use the low -level electric current signal to do work, it is much easier to let the work to be done by a pneumatic signal. The *I/P* converter gives us a linear way of translating the 4–20 mA current into a 0.2 to 1 Kg/cm^2 signal (3 to 15 psi signal). There are many designs for these converters, but the basic principle almost always involves the use of a flapper/nozzle system. Fig. 4.5 illustrates a simple way to construct such a converter. Notice that the current through a coil produces a force that will tend to pull the flapper down and close off the gap. A high current produces a high pressure so that the device is direct acting. Adjustment of the springs and perhaps the position relative to the pivot to which they are attached allows the unit to be calibrated so that 4 mA corresponds to 0.2 kg/cm^2 (or 3 psi) and 20 mA corresponds to 1 kg/cm^2 (or 15 psi).

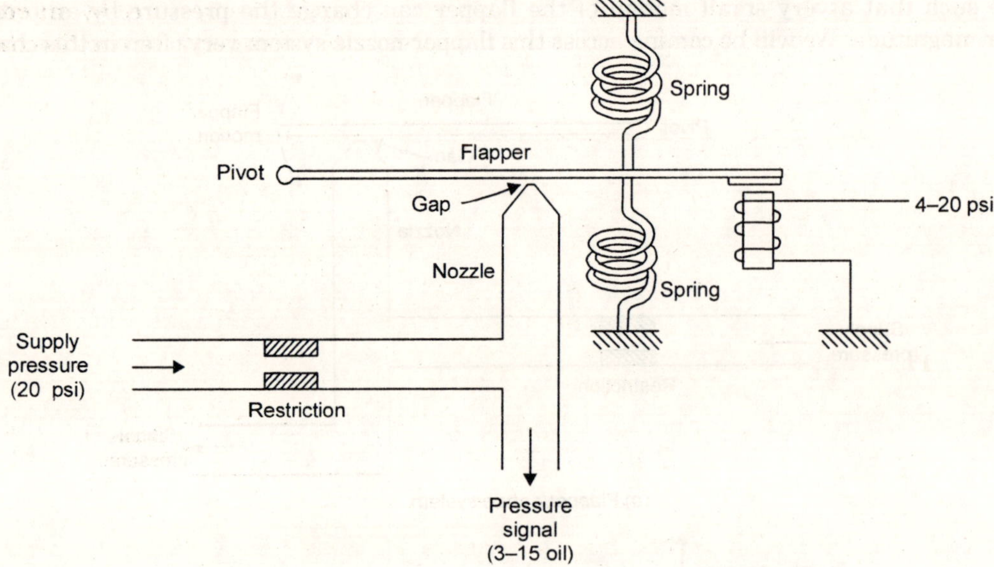

Fig. 4.5 Principles of a current-to-pressure converter

Fig. 4.6 illustrates the construction of one such converters and also lists the various electric devices with which it is commonly combined. A permanent magnet creates a field that passes through the steel body of the transmitter and across a small air gap to the pole piece. A multiturn, flexure-mounted voice coil is suspended in the air gap. The input current flows through the coil creating an electromagnetic force that tend to repel the coil and thus converts the current signal into a mechanical force. Since the total force obtainable in a typical voice coil motor with small current inputs is very less, a different approach, namely, the use of a reaction nozzle, is employed here to convert the force into a pneumatic output pressure. In this circuit, supply air flows through a restriction and out the detector nozzle. The reaction of the air jet as it impinges against the nozzle seat supplies the counterbalancing force to the voice coil motor. The nozzle back pressure is the transmitted output pressure. In order to make the transmitter insensitive to vibration, the voice coil is integrally mounted to a float submerged in silicone oil. The float is sized so that its buoyant force equals the weight of the assembly, leaving a zero net force. Zero is adjusted by changing a leaf-spring force. Span is adjusted by turning the range-adjusting screw to change the gap between the screw and the magnet, thus shunting some of the magnetic field away from the pole piece. Such converters are called 'Motion-Balance' type converters.

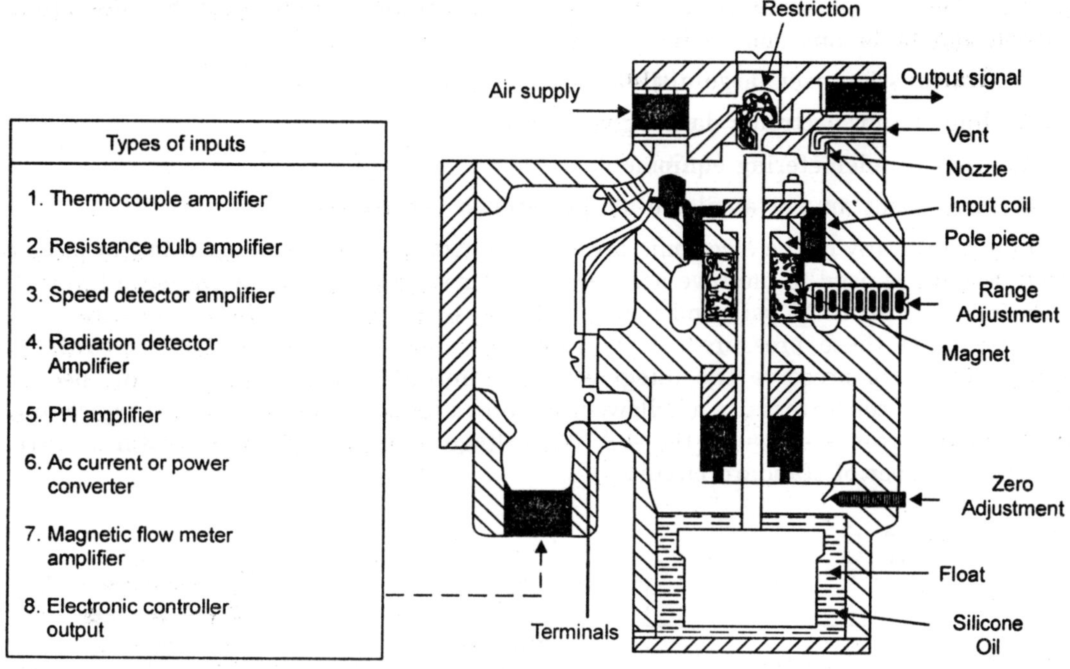

Fig. 4.6 Electric-to-pneumatic transmitter with list of typical input sources

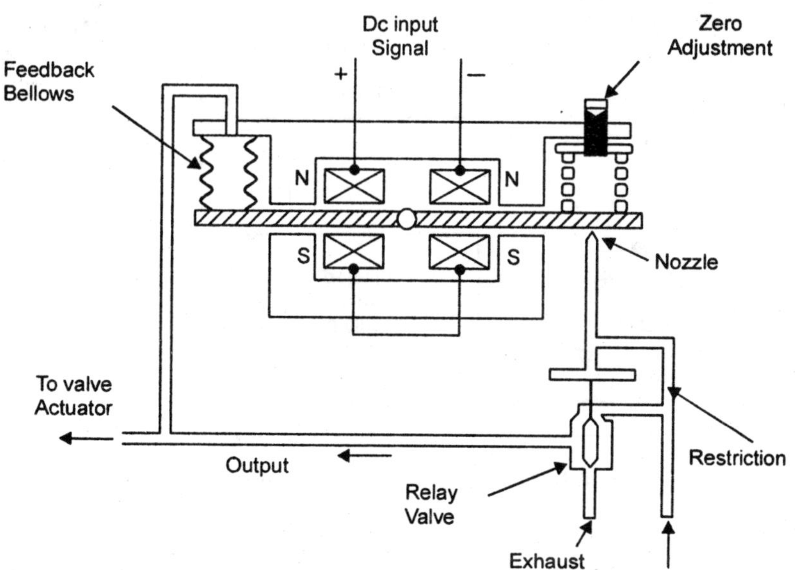

Fig. 4.7 Electropneumatic force-balance transducer

A force-balance type converter with signal feedback to improve accuracy is shown in Fig. 4.7.

4.2.3.4 Pressure-to-Current Converter : (P/I Converter)

The pressure-to-current converter, or simply *P/I* converter, is complementary to *I/P* converter. The *P/I* converter is used wherever pneumatic signals must be converted to electronic signals for any one of the following reasons.

1. Transmission over large distances.
2. Input to an electronic data-logger or computer.
3. Input to telemetering equipment.
4. Instrument air not available at the receiver controller.

In principle, any of the electronic pressure transmitters could be used, but in practice, special devices are used to improve accuracy. The air signals are at low pressure levels (0.2 to 1 Kg/cm^2, or 3 psi to 15 psi), and many of the pressure detectors are not sensitive or not linear enough at these pressures. A *P/I* converter should be atleast 0.5% accurate and preferably 0.25% to preserve the integrity of the initial signal. Because of this need for accuracy, most *P/I* converters use a bellows input and a motion balance sensor. A typical high-quality *P/E* converter is shown in Fig. 4.8. The voltage output can be converted into current values with number of standard circuits.

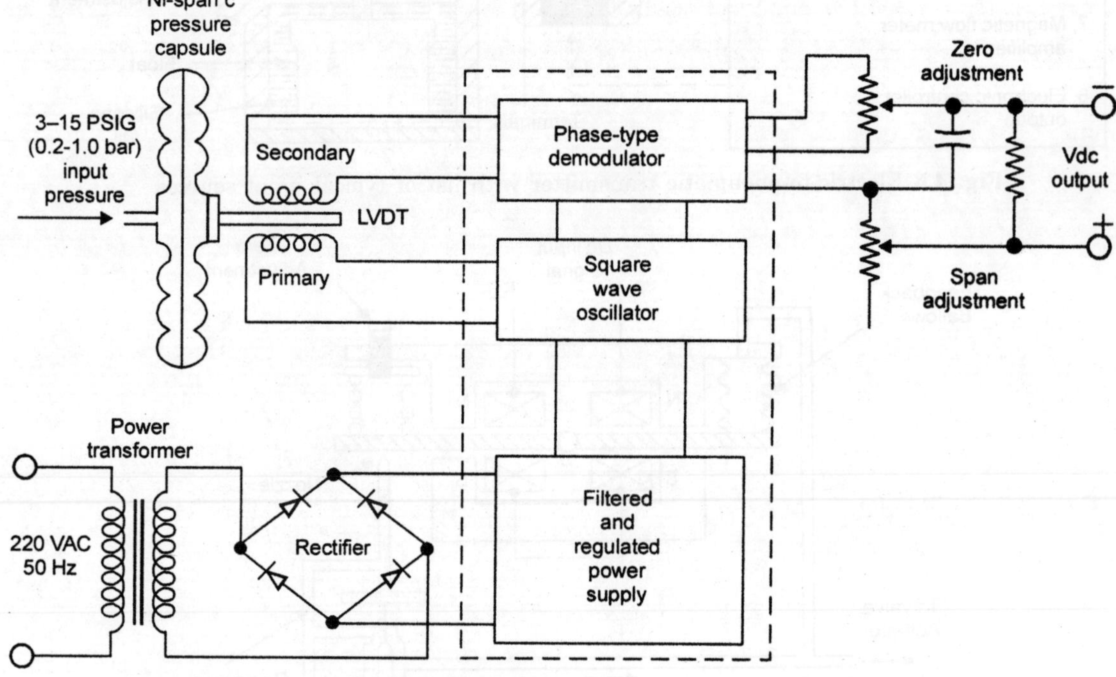

Fig. 4.8 Pneumatic-to-electronic converter

4.3 ACTUATORS

If a valve is used to control fluid flow, some mechanism must physically open or close the valve. If a heater is to warm a system, some device must turn the heater ON or OFF or vary its excitation. These are examples of the requirement of an 'actuator' in the process control loop. It occupies the intermediate position between 'signal conversion unit' and 'control valve'. Notice the distinction of this device from both the input control signal and the control element itself. Actuators take on many diverse forms to suit the particular requirements of process-control loops. We will consider several types of electrical and pneumatic actuators. We will have brief discussion on hydraulic actuators also.

4.3.1 Electrical Actuators

4.3.1.1 Solenoid

A solenoid is an elementary device that converts an electrical signal into mechanical motion, usually in a straight line. A simple solenoid consists of a coil and plunger as shown in Fig. 4.9. The plunger may be free standing or spring loaded. The coil may be operated by either dc or ac voltage. Solenoid specifications include the electrical rating and the plunger pull or push force when excited by the specified voltage. Some solenoids are rated only for intermittent duty because of thermal constraints. In this case, the maximum duty cycle (percentage on total time) will be specified. Solenoids are used when a large sudden force must be applied to perform some job. In Fig. 4.10 a solenoid is used to change the gears of a two-position transmission. An SCR is used to activate the solenoid coil. In many process control pipe lines solenoid valves (where the plunger can act as valve stem also) are used for quick closing or shut-off operations. In large pipelines shut-off valves are operated with the help of pneumatic air or hydraulic oil which are in turn controlled by solenoids valves fitted in the air line or oil line.

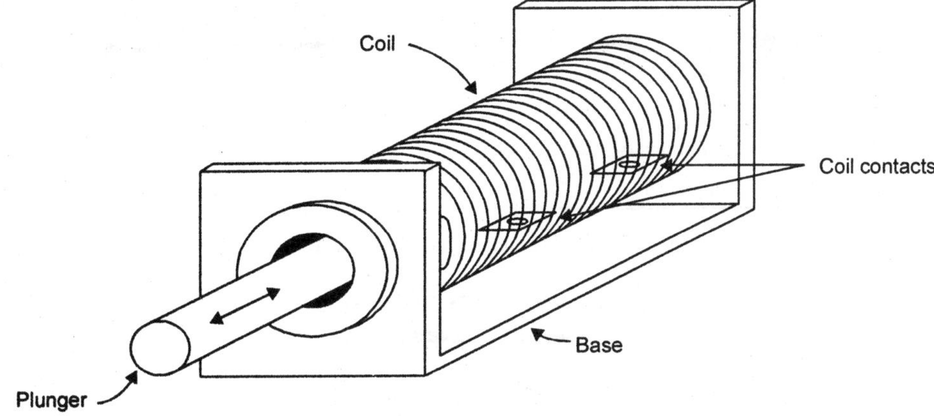

Fig. 4.9 A solenoid converts an electrical signal to a physical displacement

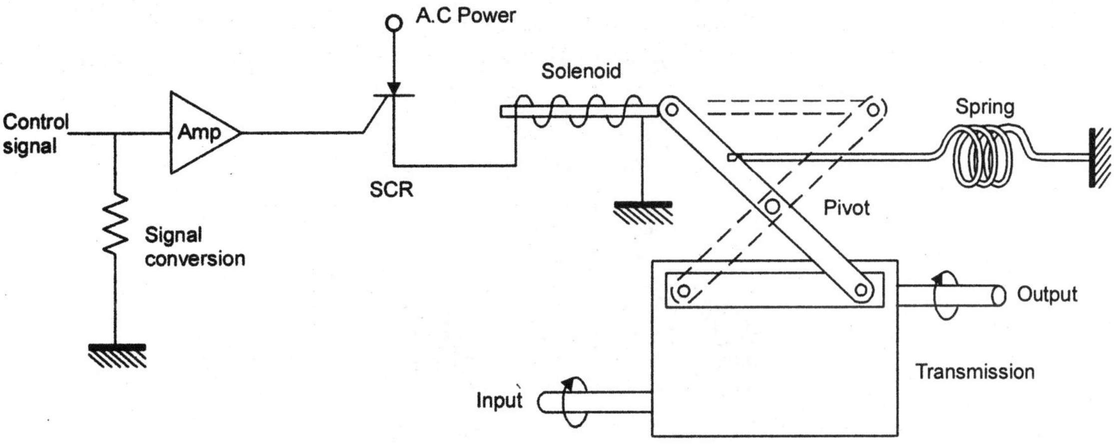

Fig. 4.10 A solenoid used to change gears

Solenoids move in a straight line and therefore require a cam or other mechanical converter to operate rotary valves. These actuators are best suited for small, short-stroke on-off valves, requiring high speeds of response. Solenoid—actuated valves can open or close in 8 to 12 milliseconds.

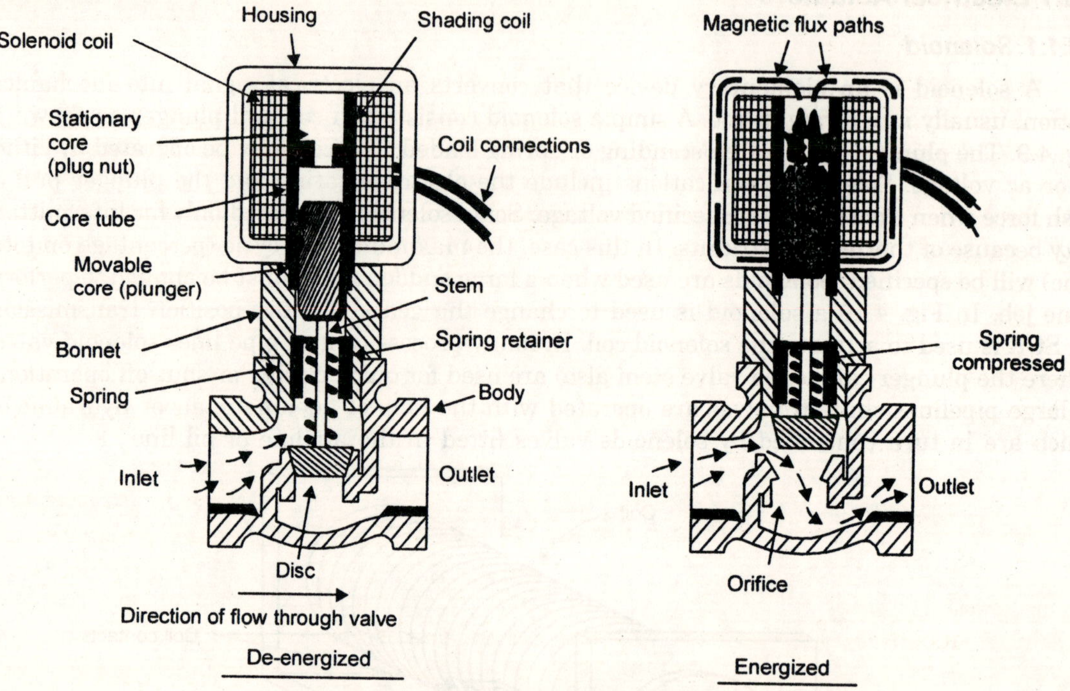

Fig. 4.11 Direct acting solenoid valve.

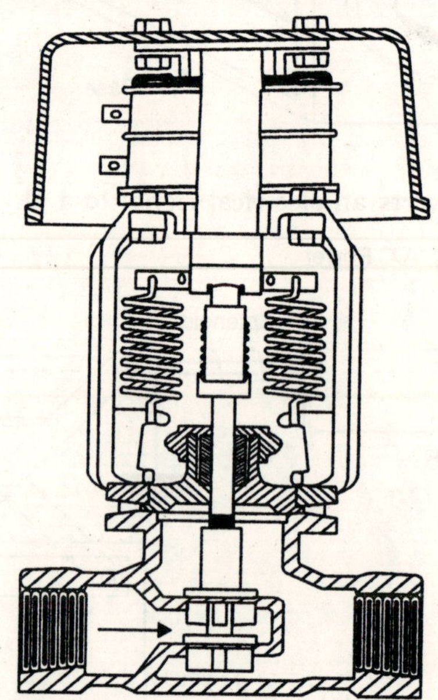

Fig. 4.12 Solenoid valve with strong return springs

A practical solenoid valve consists of the valve body, a magnetic core attached to the stem and disc, and a solenoid coil (Refer Fig. 4.11). The magnetic core moves in a tube that is closed at the top and sealed at the bottom, allowing the valve to be packless. A small spring assists the release and initial closing of the valve. The valve is electrically energized to open. The figure shows both the deenergized and energized conditions of the valve for better understanding. Stronger springs are used to overcome the friction of packing when it is required as shown in Fig. 4.12. Reversing the valve plug causes reverse action (open when deenergised).

Solenoid valves are available in two-or three-way designs, with power requirements ranging from 10 to 50 watts with 6 to 440 V ac or 6 to 220 V dc power supplies. Solenoids are reliable devices and they can provide multimillion cycles on liquid service. Solenoids are used extensively for moving valve stems. Although the force output of solenoids may not have many electrical or mechanical limitations, their use as valve actuators has economic and core (or stem) travel limitations, and they are expensive too.

Designs are available with a separate positioner which accepts 4–20 mA from controller and delivers a dc output signal to the solenoid. Valve position feedback is obtained through the use of a linear variable differential transformer (LVDT) mounted directly on the valve.

4.3.1.2 Electrical Motors

Electrical motors are devices that accept electrical input and provide a continuous rotation as a result. Motor styles and sizes vary as demands for rotational speed (rpm), starting torque, rotational torque and other specifications vary. There are numerous cases where electrical motors are employed as actuators in process control. The most common control situations where electrical motors are used may be broadly classified into two categories.

1. Motor as direct actuator: The situations where motor is driving some part of a process where speed of the motor is directly controlled to control some variable in the process. Examples : The drive of a conveyor system, variable speed pumps etc.

2. Motor along with gear boxes : (Electro Mechanical Actuators) Such motors normally run at constant speed as long as supply is available. Mostly they are reversible ones, that is, they are able to change the directions quickly. Such motors are called 'Servo Motors'. Through speed- reduction gears, torque adjustment gears and with different mechanical arrangements they can be converted into 'rotary actuators' or 'linear actuators'.

Common Varieties of Motors

DC motors : Series field, shunt-field or compound field dc motors are used as servo motors. To change the directions (open or close; forward or reverse etc) the polarity has to be changed with the help of interface units. Different types of braking designs are possible to stop the motor immediately when power is cut off or direction is changed. While changing the direction, the motor should stop running immediately and start running in the opposite direction.

A.C motors : There are many types of ac motors . Synchronous ac motors and Induction ac motors are the common ones. Sometimes single phase motors are used with special arrangements for starting and reversing them. When more torque is required three phase motors are used. Compared to dc motors ac motors are always having starting problem as they are not normally self starting and their starting torque is very low.

Stepping motors : The stepping motor (also known as stepper motor) application has increased in recent years because of the ease with which it can be interfaced with digital circuits. A stepping motor is a rotating machine that actually completes a full rotation by sequencing through a series of discrete rotational steps. Each step position is an equilibrium position in that, without further excitation, the rotor position will stay at the latest step. Thus, continuous rotation is achieved by the input of a train of pulses, each of which causes an advance of one step. It is not really continuous rotation, but discrete, step wise rotation. The rotational rate is determined by the number of steps per revolution and the rate at which the pulses are applied. A driver circuit is necessary to convert the pulse train into proper driving signals for the motor.

We will not be discussing the design and construction parts of the above motors as the details of such motors are available in standard books dealing with motors.

4.3.1.2.1 Motor as Direct Actuator

Variable speed motors are used for such application. Some of the applications where motor can be used as direct actuator are listed below :

1. Control of the speed of belt conveyors carrying coal to power plants, raw materials to metallurgical and heavy industries and carrying of components within plant units.
2. Control of the speed of pumps pumping water, acid, liquid etc in process industries.
3. Control of the speed of roller tables carrying hot metal ingots or slabs for rolling in the rolling mills.
4. Control of the speed of blowers, FD fans, ID fans etc in small power plants.

Variable voltage drives (VV), variable frequency (VF) drives, variable voltage variable frequency (VVVF) drives, pulse width modulated (PWM) drives etc are familiar motor actuators available for the above controls.

4.3.1.2.2 Electromechanical Actuators

These actuators are electrical motor driven, but coupled to mechanical gear trains. The motor runs continuously, but the resultant motion after the gear box is either rotary or linear. In the rotary actuators the motion is controlled within 90°, 180° or something less than 360°. When it is used as linear actuator, the final output spindle will be moving up and down with the travel limits fixed. Motor speed, travel limits and final shaft torque are all designed as per the process requirement. Electromechanical actuators will have the facility to operate them manually with the help of hand wheels while initially adjusting the travel limits etc. These actuators can utilize a reversible electric motor (either a.c or d.c) provided with an internal worm gear to prevent drive direction reversal by unbalanced loads. Servomotor drives can position valves in response to feedback signals from linear or rotary encoders. In other words, the control signals is compared with the feedback signal (stem position) and the error is utilized to position the valve. The drive motor will continue to run till such time the error becomes zero.

Rotary Output Actuators

The actuator shown in Fig. 4.13 is an example of the use of a double worm gear reduction to obtain output speed of about 1 rpm with an input motor speed of about 1500 rpm. The worm gear is self-locking and hence it prevents the load from moving downward by back driving the motor. Also shown is a handwheel for manual operation during power loss and initial adjustments.

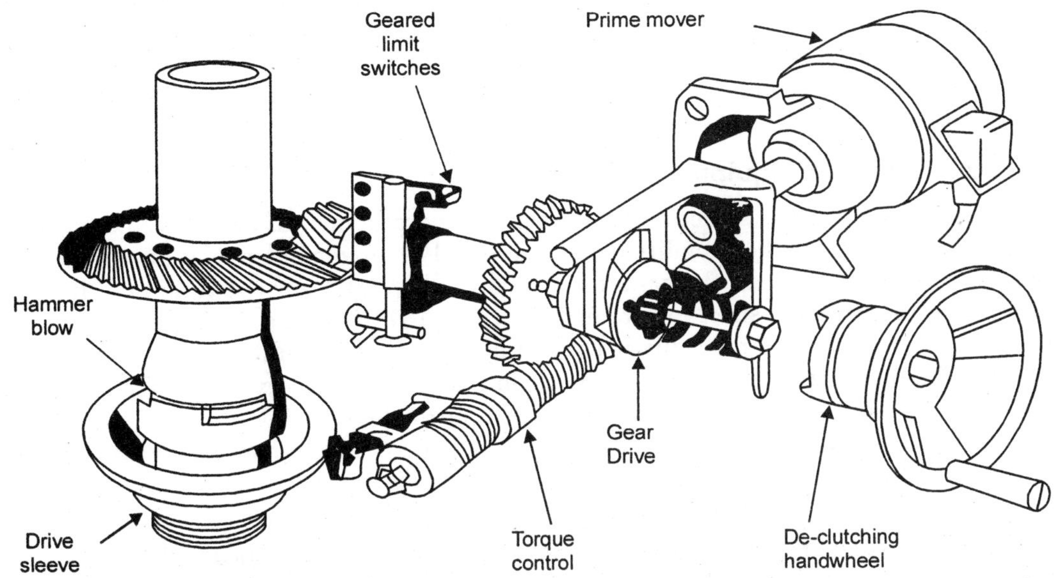

Fig. 4.13 Electric actuator with worm gear reduction

Linear Output Actuators

Fig. 4.14 shows a linear output actuator. It uses a worm and a rack and pinion to translate horizontal shaft motor output to vertical linear motion. A continuously connected handwheel, which must rotate the rotor of the motor, can be used when there is short stem travel and relatively low force output.

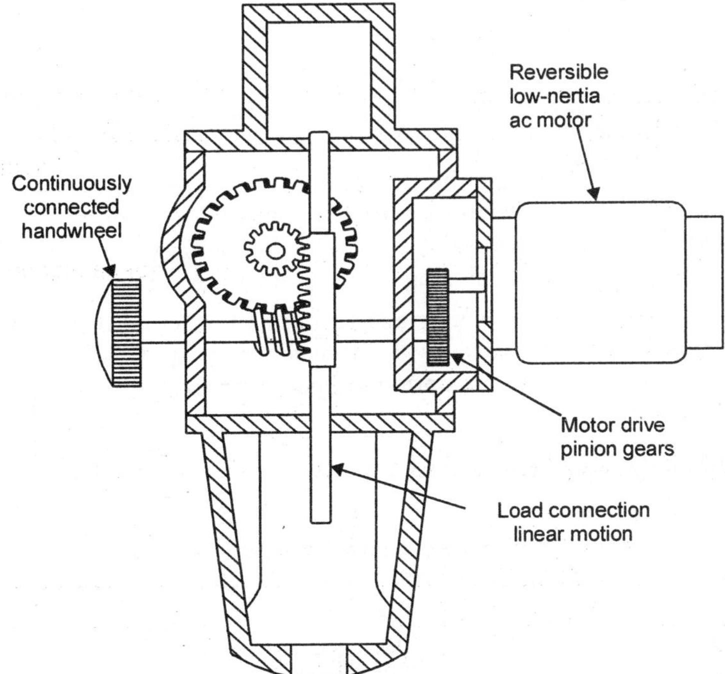

Fig. 4.14 Rack and pinion

Though we have discussed one variety each, there are number of types of rotary and linear actuators with different combination of gear arrangements are available in the market. Depending on the final control valve, speed of opening and torque requirement the selection of actuators is to be made.

Associated Electrical Equipment

Manual operation is sometimes necessary for normal operational procedures, such as start-up, or under emergency conditions. Hand wheel arrangement as we discussed earlier is useful only when you are in the actuator location in the field. For checking up full opening and closing of the actuator shaft in conjunction with valve stem and for adjusting the travel limit switches handwheel is normally used. But here what we are interested is for operating electrically the motor from the remote place (control room) via 'more' or 'less' push buttons. Hence an 'auto'-'manual' selector switch along with open-close push buttons (sometimes with actuating relays) are to be hooked up electrically. Putting the auto-manual switch in manual mode, we will be able to operate the motor with the help of push buttons. During auto control through controller this switch will remain in 'auto' position.

Limit switches In general there are two pairs of limit switches called 'travel limit switches' and 'torque limit switches'. Travel limit switches are adjusted in such a way that they break the power to the motor whenever the valve stem reaches either fully closed position ('close' limit switch operates) or fully opened position ('open' limit switch operates). Similarly 'close' torque limit switch will operate whenever the torque exerted by the motor exceeds the set value while closing the stem and 'open' torque limit switch will operate whenever the torque exceeds while opening the stem. Torque limit switches normally operate when the stem along with gears get stuck up or exceeds the travel limits because of the failures of travel limit switches.

Position Feedback

A feedback potentiometer (or encoder) is used for getting the feedback of the valve stem position. By correct gearing arrangement the transmitting potentiometer can be rotated. A position transmitter can also be used to get 4–20 mA output for positioners.

Most of the electrical equipment discussed above are run by gearing to stem rotation, so that they can be housed in the actuator unit itself. 'More' or 'less' bulbs are also incorporated to indicate in the control room whether the actuator is opening or closing. Normally green colour bulbs are used for indicating closing action and red colour ones for opening action.

Fig. 4.15 illustrates one reversible single phase a.c motor actuator's electrical connections incorporating electrical equipment discussed above for clear understanding. Nowadays the Auto/Manual selection adds one more option called computer mode (together it is called C/A/M selection).

4.3.2 Pneumatic Actuators

In general, an actuator is that portion of a valve that responds to the applied signal and causes the motion of the valve stem. For that purpose, the actuator often translates a control signal into a large force or torque as required to manipulate the control element. The 'pneumatic actuator' is most useful for such translation. The pneumatic actuators are capable of moving the valve to any position from fully closed to fully open using compressed air for power. There are two general types of pneumatic actuators: (1) The spring and diaphragm actuator and (2) The piston actuator. In a spring and diaphragm actuator, variable air pressure

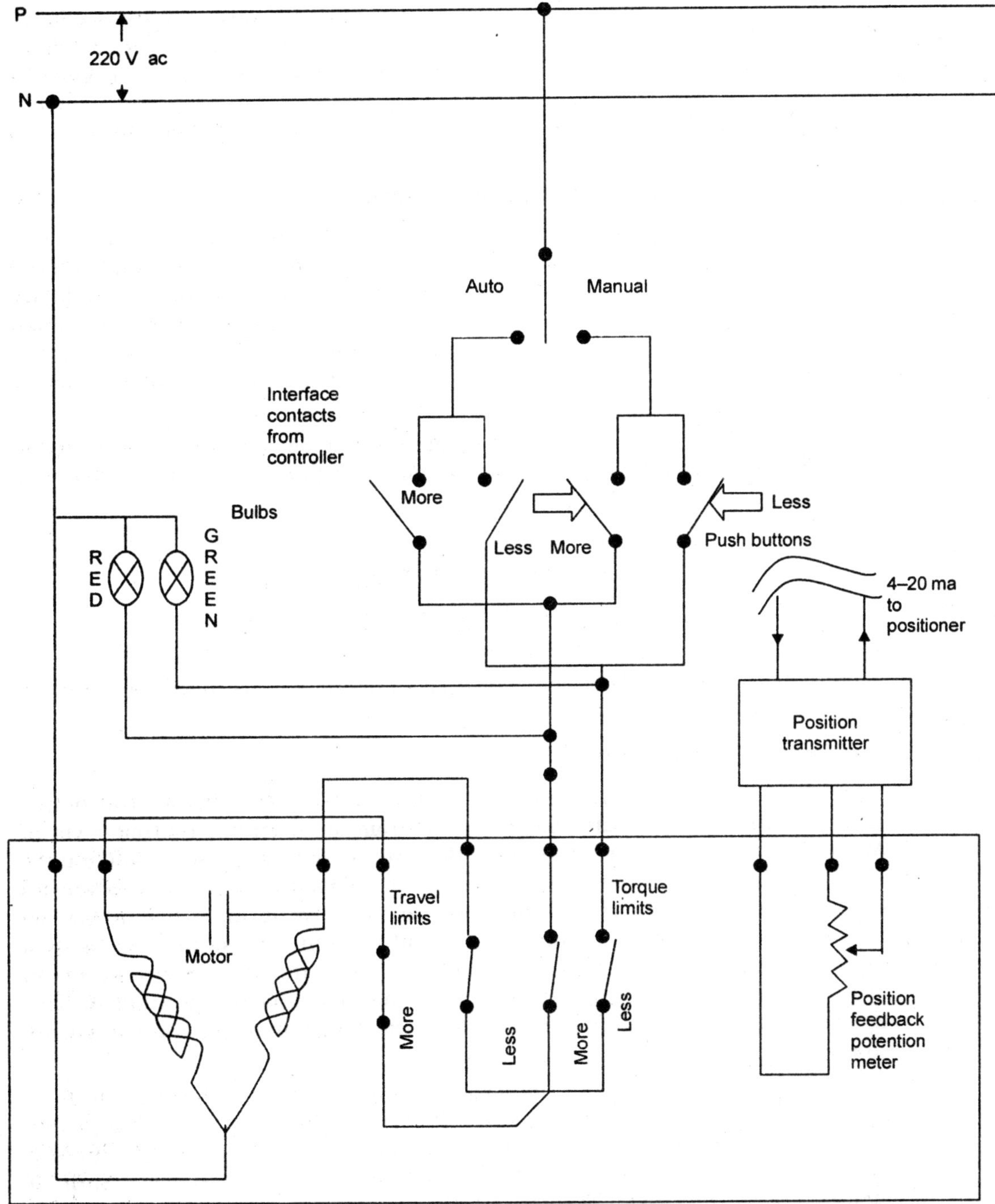

Fig 4.15 A single phase A.C motor actuator-connection diagram with electrical equipment

is applied to a flexible diaphragm to oppose a spring. The combination of diaphragm and spring forces acts to balance the fluid forces on the valve. In a piston actuator, a combination of fixed and variable air pressures is applied to a piston in a cylinder to balance the fluid forces on the valve. Sometimes springs are used, usually to assist valve closure. Excluding springs, there are two variations of piston actuators : cushion loaded and double acting. In the cushion-loaded type, a fixed air pressure, known as the cushion pressure, is opposed by a variable air pressure and is used to balance the fluid forces on the valve. In the double acting type, two opposing variable air pressures are used to balance the fluid forces on the valve.

4.3.2.1 Spring and Diaphragm Actuator : (Diaphragm Motor)

The popularity of the spring/diaphragm actuator, also called diaphragm motor, is due to its low cost, its relatively high thrust at low air supply pressure, and its availability with fail safe springs. It is available in springless designs, double diaphragm designs (for higher pressures), rolling diaphragm design (for longer strokes), and tandem designs (for more thrust). These actuators are ideal for use on valves requiring linear travel, such as globe valves. A linkage or other form of linear-to-rotary motion conversion is required to adapt these actuators to rotary valves, such as the butterfly valve.

The principle is based on the concept of pressure as force per unit area. If we imagine that a net pressure difference is applied to a diaphragm of surface area 'A', then a net force acts on the diaphragm is given by

$$F = (p_1 - p_2) A \qquad \text{...(4.1)}$$

where

$$p_1 - p_2 = \text{pressure difference (Pa)}$$
$$A = \text{diaphragm area (m}^2\text{)}$$
$$F = \text{force (N)}$$

If we need to double the available force for a given pressure, it is merely necessary to double the diaphragm area.

Direct Action Pneumatic Actuator

The action of a 'direct' pneumatic actuator is shown in Fig. 4.16. The air transmitted from a pneumatic controller (or a valve positioner) or from an electronic controller via *I/P* converter enters the upper diaphragm case, while the lower diaphragm case is vented to the atmosphere by the open hole *H*. When the top pressure increases the force acting downward also increases. This starts the valve closing. As it does so, the valve spring is compressed and the spring force increases. The valve movement will continue until the spring force is equal to the force due to the increased air pressure. Similarly, when the air pressure decreases, the valve moves upward and the spring expands until a new force balance is attained. Thus the valve stem moves through a definite distance for each change in air pressure applied to the diaphragm.

Fig. 4.16 (*a*) shows the condition in the low signal pressure state where the spring *S* maintains the diaphragm and the connected control shaft in a position as shown. Fig. 4.16(*b*) shows the case of maximum control pressure and maximum travel of the shaft. The pressure and force are linearly related, as shown in equation (4.1) and the compression of a spring is linearly related to forces acting on it. Then we see that the shaft position is linearly related to the applied control pressure.

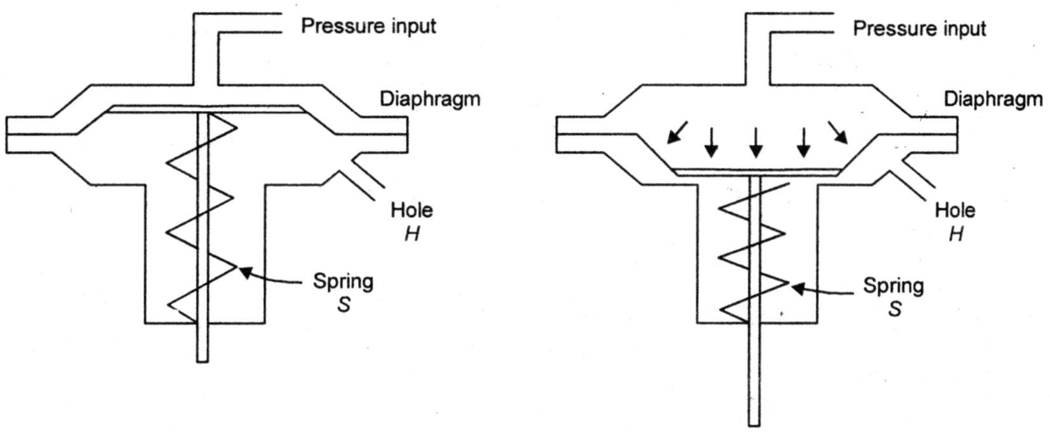

(*a*) Direct actuator in the low-pressure state

(*b*) Direct actuator in the high-pressure state

Fig. 4.16 A direct pneumatic actuator for converting pressure signals into mechanical shaft motion.

$$\Delta x = \frac{A}{K}\Delta p \quad ...(4.2)$$

where

Δx = Shaft travel(m)

Δp = Applied gauge pressure (Pa) pascals.

A = Diaphragm area (m^2)

K = Spring constant (N/m)

Fig. 4.17 shows one such direct actuator along with the control valve.

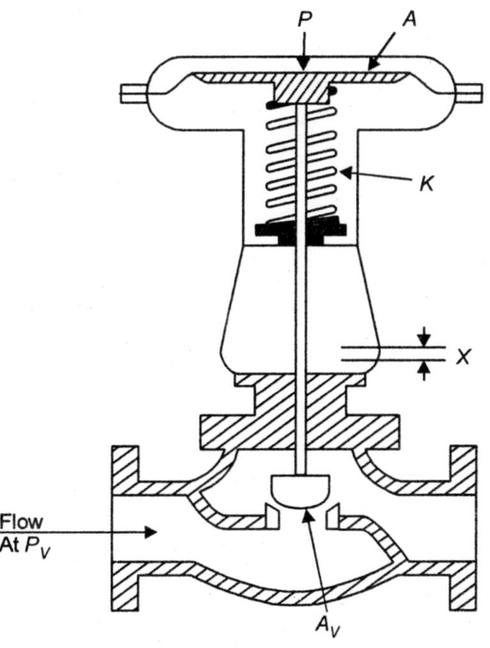

Fig. 4.17 A spring-and-diaphragm actuator cum valve.

Reverse Action Pneumatic Actuator

A reverse actuator, shown in Fig. 4.18 moves the shaft in the opposite sense from the direct actuator, but obeys the same operating principle. Thus, the shaft is pulled in by the application of a control pressure. Fig. 4.19 shows one such reverse actuator along with the control valve.

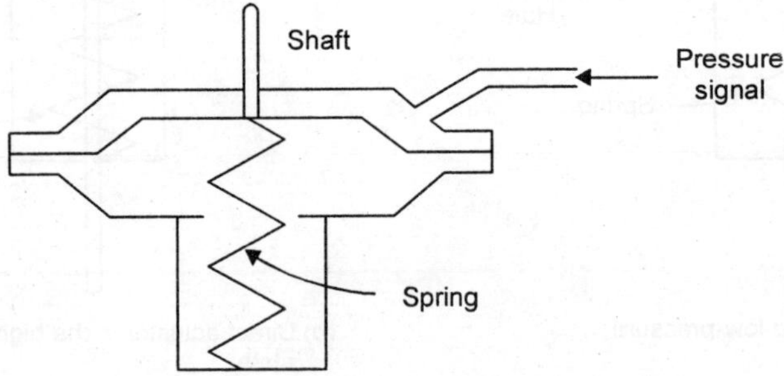

Fig. 4.18 A reverse pneumatic actuator

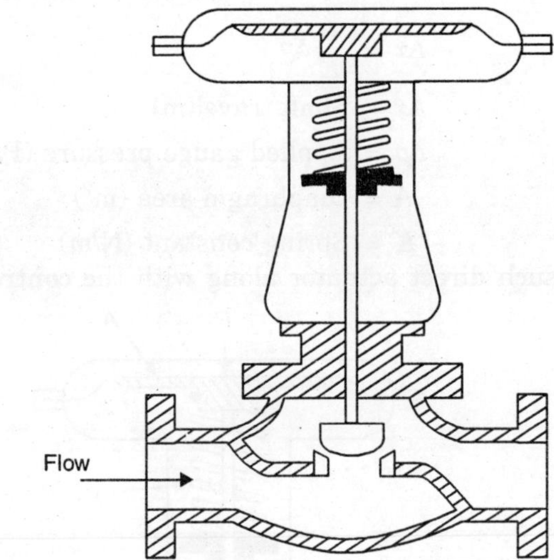

Fig. 4.19 Reverse actuating actuator cum valve

General Comments

The spring and diaphragm actuators are single-acting and are used where the controlling forces are small. The assumption is usually made that the travel of the valve stem is linear with changes in air pressure over the customary range from 3 to 15 psi (0.2 to 1 Kg/cm^2). This is not necessarily true. The non linearity is normally caused due to :

1. The nonlinear spring response
2. Nonlinear effective area of the diaphragm as it deflects.
3. The changing thrust forces that act on the movement of the plug in the controlled fluid.

To overcome this limitation, valve positioners are used whenever the positioning of the valve is opposed by non linear forces which are of more than negligible magnitude. They assure that for a definite corrective signal from the controller the valve responds with a definite amount of travel. We will be discussing about valve positioners little later.

Fail-Safe Operations : (Safe Failure Operation)

Spring actuators normally will have a definite safety position, that is, if the air pressure fails, the force of the spring will cause the valve either to open or close, depending on the least hazardous condition. In gas-heated installations, for instance, the actuator of the valve in the gas line will have a spring tending to close it, while refrigerating lines will have actuators with springs tending to open the valve, in order to guarantee heat dissipation.

Fig. 4.20 shows two types of valve plugs namely 'air-to-close' and 'air-to-open'. In the 'air-to-close' valve (Fig. 4.20 (*a*)), as the air pressure above diaphragm increases, the stem moves down and consequently the plug restricts the flow through the orifice. If the air supply above the diaphragm is lost, the valve will 'fail open' since the spring would push the stem and the plug upward. In the 'air-to-open' valve (Fig. 4.20(*b*)), as the air pressure above the diaphragm increases, the stem moves down and consequently the plug allows the flow through the orifice. If the air supply above the diaphragm is lost, the valve will 'fail closed' since the spring would pull the stem and the plug upward closing the flow through orifice.

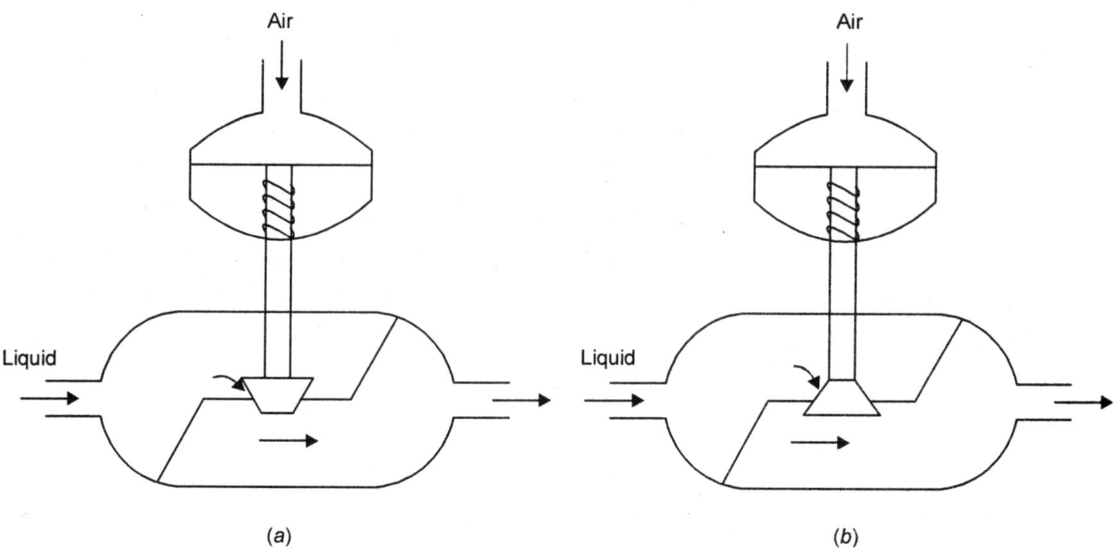

Fig. 4.20 Pneumatic valves (*a*) Fail open (*b*) Fail closed

Springless Diaphragm Actuator

The springless actuator is useful for large thrust forces. The spring of the spring/diaphragm actuator is replaced by a pressure regulator which maintains a constant pressure on the underside of the diaphragm. An air supply at a pressure of about 20 to 100 psi (1.5 to 7 Kg/cm^2) is required. The operation of the springless actuator can be explained with the help of Fig. 4.21. Assume that the cushion regulator is set to provide 9 psi (0.6 Kg/cm^2) pressure on the underside of the diaphragm. At static balance and with no thrust force on the actuator stem, the upperside pressure must be 9 psi (0.6 Kg/cm^2). Then if the input pressure increases, the nozzle back pressure increases, and the upperside pressure is raised to a high value. The actuator stem then moves downward and, as the actuator stem attains new position, the upper side pressure is returned to 9 psi. If there is an upward thrust force on the actuator stem, the underside pressure remains at 9 psi but the positioner rises the upperside pressure until static balance is achieved. For a downward thrust force the upper side pressure is reduced below 9 psi. Thus, the springless actuator can counteract a thrust force equal to approximately the underside pressure times the area of the diaphragm. This is generally from three to ten times the thrust force handled by a spring actuator.

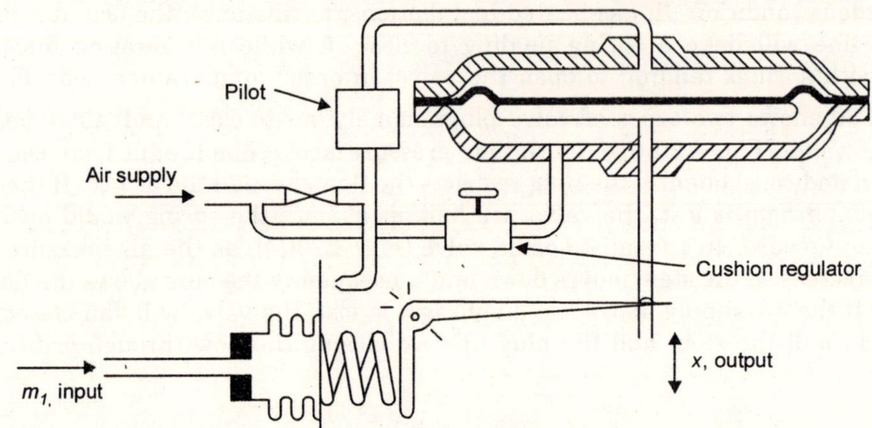

Fig. 4.21 Springless diaphragm actuator

4.3.2.2 Piston Actuator

Piston actuators are either single or double acting. The single-acting actuator shown in Fig. 4.22, utilizes a fixed air pressure, known as the cushion, to oppose the controller signal. This valve does not have spring or diaphragm area nonlinearities. In order to use such an actuator for throttling purposes, it is necessary to have a positioner. The positioner senses the actuator motion and causes the valve to move accordingly.

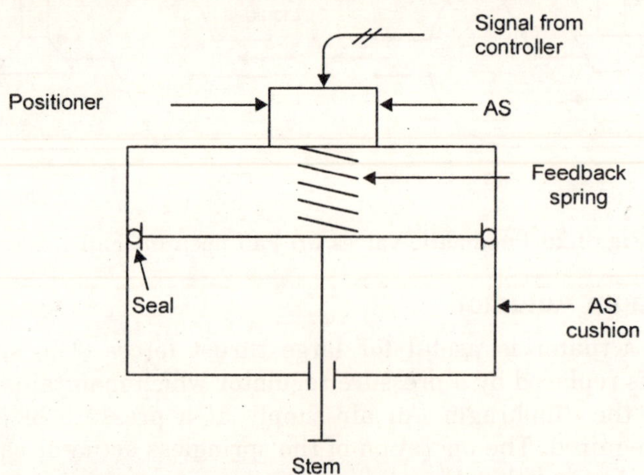

Fig. 4.22 Single-acting piston actuator

The double-acting piston actuator is one that eliminates the cushion regulator and uses a positioner with a built in reversing relay. Thus the positioner has two air pressure outlets, one connected above the piston and the other below. The positioner receives its signal and senses travel in the same manner as a single-acting positioner. The difference is in the outlets; one pressure increases and the other decreases to cause piston travel.

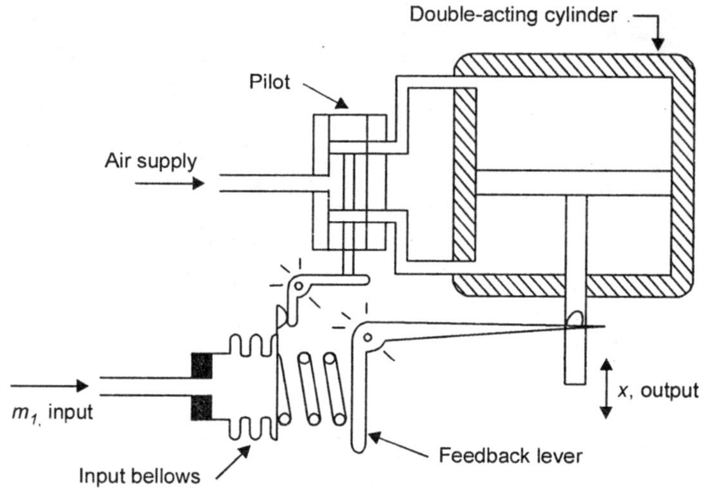

Fig. 4.23 Double-acting piston actuator

The double-acting piston actuator shown in Fig. 4.23 is employed for large thrust forces than can be handled by the single acting actuator, and the piston is used in order to obtain long stroke. The pilot is generally a spool-type diverting valve and requires an air supply of 30 to 100 psi (2 to 7 Kg/cm^2). When the input pressure increases, the bellows moves to the right and pushes the pilot spool upward. This action opens the upperside of the cylinder to the air supply and opens the lower side to the atmosphere; thus the action is to return the piston to the neutral position. Thus the position of the piston is proportional to the input pressure. A double-acting piston actuator can handle a thrust force equal to about 80 % of the supply pressure times the area of the piston.

When choosing between piston actuators and the spring diaphragm type, the fail-safe consideration may be the reason for the final selection. If properly designed, the spring is the best way of achieving fail-closed action. Fail-open action is normally less critical. Piston actuators depend upon air lock systems to force the valve closed on air failure. Such systems may work well initially, but there are many possibilities for leaks to develop in the interconnecting tubes, fittings, and check valves and such piston actuator systems are not considered reliable. Air lock systems also add to the actuator's cost. Piston actuators may also be specified with closure springs to provide positive failure positions.

4.3.2.3 Rotary Valve Actuators

When linear spring/diaphragm or piston actuators are used on rotary valves, their performance will not be linear unless a positioner is used. By the addition of a positioner, one can guarantee that the ratio between a unit change in controller signal and the resulting rotation will be uniform. Positioning a quarter-turn value (0 to 90° rotation) with a linear output actuator using a lever arm on the valve with scotch yoke design is shown in Fig. 4.24. A rack and pinion can be housed with the pinion on the valve shaft and the rack positioned by almost any linear value actuator. Fig. 4.25 shows such an actuator with dual-acting cylinder. Fig. 4.26 shows a similar actuator but with spring-loaded for emergency fail-safe operation.

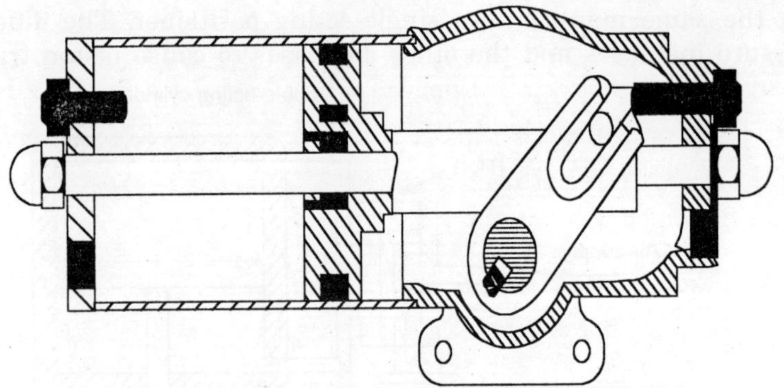

Fig. 4.24 Pneumatic cylinder with Scotch yoke

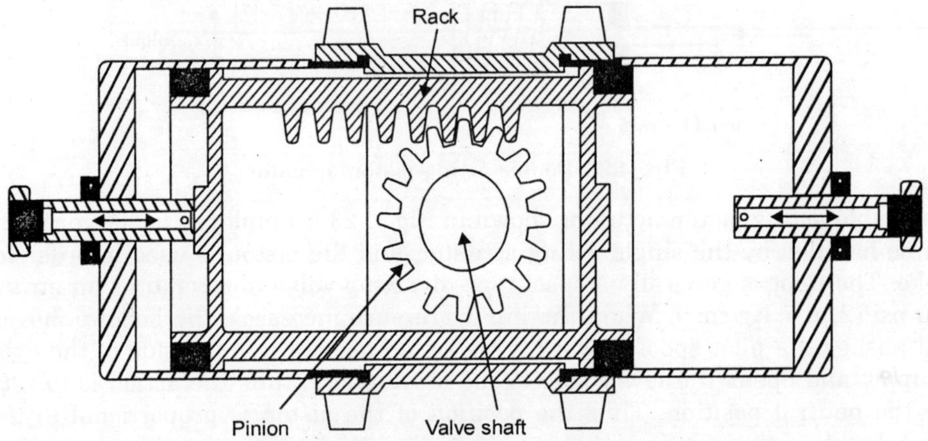

Fig. 4.25 Rack and pinion actuator with dual-acting cylinder

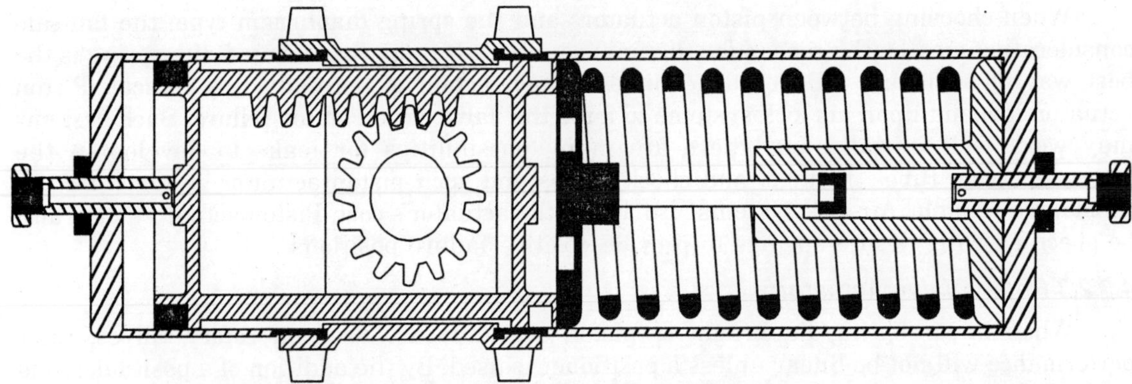

Fig 4.26 Rack and pinion actuator with spring-loaded (fail-safe) cylinder

Rotary Air Motor Actuator

Pneumatic pressure is used to power a rotary motor to drive any of the large gear motors. Control is by a four-way valve. The motor shown in Fig. 4.27 is running in one position. This will continue until the valve is repositioned or until a cam operates a shut off valve at one end of the stroke. Reversal of the four-way valve causes reverse operation. An intermediate position causes the motor to stop. The four-way valve can be operated by pneumatic or electric actuators for remote automatic control. A position transmitter will allow adaptation to closed-loop proportional control.

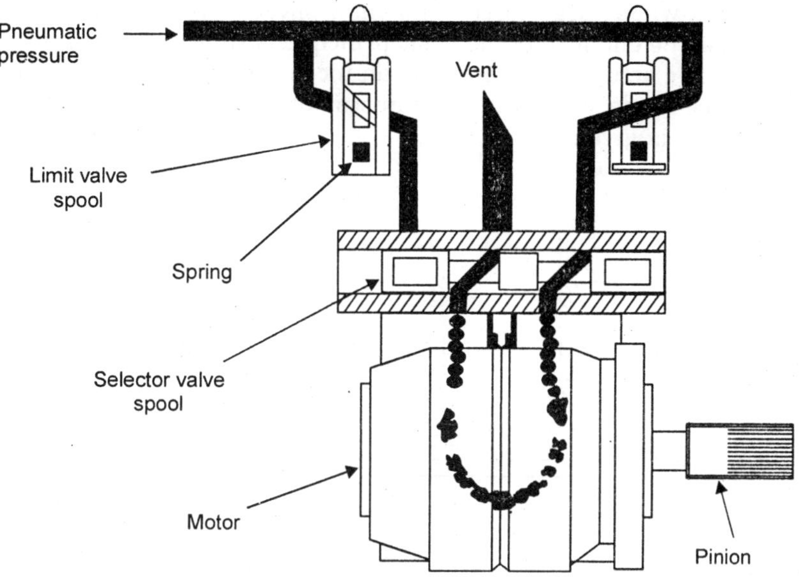

Fig. 4.27 Rotary air motor actuator

4.3.2.4 Electropneumatic Actuators

Electropneumatic actuators are pneumatically powered and electrically controlled. As we discussed earlier I/P converter can be used to convert the electrical control signal into pneumatic signal to actuate the pneumatic actuator. Fig. 4.28 combines the I/P converter in the positioner of the pneumatic actuator. The motion of the output of the actuator is related to the balance beam through the feedback lever. The output position of the actuator is therefore proportional to input electrical control signal.

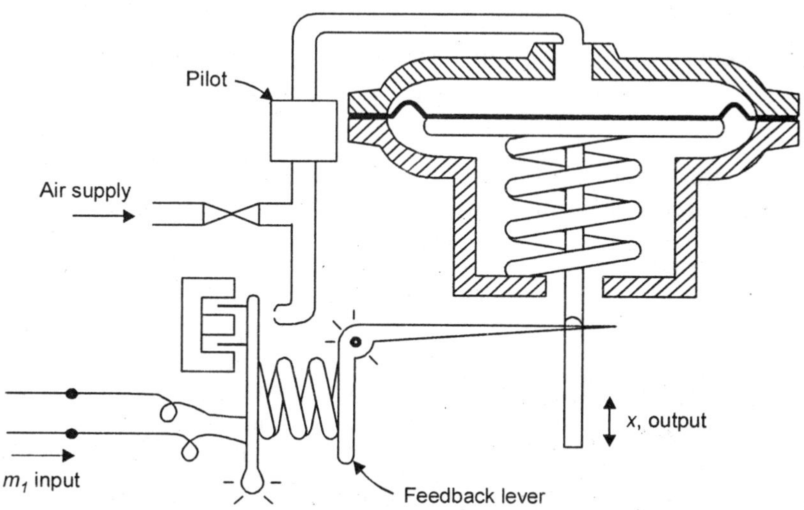

Fig. 4.28 Electro-pneumatic actuator

4.3.3 Hydraulic Actuators

Normally electrical actuators are used in non-hazardous areas and when the force or torque required to operate the control valve is moderate. Pneumatic actuators are preferred in hazardous areas like chemical process industries and more force/torque is required. Yet there are many cases when large forces and high speeds are required. In such cases, a hydraulic actuator may be employed. The basic principle is shown in Fig. 4.29. The basic idea is same as for pneumatic actuators except that an incompressible fluid is used to provide the pressure, which can be made very large by adjustment of the area of the forcing piston A_1. The hydraulic pressure is given by,

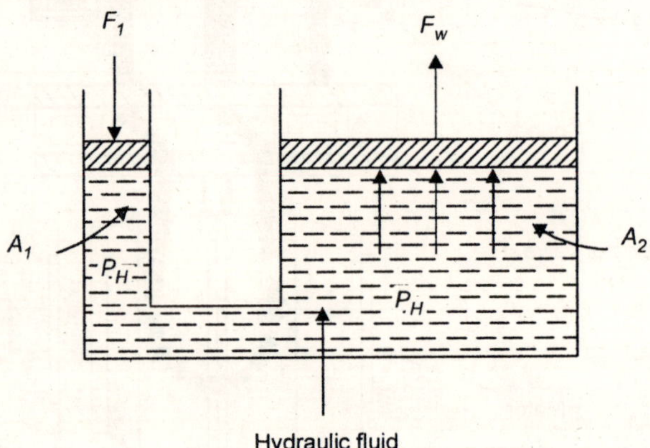

Fig. 4.29 A hydraulic actuator converts a small force F_1 into an amplified force Fw.

$$P_H = F_1/A_1 \quad \text{...(4.3)}$$

where P_H = hydraulic pressure (Pa)
F_1 = applied piston force (N)
A_1 = forcing piston area (m^2)

The resulting force on the working piston is

$$F_W = P_H A_2 \quad \text{...(4.4)}$$

where F_W = force of working piston (N)
A_2 = working piston area (m^2)

Thus, the working force is given in terms of the applied force by

$$F_W = \frac{F_1 A_2}{A_1} \quad \text{...(4.5)}$$

Hydraulic actuators can be stepping motor or servo valve driven. In the servo valve driven designs the pumps are running continuously, while in the stepping motor configurations they run only when the valve needs repositioning. A servo valve driven hydraulic piston actuator is shown in Fig. 4.30. Input, given either from electrical controller or pneumatic controller after proper signal conversion, controls the position of the vertical lever. The balance lever pivots at the bottom so that an increase of input (to the left) pushes the pilot piston to the left. This action opens the left end of the piston to supply pressure and opens the right end of the piston to drain. The large power piston, therefore, moves to the right until, as the balance lever rotates about the top most end, the pilot piston is returned to centre, the motion of the output x_1 is therefore proportional to the input motion m_1. The hydraulic actuator requires a continuously running electric motor and pump to provide a source of high-pressure oil, and a drainer or sump to collect the return.

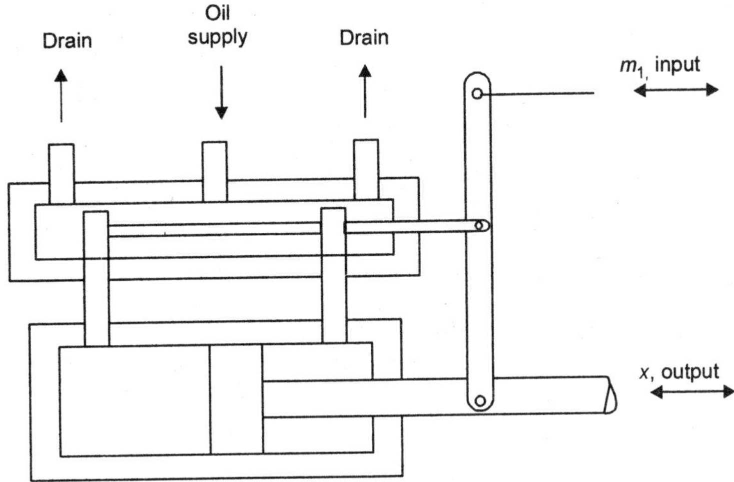

Fig. 4.30 Hydraulic piston actuator

In some cases it is desired to control the position of very large loads as part of the control system. This often can be done by using the low-energy controller output as the set point input to a hydraulic control system as illustrated in Fig. 4.31. In this system, high-pressure hydraulic

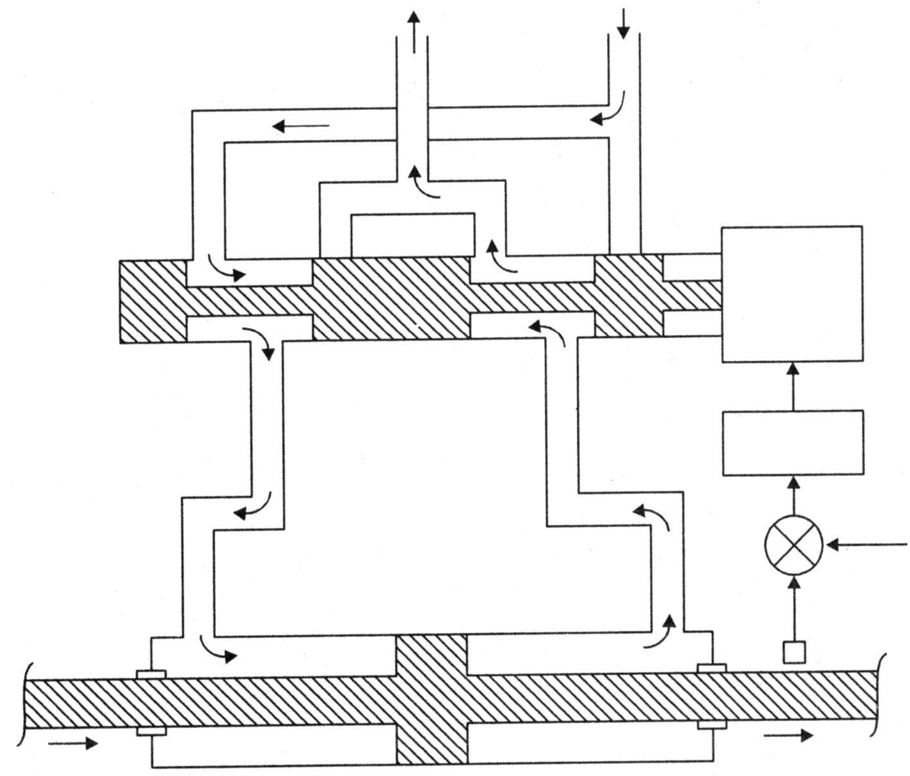

Fig. 4.31 A hydraulic servo system. The process-control system provides the set point of the servo system

fluid can be directed to either side of a force piston, which causes motion in either direction. The direction is determined by the position of a control valve piston in the hydraulic servo valve. The position of this valve piston is controlled by a linear motor driven by the output of an amplifier and error detector. The inputs to the error detector are the process controller output, which forms the set point of the hydraulic servo, and a feedback from the force piston shaft. Thus, the amplifier will drive the hydraulic servo until the feedback matches the setpoint input.

4.3.4 Comparison between Actuators

Table 4.1 gives a summary of advantages and disadvantages for the three types of actuators namely electrical, pneumatic and hydraulic.

Table 4.1 Comparison between Actuators

Sr. No	Actuator types	Advantages	Disadvantages
1.	Electrical (servomotor or stepping motor)	Direct interface with computer system. Simple design.	Low thrust. Slow speed No mechanical fail safe hazardous
2.	Electromechanical (Motors combined with gear boxes)	High thrust High stiffness coefficient Flexible adaptation.	Complex design No mechanical fail safe Large, heavy structure Hazardous.
3.	Hydraulic and Electro hydraulic	High thrust Fast speed High stiffness coefficient Self lubrication.	Complex design Large, Heavy structure Hazardous Fluid viscosity sensitive.
4.	Pneumatic and Electro-pneumatic	Low cost Mechanical fail safe Simple design Small package Suitable for highly hazardous areas also Good control with control device	Slow speed Lack of stiffness Instability Moderate thrust Quality air requirement.

4.4 VALVE POSITIONER

The main purpose of having a valve positioner is to guarantee that the valve does move to the position where the controller wants it to be. By adding positioner, one can correct for many variations, including changes in packing friction due to dirt, corrosion, or lack of lubrication; variations in the dynamic forces of the process; sloppy linkages (dead band); or non linearities in the valve actuator. The effective dead band of a valve/actuator combination can be as much as 5%; with the addition of a positioner it can be reduced to less than 0.5 %. It is the job of the positioner to protect the controlled variable from being upset by any of the variations.

In addition, the positioner can be used for split-ranging the control signal between more than one valve, for increasing the actuator speed (by increasing the air pressure/volume incase of pneumatic actuators) for modifying the valve characteristics by cams or electronic function generators. But these reasons do not necessitate the use of positioners as they can be achieved by other means without using positioner also.

When the valve is remote manual (open loop) operation, it will always benefit from the addition of a positioner, because it will reduce the valve's hysteresis and dead band while increasing its response. When the valve is under automatic (closed loop) control, the positioner will be helpful when the loop response is not very fast (analysis, temperature, liquid level, blending, slow flow, large volume gas flow etc), while the positioner will degrade loop response, contribute to proportional offsets, and limit cycling in fast loops (fast flow, liquid pressure, small volume gas pressure etc). Pneumatic actuators without springs always require valve positioners.

The valve positioner is a high-gain plain proportional controller which measures the valve stem position, compares that measurement to its set point (the controller output signal), and, if there is a difference, corrects the error. The open-loop gain of positioners ranges from 10 to 200 (proportional band of 10 % to 0.5%), and their periods of oscillation range from 0.3 to 10 seconds (frequency response of 3 to 0.1 Hz). In other words, the positioner is a very sensitively tuned proportional only controller.

The positioner in effect is the cascade slave of the loop controller. In order for a cascade slave to be effective, it must be faster than the speed at which its set point, the master output signal, can change. The rules of thumb used in this respect suggest that the time constant of the slave should be ten times shorter (open-loop gain ten times higher) than that of the master and the period of oscillation of the slave should be three times shorter (free response three times higher) than that of the primary. It is recommended not to use positioners if the positioned valve is slower than the process variables it is assigned to control.

4.4.1 Electronic Valve Positioner (For Electric Actuators)

Electronic valve positioner is sometimes called as interface unit. It is a high-gain proportional -only controller. It may be considered as a slave controller of a cascade loop in which the master controller is the primary controller itself. The master or primary controller's output signal is the set point for the slave controller (positioner here). The feedback signal (valve position) is from the position transmitter. Feedback potentiometer forms an integral part of the electrical actuator (as discussed under electrical actuators section. Refer Fig. 4.15). The resistance variation is converted into a linear current variation with the help of position transmitter and fed to interface unit in the form of current representing the stem position. Fig. 4.32 shows the complete loop for positioning the control valve as required by the primary controller signal. The output of the secondary controller is normally $\pm$ V dc for dc motor or relay contacts (more or less contacts) for ac motors.

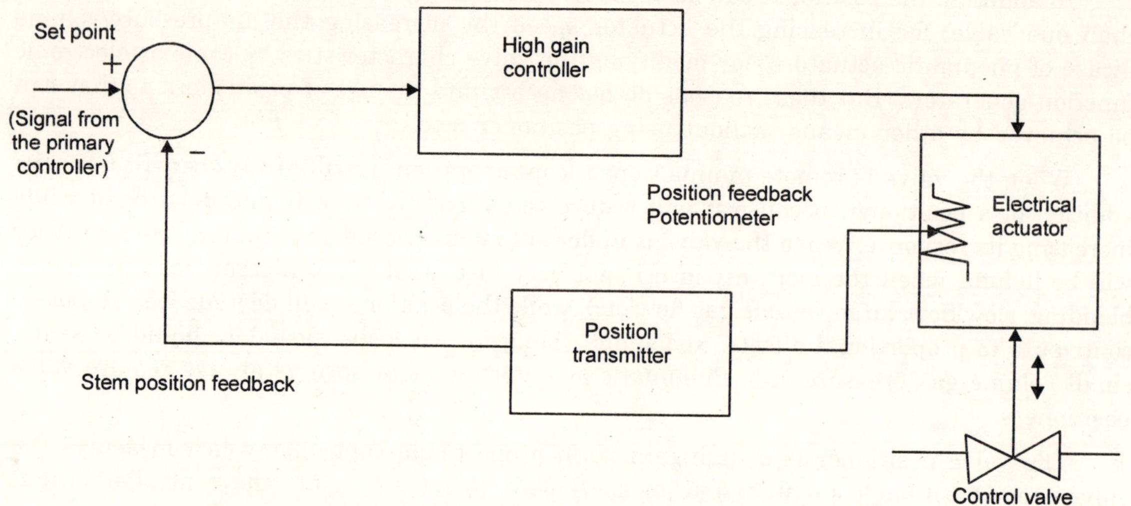

Fig. 4.32 Valve positioner for electrical actuators (interface)

4.4.2 Valve Positioner for Pneumatic Spring Actuator

A motion-balance positioner used along with spring and diaphragm actuator is shown in Fig. 4.33. The positioner consists of an input bellows, a nozzle and amplifying pilot, and the feedback levers and spring. An air supply of 20 to 100 psi must be provided. When the input air pressure m_1 increases, the input bellows moves to the right and causes the flapper to cover the nozzle. The nozzle back-pressure change is amplified by the pilot and is transmitted to the diaphragm. The diaphragm moves down and the feedback lever compresses the spring to return the flapper to a balanced position. Thus the actuator stem assumes a position dictated by the input air pressure. The spring actuator becomes a power means and the characteristics of the spring and the diaphragm are relatively less important. The use of the positioner results in several improvements in performance which is true in general for all valve positioners.

1. Hysteresis is reduced and linearity is usually improved because the static operation is governed by the feedback spring and input bellows.
2. The actuator can handle much higher static friction forces because of the amplifying pilot.
3. Variable thrust forces on the motor stem do not disturb the stem position to any great extent.
4. Speed of response is generally improved because the pneumatic controller must only supply sufficient air to fill the small input bellows rather than the large actuator chamber.

It is to be noted that the use of a positioner with a spring actuator does not improve the ability of the actuator to handle large inertia or thrust forces unless special adjustments of motor operating range are made.

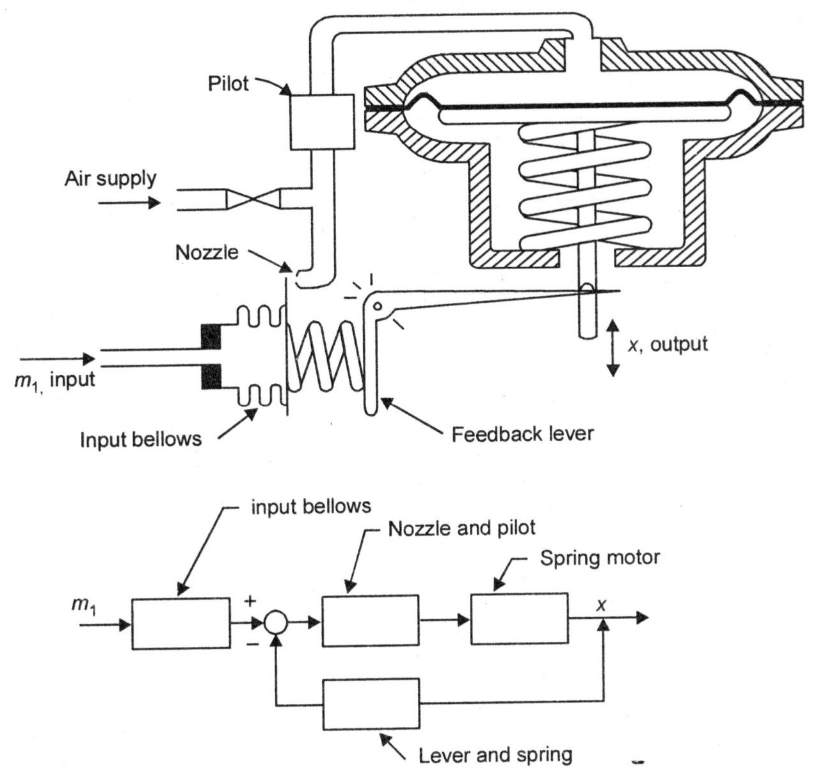

Fig. 4.33 Spring and diaphragm motor with positioner

4.4.3 Pneumatic Motion-Balance Positioner : (Refer Fig. 4.34)

This is same as the one explained in the section 4.4.2 but for the flapper arrangement . The motion-balance positioner compares the motion of an input bellows or diaphragm with linkage attached to the valve stem. Bellows-type input elements are generally thought to be more accurate than simple diaphragms.

4.4.4 Pneumatic Force-Balance Positioner

The force-balance positioner (Refer Fig. 4.35) has an element that compares the force generated by the input signal with the force generated by the feedback spring connected to the valve stem. If the position of the valve stem differs from that called for by the input pressure, the spool valve moves and air is sent to the actuator until the correct position is achieved. The positioner reduces the effects of stem position and unbalanced pressure forces on the valve plug. It also increases the speed of response of the system, since the air entering the actuator does not have to flow through a long transmission line.

4.4.5 Electronpneumatic Force-Balance Positioner

Fig. 4.36 illustrates the functioning of an electropneumatic force-balance valve positioner. The force generated by the electrical input signal through electromagnetic coil is compared with the force generated by the feedback spring connected to the valve stem. The pivot acts as a balancing media and the position of the electromagnetic coil attached with flapper decides the air inlet to the actuator. It is also possible to have a *I/P* converter in between and use the same positioner as discussed in section 4.4.4.

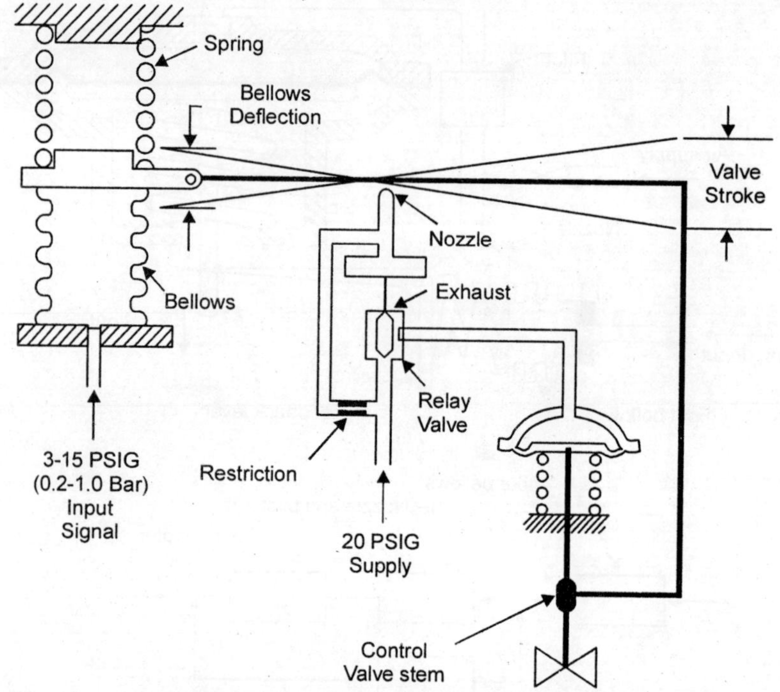

Fig. 4.34 Motion-balance positioner

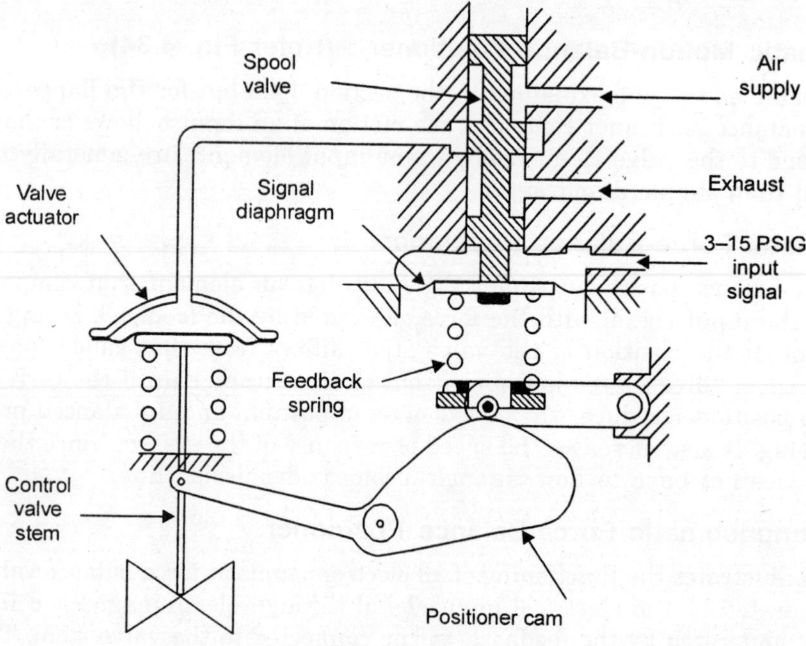

Fig. 4.35 Force-balance positioner

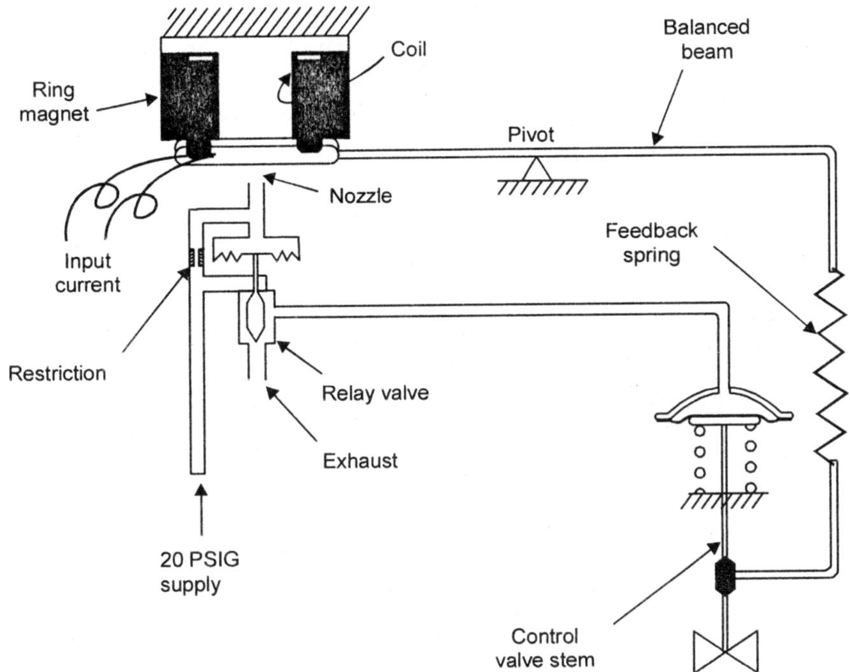

Fig. 4.36 Electropneumatic force-balance positioner

4.4.6 Electrohydraulic Positioner : (Refer Fig. 4.37)

The standard command signal controls a torque motor or voice coil to position a flapper or other form of variable nozzle. This positions a spool valve to control the hydraulic positioning of a high-pressure second-stage valve. The second-stage valve directs operating pressure to the cylinder for accurate positioning. Closing the loop requires mechanical or electrical feedback to compare the piston position with the controller output (command) signal.

4.5 CONTROL ELEMENTS

In the beginning of this chapter we have mentioned that the final control element system often consists of three parts namely (1) Signal conversions (2) Actuators, and (3) Control elements. We have already discussed in detail about signal conversions in section 4.2 and about different types of actuators in section 4.3. Also as a part of actuators, we have discussed about valve positioners in section 4.4 although it is applicable only for control valves. In this section we will be seeing briefly about direct mechanical and electrical control elements and then discuss in some detail about control valves because control valves play an very important role as control element in process industries, especially in chemical process industries.

4.5.1 Mechanical Control Elements

Control elements that perform some mechanical operation in a process (by virtue of operations) are called mechanical control elements. To understand better we will see two such examples.

1. Solid material hopper valves : consider the grain supply bin shown in Fig. 4.38. The control system is to maintain the flow of grain from the storage bin to provide a constant flow rate on the conveyor. This flow depends on the height of grain in the bin, and hence the hopper valve must open or close to compensate for the variation. In this case an actuator operates a vane-type valve to control the grain flow rate. The actuator could be a motor to adjust shaft position (electrical), a pneumatic spring/diaphragm type or a hydraulic cylinder.

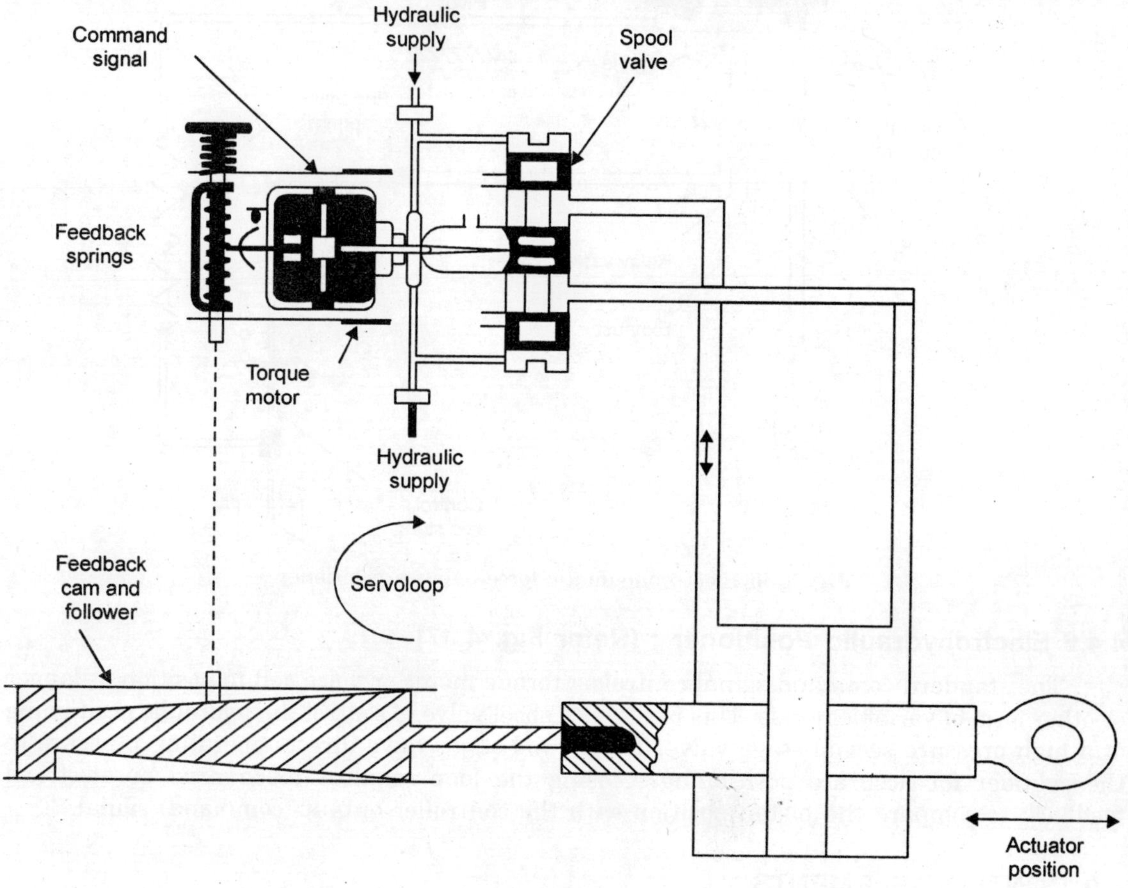

Fig. 4.37 Two-stage servo-valve with mechanical feedback

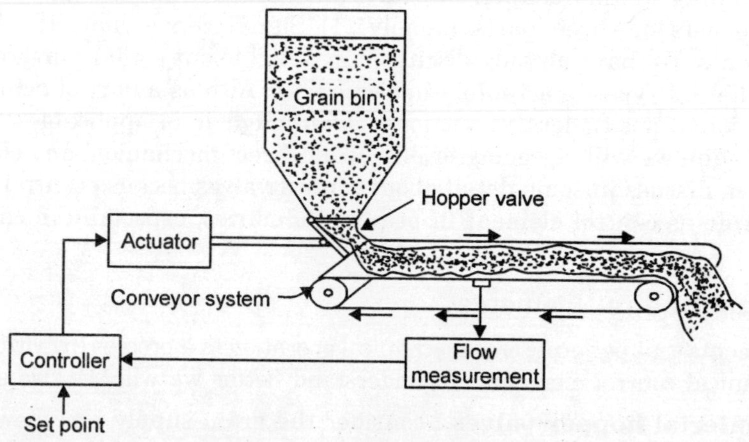

Fig. 4.38 An example of mechanical control element in the form of a hopper valve

2. Controlling of paper thickness : The essential features of a system for continuously controlling paper thickness are shown in Fig. 4.39. The paper is in a wet fiber suspension and is passed between rollers. By varying the roller separation, paper thickness is regulated. The mechanical control element shown is the movable roller. The actuator could be electrical, pneumatic, or hydraulic, and adjusts roller separation based on a thickness measurement.

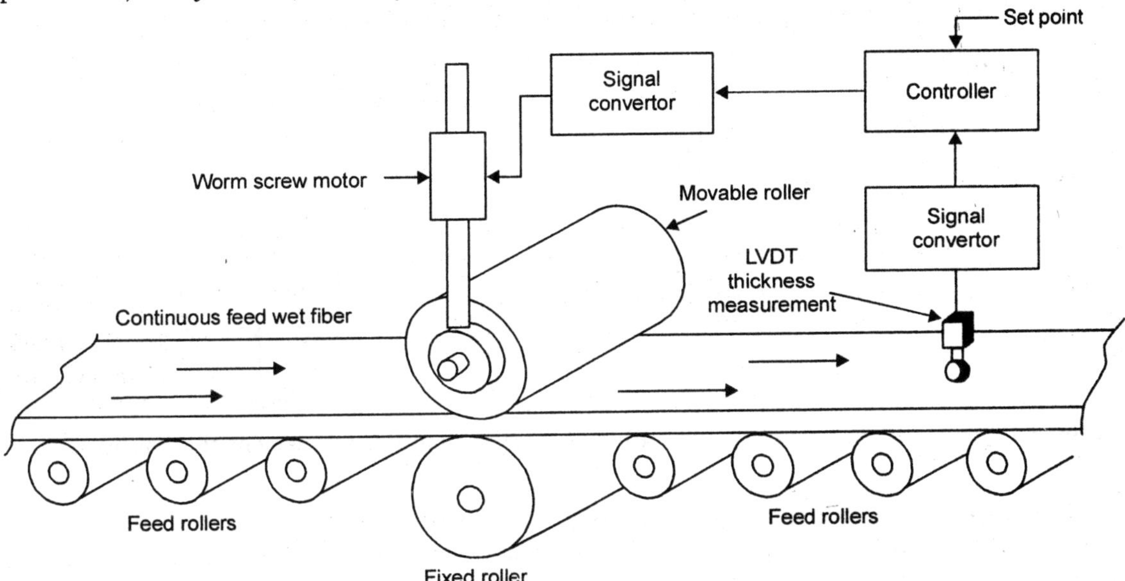

Fig. 4.39 A continuous operation paper thickness controlling system using the mechanical final control elements

4.5.2 Electrical Control Elements

There are numerous cases where a direct electrical effect is impressed in some process-control situation. We will see three typical cases of electrical control elements.

1. DC motor speed control : The speed of large electrical motors depends on many factors, including supply voltage level, load, and others. A process control loop regulates this speed through direct change of operating voltage or current, as shown in Fig. 4.40 for a dc motor.

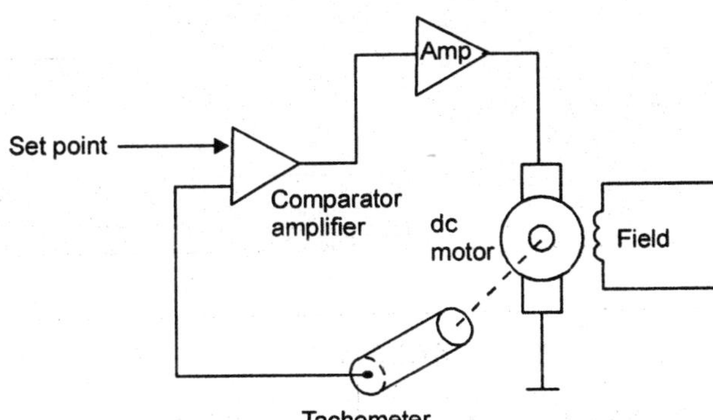

Fig. 4.40 Electrical final control as found in the control of a dc motor speed.

2. Rotational rate control of a kiln : Here the motor speed control is an intermediate operation in a process control application. In the operation of a Kiln for solid chemical reaction, the rotation rate may be varied by motor speed control based on reaction temperature as shown in Fig. 4.41.

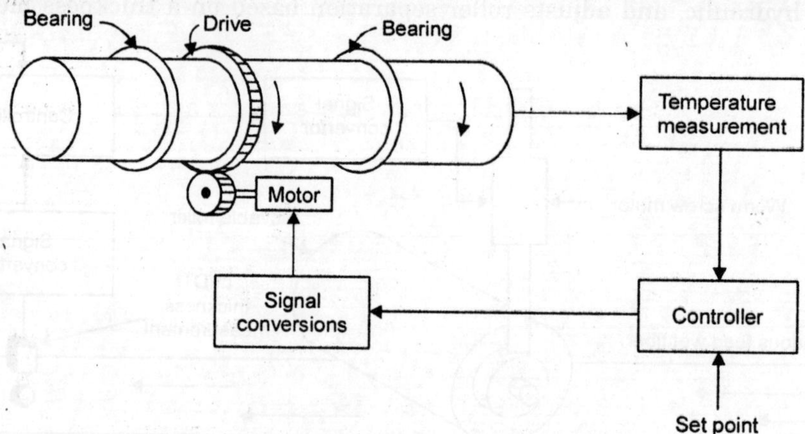

Fig. 4.41 An electrical control system with an electrical final control element that varies the rotational rate of a reactions Kiln.

3. Temperature control using heaters : Temperature often is controlled by using electrical heaters in some application of industrial control. Thus, if heat can be supplied through heaters electrically in an endothermic reaction, then the process control signal can be used to ON/OFF cycle a heater or set the heater within a continuous span of operating voltages as in Fig. 4.42. In this example, a reaction vessel is maintained at some constant temperature using an electrical heater. The process-control loop provides this by smoothly varying excitation to the heater.

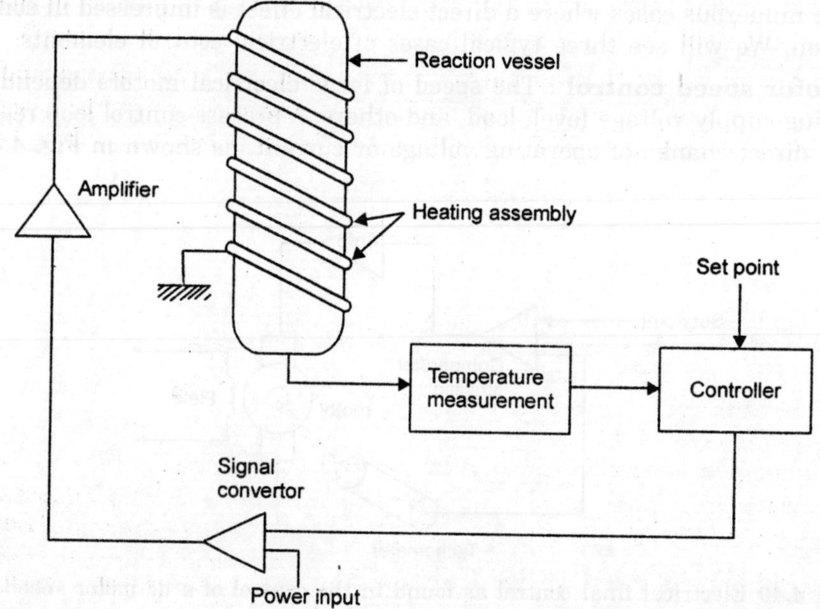

Fig. 4.42 Control of heat to a reaction vessel can be provided by purely electrical means

4.5.3 Control Valves as Control Elements

The chemical and petroleum industries, and many other industries depend in part on operations that involve fluids and the regulation of fluid parameters. The word 'fluid' here represents either gases, liquids or vapours. Many principles of control can be equally applied to any of these states of matter with only slight corrections. Many fluid operations require regulation of such quantities as density and composition, but by far the most important control parameter is 'flow rate'. A regulation of flow rate emerges as the regulatory parameter for reaction rate, temperature, composition, or a host of other fluid properties. We will consider in some detail that process-control element specifically associated with flow—'the control valve'.

Control valve principles : Flow rate in process control is usually expressed as volume per unit time. If a mass flow rate is desired, it can be calculated from the particular fluid density. If a given fluid is delivered through a pipe, then the volume flow rate is

$$Q = A \,.\, v \qquad \text{...(4.6)}$$

where Q = flow rate (m^3/sec)

A = pipe area (m^2)

v = flow velocity (m/sec)

A control valve regulates the flow rate in a fluid delivery system. In general, a close relation exists between the pressure along a pipe and the flow rate so that if the pressure is changed, then the flow rate is also changed. A control valve changes flow rate by changing the pressure in a flow system because it introduces a constriction in the delivery system. In Fig. 4.43, the placement of a constriction in a pipe introduces a pressure difference across the pipe. We can show that the flow rate through the constriction is given by

$$Q = K\sqrt{\Delta p} \qquad \text{...(4.7)}$$

where K = Proportionality constant (m^3/sec/Pa ½)

$\Delta p = p_1 - p_2$ = Pressure difference (Pa)

= Pressure drop across the constriction.

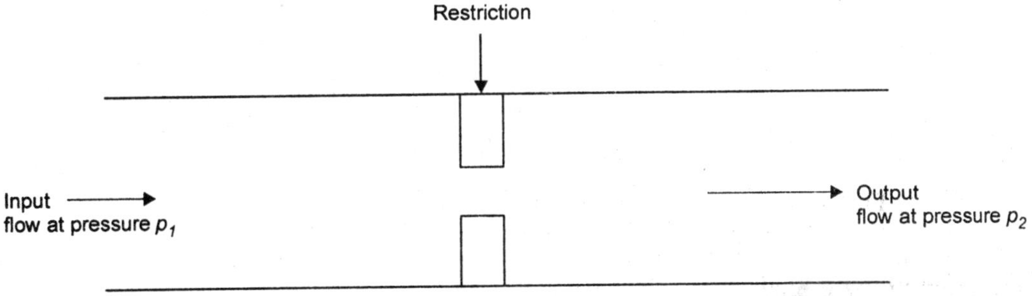

Fig. 4.43 Flow rate through a restriction in a line is a function of the pressure drop across the restriction

The constant K depends on the size of the valve, the geometrical structure of the delivery system, and, to some extent, on the material flowing through the valve. Now the actual pressure of the entire fluid delivery (and sink) system in which the valve is used (and, hence, the flow rate) is not a predictable function of the valve opening only. But because the valve opening does change flow rate, it provides a mechanism of flow control.

4.5.3.1 Control Valve Types-Characteristics of Control Valves

The different types of control valves are classified by a relationship between the valve stem position and the flow rate through the valve. Control valves exhibit an 'inherent characteristic' and an 'installed or effective characteristic'.

Inherent characteristics : This control valve characteristics is assigned with the assumptions that the stem position indicates the extent of the valve opening and that the pressure difference is determined by the valve alone.

Installed or effective characteristics : The control valve when installed in a process with pipe lines, downstream and upstream equipment will exhibit a different flow rate—stem position relation and is called installed or effective characteristics.

4.5.3.1.1 Inherent characteristics

Based on inherent characteristics, there are three basic types of control valves, whose relationship between stem position (as percentage of full range) and flow rate (as a percentage of maximum flow rate) is shown in Fig. 4.44.

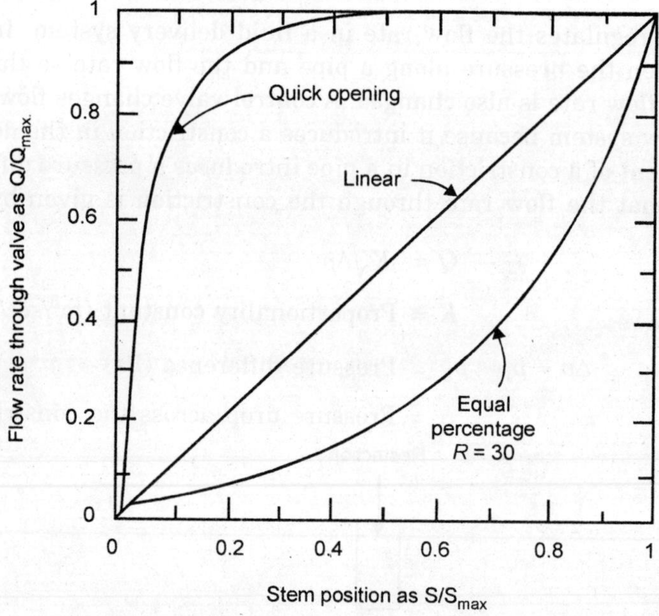

Fig. 4.44 Different responses of the three main types of control valves with respect to stem position

The assumptions made are :

1. The actuator is linear (valve travel is proportional with controller output)
2. The pressure difference across the valve is constant.
3. The process fluid is not flashing, cavitating, or approaching sonic velocity (choked flow)

1. Quick opening valve

This type of valve is used predominantly for full on/full off control applications. The valve characteristic in the Fig. 4.44 shows that a relatively small motion of valve stem results in maximum possible flow rate through the valve. Such a valve, for example, may allow 90% of maximum flow rate with only a 30% travel of the stem.

Such a valve is also called as 'decreasing sensitivity type valve'. The valve sensitivity ($\Delta Q/\Delta S$) at any flow decreases with increasing flow. The maximum port area is sufficiently large that pressure losses elsewhere than at the valve port may restrict the maximum flow.

2. Linear valve

This type of valve, as shown in Fig. 4.44, has a flow rate that varies linearly with the stem position. It represents the ideal situation where the valve alone determines the pressure drop. The relationship is expressed as

$$\frac{Q}{Q_{max}} = \frac{S}{S_{max}} \quad ...(4.8)$$

where

Q = Flow rate (m^3/sec)

Q_{max} = Maximum flow rate (m^3/sec)

S = Stem position (m)

S_{max} = Maximum stem position (m)

The valve sensitivity ($\Delta Q/\Delta S$) is more or less constant at any flow.

3. Equal percentage valve

A very important type of valve employed in flow control has a characteristic such that a given percentage change in stem position produces an equivalent change in flow, that is, an equal percentage. Generally, this type of valve does not shut off the flow completely in its limit of stem travel. Thus, Q_{min} represents the minimum flow when the stem is at one limit (closing limit) of its travel. At the other extreme, the valve allows a flow Q_{max} as its maximum, open valve, flow rate. For this type, we define the 'rangeability' R as the ratio

$$R = \frac{Q_{max}}{Q_{min}} \quad ...(4.9)$$

The curve in the Fig. 4.44 shows a typical equal percentage curve that depends on the rangeability for its exact form. The curve shows that increase in flow rate for a given change in valve opening depends on the extent to which the valve is already open. This curve is typically exponential in form and is represented by

$$Q = Q_{min} R^{S/S_{max}} \quad ...(4.10)$$

Such a valve is also called as 'increasing sensitivity type valve'. This is termed the equal-percentage, logarithmic, parabolic, or 'characterised' type of valve because the flow-lift curve plotted on semi logarithmic coordinates is approximately a straight line. The valve sensitivity ($\Delta Q/\Delta S$) increases with increasing flow rate. The valve sensitivity at any given flow rate is a constant percentage of the given flow rate. Thus, the term equal-percentage.

4. Special characteristic valves

In addition to the already discussed three basic types of valves namely quick opening, linear and equal percentage, we will see some valves with special characteristics. The characteristic curve depends on the geometrical shape of the plug's surface.

The equation 4.7 may be rewritten, taking flow characteristics [$f(x)$] and density of the fluid into account, as below :

$$Q = K\, f(x) \sqrt{\frac{\Delta p}{\rho}} \qquad ...(4.11)$$

where $f(x)$ = value flow characteristic.

ρ = density/specific gravity of the fluid.

Types of valve plugs for different inherent characteristics are shown in Fig. 4.45. Fig. 4.46 shows the flow capacity characteristics for various valves including those three basic types of valves. $f(x)$ for different types of values are given below:

Linear: $f(x) = x$

Square root: $f(x) = \sqrt{x}$

Equal percentage: $f(x) = \alpha^{x-1}$

Hyperbolic: $f(x) = \dfrac{1}{\alpha - (\alpha - 1)x}$

(α can be varied)

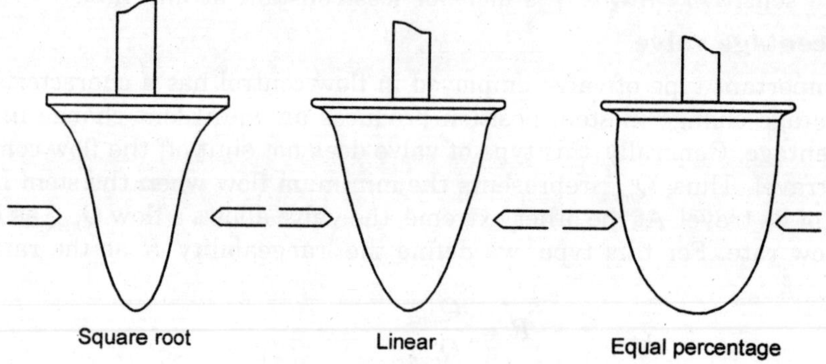

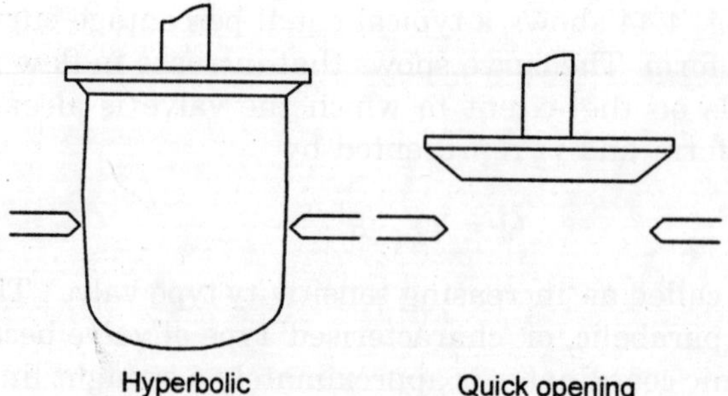

Fig. 4.45 Types of plugs for control valves

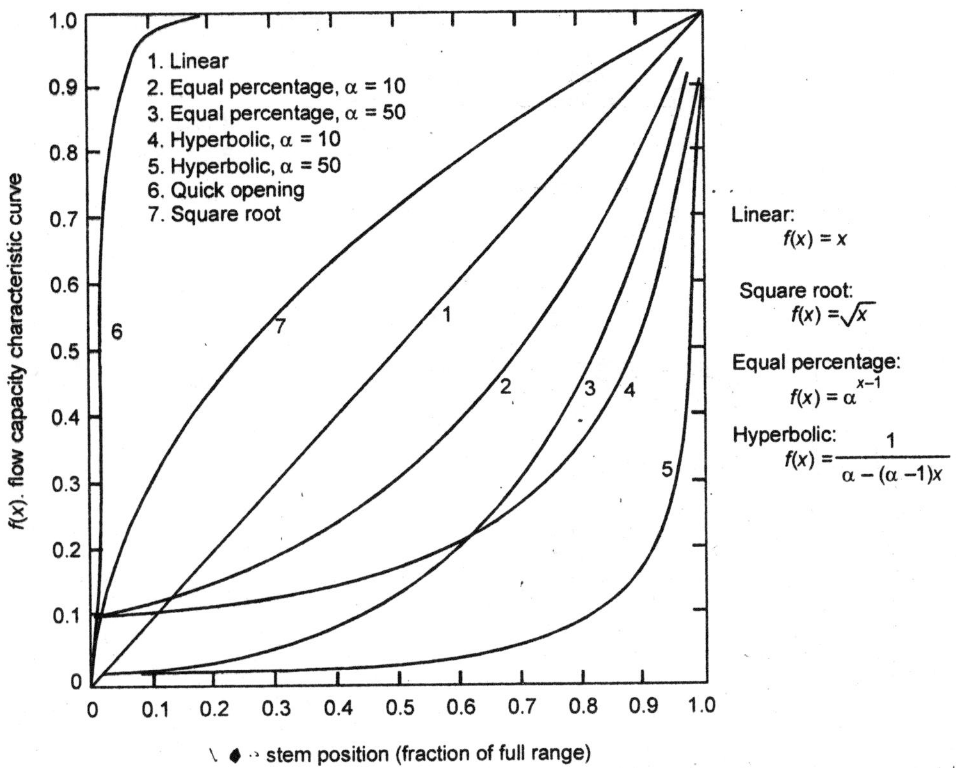

Fig. 4.46 Flow capacity characteristics for various valves.

4.5.3.1.2 Installed (or effective) valve characteristics

The flow characteristic of a valve in a process control system depends on the inherent characteristic and on the change in valve pressure drop with flow rate. When the control valve is installed as part of a process plant, its flow characteristics are no longer independent of the rest of the system. The fluid flow through the valve is subject to frictional resistances in series with that of the valve. Consider a simple system with pump, valve and connected pipe lines as shown in Fig. 4.47 and the consequence of distortion in the characteristics illustrated both for linear and equal percentage valves in the curves below. From these curves, one can conclude that the particular installation involved can have a very substantial effect on both flow characteristics and 'Rangeability'. Clearance flow alone can increase as much as ten fold, and equal-percentage characteristics can be distorted toward linear or even quick opening under conditions of excessive distortion. The distortion coefficient D_C used here is given by the equation :

$$D_C = \frac{(\Delta P_t)_{min} \Delta P}{(\Delta P_t)_{max} \Delta Ps} \quad ...(4.12)$$

where ΔP = Pressure drop across valve.

ΔPs = Pressure drop across system

ΔPt = Total pressure drop = $\Delta P + \Delta Ps$

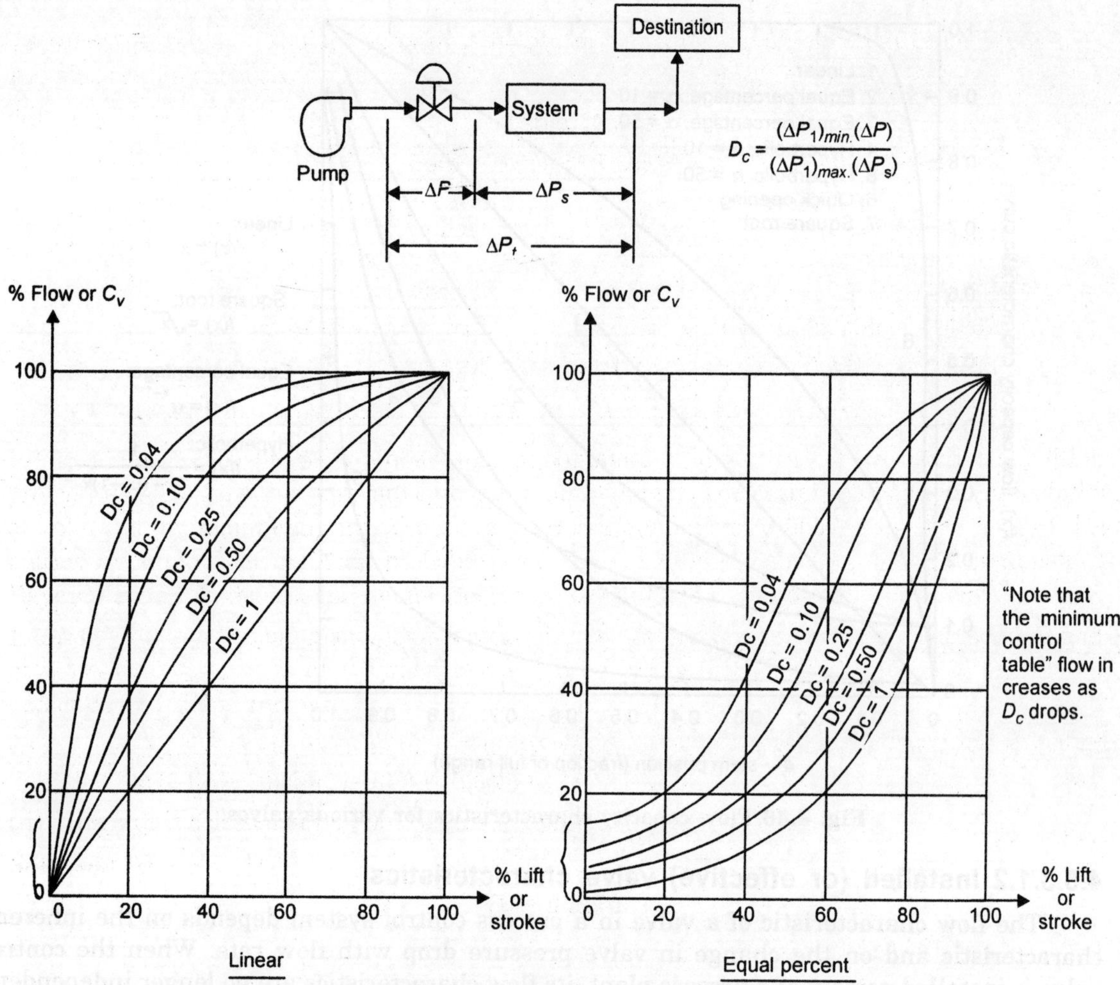

Fig. 4.47 The effect of the distortion coefficient (dc) on inherently linear and equal-percentage valves.

The predictability of installed valve behaviour is further reduced by other factors, such as the following :

1. Deviation in inherent valve characteristics.
2. Actuators without positioners will introduce non-linearities.
3. Pump curves will also introduce nonlinearities.

It should also be recognized that in order to learn the true requirements for valve characteristics, a full dynamic analysis is required.

4.5.3.2 Valve Body and Commercial Valve Bodies

The control valve is essentially a variable resistance to the flow of a fluid, in which the resistance and therefore, the flow can be changed by a signal from a process controller. The control valve itself is divided into the body and the trim.

The body consists of a housing for mounting the actuator and flange connections for attaching/installing the valve in a pipeline. The trim, which is enclosed within the body, consists of a plug, a valve seat and a valve stem. The actuator moves the valve stem and inturn the stem moves the plug in a valve seat in order to change the resistance to flow through the valve. That is, the cross-sectional area between the plug and the seat is changed to change the flow rate.

Fig. 4.48 shows the principal parts of a double-seated control valve whereas Fig. 4.49 shows that of a single-seated valve. The bonnet assembly is attached to the valve body. The body stem moves through the bonnet which contains a means for sealing against leakage such as a stuffing box assembly with suitable packaging or a sealing bellows. The blind head may be with or without guide bushings. The valve plug has extensions on top and bottom which are the valve plug guides. These guides keep the valve-plug motion in alignment. The yoke is the structure which is supported rigidly on the bonnet assembly and carries the diaphragm actuator.

Valve trim consists of those internal components within the valve body which come in contact with the process fluid passing through the valve. Valve trim includes components such as seat rings, valve stems and valve plugs.

Valve bodies are generally cast. The most frequently used materials are cast iron, cast steel and bronze. For corrosion service, stainless steel, nickel, carbon-molybdenum etc are used.

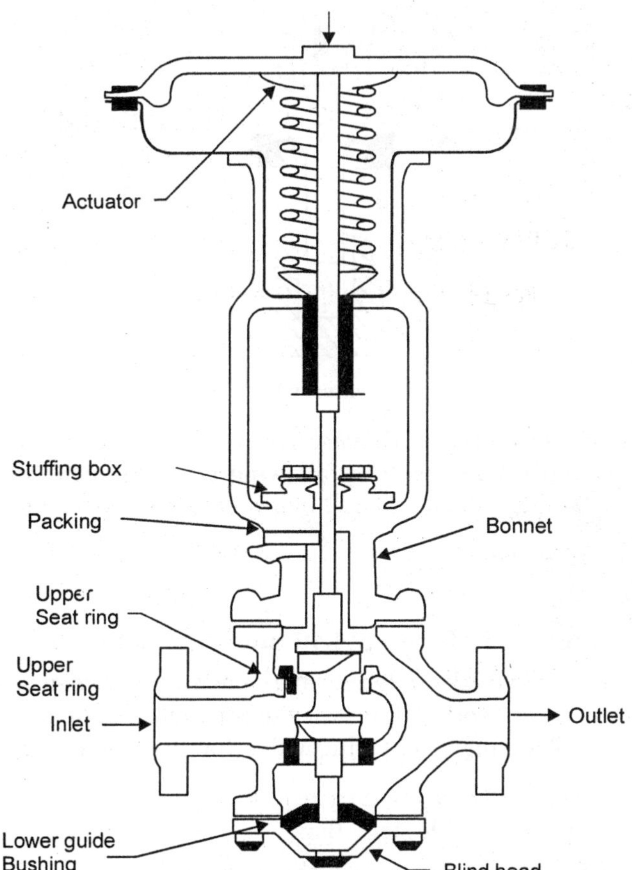

Fig. 4.48 Double-seated pneumatic control valve

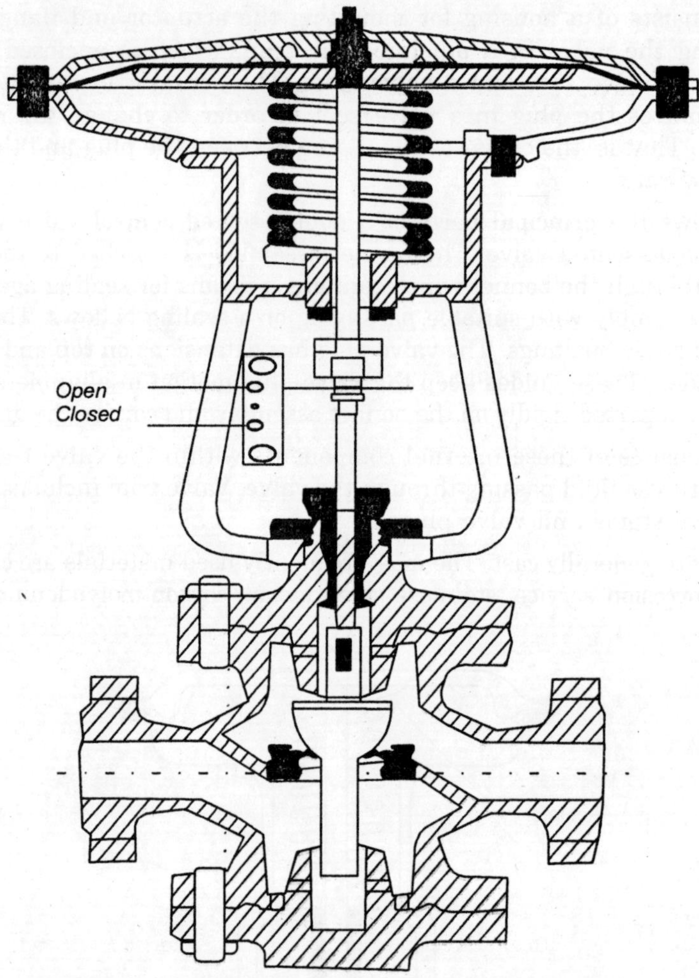

Fig. 4.49 Single-seated two-way control valve with single-acting motor (air-to-close action)

4.5.3.2.1 Sliding-Stem Control Valves: The valve body discussed above is meant for operating sliding -stem control valves. The following types of such valves are available :

1. Single-seat plug valves
2. Double-seat plug valves
3. Lifting gate valves

1. Single-seat plug valves : (Refer Fig. 4.49 and Fig. 4.52)

The single-seat plug valve has only one port opening between seat and plug and the entire flow passes through this port. A few types of plugs for single-seat valves are shown in Fig. 4.50 and Fig. 4.45 . The single-seat plug valve has the following features :

1. It is simple in construction
2. It can be shut off to provide zero flow.
3. There is a large force acting across the port and seat area.

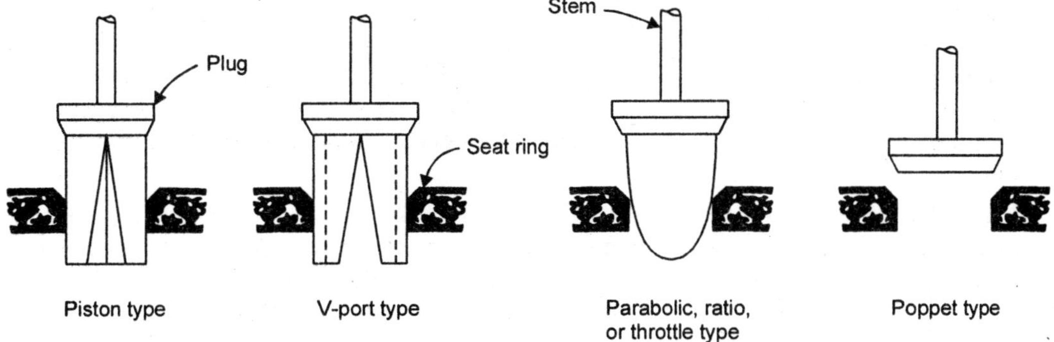

Fig. 4.50 A few types of single-seat valve plugs.

2. Double-seat plug valves : (Refer Fig. 4.48 and Fig. 4.52)

The double-seat valve has two port openings and two seats and two plugs. The port openings are not usually identical in size. Two types of double-seat valve plugs are shown in Fig. 4.51. This type has the following features :

(*a*) Net force acting on the valve stem is generally small and therefore pressure balanced.

(*b*) It cannot be shut off tightly because of differential temperature expansion of valve plug and valve body.

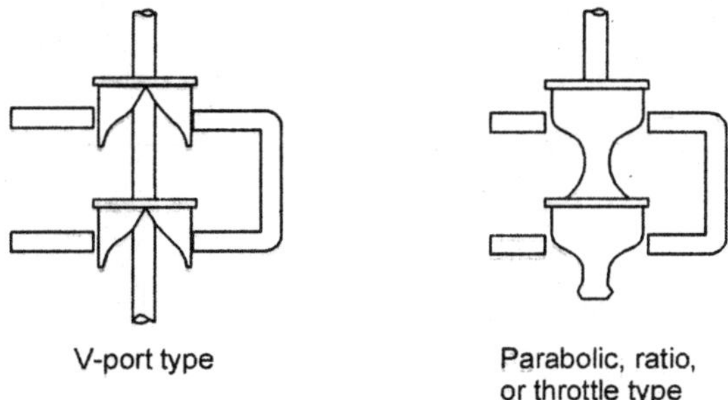

Fig. 4.51 Two types of double-seat valve plugs

The single-seat and double-seat control valves are shown in Fig. 4.52 for clarity.

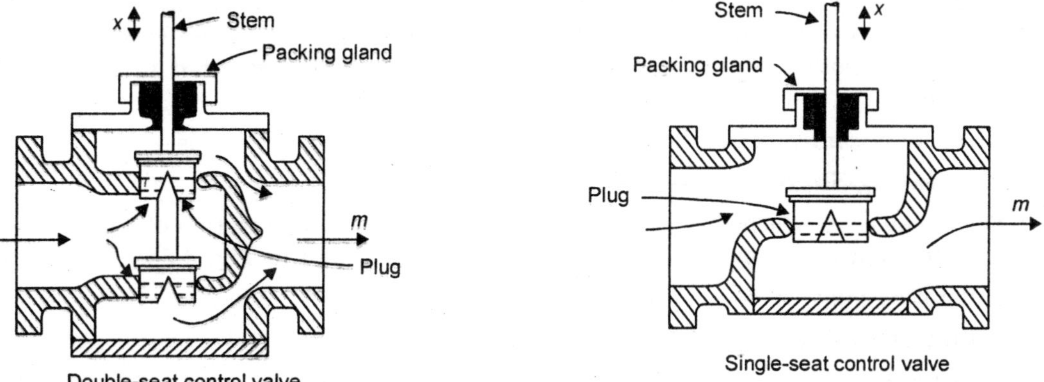

Fig. 4.52 Sliding-plug control valve

Valve plugs : The piston type plug has one or more grooves along its length and the flow passes vertically in the groves between the plug and seat ring. The V-port-type plug is open on the inside and the flow passes horizontally through the triangular shaped area over the seat ring. The parabolic plug presents an annular area to flow between the plug and the seat ring. The poppet type plug offers a cylindrical-shaped flow area and is used with small total lift.

3. Lifting gate valves : The gate valve in Fig. 4.53 is often used for fluids containing solid matter, because it presents an open area directly to the flow of fluid and does not involve a change of direction of flow stream. A gate valve can usually be shut-off tightly by wedging into the seat. The chopping action at shut-off is very useful for stringy materials such as paper pulp.

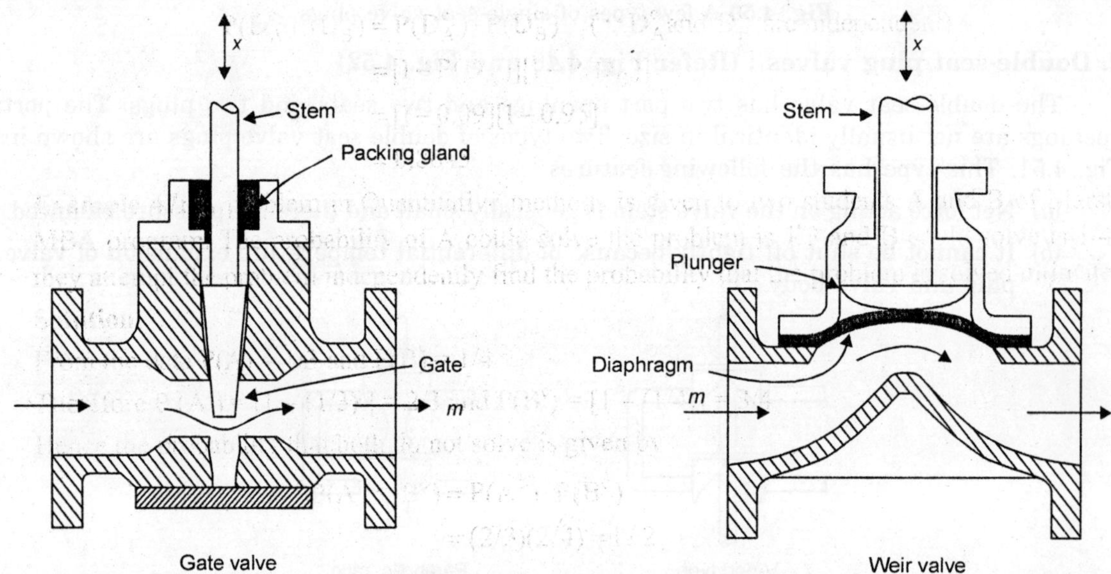

Fig. 4.53 Single-seat valves with a lifting-gate action

The wier valve in Fig. 4.53 is particularly suited to certain chemical fluids, because it has a smooth contour inside the body with no 'pockets' for solid matter, and because it has no packing gland around the stem. The flexible diaphragm of rubber or other non metallic material is positioned by the plunger and stem. Fluid pressure inside the valve body holds the diaphragm smoothly against the plunger. This valve is also referred to as 'Saunders Diaphragm Valve'.

Sliding stem control valves are in general classified as 'Globe valves'

4.5.3.2.2 Rotating-Shaft Control Valves

Control valves in which the restriction is accomplished by the rotation of a plug or vane may be called rotating shaft type. Some of them are illustrated in Fig. 4.54.

1. Rotating-plug valves
2. Butterfly valves
3. Louvers.

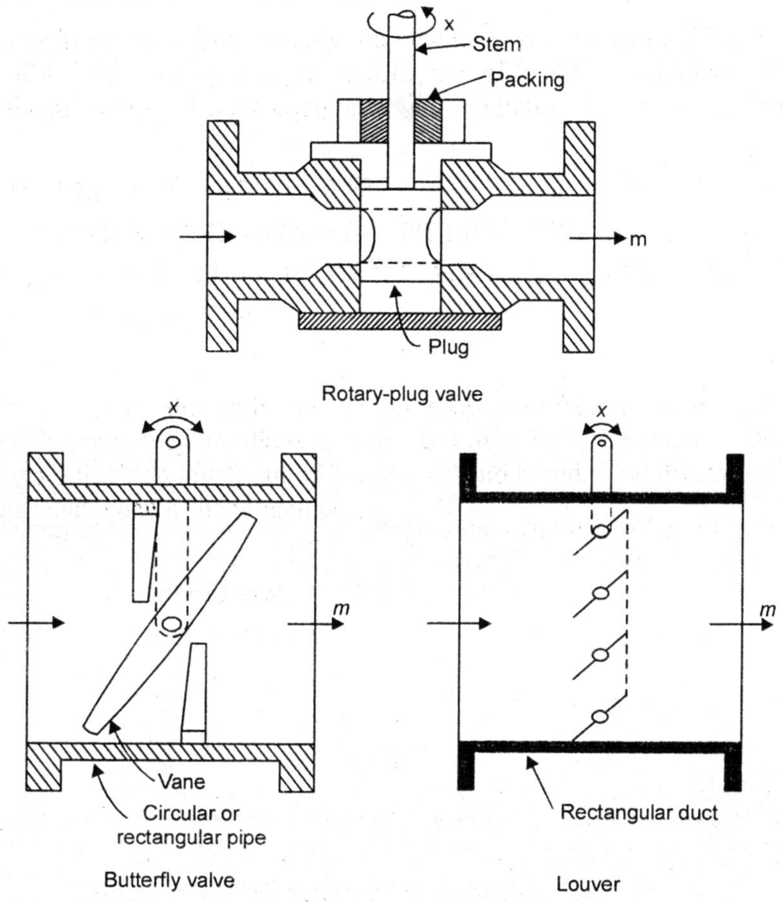

Fig. 4.54 Several rotating-shaft valves

1. Rotating-plug valves : The plug is a cylindrical or conical element with a transverse opening. It is rotated in the valve body by an external lever so that the opening on one side of the plug is gradually covered or uncovered. The shape of the opening or part may be circular, *V*-shape, rectangular, or any form that is desired to produce a given flow-angle characteristic. A rotating-plug valve having a conical plug can generally be closed tightly and has high rangeability. This type of valve is normally employed for throttling the flow of oil to burner system. The valve with a spherical plug that controls the flow of fluid through the valve body is called as 'ball valve'.

2. Butterfly valves : The butterfly valve consists of a single vane rotating inside a circular or rectangular pipe or casing. The shaft projects through the casing and operated externally. The total rotation of the vane is restricted to about 60 degrees, because the additional 30 degrees does not produce much further increase in flow (Theoretical rotation of a butterfly vane is 90 degrees). The *V*-port butterfly valve incorporates a *V*-slot in the body so that rotation of the vane opens a portion of the *V*-slot. The rangeability of butterfly valves may vary from 5 to 50 and tight shut-off may be obtained with special design. The butterfly valve is most often employed for the control of air and gas. Vane positions of butterfly valve at three extreme positions are shown in Fig. 4.55 for clear understanding.

Fig. 4.55 Vane positions of butterfly valve

3. Louvers : The louver consists of two or more rectangular vanes mounted on shafts one above the other and interconnected so as to rotate together. The vanes are operated by an external lever. In the unirotational louver the vanes remain parallel at all positions. In a counter rotational louver alternate vanes rotate in an opposite direction. Flow guides are sometimes installed between adjacent vanes in order to improve the effectiveness of throttling. A louver cannot provide tight shut-off because of the long length of seating surfaces. Louvers are used exclusively for control of air flow (draft) at low pressure.

The flow-angle characteristics of rotating-shaft control valves are shown in Fig. 4.56. The flow-angle characteristic of rectangular port rotary plug is almost linear. The flow-angle characteristic for butterfly valve is for a 60° butterfly valve. It may be seen from the flow-angle characteristic of a 90° unirotational louver that the sensitivity is very high at mid flow and that the last 30° of rotation is relatively ineffective.

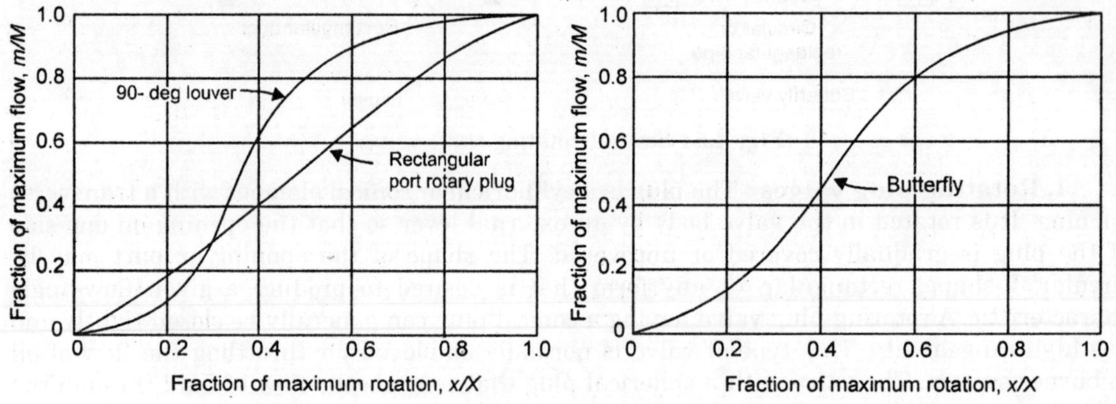

Fig. 4.56 Actual flow-angle characteristics of rotary-shaft control valves

4.5.3.2.3 Special Control Valves: There are some more types of valves in industrial use than the ones discussed above. A 'Cock valve' consists of a tapered plug which fits closely into a tapered hole in the valve body. The plug and the mating hole are ground together to make a leakproof fit.

'Bellows-seal-valves' are used in pipelines handling toxic or valuable fluids. Bellows-seal is used instead of ordinary packings to isolate valve stem action from the atmosphere.

'Three-way valves' are either for diverting a stream into two separate streams or for mixing two fluids in controlled proportions.

'Pinch or clamp valves' are used in such industries as mixing, water treatment, sewage and waste disposal, chemical, food, cosmeticals, and others. Traditionally pinch-valves are used throughout the mixing and ore processing industries for handling of all types of slurries. In the water and waste treatment industry, the pinch valve handles many difficult fluids such as lime slurry, raw sewage, recycle sludge, grit and garbage particles, grease and other equally obnoxious materials. In the chemical industry, pinch valves are used on many slurries, erosive or corrosive or both.

4.5.3.3 Rangeability

The rangeability of a control valve is defined as the ratio of maximum controllable flow to minimum controllable flow (as already touched upon in section 4.5.3.1 under equal percentage valve).

$$R = \frac{Q_{max}}{Q_{min}} \quad ...(4.9)$$

where R = Rangeable number.

Q_{max} = Maximum controllable flow.

Q_{min} = Minimum controllable flow.

The minimum controllable flow is defined as the flow below which the valve tends to close completely. It is not the leakage flow (which occurs when the valve is closed), but the minimum flow that is controllable in the sense that it can be changed up or down as the valve stroke is changed. Using this definition, manufacturers of valves usually claim 50:1 rangeability for equal percentage valves, 33:1 for linear valves, and about 20:1 for quick -opening valves. These claims suggest that the flow through these valves can be controlled down to 2%, 3% and 5% of their valve flow coefficient (C_V) (We will discuss in little more detail about C_V while discussing about control valve sizing)

The above definition of rangeability is based on the inherent C_V determined during testing. Installed characteristics may be different depending on the actual conditions. From Fig. 4.47, it can be seen that the minimum controllable flow rises as the distortion coefficient (D_C) drops. At a D_C value of 0.1, for example, the 50:1 rangeability of an equal-percentage valve has dropped to closer to 10:1. This is because the valve pressure drop is much higher at lower flows, and therefore the minimum valve opening will pass much more flow. Thus, the required rangability should be calculated as the ratio of the C_V required at maximum flow (and minimum pressure drop) and the C_V required at minimum flow (and maximum pressure drop).

Turndown

Turndown is a similar concept of rangeability and is defined as the ratio of normal maximum flow to minimum controllable flow.

$$T = Qn_{max}/Q_{min} \quad ...(4.13)$$

where T = Turndown

Qn_{max} = Normal maximum flow

Q_{min} = Minimum controllable flow

Normal maximum flow is generally taken as 70 % of maximum controllable flow so that

$$T \cong 0.7\,R \qquad \text{...(4.14)}$$

The importance of rangeability and turndown lies in the application of the control valve. For example, if the design of an oil burner and furnace requires a 30 to 1 range of oil flow to accommodate various loads on the furnace, the turndown must be atleast 30 and the rangeability must be atleast 43 (= 30/0.7).

4.5.3.4 Control Valve Sizing

The proper sizing of control valve is important because of the effect on the operation of the automatic controller. If the control valve is over size, for example, the valve must operate at low lift and the minimum controllable flow is too large. In addition, the lower part of the flow-lift characteristic is most likely to be non uniform in shape. On the other hand if the control valve is undersize, the maximum flow desired for operation of a process may not be provided.

Factors that influence sizing of control valves: Pressure drop across the control valve, flow rate through the valve and specific gravity (or specific weight) are the main determining factors in selecting a suitable size for control valves. Other factors such as type of fluid, gas or liquid, critical flow conditions for gases and vapours, and viscosity of liquids influence valve size. Before selecting valve size, valve and process characteristics must match to compensate for non linearities in the control valve and process.

Flow coefficient: One of the most useful factors to determine the size of a control valve is the 'flow coefficient' or C_V factor (or K_V factor). Practically all control valve manufacturers supply C_V factors for their valves. These factors form the basis for all calculations. The flow coefficient indicates the amount of flow the control valve can handle under a given pressure drop across the control valve.

C_V factor : The flow coefficient (C_V) is defined as the flow rate of water in gallons per minute at 60°F through a valve at maximum opening with a pressure drop (or pressure differential) of 1 psi measured in the inlet and outlet pipes directly adjacent to the valve body.

K_V factor : Whenever the flow coefficient is mentioned in metric units, it is denoted by the symbol K_V which is defined as the flow rate of water in m^3/hour at about 30°C flowing through the fully opened control valve at a pressure drop of 1 kg/cm^2 across the control.

The following relationships between C_V and K_V can generally be used.

$$C_V = 1.17\,K_V \qquad \text{...(4.15)}$$

$$K_V = 0.86\,C_V \qquad \text{...(4.16)}$$

The flow coefficient is determined by the manufacturer for various types and sizes of valves by actual experiments with water. The flow coefficient for 100% valve opening is termed as C_V (or K_V) of the particular valve size and the variation of C_V (or K_V) at different valve openings is given in the form of a graph, which is termed as valve characteristic.

Flow Rate *vs* Flow Coefficient

Combining Bernoulli's theorem for the conservation of energy, and the continuity equation for the conservation of mass, the ideal flow rate through a pipe restriction such as a control valve can be expressed as :

$$Q = C_V \sqrt{\frac{\Delta_p}{S_G}} \quad \text{...(4.17)}$$

where, Δp = Pressure drop across control valve

S_G = Specific gravity of the liquid

It is important to note that C_V is not a dimensionless coefficient.

Fluid flowing though a control valve obeys the basic laws of the conservation of mass and energy. The pressure profile through a typical control valve is illustrated in Fig. 4.57.

where P_1 = Upstream pressure

P_2 = Downstream pressure

P_{vc} = Pressure at vena contracta (The minimum pressure point down stream from the throttling point is known as the 'vena contracta')

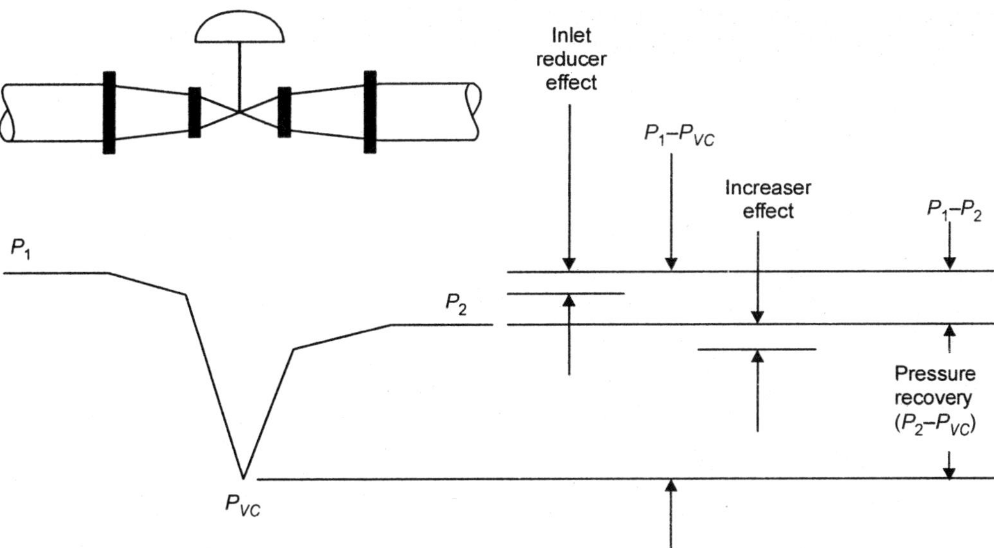

Fig. 4.57 Pressure profile through a valve

Fluids that pass through a valve may be either liquids, steam or gases (including vapours). The same basic flow equation (4.17) holds for all the fluids. Practical equations are in common use for each fluid. Normally it is not necessary to take viscosity of the medium into account. Only when the viscosity of the medium is considerably higher than that of water, correction has to be applied.

Guidelines for Sizing of Control Valves

1. The valve shall be sized for the actual flow condition and not for the ultimate design capacity of the system. Normal maximum flow rate is normally about 70% of the ultimate design capacity.
2. Most of the pressure drop of the system should be across the control valve. As a general rule around 70% of the system drop should be across the control valve.

3. When the pipe line is dimensioned with normal allowable velocities (low pressure loss) the control valve will be a few sizes smaller than the pipeline. Only in extreme cases where very high velocities have been used in the pipe line, the size of the control valve will be same as that of the pipe line.
4. The final selection must be done such that the calculated C_V is attained at about 75 to 80% of the full valve travel. In case of high pressure gases and steam where expansion takes place after the control valve, calculated C_V must be attained at about 50 to 60% of the valve travel.
5. Regardless of the application such as flow control or pressure control the valve sizing is done on the basis of flow coefficient C_V.

4.5.3.5 Cavitations and Flashing

4.5.3.5.1 Cavitations

The phenomenon of cavitation is related to Bernoulli's theorem, which describes the pressure profile as fluid flows through a pipe or passes through a narrower passage, restriction, orifice or control valve (Refer Fig. 4.58). As the fluid accelerates, some of the pressure head is converted into velocity head. This transfer of static energy is needed to push the same mass flow through the smaller passage. The fluid accelerates to its maximum velocity, which is also the point of minimum pressure (vena contracta), and then gradually slows down as it again expands back to the full pipe area. The static pressure also recovers but part of it is lost due to friction. If the static pressure head drops below the liquid vapour pressure (P_V) at that temperature, then vapour bubbles will form downstream of the restriction. As the static pressure recovers to a point greater than the vapour pressure, the vapour bubbles collapse back into their liquid phase. The collapse of the bubbles produces high-energy implosions which is called 'Cavitation'. These implosions generate noise, fluid shock cells, and gets that impinge upon the trim metal parts. It is thought that this phenomenon generates a tremendous and concentrated impact force that destroys the metal as it fractures out tiny metal particles. Cavitation damage gives a very distinctive appearance which is like sandblasting.

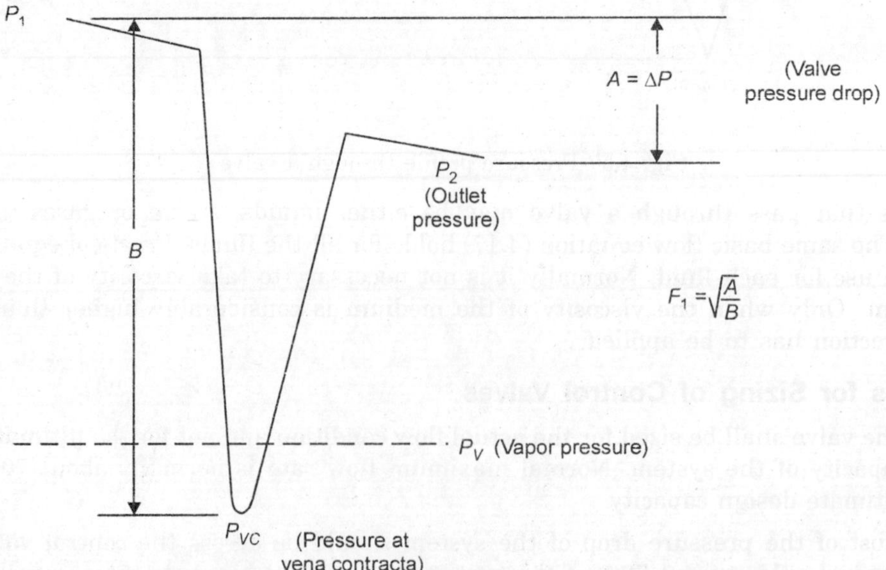

Fig. 4.58 Pressure profile: Single-seat valve experiencing cavitation

Cavitation damage always occurs downstream of the vena contracta when pressure recovery in the valve causes the temporary voids to collapse. Destruction is due to the implosions, which generate the extremely high pressure shock waves in the substantially non compressible stream. When these waves strike the solid metal surface of the valve or downstream piping, the damage gives a cinder-like appearance. Cavitation is usually coupled with vibration and a sound like rock fragments or gravel flowing through the valve.

Elimination of Cavitation

No known material will withstand continuous cavitation without damage and eventual failure. The length of time it will take is a function of the fluid, metal type and severity of the cavitation. Without special trim geometry, some of the mitigating actions possible were to use extremely hard trim materials or overlays to increase the downstream back pressure, or to limit the pressure drop by installing control valves in series to distribute the drop and to reduce the vena contracta pressure in each valve.

Cavitation damage also varies greatly with the type of liquid flowing. The greatest damage is caused by a dense pure liquid with high surface tension. (water, mercury etc). Density governs the mass of the microjet stream, and surface tension governs the most important jet velocity (Refer Fig. 4.59). Because no known material can remain indefinitely undamaged by severe cavitation, the only sure solution is to eliminate cavitation completely. Some of the methods by which cavitation can be reduced or eliminated are listed below.

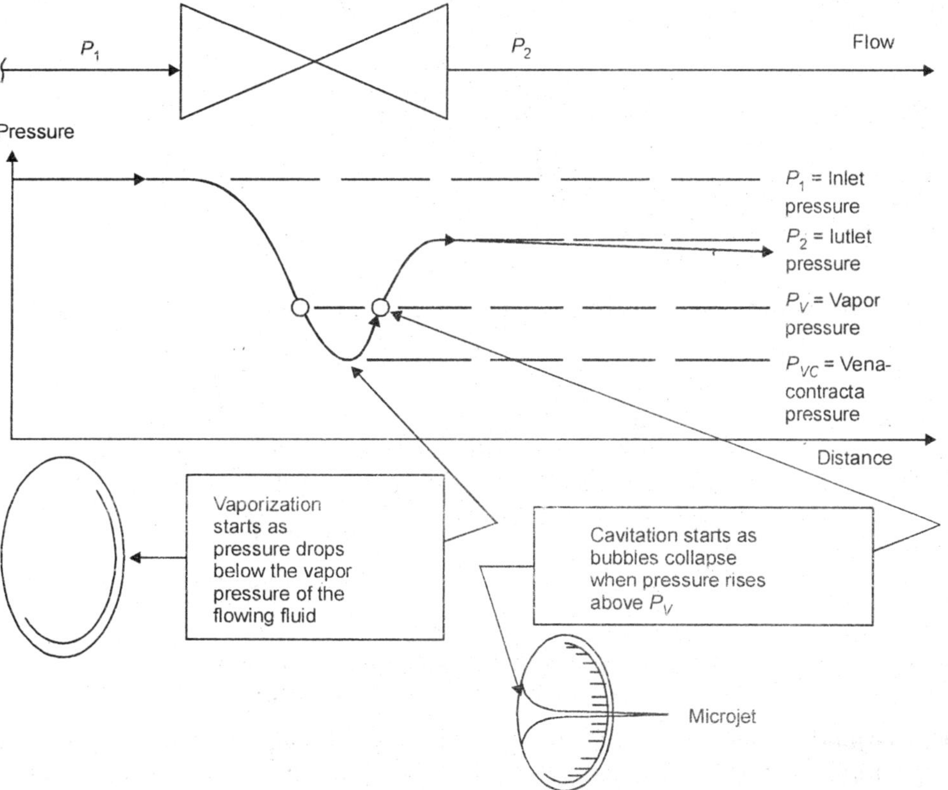

Fig. 4.59 Cavitation : Downstream of the vena contracta the pressure rises. When vapour pressure is reached, the vapour bubbles implode, releasing microjets that will damage any metallic surface in the area.

1. Revised process conditions : A reduction of operating temperature can lower the vapour pressure sufficiently to eliminate cavitation. Similarly, increased upstream and downstream pressures, with ΔP unaffected, or a reduction in the ΔP can both relieve cavitation. Therefore, control valves that are likely to cavitate should be installed at the lowest possible elevation in the piping system and operated at minimum ΔP. Moving the valve closer to the pump will also serve to elevate both upstream and downstream pressures. If cavitating conditions are unavoidable, then it is preferred to have some permanent vaporization (Flashing) through the valve.

2. Revised valve : The valves most likely to cavitate are the high recovery valves (ball, butterfly, gate etc) having low liquid pressure recovery factor (F_L) and low cavitation coefficient (K_C). (The cavitation coefficient K_C is the ratio between the valve pressure drop at which cavitation starts and the difference between the inlet and the vapour pressure of the application. The liquid pressure recovery factor F_L is related to the ratio between the valve pressure drop and the difference between the inlet and the vena contracta pressure.) Fig. 4.60 illustrates some of the ways available to eliminate cavitation.

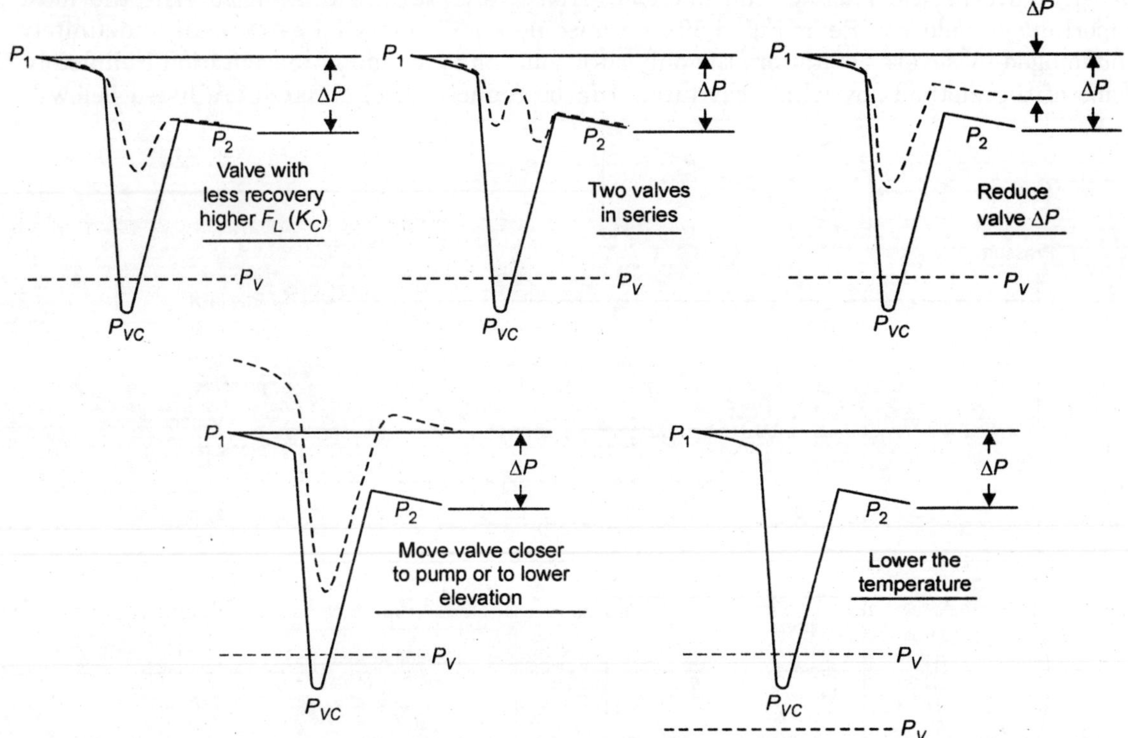

Fig. 4.60 The pressure profile shown in dotted lines illustrate some of the options available to the process control engineer to eliminate cavitation.

3. Gas injection : Another valve design variation that can alleviate cavitation is based on the introduction of non-condensible gases or air into the region where cavitation is expected. The presence of this compressible gas prevents the sudden collapse of the vapour bubbles as the pressure recovers to values exceeding the vapour pressure, and, instead of implosions, a more gradual condensation process occurs. The gas or air may be admitted through the valve shaft or through downstream taps on either side of the pipe, in line with the shaft and as close to the valve as possible as shown in Fig. 4.61. Since the fluid vapour pressure is usually less than atmospheric pressure, the gas or air need not be under pressure.

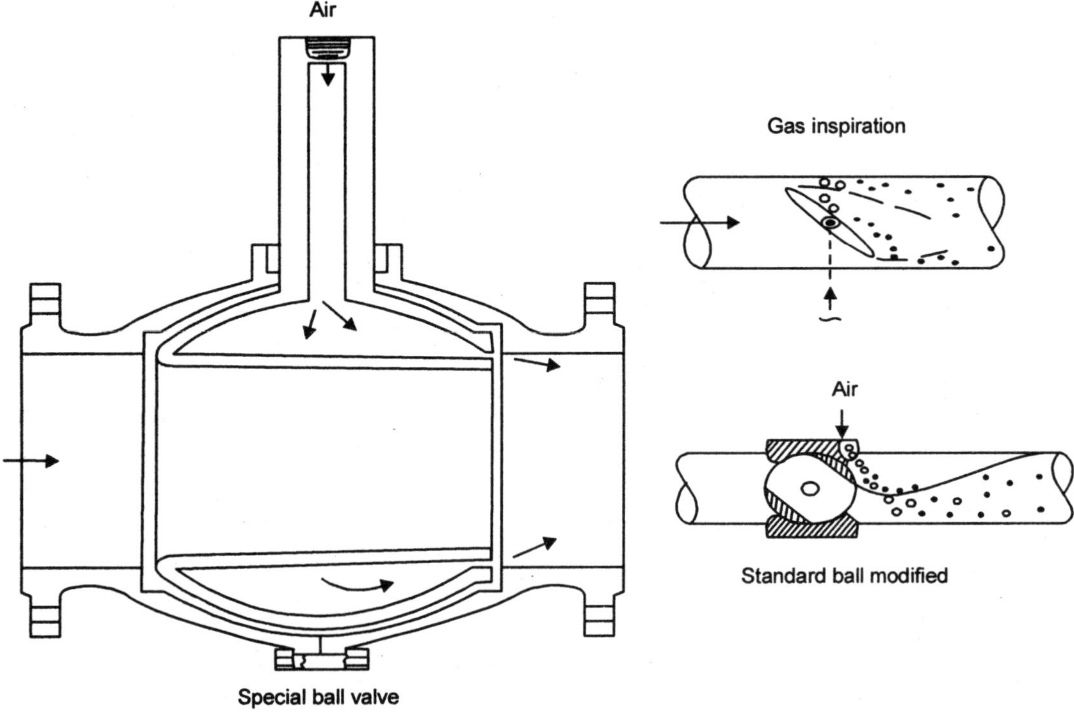

Fig. 4.61 Valve design variations to alleviate cavitation through the admission of air into the flowing stream

4. Revised installation : In order to eliminate cavitation, it is possible to install two or more control valves in series. Cavitation problem can also be alleviated by absorbing some of the pressure drop in restriction orifices, chokes, or in partially open block valves upstream or downstream to the valve.

Cavitation can cause erosion, noise, and vibration in piping systems and therefore should be avoided. Excessive cavitation can also cause choked flow conditions in the valve. Sizing a valve for choked flow allows one to determine its maximum flow capacity, but this is of little value, because valves should not be operated under cavitating conditions.

4.5.3.5.2 Flashing : (Refer again the Fig. 4.59)

Cavitation occurs when $P_2 > P_V$, while flashing takes place when $P_2 < P_V$. When the valve outlet pressure, P_2, is less than or equal to the vapour pressure of the process liquid, some of the liquid 'flashes' into vapour and stays in the vapour phase as it enters the downstream piping. The specific volume increases as liquid changes to vapour (and hence large increase in volume), which in turn causes an increase in the fluid velocity. If enough vapour is formed, the resulting high velocities can erode metals. In this circumstance the piping downstream of a valve needs to be much larger than the inlet-piping in order to keep the velocity of the two-phase stream low enough to prevent erosion. In many cases, flashing is a normal part of the process; it cannot be avoided, and special system and valve designs are required to accommodate it. The ideal valve to use for such applications is an angle valve with an oversized outlet connection. Erosion caused by high exit velocities can also cause corrosion problems. The preferred arrangement for flashing service is to use a reduced port angle valve discharging directly into a vessel or flash tank.

The heat of vapourisation comes from the process liquid, causing its temperature to decrease. The relative masses of liquid and vapour will thereby approach thermodyanamic equilibrium. The amount of flashing can be calculated from an energy balance. Even small amounts of flashing (1 to 3% by weight) can significantly affect a valve's capacity, sizing, and selection; therefore, flashing should be stated in the valve specification data sheets. Large amount of flashing (10 to 15% by weight) require special valve designs, such as oversized outlets, replaceable throats, and special trims. In order to select the right valve, it is necessary to know the fraction of the liquid which will flash to vapour and the velocity of the vapourised mixture. The calculations part is not dealt with here as it is beyond the scope of this book.

4.5.3.6 Selection of Control Valves

Selection of a control valve for a particular application is a very important function for a design engineer. It is really a complex process as there are lot of things to be considered before selecting a suitable control valve. Load changes, pressure drop, rangeability, process gain, flow capacity and characteristic, process itself, surroundings, safety and hazardous conditions, cavitation and flashing details, valve sizing, trim material and cost are some of the important parameters which will finally decide the selection of a control valve. In this section we will discuss briefly the steps involved in selection of control valves.

1. Need for a control valve : Before proceeding through the steps of selecting a control valve, one should evaluate if a control valve is really needed in the first place, or if a simpler and more elegant system will result through some other means. (e.g., variable speed centrifugal pump can be used instead of constant speed pump and a control valve).

2. Collection of process data : One must fully understand the process that the valve controls. It means not only understanding normal operating conditions, but also the requirements that the valve must live upto during start-up,shut-down, and emergency conditions. Therefore all anticipated valves of flow rates, pressures, vapour pressures, densities, temperatures, viscosities etc must be identified in the process of collecting the data for valve sizing in addition to quality and safety requirements.

3. Assigning valve pressure drop : Assigning the sizing pressure drop for the valve is more complex than picking a number like 25% of the total system drop or a number like 25 psi. It requires an understanding of the interrelationships that exist in pumping, fan or compressor systems. The proper approach to the selection of valve pressure drop is to first determine the total friction drop of the system at normal flow and assign 50% of that to valve pressure drop. Based on that assignment, one should next determine the resulting valve drop at minimum and maximum flows and select a valve which can handle the required C_V rangeability.

4. Control valve performance : Good control valve performance usually means that the valve is stable across the full operating range of the process, it is not operating near to one of its extreme positions, it is fast enough to correct for process upsets or disturbances, and it will not be necessary to retune the controller every time the process load changes. In order to meet the above goals one must consider the following factors :

(*a*) *Valve characteristics :* Inherent (quick opening, linear or equal percentage) and installed characteristics.

(*b*) Gain of control loop components like process, sensor, controller and valve gain.

(*c*) Process nonlinearity.

(*d*) Valve rangeability.

(*e*) Control valve sequencing.

(*f*) Split-ranging or floating.

5. Control valve sizing : One should first determine both the minimum and maximum C_V requirements for the valve, considering not only normal but also start-up and emergency conditions. The selected valve should perform adequately over a range of 0.8 [C_V(min)] to 1.2 [C_V (max)]. If this results in a rangeability requirement which exceeds the capability on one valve, use two or more valves.

6. Valve actuator selection : Knowing the applications and relative advantages of different actuator designs, one should be able to select proper actuator for an application. The following factors are to be kept in mind.

(*a*) Whether electrical, pneumatic or hydraulic.

(*b*) Actuator speed of response.

(*c*) Actuator power or torque.

(*d*) Valve failure position (Fail-safe operation).

7. Valve positioner : The following factors are to be considered with respect to valve positioner.

(*a*) When not to use positioner : one should be clear.

(*b*) To eliminate dead band.

(*c*) Split range operation.

8. Process application considerations : In selecting control valves, the properties of the process liquid must be fully considered. The following factors are to be taken into account.

(*a*) High-pressure service

(*b*) High differential pressure usuage.

(*c*) Vaccum service

(*d*) High temperature service : Limitations of metallic parts, packing designs, Jacketed valves etc to be taken into account here.

(*e*) Low-temperature service : Cold box and cryogenic valves.

(*f*) Cavitation and erosion.

(g) Flashing and erosion.

(h) Viscous and slurry service.

(i) Leakage.

(j) Small-flow values.

(k) Control valve noise.

(l) Piping and installation considerations.

(m) Climate and atmosphere corrosion.

9. Control valve specification form : Compiling the information necessary to specify a control valve is best done with the aid of tabulation sheet. Many companies have their own customized forms.

10. Test report and test certificate : This is to be obtained from the manufacturers of control valves for future record, ofcourse after selecting and getting a proper control valve for the required application.

4.5.3.7 Signal Transmission Lines

Signal transmission lines are used to carry the measurement signal from the measuring device to the controller and to carry the controller output signal to the final control element. In the past, transmission lines were pneumatic (compressed air or compressed liquids) but with the advent of electric analog controllers and especially the expanding use of digital computers for control, transmission lines carry electrical signals.

4.5.3.7.1 Electrical Signal Transmission

Electric systems have a great advantage of nearly unlimited distances of transmission either by wire-carried signal or by radio linkage. When a control loop has been implemented using analog electrical signals, it is most common to transmit the analog signal as a current signal. Whenever a process control loop is designed, some operating range is specified for the controlled variable to be regulated. It may be necessary to define a set point anywhere within the range. The presently followed standard specifies that the signal conditioning be such that a 4–20 mA dc current (or 1–5V dc) range on the signal transmission wires represents the specified range of the variable. There are three significant points regarding the use of current transmission to represent the controlled variable.

1. Load impedance : By using a current to carry analog information about the variable, we avoid errors introduced by attaching different loads to the transmitting circuit. Thus, changing lead resistance or inserting a series resistance in the leads will not change the current delivery. Generally, the transmitting circuits are designed to work into any load from 0 ohms to about 1000 ohms. (Load independency upto 1000 ohms).

2. Interchangeability : By using a specified current range to represent the variable range, we provide for interchangeability of the controller in the process-control loop. Once the dynamic variable range has been translated to a fixed current range, then all control can be based on a set point and deviation as some percentage of this range. Thus, a controller only sees a 4–20 mA signal, for example, and it does not matter what specific dynamic variable this represents.

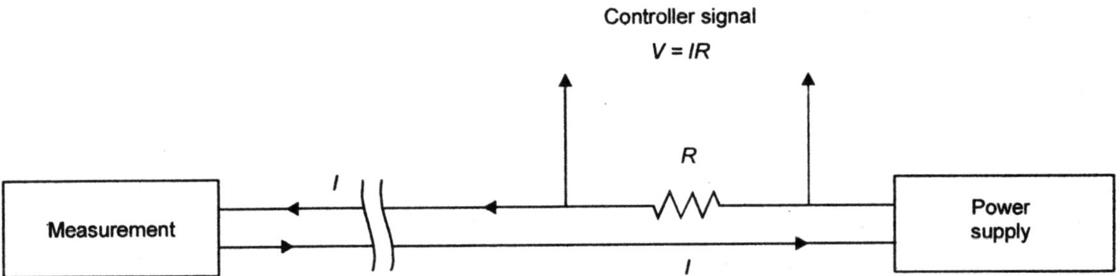

Fig. 4.62 A two-wire system can be used for both power leads to a measurement system and also carry information on the measurement through a current

3. Measurement/power supply : Generally speaking, the current signal lines are also the power delivery lines to energise the transducer and local signal conditioning. Thus, only two wires are necessary to connect a transducer and signal conditioning measurement system to the rest of the loop. The signal conditioning is designed so that the circuit draws more or less current from the power source in proportion to the value of the dynamic variable. Fig. 4.62 shows such a two-wires transmission circuit which also includes a series resistor in the line (normally 250 Ω conditioning resistor) to convert 4–20 mA into 1–5V signal..

4.5.3.7.2 Pneumatic Signal Transmission

In a pneumatic control loop or in a electric conrol loop with the final control element as a pneumatic device, the standard transmission signal is an air pressure level of a range 3-15 psi (or 0.2 to 1 kg/cm^2). In the second case, a current-to-pressure converter (*I/P* Converter) is employed to scale the 4–20 mA signal to a 3–15 psi signal. Unlike the electric transmission, the pneumatic transmission has always transmission lags. Distances up to 150m can be operated with success. The longer distances can be attained by specifying such requirements as higher air capacities, use of volume boosters, bigger *OD* tubing, and, as a last resort, field-mounted controllers.

The pneumatic transmission system shown in Fig. 4.63 may be used for distances upto 500 feet (≅150 m). The controlled variable is converted into an air pressure at the transmitter *T*. The air pressure is then conducted through a single tube to the receiver *R* where it is transduced to a position or force for operation of the controller. The details of a pneumatic transmission system are illustrated in Fig. 4.64 for a particular system transmitting a measured pressure of range 0 to 100 psi.

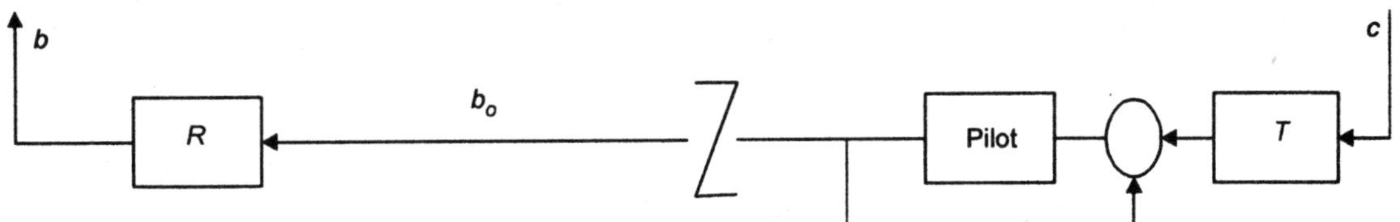

Fig. 4.63 Pneumatic transmission

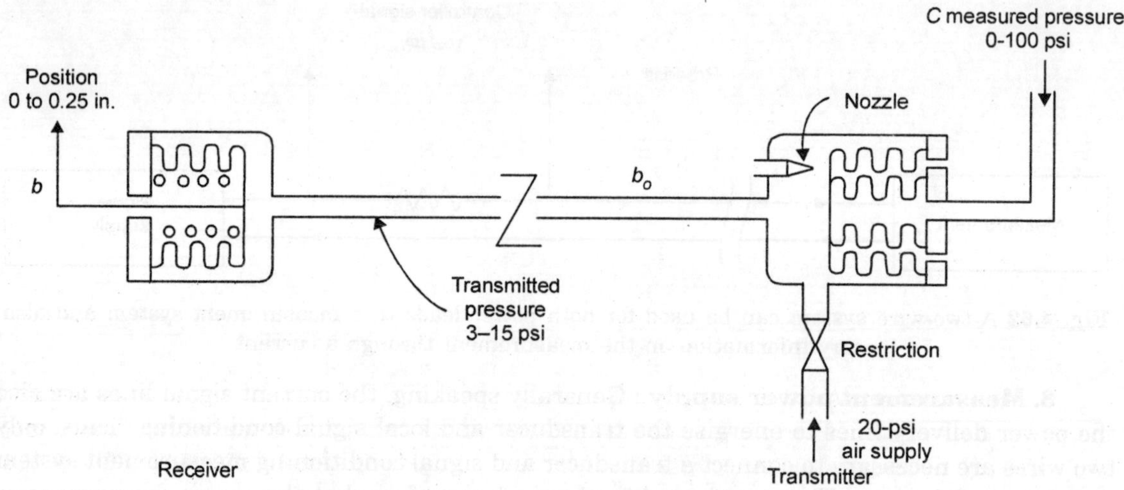

Fig. 4.64 A pneumatic transmission system

The connecting tube which carries the transmitter pressure to the receiver is almost always one-quarter inch *OD* (≅ 6 mm *OD*) standard copper, aluminium or plastic tubing. Practically the only limitation of distance of pneumatic transmission is the lag of transmitting the signal through a long connecting tube. The capacitance of the tube is caused by the volume of the tube and it is distributed along the length of the tube. The resistance of the tube is due to fluid friction and is likewise distributed along the tube. Because of the distributed parameters, the calculation of the lag is extremely difficult and it is necessary to rely upon experimental tests to determine the lag.

Unless the process changes very fast or the transmission lines are very long, the dynamic behaviour of a pneumatic transmission line can be neglected from consideration. When the assumptions above do not hold, it has been found that the following transfer function correlates successfully the pressure at the outlet (P_O) to the pressure at the inlet (Pi) of a pneumatic transmission line :

$$\frac{\bar{P}_o(s)}{\bar{P}_i(s)} = \frac{e^{-\tau_d s}}{\tau_p s + 1} \quad ...(4.18)$$

with $\tau_d/\tau_p \cong 0.25$.

4.5.3.8 Methods of Fluid Control

A source of fluid head is required to provide the transport of fluids through processing equipment. Several arrangements of the fluid source may be used depending upon the type of fluid and processing equipment. A pump, compressor, blower, fan or overhead tank is ordinarily used to provide the source of fluid head or pressure. Six arrangements of fluid sources and a control valve or motor operator are shown in Fig. 4.65. Fig. 4.65(*a*) illustrates 'series throttling' method, Fig. 4.65(*b*) that of 'by pass' arrangement and Fig. 4.65(*c*), the 'variable delivery' means.

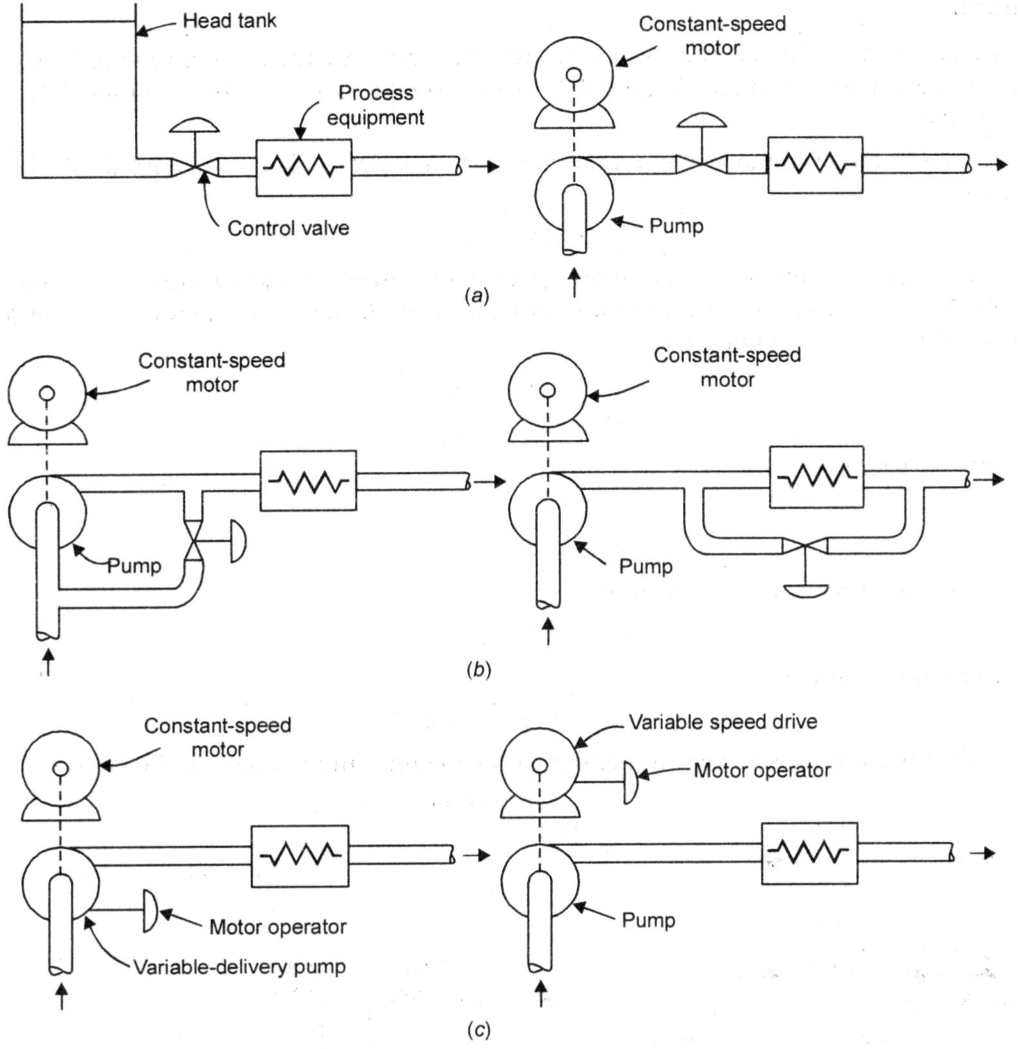

Fig. 4.65 Methods of fluid control
(*a*) Series throttling (*b*) Bypass (*c*) Variable delivery

The choice of any one arrangement of a fluid source depends upon a great number of factors such as the type of fluid, the size of the installation, and the efficiency. The arrangement of the fluid source is very important in automatic control, and it is probably safe to say that most difficulties in automatic control arise because of poorly installed or maintained fluid sources.

4.6 PROBLEMS AND SOLUTIONS

Problem 4.1

An interface unit requires a 5–10 V input signal from 4–20 mA control signal. Design a signal conversion system to provide this relationship.

Solution :

We may convert the current into a voltage, and then provide the required gain and bias. We can get a standard voltage signal 1–5 V using a conditioning resistor of value 250Ω in the current line.

Then we can design an opamp amplifier system which will give an output given by the equation

$$V_{out} = KV_{in} + V_B$$

where K is the gain and V_B is the appropriate bias voltage. We know that 1 V input must provide 5V output and 5 volts input must provide 10 V output. This allows us to find K and V_B using simultaneous equations as

$$5 = 1\,K + V_B$$

$$10 = 5\,K + V_B$$

subtracting, we get

$$5 = 4\,K$$

$$\therefore \qquad K = 5/4 = 1.25.$$

that we use in either equation to find

$$V_B = 3.75\ V$$

Thus, the result is

$$V_{out} = 1.25\ V_{in} + 3.75$$

With this information, we can implement an opamp configuration as shown in Fig. 4.66.

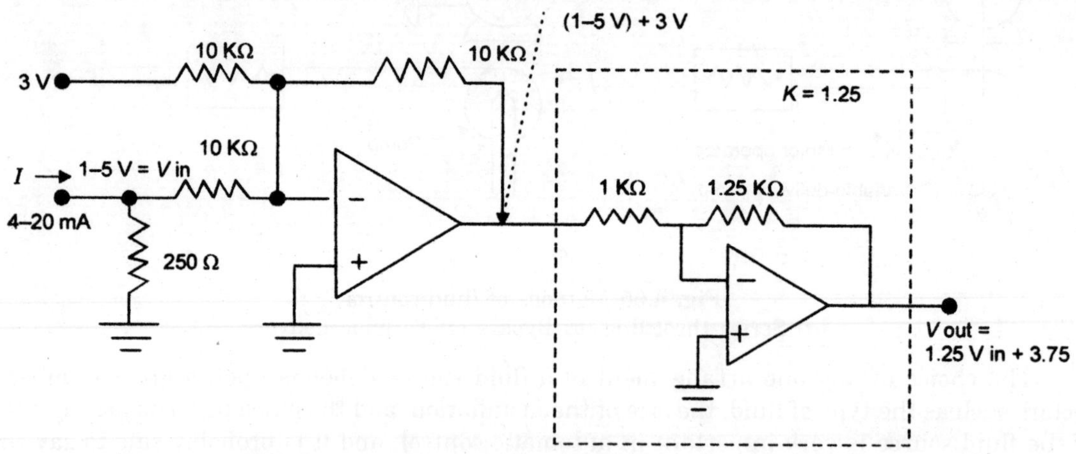

Fig. 4.66 An op amp circuit designed for problem-4.1

Problem 4.2

A stepper motor has 6° per step and must rotate at 300 rpm. What input pulse rate, in pulses per second, is required ?

Solution :

A full revolution has 360° so that with 6° per step it will take 60 steps to complete one revolution. Thus

$$\left(360\frac{\text{revolutions}}{\text{minute}}\right)\left(60\frac{\text{pulses}}{\text{revolution}}\right) = 18{,}000 \text{ pulses/minute.}$$

$$= \frac{18{,}000}{60} \text{ pulses/sec.}$$

$$\text{Required input pulse rate} = 300 \text{ pulses/sec.}$$

Problem 4.3

Suppose a force of 600 N must be applied to open a valve using pneumatic diaphragm actuator. Find the diaphragm area if a control gauge pressure of 0.7 kg / cm² must provide this force.

Solution :

We may calculate the area from the formula

$$\text{Force} = \text{Pressure} \times \text{Area} \quad (F = PA)$$

$$\text{Area} = \frac{\text{Force}}{\text{Pressure}} = \frac{600 \text{ N}}{0.7 \text{ Kg/cm}^2}$$

$$= \frac{600 \text{ N}}{7 \text{ N/cm}^2} = 85.7 \text{ cm}^2$$

or about 10.5 cm in diameter.

Problem 4.4

Find the working force resulting from 250 N applied to a 1.25 cm radius forcing piston in a hydraulic actuator (a) if the working piston has a radius of 10 cm. Then (b) find the hydraulic pressure.

Solution :

(*a*) We can find the working force from

$$F_W = \frac{A_2}{A_1} F_1 \qquad \text{...(4.5)}$$

or

$$F_W = \left(\frac{R_2}{R_1}\right)^2 F_1 = \left(\frac{10 \text{ cm}}{1.25 \text{ cm}}\right)^2 (250 \text{ N})$$

$$F_W = 16{,}000 \text{ N}$$

(*b*) Thus, the 250 N force provides 16,000 N of force. Then the hydraulic pressure is

$$P_H = \frac{F_W}{A_2} = \frac{16000 \text{ N}}{\pi\left(10 \times 10^{-2} m\right)^2} = 5.1 \times 10^5 \text{ N/m}^2$$

$$= 5.1 \text{ Kg/cm}^2$$

Problem 4.5

An equal percentage valve has a maximum flow of 45 m^3/s and a minimum of 1.5 m^3/s. If the full travel is 4 cm, find the flow at a 2 cm opening.

Solution :

$$\text{The rangeability } R = \frac{Q_{max}}{Q_{min}} = \frac{45 \text{ m}^3\text{/sec}}{1.5 \text{ m}^3\text{/sec}} = 30$$

Then the flow at a 2 cm opening is

$$Q = Q_{min} R^{S/Smax} \qquad ...(4.10)$$

$$= (1.5 \text{ m}^3)(30)^{2/4} = 1.5\sqrt{30}$$

$$Q = 8.2 \text{ m}^3\text{/sec.}$$

Problem 4.6

Find (a) the proper C_V for a valve that must allow 150 gallons of ethyl alcohol per minute with a specific gravity of 0.8 at a maximum pressure drop of 50 psi,. and (b) the required valve size making use of the valve flow coefficient (Kv) table given below.

Valve size cms	*Kv*	*Valve size Cms*	*Kv*
0.75	0.25	7.50	95
1.25	2.50	10.00	150
2.50	12.0	15.00	350
3.75	30.0	20.00	625
5.00	50.0		

Solution :

(*a*) We find the C_V from

$$Q = C_V \sqrt{\frac{\Delta P}{S_G}} \qquad ...(4.17)$$

Then

$$C_V = Q\sqrt{\frac{S_G}{\Delta P}} = \left(150 \frac{\text{gal}}{min}\right)\sqrt{\frac{0.8}{50\text{psi}}} = 18.97$$

(*b*) To find the required valve size, we have the characteristic table available in K_V only. To convert Cv into Kv we have

$$K_V = 0.86\ C_V \qquad ...(4.16)$$

$$= 0.86 \times 18.97 = 16.31$$

Since the valve size has to be selected out of the sizes available, we may have to select the one which is having K_V next to the calculated K_V =16.31 as 30.00 with valve size of 3.75 cm.

Problem 4.7

A springless pneumatic actuator has a diaphragm of 500 cm^2 area. Its positioner operates from 0.2 to 1 Kg/cm^2. The under-side cushion pressure is set at 0.3 Kg/cm^2. What range of thrust load can be accommodated ?

Solution :

$$\text{Maximum upward thrust} = (1\ \text{Kg/cm}^2 - 0.3\ \text{Kg/cm}^2) \times 500\ \text{cm}^2 = 350\ \text{Kg.}$$

$$\text{Maximum downward thrust} = (0.3\ \text{Kg/cm}^2 - 0.2\ \text{Kg/cm}^2) \times 500\ \text{cm}^2 = 50\ \text{Kg}$$

Problem 4.8

Flow through a linear valve (constant sensitivity type characteristic) is given by

$$\frac{Q}{Q_{max}} = \frac{1}{R}\left[1+(R-1)\frac{S}{S_{max}}\right]$$

where Q is the flow at any lift 'S', Q_{max} is the maximum flow at maximum lift S_{max}, and R is the rangeability .

If the valve passes 100 m^3/m of water at a maximum lift of 5 cm, and the rangeability is 20, compute the valve sensitivity.

Solution :

From the above given equation we may get the valve sensitivity as

$$\frac{dQ}{dS} = \frac{(R-1)}{R}\cdot\frac{Q_{max}}{S_{max}} = \frac{20-1}{20}\times\frac{100}{5} = 19\ \text{m}^3/\text{cm.}$$

$\therefore$ Valve sensitivity = 19 m^3/cm.

Problem 4.9

A heating furnace requires a control valve passing 10 gpm preheated light fuel oil ($S_G = 0.8$) at full load and only 0.2 gpm at the smallest heating load. The pressure differential at wide open in 20 psi. Calculate (a) Turn down (b) Rangeability (c) Cv and d) Kv.

Solution :

(*a*) $$\text{Turn down} = T = \frac{Q(\text{normal maximum})}{Q(\text{minimum controllable})} = \frac{10}{0.2} = 50$$

(*b*) $$\text{Rangeability} = R = \frac{Q(\text{maximum controllable})}{Q(\text{minimum controllable})}$$

Normal maximum flow is generally taken as 70% of the maximum controllable flow.

$$\therefore \quad Q(\text{maximum controllable}) = \frac{Q(\text{normal maximum})}{0.7}$$

$$\therefore \quad R = \frac{T}{0.7} = \frac{50}{0.7} \cong 70.$$

(c) $$Cv = Q\sqrt{\frac{S_G}{\Delta P}} = 10\sqrt{\frac{0.8}{20}}$$

$$Cv = 2.$$

(d) $$Kv = 0.86\ Cv.$$

$$= 0.86 \times 2$$

$$Kv = 1.72$$

Problem 4.10

The vapour pressure of the liquid flowing through a control valve is given as 0.6 Kg / cm^2 If the down stream pressure is 0.5 Kg / cm^2, is there a possibility (a) for cavitation to occur (b, for flashing to occur (c) What should be the upstream pressure for making the condition tc switch over from one state to another state if the pressure recovery ratio is 0.8.

Solution :

(*a*) Cavitation occurs when

Downstream pressure P_2 > vapour pressure Pv.

$$Pv = 0.6\ \text{Kg/cm}^2.$$

$$P_2 = 0.55\ \text{Kg/cm}^2.$$

Since $P_2 < Pv$, there is no possibility for cavitation to occur.

(*b*) Flashing takes place when

Downstream pressure P_2 < vapour pressure Pv.

$$Pv = 0.6\ \text{Kg/cm}^2,\ P_2 = 0.55\ \text{Kg/cm}^2.$$

$$\therefore \quad P_2 < Pv.$$

Hence there is a possibility for flashing to occur.

(*c*) We have to find out the upstream pressure which will make the situation for 'cavitation' instead of 'flahsing' as found out above.

We should achieve $P_2 > Pv$.

$\therefore$ *i.e.*, $$P_2 \geq \frac{0.6}{0.8} \geq 0.75\,.$$

If upstream pressure is lifted above 0.75 Kg/cm^2, the down stream pressure is expected to go more than 0.6 Kg/cm^2, the vapour pressure. Hence there will be possibility of cavitation to occur.

4.7 PROBLEMS AND QUESTIONS

1. What is the purpose of final control element ?
2. Explain the function of signal transmission lines ?
3. Compare electrical and pneumatic signal transmission lines ?
4. Explain the importance of control valve sizing.
5. Write short notes on (*a*) Cavitation and (*b*) Flashing.
6. Explain briefly about the construction and operating principles of different types of pneumatic valves and their flow capacity characteristics.

7. Discuss about the important factors before selecting (*a*) air-to-close and (*b*) air-to-open pneumatic control valve
8. Describe the function of a control valve.
9. Write about the factors involved in the selection of a control valve.
10. Define the flow coefficient of a control valve. How *Cv* and *Kv* differs ?
11. What is the need for a valve positioner ? Explain about various electrical, pneumatic and hydraulic valve positioners.
12. List the factors that decide the lag in a pneumatic signal transmission line.
13. Describe the working of a pneumatic actuator with positioner.
14. Describe the working of a electrical actuator with positioner.
15. Describe the construction of control valves.
16. Explain about the different types of control valve bodies.
17. Give two examples of electric actuators.
18. Why installed- characteristics of control valve is different from inherent characteristics?
19. What are the three different inherent characteristics of a control valve ?
20. What are the effects of oversizing and undersizing of a control valve?
21. When a valve is called as equal percentage valve and why ?
22. Why a spring actuator often requires a positioner ?
23. Differentiate spring and springless pneumatic actuators .
24. A spring actuator is to be used for positioning a dead weight of 300 lbs on its stem. The pressure input is 3 to 15 psi and the stroke is 2 inches. What is the minimum diaphragm area ?
25. What are the different methods of fluid control ? Explain about bypass fluid control with diagrams.
26. What is called 'decreasing sensitive type' characteristic ?
27. Write short notes on (*a*) Rangeability and (*b*) Turn down of a control valve.
28. Name three types of 'rotating-shaft' control valves and explain about them.
29. The area of opening of a valve versus lift is given by $A = a + bx^2$. Derive the flow versus lift characteristics for this parabolic curve.
30. What is called piston actuators ? How does it differ from spring and diaphragm actuator ?
31. What are the different shapes of plugs used in a control valve ?
32. How effective valve characteristics (installed characteristics) differ from inherent characteristics?
33. The temperature of an air stream is controlled by sending part of the stream to a steam-heated exchanger and bypassing the rest of the stream through a control valve. What type of valve should be used to give the best control over a range of inlet air temperatures ? Assume that the total flow of air is constant, that the control point is 60°C, and that the exchanger could heat the entire stream to 90°C with steam at 150°C.
34. A stepping motor has 7.5° per step. Find the rpm produced by a pulse rate of 2000 pps on the input.
35. An equal percentage control valve has a rangeability of 32. If the maximum flow rate is $100 m^3/hr$, find the flow at $(2/3)^{rd}$ and $(4/5)^{th}$ open settings.

5

PROCESS CONTROL SYSTEMS—SELECTED UNIT OPERATIONS

5.0 INTRODUCTION

In the first 4 chapters of this book we have discussed about mathematical modeling of first order and higher order systems, process with dead time, continuous and batch process, different controller characteristics and tuning, control systems with multiple loops and final control elements. Then and there we have discussed many of the practical applications. Especially in chapter-3, we have discussed cascade control, ratio control, feed forward and feed back control, selective control, split-range control, adaptive control and inferential control systems with many applications from boilers and chemical industries. In addition we will be discussing in this chapter some of the important process control applications in Boilers as well as Chemical industries. Reactors, Mixers, Evaporators, Dryers, Heat exchangers and Distillation column in chemical industries will be dealt with. We will not be able to cover the complete range of control systems in the above units, but will surely cover the important ones in those units. The basic operation of the unit considered will be discussed first for the understanding of the readers and then few connected important control systems in the unit will be discussed. Interaction of control loops will be covered wherever required.

5.1 BOILER

5.1.1 Boiler Operation

Steam boilers are used industrially both for electric power generation and for supplying process steam. They consist of a furnace (combustion chamber) where air and fuel are combined and burnt to produce combustion gases and a water-tube system, the contents of which are heated by these gases. The heat exchangers in Boilers are available in two basic designs : Fire-tube where bundle of tubes carry combustion gases and water-tube where the tubes carry feedwater. In our discussions we will be assuming water-tube boilers. The tubes are connected to the steam drum, from where the generated water vapour is withdrawn. The steam from the drum is passed through the superheater tubes, which are exposed to the combustion gases. The superheated steam from different boilers of a station is connected to a common steam bus and then distributed either to turbines of turbo-generators to produce electrical power or to process steam line for supplying process steam to various processes of chemical industries.

The whole operation of the boiler may be divided into two systems namely feed water-steam-condensed water-system and Fuel-air-waste gas system.

Feed water-steam-condensed water-system

Makeup water (treated from raw water at water treatment plant) combines with condensed water to form feed water which passes through heat exchanger tubes, boiler drum and super heater before getting supplied to turbine as steam. Steam after utilizing its energy for rotati g the turbine (and hence the alternator) gets condensed in the condenser before getting pumped as condensed water to get added to makeup water and form feed water. The most important controls in this system are :

1. Drum water level control
2. Superheated steam temperature control
3. Steam pressure control.

Fuel-air-waste gas system

Fuel supplied as per boiler steam demand is burnt with air supplied by Forced Draft Fan (FD Fan) in the combustion chamber. The heat energy of combustion gases is exchanged with water system through water tubes and superheater tubes. After this primary heat exchanging process, the combustion gases are called flue gas. The remaining heat energy of the flue gas is utilized to preheat feedwater in economizer and to preheat combustion air in air preheater. Flue gas coming out of air preheater is sucked and thrown out as waste gas with the help of Induced Draft Fan (ID Fan) and chimney. The most important controls in this system are :

1. Combustion air control (FD fan control)
2. Combustion control (air-fuel ratio control)
3. Furnace draft control (ID fan control)

The main in-line Instruments used in a boiler is shown in Fig. 5.1 for better understanding.

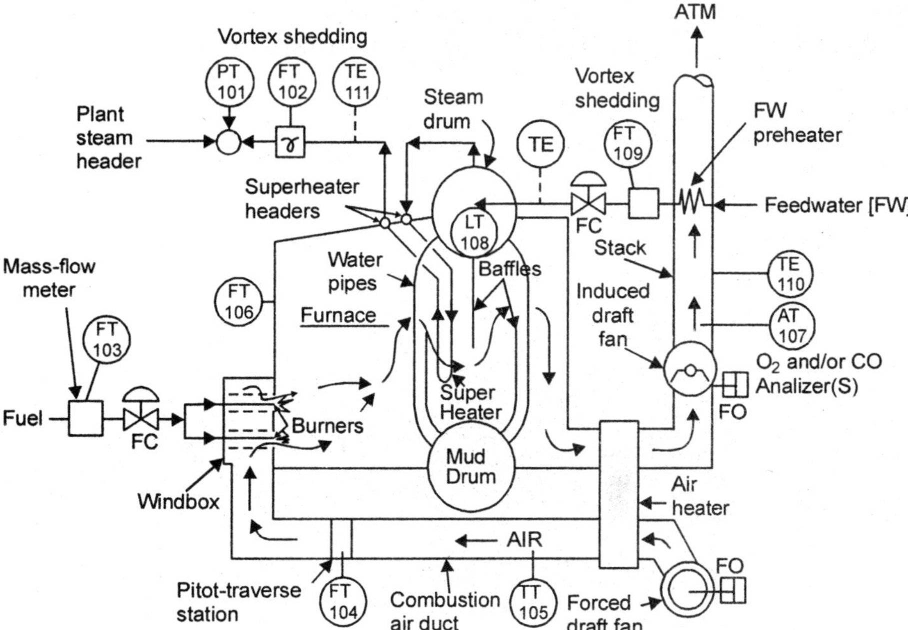

Fig. 5.1 The main in-line instruments are shown here for a drum-type boiler.

5.1.2 Boiler Controls

Drum level control : This has already been discussed in Chapter-3 (Fig. 3.12) as an example for feed forward control system. A simple feedback control would do with level as controlled variable and feedwater flow as manipulated variable. As the boiler steam load variation is dependent on power station load requirement and its variation may take little more time to reflect on the level, steam flow rate signal is taken for feed forward control with feed water flow rate as manipulated variable. This improves the performance of the overall loop.

Fuel-air ratio control : This ratio control has also been discussed in Chapter-3 (Fig 3.24) as an example for ratio control. For safety reasons firing rate demand as a set point is given parallely to fuel controller as well as air controller. The air controller receives the set point through a ratio setter so that the required Fuel-Air ratio is maintained for proper combustion.

Steam header pressure control : When a power station is operating with number of boilers and a common steam bus, it becomes necessary to have a split-range control to share the pressure requirements among boilers. The split-range control used for this purpose is already discussed in Chapter-3 with reference to Fig. 3.31.

In addition to the above three critical controls we will be discussing two more important controls often used in boilers operation namely (1) Parallel control of FD and ID fans and (2) Superheated steam temperature control.

5.1.3 Parallel Control of Inlet and Outlet Dampers : (To reduce Interactions)

5.1.3.1 Inlet Damper (FD Fan) Control

Combustion air for steam boiler is supplied by forced draft fan. Air requirement set point is given from Fuel-Air ratio station and controlled through a control valve in the air line as discussed already under Fuel-air ratio control. Here the controlled variable is combustion air

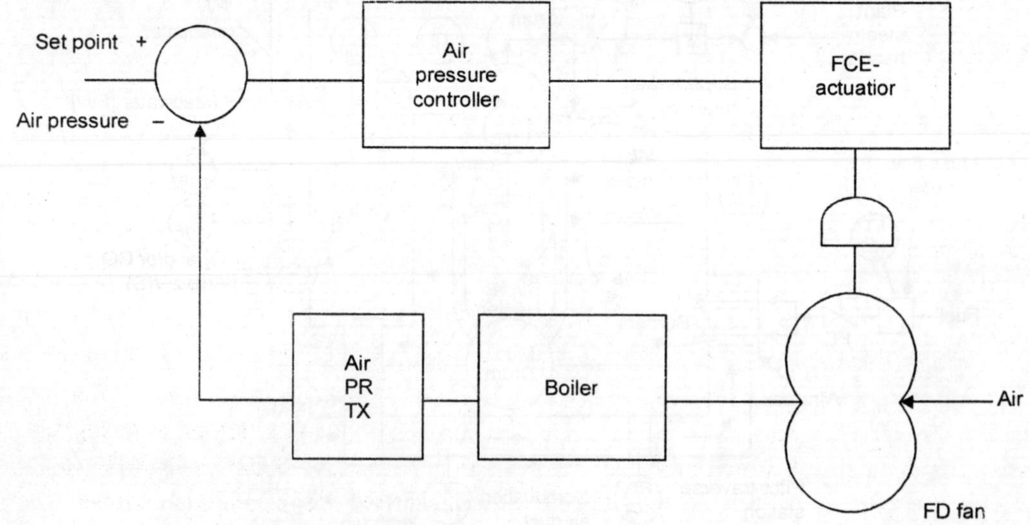

Fig. 5.2 F.D fan control.

flow rate. It is always required to control the air line pressure so that the air requirement is always met. For controlling the combustion air inlet pressure, the control elements are either damper or vane, or louver or butterfly valves. Controlling the speed of the FD fan motor (speed control) is also a method to control air pressure. A simple feedback control loop is shown in Fig. 5.2.

5.1.3.2 Outlet Damper (ID Fan) Control

The combustion chamber pressure (Furnace Draft) should neither be too negative nor be too positive. If it is too negative, cold air from atmosphere may enter the chamber through inspection holes, grate or some openings due to refractory wall damage etc. This excess cold air infiltration will disturb the thermal equilibrium and also affect the combustion efficiency. If it is too positive, hot combustion gas leakage will be there. This leakage not only causes loss of heat energy but also damages equipment outside the furnace. Hence always a slight negative pressure is maintained with the help of Induced Draft (ID) fans and chimney. The main purpose of ID fan and chimney is to push out the waste gas into the atmosphere. Hence by controlling damper position on the waste gas line or controlling the speed of the ID fan motor both the objectives of throwing out waste gases and maintaining a slight negative pressure in the combustion chamber are achieved. A simple feedback control to achieve the necessary furnace draft by throttling the ID fan is shown in Fig. 5.3. Pressure tapping position for measuring furnace pressure may be located some distance below top and the middle across the widths.

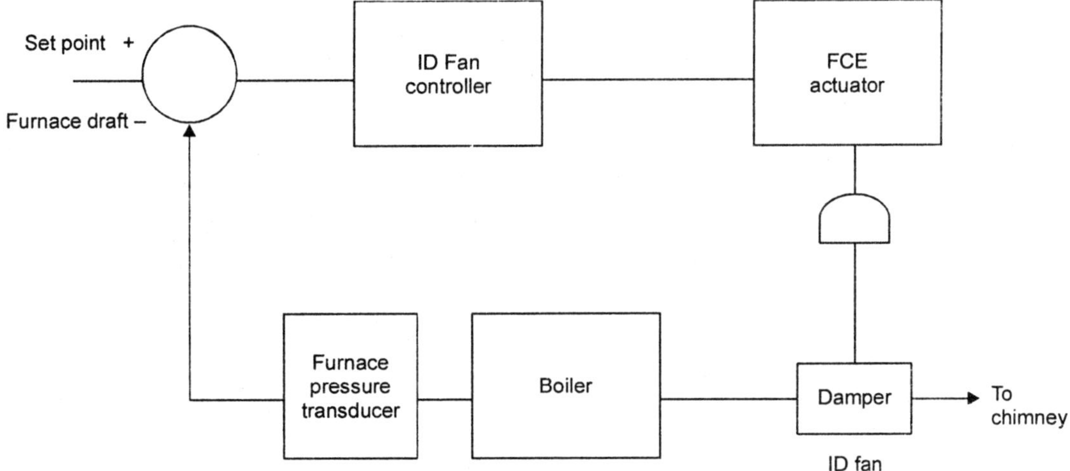

Fig. 5.3 ID fan control.

5.1.3.3 Parallel Control of Inlet and Outlet Dampers

The operation of both FD fan control and ID fan control together in a boiler is called as a balanced draft operation. Whenever both forced draft and induced draft are used together, at some point in the system the pressure will be the same as that of the atmosphere. Balanced-draft boilers are not normally designed for positive furnace pressure. In the case of a balanced draft boiler the maintenance of constant furnace pressure or draft (which is always slightly negative) keeps the forced and induced draft in balance.

Interaction between air flow and furnace pressure may occur because the change in air flow may affect furnace pressure and vice versa. Additionally, stability problems and interactions may occur in the overall system also. Though theoretically either of the forced draft fan or induced draft fan can be used for either of the controls, normally induced draft fan is used to control the furnace draft, with forced draft fan performing basic air flow control function. Interaction cannot be completely eliminated between these two loops, but it can be minimized by system designs. One such design is shown in Fig. 5.4.

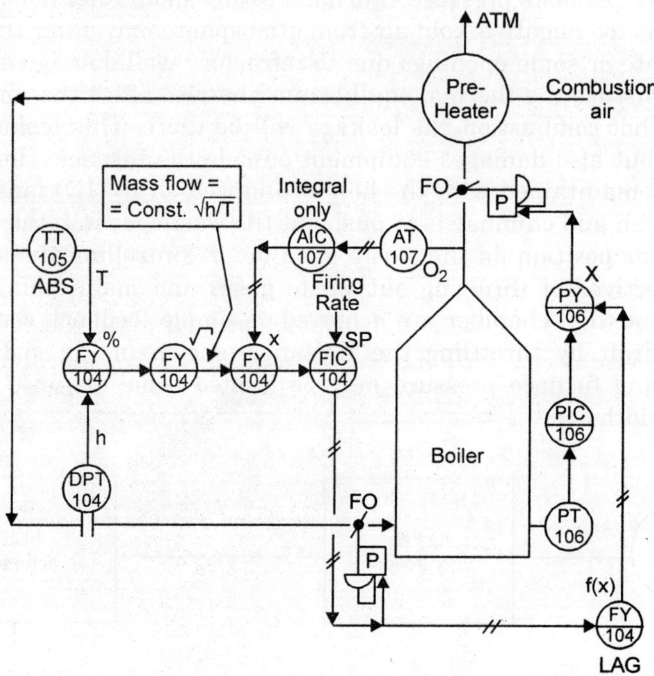

Fig. 5.4 Parallel control of inlet and outlet dampers reduces interactions.

Legend

FIC-104 : Air flow controller (FD fan controller).

DPT-104 : Air flow transmitter.

TT-105 : Air temperature transmitter.

FY-104% : Temperature corrector

FY-104√ : Square root extractor.

FY-104 X : Oxygen trim multiplier.

AT-107 : Oxygen percentage in flue gas analyzer.

AIC-107 : Oxygen trim controller.

FY-104 f(x) : Dynamic lag.

PT-106 : Furnace pressure transmitter.

PIC-106 : Furnace draft controller (ID fan controller).

PY-106 : Air flow trim for furnace draft control.

The common rule is that air flow should be measured and controlled on the same side (air or combustion gas) of the furnace to minimize interaction between the flow and pressure loops.

If the combustion air is preheated, then its temperature will vary substantially and compensation is needed. The mass flow of air is related to $\sqrt{h/T}$, where 'h' is a differential across a restriction and T is absolute temperature. This loop is shown in Fig. 5.4 together with an excess oxygen trim on the air flow controller. Excess oxygen trim is required as an optimization control since it is based on the oxygen percentage in the flue gas. More oxygen indicates excess air in the combustion process and less oxygen indicates starvation of air. Though this measurement cannot be used for direct control for air flow but can be very well used as trim control.

In the Fig. 5.4 it is attempted to reduce interaction by connecting the two dampers (or fans) in parallel and using the furnace pressure as a trimming signal. This can also be used to overcome problems resulting from noisy furnace pressure signals and slow response caused by the series relationship between flow and pressure loops. In this arrangement air flow controller moves both dampers equally, and the furnace pressure corrects ID fan damper for any mismatch. The furnace pressure might respond faster to a change in the downstream damper opening and therefore a dynamic lag (FY-104) is provided. The air flow controller throttles the speed or blade pitch of the forced draft fan and the signal to the induced draft fan is given in a feedforward manner. Thus, as soon as the air flow to the furnace is changed, the outflow will also start changing. This improves the control of the furnace draft as the feedback controller (PIC-106) will need only to trim the feedforward signal at PY -106 to account for measurement and other errors.

5.1.4 Super-Heated Steam Temperature Control

The purpose of superheated steam temperature control is to improve the thermal efficiency of the steam turbine. The basic methods of superheated steam temperature control are :

1. Excess air control.
2. Combustion gas bypass control.
3. Adjustable burner (tilting) control.
4. Using attemperation (De-superheater).

By far, the most commonly used method of control is by use of attemperators (De-superheaters). De-superheater is an equipment used to reduce the temperature and heat content of superheated steam. The De-superheater may be located either between saturated steam outlet of the boiler and the superheater (condensing attemperator) or between two successive sections of the superheaters (interstage attemperators) or at the superheater outlet (after cooler attemperator). Construction wise, the attemperators could be either spray type or shell type (non-contract-type). In spray-type, direct contact heat exchanger is provided whereby feedwater which has bypassed the economizer is sprayed into the steam flow and by absorbing latent heat, cools the steam. The water joins the main steam flow and evaporates. The shell type (non-contact) attemperators are shell and tube heat exchangers. The water on its way from the economizer to the boiler is passed through finned U-tubes and part of steam passed over the tubes. The rest of the steam by-passes the exchanger. The basic three methods of using desuperheater (attemperator) for superheated steam temperature control are shown in Fig. 5.5.

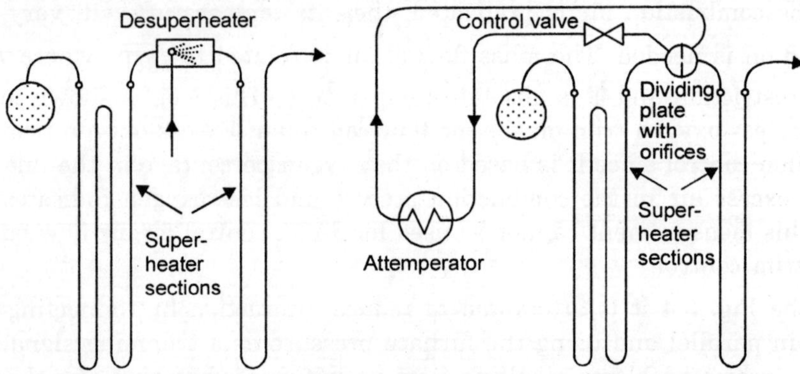

(*a*) Pure water sprayed into the superheated steam vaporizes; degree of superheat is reduced by heat of vapourisation

(*b*) In an effort to avoid steam contamination, desuperheating can also be done through use of shell-and-tube heat exchanger

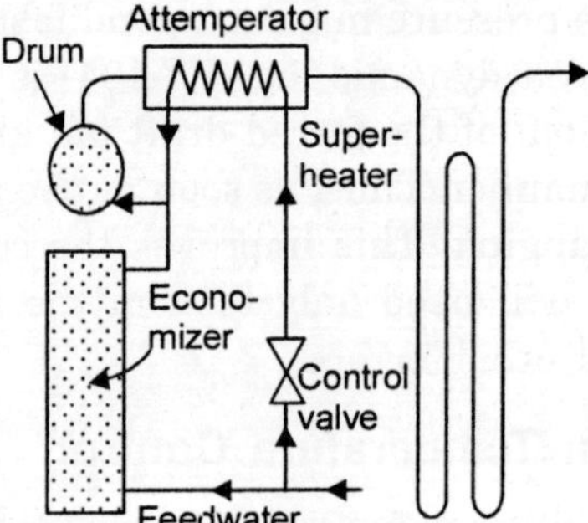

(c) In this hookup the saturated boiler steam is partially condensed by feedwater, controlling the steam temperature

Fig. 5.5 Basic methods of desuperheating.

Temperature control by attemperation is more responsive and can be used as single stage or two stage controls. Single stage is a simple feedback controller with final temperature (after de-superheater) as controlled variable and injection spray water flow as manipulated variable. A two stage control will have two controllers in series and the controlling strategy is shown in Fig. 5.6 below.

In bigger size boilers, the total saturated steam is divided into two parts and controlled in two streams with two stage controllers in each stream and added together in the end. The simplified arrangement is shown in Fig. 5.7.

To use desuperheater spray for steam temperature control, the boiler will normally be provided with added superheater area. The desuperheater characteristics in a boiler is shown below in Fig. 5.8. With all the above basic information in mind, we will be discussing two strategies of superheated steam temperature control.

1. Cascade configuration with feed-forward correction.

1. Cascade configuration with feed-forward correction.
2. Feed-forward-plus-feedback control strategy with an additional override control.

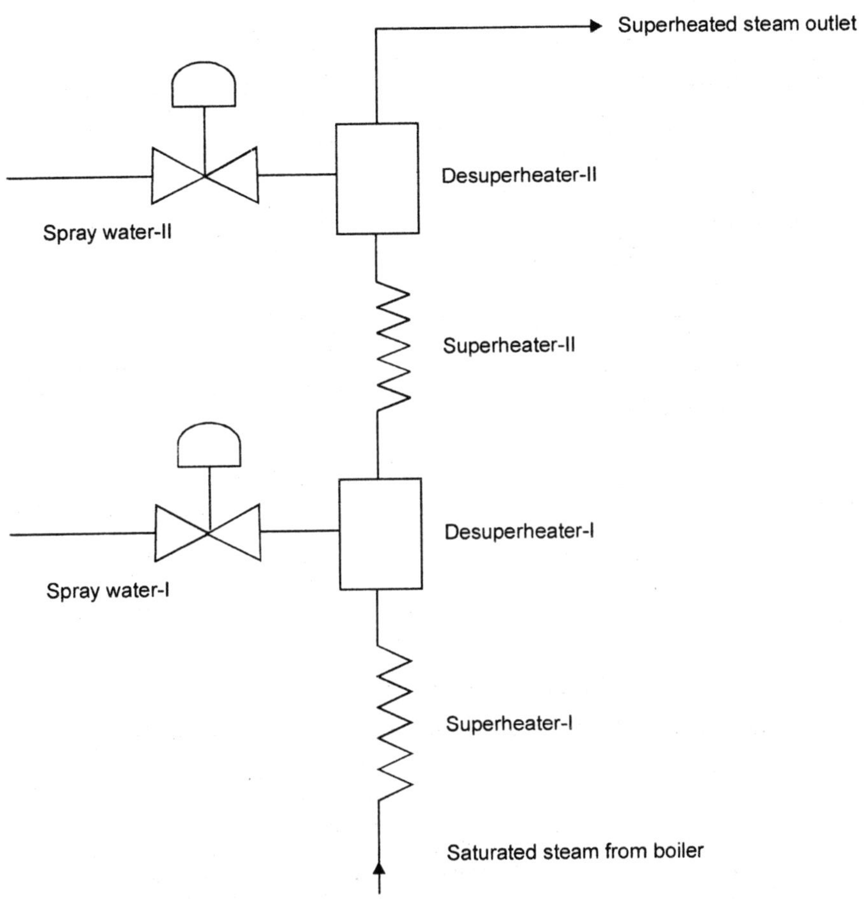

Fig. 5.6 Two stage superheated steam temperature control.

5.1.4.1 Cascade configuration for Superheated Steam Temperature Control

The desuperheater control loop can be configured as a cascade loop, as illustrated in Fig. 5.9. Here the saturated steam from the steam drum is returned back into the furnace where it is superheated. If the amount of superheat is excessive, then water is sprayed into the steam and it is returned once more to the furnace to make sure that all water is vapourised. The slave temperature controller (TIC-112) is placed right after the spray attemperator, while the cascade master (TIC-111) is located at the steam outlet. The feedforward correction is based on steam flow (FT-102), and the relationship between load and uncontrolled temperature (Fig. 5.8) is predicted by the function unit (FY 102). The simplified control loop diagram is shown in Fig. 5.10 for better understanding.

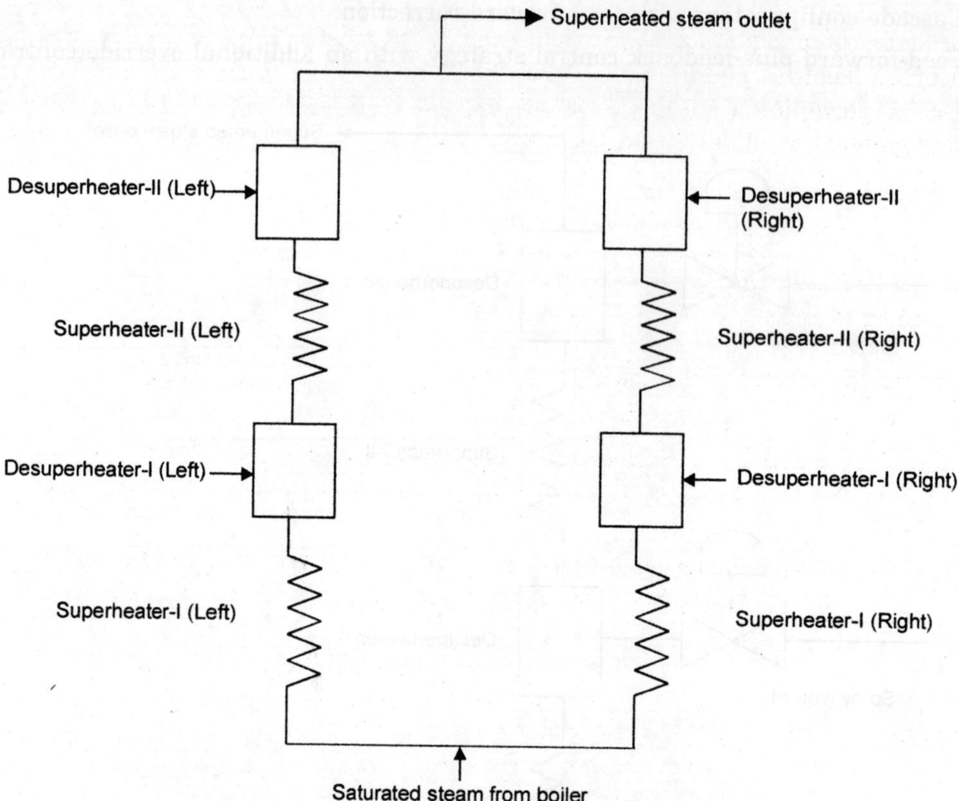

Fig. 5.7 Superheated steam temperature control-Large boiler.

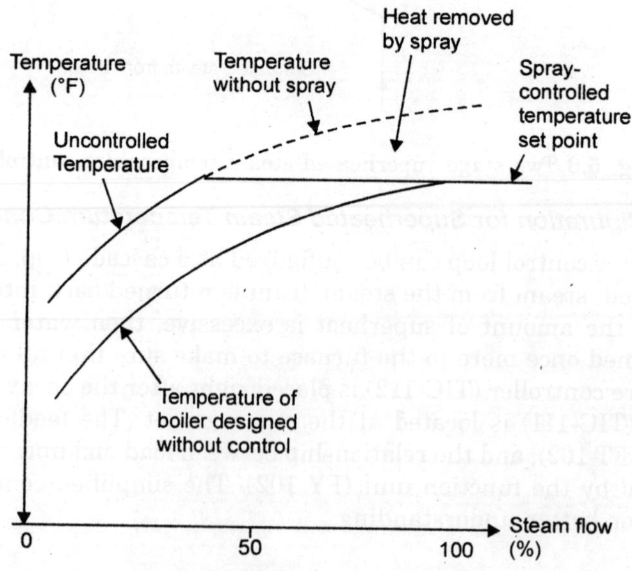

Fig. 5.8 Desuperheater characteristics.

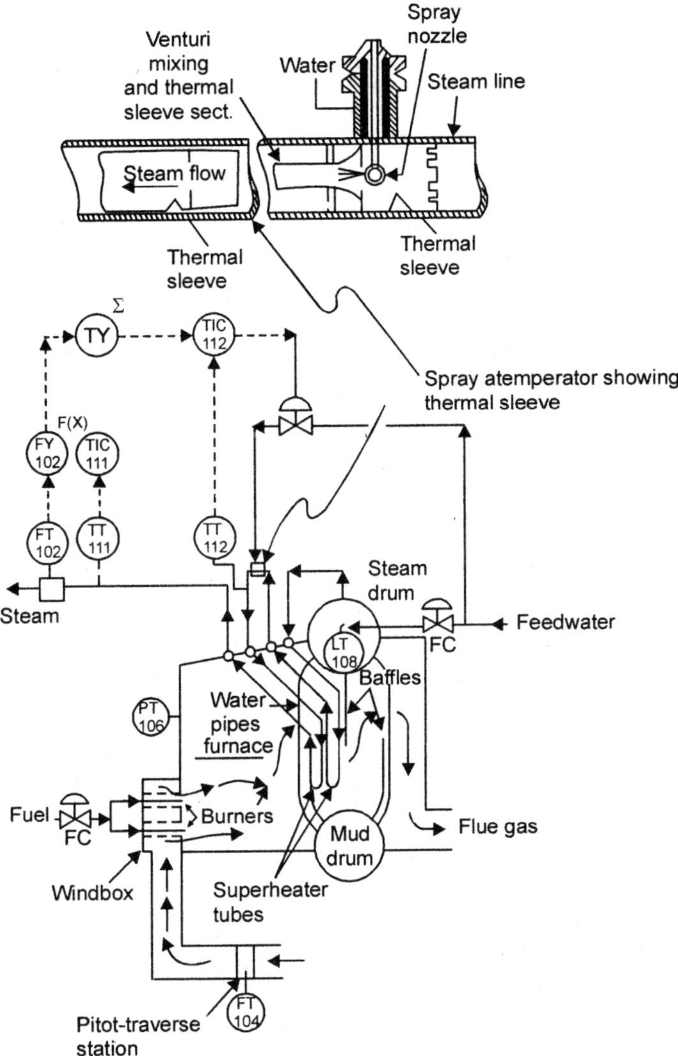

Fig. 5.9 Cascade configuration for controlling the desuperheater control valve.

5.1.4.2 Feed Forward-Plus-Feedback Control Strategy

Fig. 5.11 shows the application of a feedforward-plus-feedback strategy. Since air flow rate is an index of firing rate and excess combustion air, the air flow measurement is used as the anticipatory (feedforward) signal. This signal is combined with the output of the steam temperature feedback controller. The output of this summer provides a signal for the spray water flow control valve. Note that the feedback controller is provided with an override controller that provides a minimum output value tracking signal for the feedback controller. This is included so that when the boiler load is below that of the steam temperature control range, the output of the feedback controller will be the signal necessary so that the output of the summer will provide a 'just closed' position of control valve. This function is necessary for good control, since on increasing or decreasing steam flow rates, the steam temperature may be at the design temperature level with different firing and air flow rates. This relationship

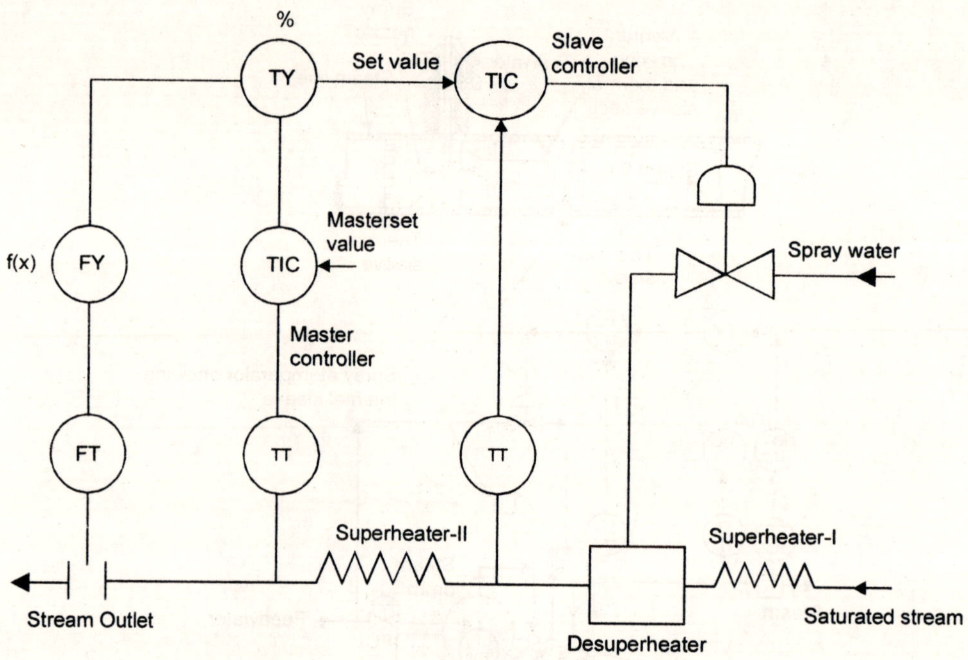

Fig. 5.10 Superheated steam temperature control-cascade.

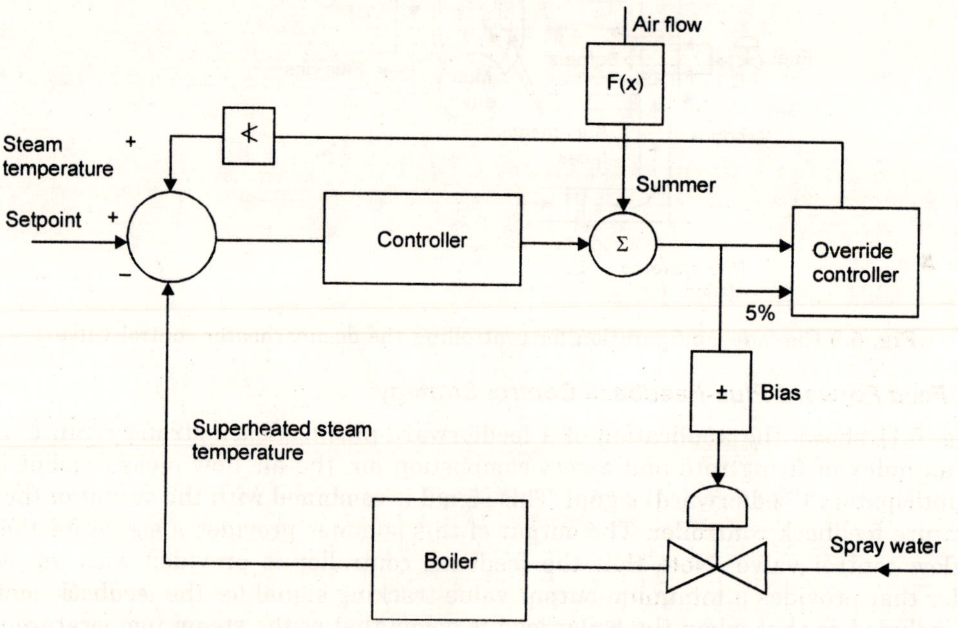

Fig. 5.11 Feed forward-plus-feedback control of superheat spray.

is also affected by the rate of load change. The bias logic is provided to obtain a positive value signal of the output of summer when the signal to the control valve is reduced to zero percent. As shown this is a 5 percent bias. This allows a set point of the override controller to be a value of 5 percent with control action above and below this signal level. The $f(x)$ logic allows for a nonlinear relationship between the measured air flow signal and the position demand signal to the spray water control valve.

5.2 REACTOR

5.2.1 Reactor Operation

Reactors or reaction vessels form the heart of any chemical plant. These vessels are used either for carrying out unit processes (reactions) or can be used for carrying out unit operations such as blending, dispersion, gas absorption, dissolution, batch distillation etc under control conditions. Depending on the process operations the vessels may require heating, cooling, agitation and may be required to operate under varying pressures.

These vessels are classified into three main groups.

1. **Batch reactor :** They are also called a Batch Stirred Tank Reactors (BSTR). These are used for exclusively liquid phase reactions. The reactants are added to the empty vessel and the controls are removed after the completion of the reaction. In this system, temperature, pressure and composition may vary with time requiring controls.
2. **Continuous flow reactors :** They are also called as Continuous Stirred Tank Reactors (CSTR). In these reactors the reactants flow continuously into the reactor and the products flow out continuously. Under ideal conditions, in a well agitated system, an uniform concentration is maintained throughout the vessel.
3. **Semi batch reactors :** In these reactors one of the reactors is initially charged batch wise, while the other reactant is fed into the reactor continuously.

Agitation

Agitation of the reaction vessel contents is a requirement in a number of processing operations. Both heat and mass transfer are greatly influenced by agitation or mixing. An agitator, also known as stirrer, produces high velocity liquid streams which move through the vessel resulting in intense mixing.

Heating and Cooling Systems

Chemical reactions are accompanied by either absorption or liberation of heat. The reaction vessel must, therefore be provided with the means for supplying or removal of heat. Heating systems can be either direct or indirect. Electrical heating systems heat the vessels directly, but are low efficient. Indirect heating systems are most widely used. The heat is received from fluids such as steam, hot oil, hot water or air, molten salt mixtures etc. In direct cooling systems fluids employed are air, liquid ammonia, water or brine. In indirect heat transfer systems the fluid is supplied in a jacket, which surrounds the vessel wall.

A jacketed batch reactor is shown in Fig. 5.12 which gives an overall idea of the reactor. Low carbon steel, stainless steel or other alloy steels are used for construction of reactor vessels.

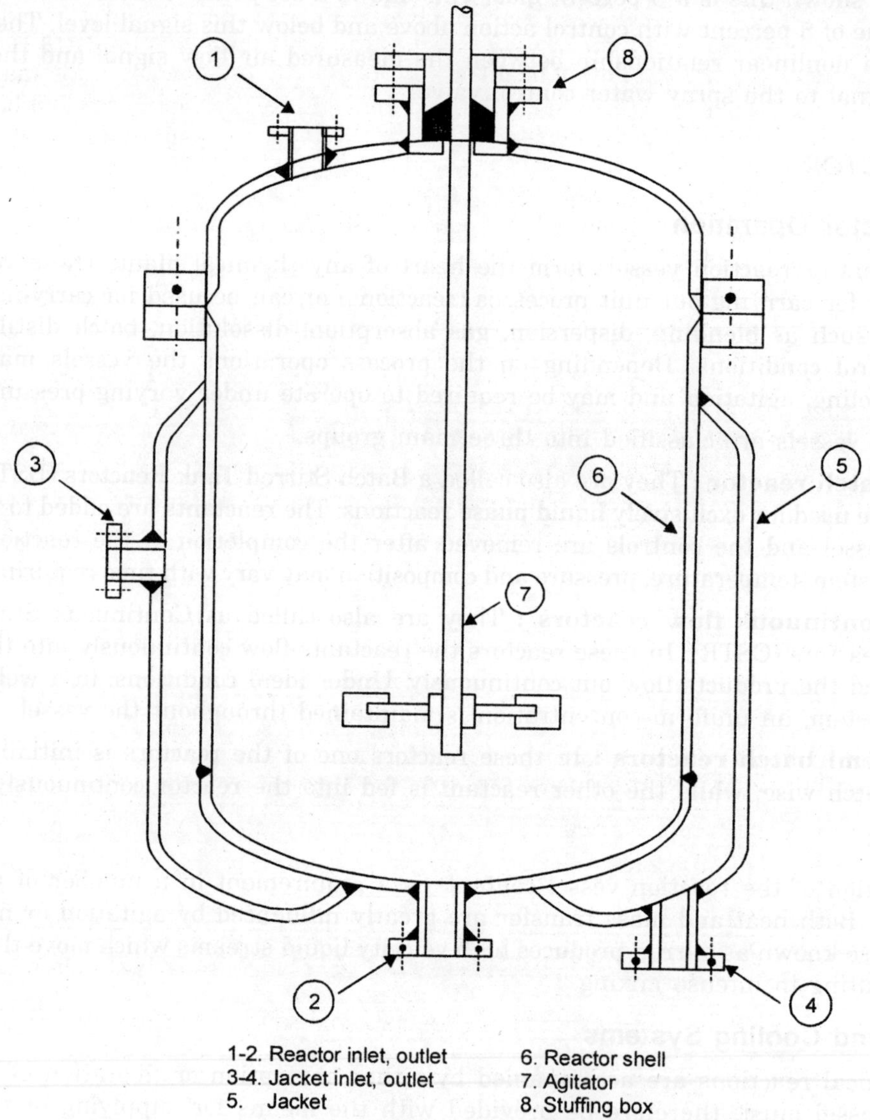

Fig. 5.12 Jacketed Batch Reactor (Section lines not shown).

5.2.2 Reactor Controls

The important controls in a reactor can be identified as follows :

1. Reaction temperature control.
2. Endpoint control-using analysers
3. Reactor pressure control.

Out of these controls we will be discussing about reaction temperature control as it is the most important. Depending on the heat transfer medium used, the control strategies may change. We will take three such temperature control strategies which use (1) Cascade, (2) Two-directional cascade and (3) Split-range sequential controls.

5.2.3 Cascade Control of Reactor Temperature

A simple temperature control scheme may have a feedback controller with the reaction temperature measured, and the flow of heat transfer medium to the reactor Jacket manipulated. [Refer Fig. 5.13]. In order to control reaction temperature, the released heat must be removed from the system as it is liberated by the reactants. The coolant media used, may be water, is passed through once and left out in the above scheme. This 'Once-through' method of cooling is undesirable because the coolant temperature is not uniform. This can cause cold spots near the inlet and hot spots near the outlet. Another disadvantage of this configuration is the variable residence time of the coolant within the jacket as the flowrate changes. This causes the deadtime of the jacket to vary, which in turn necessitates the modification of the control loop tuning constants as the load varies.

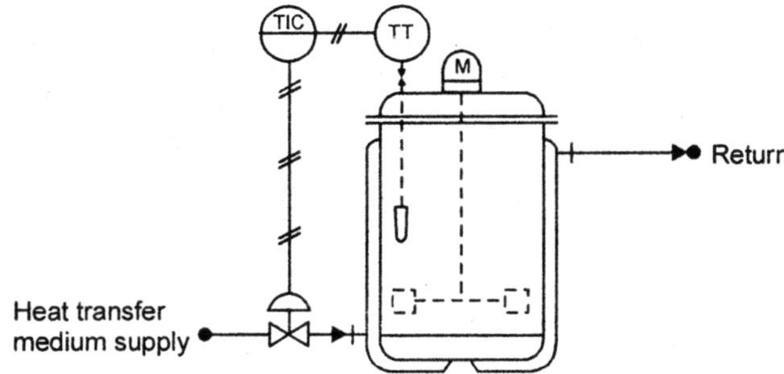

Fig. 5.13 In once-through cooling of chemical reactors, the coolant temperature is not uniform.

The recirculated cooling water configuration shown in Fig. 5.14 is more desirable, because it guarantees a constant and high rate of water circulation. This keeps the jacket dead time constant, the heat transfer coefficient high, and the jacket temperature uniform, thereby eliminating cold and hot spots.

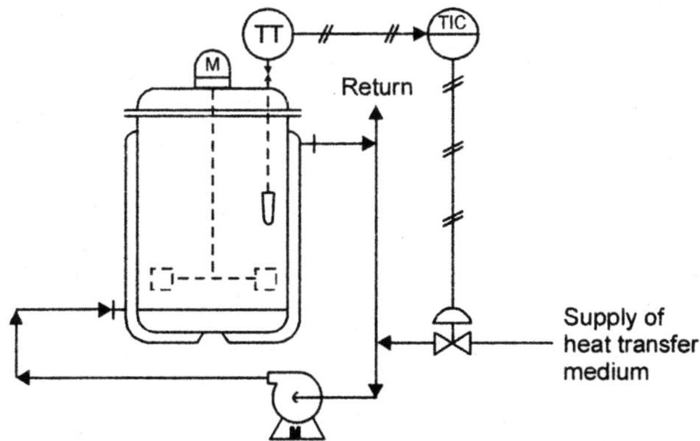

Fig. 5.14 Recirculated cooling of chemical reactors guarantees a constant and high rate of water circulation.

A superior method of reactor temperature control, a cascade loop, is illustrated in Fig. 5.15. Here the controlled process variable (reactor temperature), whose response is slow to changes in the heat transfer medium flow (manipulated variable), is allowed to adjust the setpoint of a secondary loop, whose response to coolant flow changes is rapid. In this case, the reactor temperature controller (Master) varies the set point of the jacket temperature controller (Slave). The purpose of the slave loop is to correct for all outside disturbances, without allowing them to affect the reaction temperature. For example, if the control valve is sticking or if the temperature or pressure of the heat transfer media changes, this would upset the reaction temperature in previous two cases, but not in this control because the slave would notice the resulting upset at the jacket outlet and would correct for it before it had a chance to upset the master. It is preferred that the slave controller maintain the jacket outlet (and not inlet) temperature, because this way the jacket and its dynamic response is included in the slave loop. Another advantage of this configuration is that it removes the principal non linearity of the system from the master loop, since reaction temperature is linear with jacket-outlet temperature. The non-linear relationship between jacket-outlet-temperature and heat-transfer-medium flow is now within the slave loop, where it can be compensated for by an equal-percentage valve, whose gain increases as the process gain drops. In most instances the slave will operate either with PD control or proportional only control.

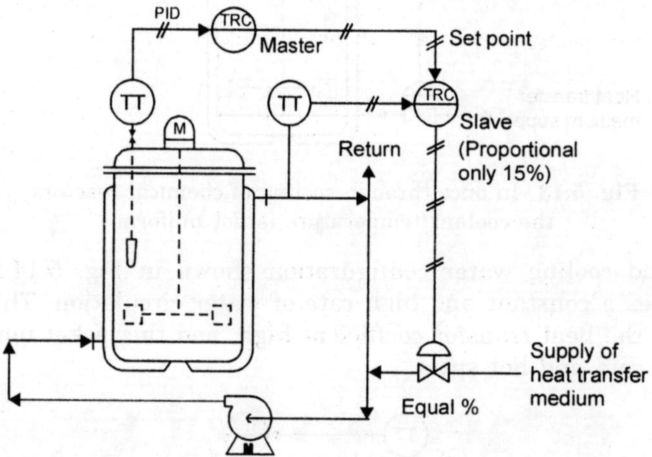

Fig. 5.15 Cascade control of a reactor with recirculation reduces the period of oscillation of the master loop.

5.2.4 Two-Directional Cascade Control of Reactor Temperature

If heat needs to be added in some phases of the reaction while in other phases it must be removed, the control must be configured in a two-directional manner. Fig. 5.16 depicts a cascade temperature control system with provisions for batch heat-up. The heating and cooling-medium control valves are split-range controlled, such that the heating-medium control valve operates between the air signal values of 9 and 15 psi (0.6 and 1 kg/cm^2) and the cooling medium control valve operates between 3 and 9 psi (0.2 and 0.6 kg/cm2). It is important to match the characteristics of the valves and to avoid non linearity at the transition, which can result in cycling. It is equally undesirable to keep both valves open simultaneously, because it results in energy waste. This control system is a fail-safe arrangement, because in case of air failure the heating valve is closed and the coolant valve is opened.

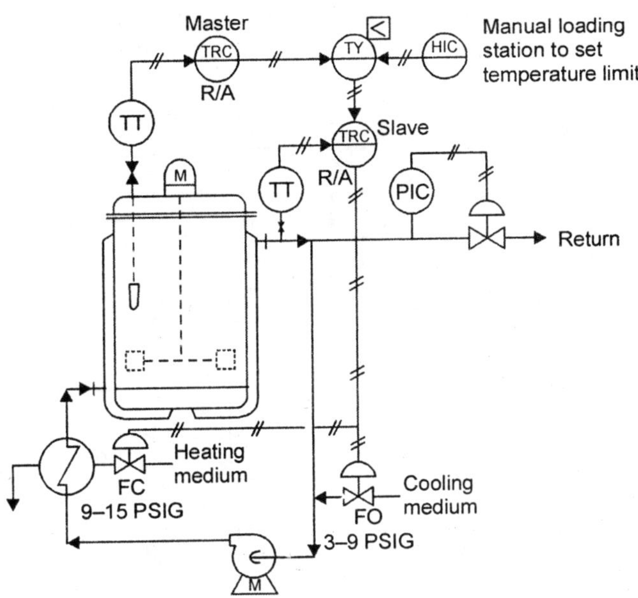

Fig. 5.16 Two-directional cascade loop with a maximum limit on jacket temperature allows heat to be added in some phases of the operation and removed in others.

An arrangement is provided whereby an upper temperature limit is set on the recirculating heat transfer medium stream. This is an important consideration if the product is temperature-sensitive or if the reaction is adversely affected by high reactor well temperature. In this particular case, the set point to the slave controller is prevented from exceeding a preset high temperature limit. Another feature shown is a back pressure control loop in the heat-transfer medium return line. This may be needed to impose an artificial back pressure, so that during the heat-up cycle no water leaves the recirculation loop and therefore the pump does not experience cavitation problems.

5.2.5 Split-Range Sequencing with Multiple Coolants

The use of a single coolant and single heating media as discussed above is often insufficient or uneconomical. If one type of coolant (or heating media) is less expensive than another, it is desirable to fully utilize the less expensive one before starting to use the expensive one.

For example if provision is there for using cold water and chilled water as coolants, usage of cold water which is less expensive is advised. When the condition exceeds the limit that cold water alone will not do, then chilled water usage may be started. Similarly hot water and steam can be compared for heating media. Here we will discuss a case where steam alone is used as heating media, but cold water as well as chilled water are used as cooling media for the sake of simplicity.

Fig. 5.17 illustrates such a scheme of reactor temperature control providing split-range sequencing with multiple coolants to minimize cost. It is desirable to fully utilize the cold water range before starting to use chilled water. Also the destination of the returning water should not be selected on the basis of the origin of that water, but rather should be based on the temperature of that water. This will reduce the upset caused in the plant utilities when a reactor switches from heating to cooling. Three way valves are used for selecting either cold water or chilled water as per the controller requirement. Similarly for the return water the destination is selected via a three-way valve operated by a thermal switch.

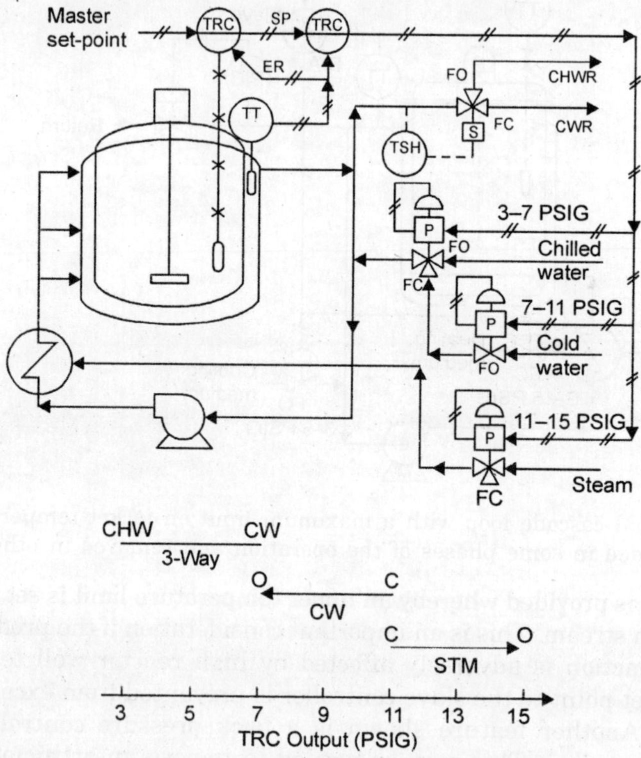

Fig. 5.17 Split-range sequencing with multiple coolants can be used to minimize cost.

When the slave controller output is 15 psi, the steam valve is fully opened supplying heat energy to the reactor. As the temperature of the reactor starts increasing the heat energy out of reaction is also getting added. The controller output starts getting reduced from 15 psi to 11 psi and hence the steam valve getting closed from 100% open to full closed. Further temperature rise (now it is only due to reactions in the reactor) makes the controller output to decrease from 11 psi and the cold water as coolant is opened. From 11 to 7 psi, the cold water valve is in operation and fully opened when the output becomes 7 psi. Further increase in temperature will further reduce the controller output from 7 psi towards 3 psi. During this period, the cold water valve remains fully opened and the chilled water valve operates from fully closed to open. When the temperature drops the reverse action takes place.

5.3 MIXING CONTROLS

5.3.1 Mixing Peration

Automatic, continuous, in-line mixing (blending) systems provide control of gases, liquids, and solids in predetermined proportions at a desired total mixed flow rate. The mixing systems consist of flow transmitters (to detect controlled variables), ratio stations/relays (to set proportions), controllers, and final control elements (to complete the closed loop control). Mixing applications provide continuous control of the flow of each component with fixed ratios between components, so that when the streams are combined to form the finished blend at a fixed through-put rate, the composition of the finished product is within specifications. Blending/mixing systems are applied to a variety of materials in a number of industries : Solvents, paints, reactor feeds, foams, fertilizers, soaps, and liquid cleaners in the chemical industry; gasoline, asphalt, lube and fuel oils, and distillates in the petroleum industry; wine, beer, candy, soups, ice cream mix, and cake mixers in the food industry; cement, wire insulation and asbestos products in the building industry; iron ore, coke, lime and dolomite in ferrous industry.

The mixing operations can be classified into three categories in general : (1) Mixing and blending of solids : Example : In an iron and steel industry different grades of iron ores are mixed and sent to steel plant before they are charged in blast furnace. (2) Mixing of liquids : very common in chemical industry. Two streams of liquids are measured and mixed in a desired ratio before entering into the vessel. Thorough mixing of them is possible with the help of stirrer before the final mixture is taken out. (3) Mixing of two or more gases : Three fuel gases of different calorific values can be mixed and stored in a gas holder to have uniform calorific value of the mixed gas before it is used in furnaces.

5.3.2 Mixing Controls

In most of the mixing controls, the 'ratio' of two streams (reactants) must be controlled. In this case, one of the flow rates is measured but allowed to float, that is, not regulated, and the other is both measured and adjusted to provide the specified constant ratio. An example of such a ratio control is already discussed in Chapter-3 (Fig. 3.22).

One may also encounter number of feedback control loops like temperature control, level control, composition control etc when two reactants are mixed, heated and final product taken out at a desired composition. Sometimes, the flow rates of two streams to the mixer can be controlled by means of other than ratio control. One stream flow may be controlled based on the output product rate and the other stream flow to have the required composition.

We may have to encounter split range controls and override controls also in mixing process when number of streams to be mixed is more than two.

We will be discussing the four following basic applications of mixing control covering almost all possible combinations.

1. Mixing of solids.
2. Common feedback loops of a mixing process.
3. Outlet flow rate and composition control.
4. Gas mixing station with three streams.

5.3.3 Mixing of Solids

Let us take a simple example of mixing two grades of iron ores in a predetermined proportion and stock them for further use in blast furnace. Fig. 5.18 illustrates one such application. Bunker-1 contains Grade-1 ore and Bunker-2 contains grade-2 ore. The quantity of ore from Bunker-2 is normally controlled through manual operation of the bin-gate. Then it is fed through feeder-2 to the hopper where Grade-1 ore is mixed with. Hopper output is fed to the common bunker through common feeder for further use. Feeding rates of both bunkers are measured by feeder weighing machines (normally load cells are used) W_1 and W_2.

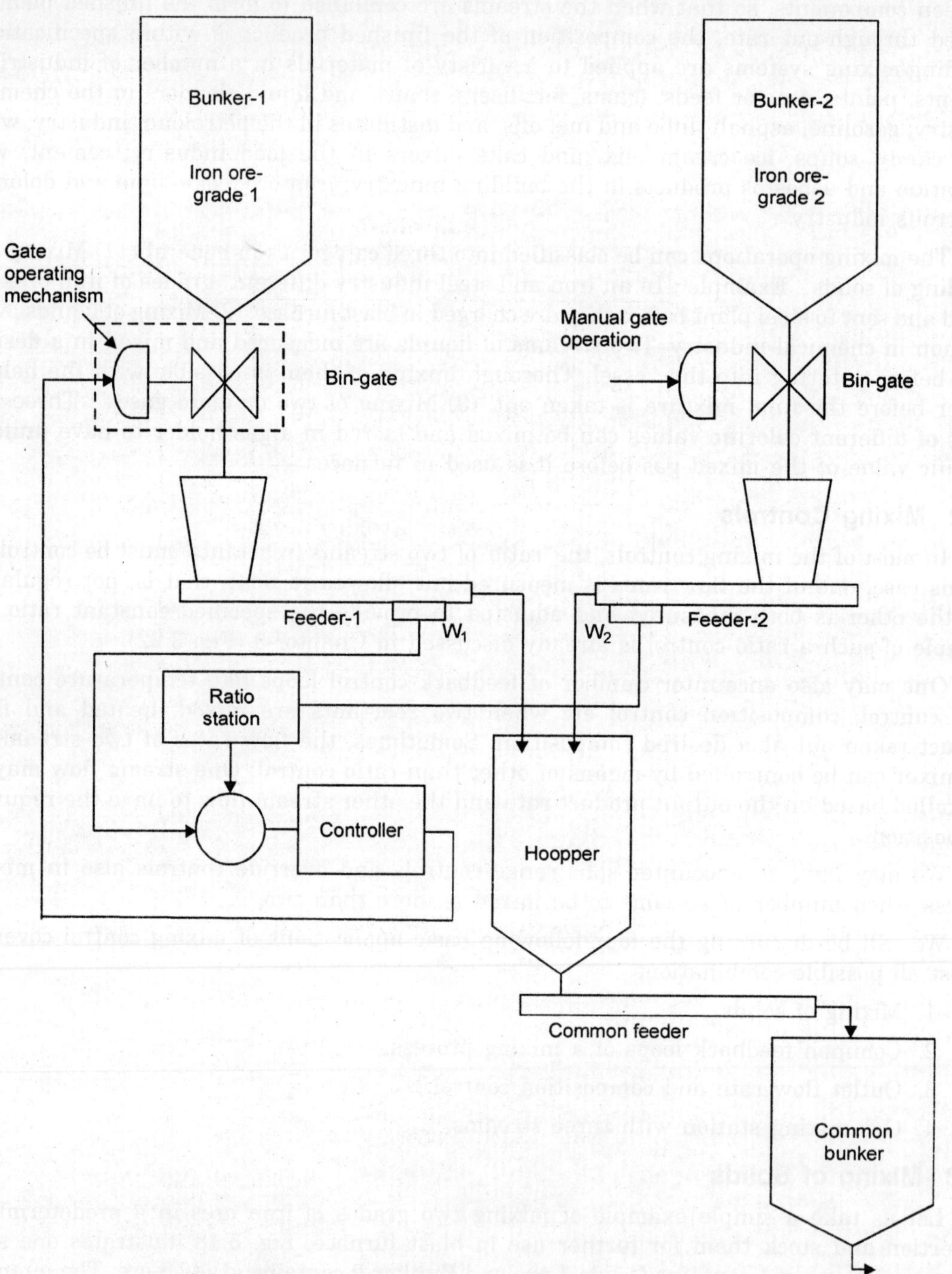

Fig. 5.18 Mixing of solids-Ratio control.

W_2, the measured weight flow rate of feeder-2 is given to a ratio station where the ratio desired between W_2 and W_1 is set. The controller adjusts the gate operating mechanism in bunker-1 to control the weight flow rate of feeder-1 to satisfy the ratio requirement.

5.3.4 Common Feedback Loops in a Mixing Process

Let us consider a case where two streams 1 and 2 are being mixed together producing a output stream. The tank can be a well-stirred one so that the mixing of both streams is homogeneous. As streams may have different temperatures and composition, a temperature control system and a composition control system may form parts of the total mixing control process. There can be another loop to control the level of the mixing tank.

Refer Fig. 5.19 where mixing process is controlled by three feedback loops. Stream-2 flow rate is neither measured nor controlled. Stream-2 may be an outlet from the preceding section and the stream-1 whose quantity has to be controlled to maintain the composition of the mix. Hence the composition (or sometimes pH value of the mix) has to be measured by an analyzer and controlled through a feedback controller with stream-1 flow rate as manipulated variable. Temperature of the mix has to be measured and controlled with the help of heat exchanging medium (steam), the flow rate of which becomes the manipulated variable. Level of the tank can be controlled by having a control element in the outlet stream.

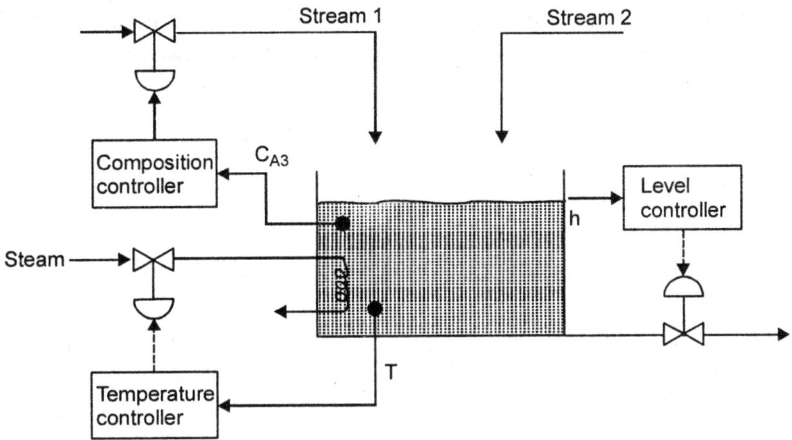

Fig. 5.19 Feedback loops of a mixing process.

In the above system, the possible three control objectives are :

1. Keep the volume (or level) of the mixture in the tank at a desired value.
2. Maintain the temperature of the mixture and hence the temperature of the effluent stream (outlet stream) constant at a desired set point.
3. Keep the composition (or pH) of the mixture and hence that of the outlet stream at a desired value.

5.3.5 Outlet Flow Rate and Composition Control

Here we will discuss a mixing process of two streams. The requirements are to regulate the product composition and outlet flow rate. The input stream flow rates are manipulated variables. Selection of loops and the interaction between loops in the mixing process are also to be considered. Let us consider the following example and try to know the consequences with sample calculation.

Two streams with flow rates F_1 and F_2 and compositions (mole percent) $x_1 = 80\%$ and $x_2 = 20\%$ are mixed in a vessel [Fig. 5.20(*a*)]. We would like to form two control loops to regulate the product composition x and flow rate F. Let $F \equiv y_1$ and $x \equiv y_2$ be the two controlled outputs, while $F_1 \equiv m_1$ and $F_2 \equiv m_2$ are the two available manipulated variables. There are two possible control configurations with different pairings between the inputs and outputs, and they are shown in Fig. 5.20(*b*) and (*c*). The question now is, which one should we prefer ?

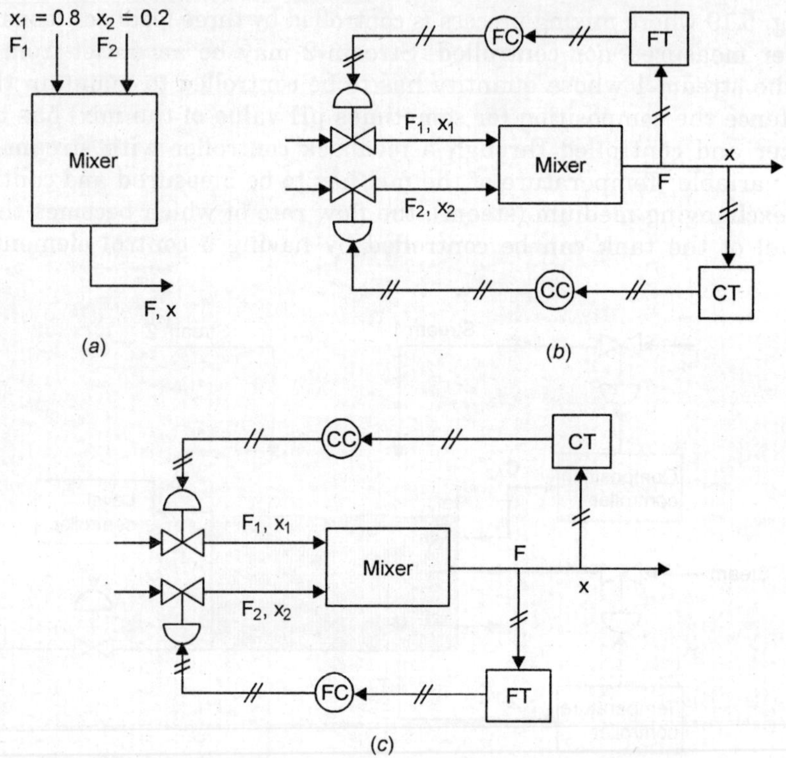

Fig. 4.20 (*a*) Mixer; (*b*), (*c*) alternative loop structures.

The steady state mass balances yield

$$F = F_1 + F_2 \quad ...(5.1)$$

$$Fx = F_1x_1 + F_2x_2 \quad ...(5.2)$$

Assuming the desired steady state for operational purposes as

$$F = 200 \text{ mol/hr and } x = 60\% \text{ (by moles)}$$

We find the steady state solutions for equations (5.1) and (5.2):

$$F_1 = 133.4 \text{ and } F_2 = 66.6$$

To compute the relative gain between F_1 and F, do the following :

1. Change F_1 by one unit (*i.e.*, $F_1 = 134.4$) while holding $F_2 = 66.6$ (same). By solving the equations (5.1) and (5.2) for F and x, we find the following new steady states :

   $F = 201$ and $x = 0.6012$ (60.12%)

   Therefore,

$$\left(\frac{\Delta F}{\Delta F_1}\right)_{F2} = \frac{1}{1} = 1, \left(\frac{\Delta x}{\Delta F_1}\right)_{F2} = \frac{0.0012}{1} = 0.0012$$

2. Change F_1 by one unit (*i.e.*, $F_1 = 134.4$) while holding $x = 60\%$ constant. By solving the equations (5.1) and (5.2)

   We get, $F = 201.67$ and $F_2 = 67.27$

   Therefore,

$$\left(\frac{\Delta F}{\Delta F_1}\right)_x = \frac{1.67}{1} = 1.67$$

   Consequently, the relative gain between F and F_1 is

$$\lambda_{11} = \frac{(\Delta F/\Delta F_1)_{F2}}{(\Delta F/\Delta F_1)_x} = \frac{1}{1.67} = 0.6$$

   It follows easily that the complete relative-gain array is

$$\begin{array}{cc} & F_1 \quad F_2 \\ A = & \begin{bmatrix} 0.6 & 0.4 \\ 0.4 & 0.6 \end{bmatrix}_{x}^{F} \end{array}$$

We can draw two main conclusions now :

1. The two loops with minimum interaction are formed when we couple F with F_1 and x with F_2. [Fig. 5.20(*b*)]
2. Although the interaction between the two selected loops are smaller than that of the alternate configuration [Fig. 5.20(*c*)], it is still significant. Thus any control action to regulate F will seriously disturb x, and vice versa.

5.3.6 Gas Mixing Station with Three Streams

In many chemical and metallurgical industries, there are requirements for two or three gases generated during the processes to be mixed to get a mixed stream (normally called as 'mixed gas'). Such a mixed gas is used as heating source for the furnaces elsewhere in the same industry.

As an example, let us consider a case in iron and steel plant where coke-oven gas, blast furnace gas and steel melting converter gas are produced while processing. These three gases are mixed and stored in a gas holder for further use in heating furnaces and annealing furnaces. The three gases will have different calorific values.

The desired requirement here are :

1. To get a product outlet stream (mixed gas) at a constant pressure.
2. To have the constant calorific value.
3. To meet the required outlet flow rate.

A mixing process scheme may be designed to achieve the above requirements, assuming some nominal values of calorific values for the streams.

Cv_1 = Calorific value of stream1 (coke-oven gas) = 4000 Kcal/Nm^3.

Cv_2 = Calorific value of stream2 (blast furnace gas) = 1000 Kcal/Nm^3.

Cv_3 = Calorific value of stream3 (steel melting shop gas) = 2000 Kcal/Nm^3.

Cv = Calorific value of mixed gas = 3000 Kcal/Nm^3.

If we assume the flow rates of three input streams as C, B and S and the flow rate of outlet as M, we may be able to write the following steady-state equation.

$$4C + B + 2S = 3M \quad \text{and} \quad C + B + S = M$$

or per unit volume of output stream, we can write

$$4C + B + 2S = 3 \quad \text{and} \quad C + B + S = 1$$

Assuming any one of the streams' flow rate, the other two can be calculated. The assumption should be such that no negative values are allowed.

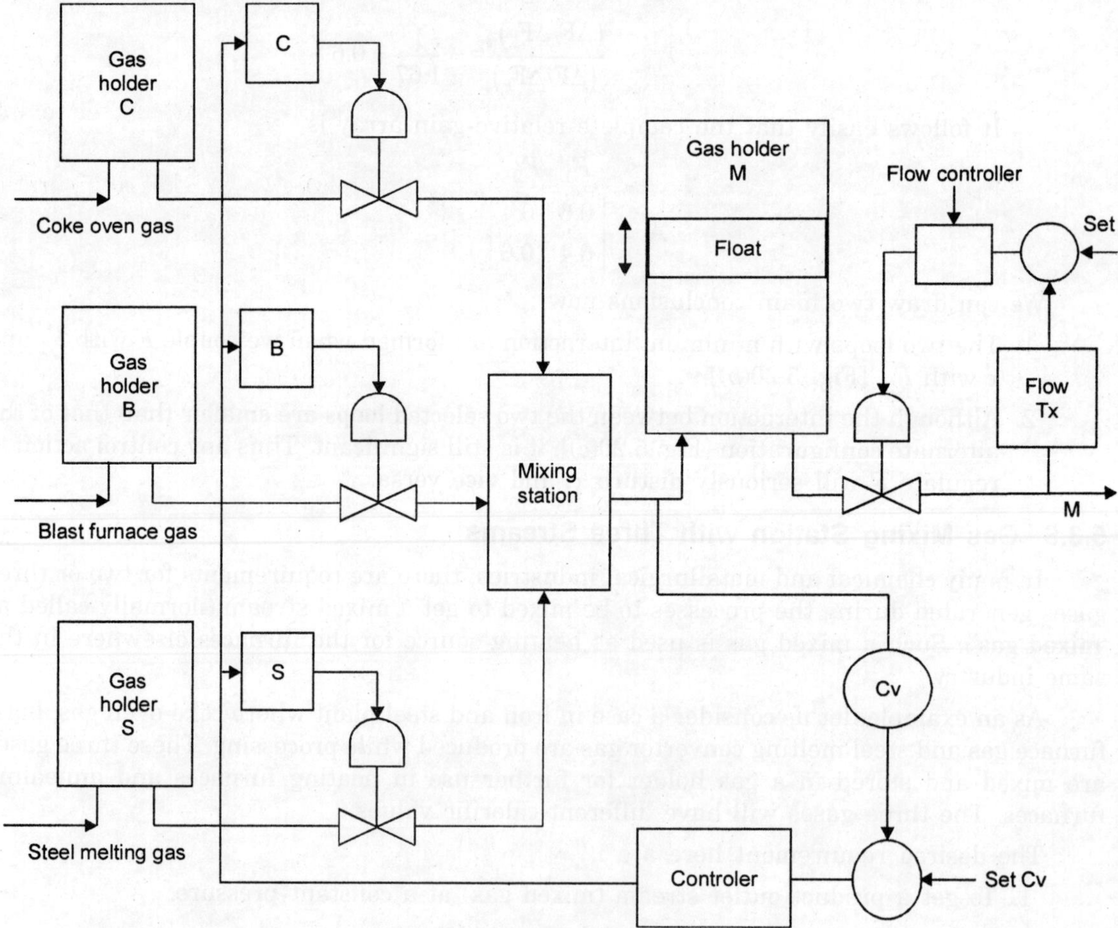

Fig. 5.21 Gas mixing station.

For example if we assume $C = 0.6$, we get $B = 0.2$ and $S = 0.2$. This means, if we mix 60% of output stream flow rate as coke oven gas, 20% as blast furnace gas and 20% as steel melting gas, we are to get the mixed gas of 3000 Kcal/Nm^3 calorific value. We can have theoretically infinite number of such combinations.

Refer Fig. 5.21 for the mixing process discussed so far.

The gas holder-M practically serves two purposes :

1. To store the excess gas available than consumed.
2. To maintain the outlet pressure constant.

The weight of the float decides the pressure. Whenever the pressure goes high, the float moves up allowing the gas to occupy more volume and hence loose pressure and vice versa. The outlet pressure will be always equal to the weight of the float divided by cross sectional area of the gas holder ($P = F/A$).

Our other requirement of maintaining calorific value constant at the outlet is met by measuring the calorific value at the outlet of the mixing station and controlling the flow rates of the input streams. There are number of ways of doing this. We may keep the flow rate of one stream constant and control the other two loops using ratio control. After fixing priority for the mixing streams we can have a split-range operation also. What is shown in the Fig. 5.21 adopts sharing of controller output among themselves as per the preset ratios which are to be further tuned to the requirement. The control can be made further complex by measuring calorific values of each stream and automatically providing the ratios.

Three-way valves : Three way valves, a form of specialized valve body configuration, are used for mixing service. Two fluid streams can be combined and passed through a common outlet port (Fig. 5.22).

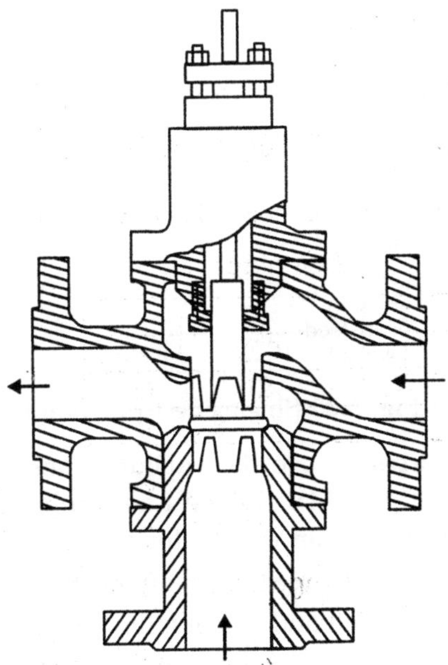

Fig. 5.22 Three-way valve for mixing service.

5.4 EVAPORATION

5.4.1 Evaporator Operation

Evaporation is one of the oldest unit operations. It is the process of concentrating aqueous solutions in a closed vessel or group of vessels in which the concentrated solution is the desired product and indirect heating (usually steam) is the energy source. Evaporation is an energy-intensive process. Its efficiency can be improved by increasing the number of evaporators in series (effects), and some of the energy can be recovered by vapour recompression. Evaporators can be arranged in forward-feed, reverse-feed, or parallel feed configurations with each stage being heated by the vapours of the previous stage. The product concentration can be measured by a variety of analysers, including density, conductivity, refractive index, percent solids, turbidity, and boiling or freezing point analysers.

'Single-effect evaporation' occurs when a dilute solution is contacted only once with a heat source to produce a concentrated solution and an essentially pure water vapour discharge. The operation of such an evaporator is shown in Fig. 5.23.

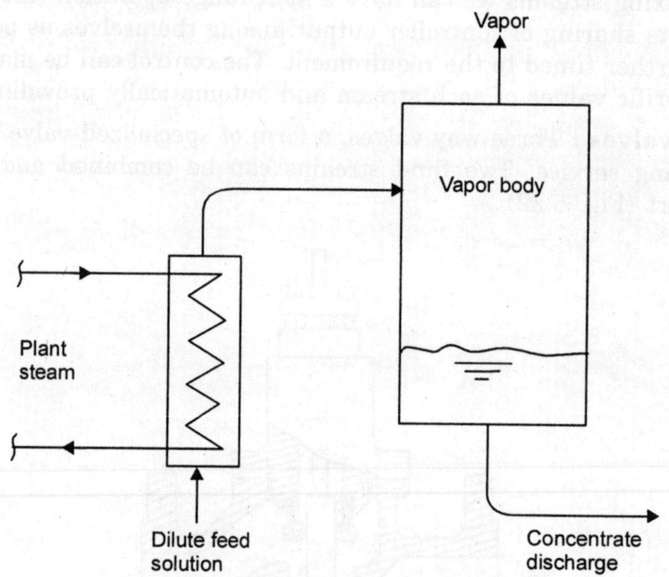

Fig. 5.23 Single-effect evaporator.

'Multiple-effect evaporations' use the vapour generated in one effect as the energy source to an adjacent effect. (Fig. 5.24). Double and triple-effect evaporators are the most common; however, six-effect evaporation can be found in paper industry where kraft liquor is concentrated, and as many as twenty effects can be found in desalinization plants.

In 'Co current operation', the feed and steam follow parallel paths through the evaporator train. In 'Counter current operation', the feed and steam enter the evaporation train at opposite ends.

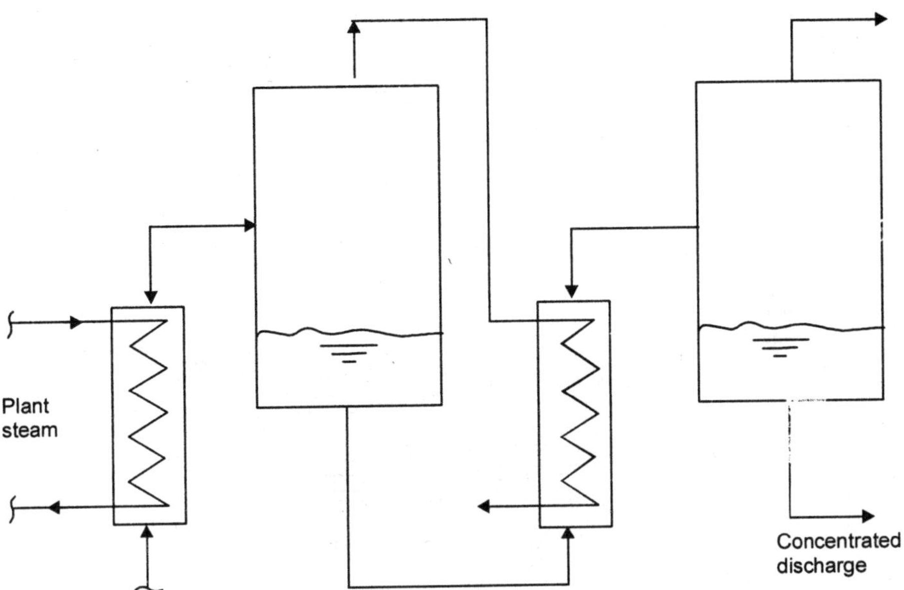

Fig. 5.24 Multiple-effect evaporator.

5.4.1.1 Types of Evaporators

The following six types of evaporators are used for most applications, and the length and orientation of the heating surfaces determine the name of the evaporator.

1. Horizontal-tube evaporator
2. Forced circulation evaporator
3. Short-tube vertical evaporator
4. Long-tube vertical evaporator
5. Falling-film evaporator
6. Agitated-film evaporator

5.4.1.1.1 Horizontal-tube evaporator

A typical horizontal-tube evaporator is shown in Fig. 5.25. These evaporators were among the earliest types. Nowadays, they are limited to preparation of boiler feedwater, and in special construction at high cost, for small volume evaporation of severely scaling liquids, such as hard water. In their standard form they are not suited to scaling or salting liquids and are best suited in applications requiring throughputs.

5.4.1.1.2 Forced circulation evaporator

Forced circulation evaporators (Fig. 5.26) have the widest applicability. Circulation of the liquor past the heating surfaces is assured by a pump, and consequently these evaporators are frequently external to the flash chamber so that actual boiling does not occur in the tubes, thus preventing salting and erosion. The external tube bundle also lends itself to easier cleaning and repair than the internal heater shown in Fig. 5.26. Disadvantages include high cost, high residence time, and high operating cost due to the power requirements of the pump.

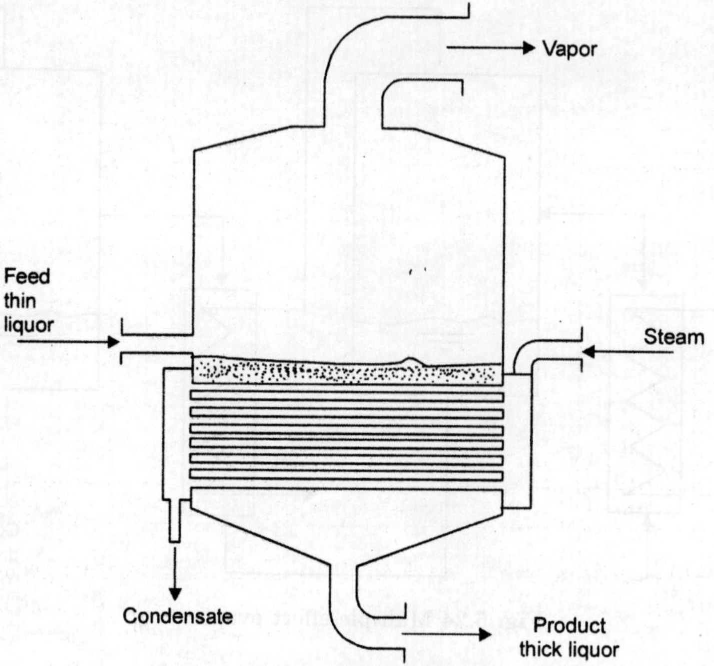

Fig. 5.25 Horizontal-tube evaporator.

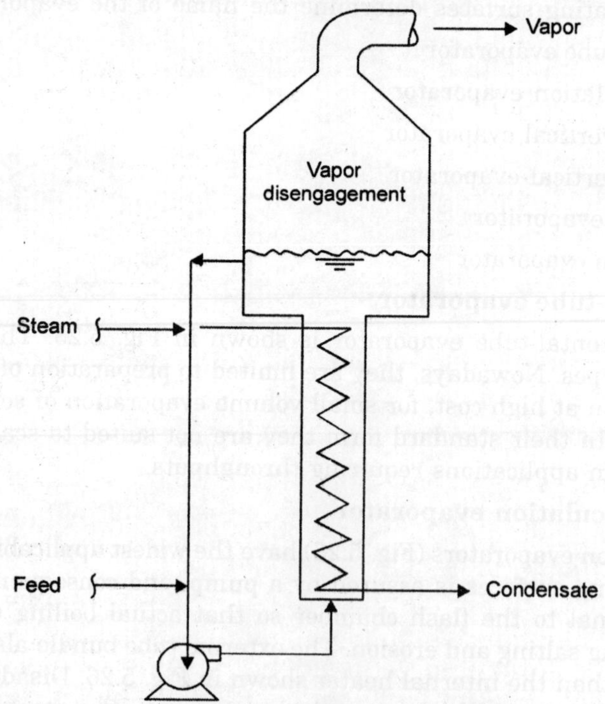

Fig. 5.26 Forced-circulation evaporator.

5.4.1.1.3 Short-tube vertical evaporator

A short-tube vertical evaporator shown in Fig. 5.27 is common in sugar industry for concentrating sugar juice. Liquor circulation through the heating element (tube bundle) is by natural circulation (Thermal convection) since the mother liquor flows through the tubes, they are much easier to clean. This evaporator is suitable for mildly scaling applications. Level control is important because if the level drops below tube ends, excessive scaling results. Ordinarily, the feed rate is controlled by evaporator level to keep the tubes full. The disadvantage of high residence time in the evaporator is compensated for by the low cost of the unit for a given evaporator load.

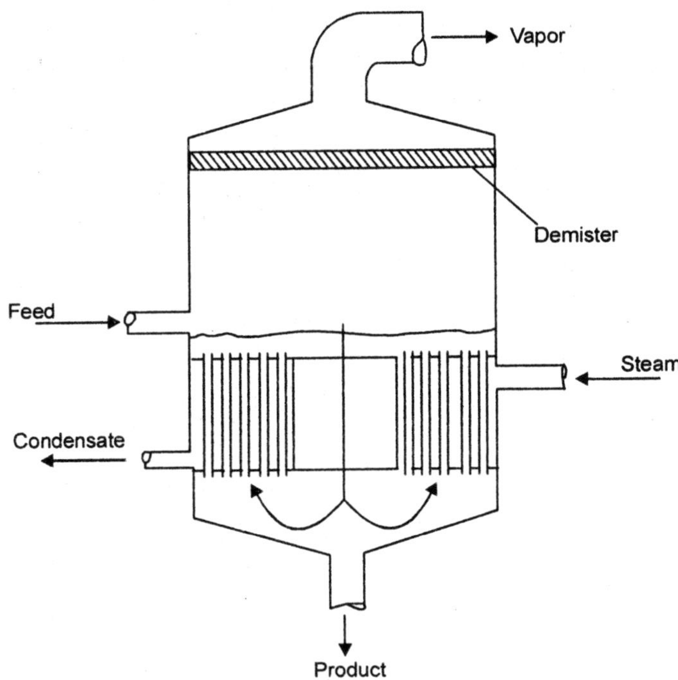

Fig. 5.27 Short-tube vertical evaporator.

5.4.1.1.4 Long-tube vertical evaporator : (Refer Fig. 5.28)

The cost per unit capacity of long-tube vertical evaporator is low. It is also called as 'Rising Film Concentrator(RFC)'. Typical application include concentrating black liquor in the pulp and paper industry and corn syrup in the food industry. Most of these evaporators are of the single-pass variety, with little or no internal circulation. Thus, residence time is minimized. Level control is important in maintaining the liquid seal in the flash tank. They offer low cost per unit weight of water evaporated and have low hold up times but tend to be tall, requiring more head room than other types. These units are sensitive to changes in operating conditions making them difficult to control.

5.4.1.1.5 Falling-film evaporator

A falling-film evaporator, shown in Fig. 5.29, is commonly used with heat-sensitive materials. Physically, the evaporator looks like a long-tube vertical evaporator, except that feed material descends by gravity along the inside of the heated tubes, which have large inside diameters.

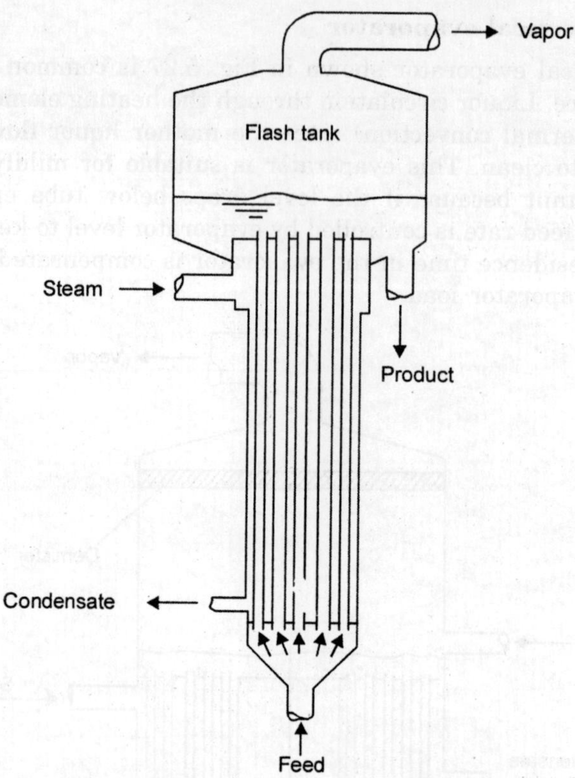

Fig. 5.28 Long-tube vertical evaporator.

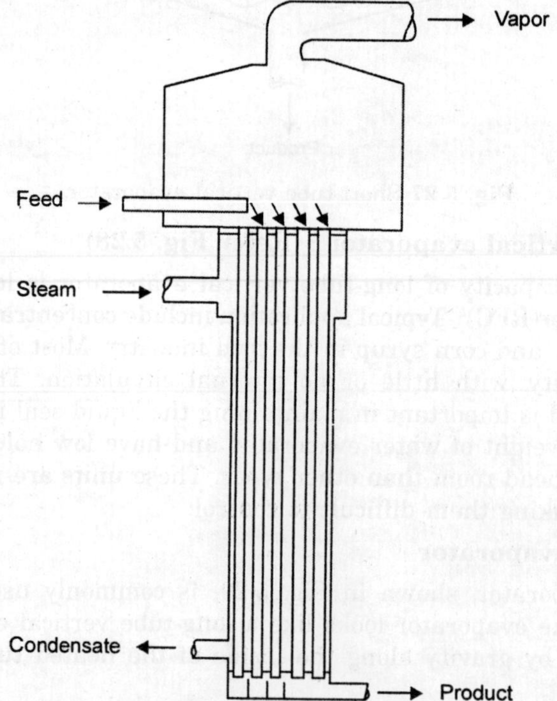

Fig. 5.29 Falling-film vertical evaporator.

5.4.1.1.6 Agitated film evaporator

Like the falling-film evaporator, an agitated film evaporator is commonly used for heat sensitive and highly viscous materials. It consists of a single large diameter tube with the material to be concentrated falling in a film down the inside where a mechanical wiper spreads the film over the inside surface of the tube. Thus, large heat transfer coefficients can be obtained, particularly with highly viscous materials.

5.4.2 Evaporator Controls

The control systems are mainly to achieve the final product concentration. Perhaps one of the most controversial issues in any evaporator control scheme is the method used to measure the product density. Common methods include (1) temperature rise, boiling point rise; (2) conductivity; (3) differential pressure; (4) gamma gauge; (5) U-tube densitometer; (6) buoyancy float and (7) refractive index. Each method has its strengths and weaknesses. In all cases, however, care must be taken to select a representative measurement location to eliminate entrained air bubbles or excessive vibration. The relative location of the product density transmitter with respect to the final effect should also be considered. Measurement details are not considered in this book, but can be found in process measurement books.

The control system to be considered in achieving final product concentration include (1) Feedback, (2) Cascade, (3) Selective and (4) Feed forward. For ease of illustration, a double-effect, cocurrent flow evaporator is considered for our further discussions. Extension to more or fewer effects will not change the basic control system configuration.

The choice of system should be based on the needs and characteristics of the process. Evaporators as a process class tend to be capacious (mass and energy storage capability) and have significant dead time (30 seconds or greater). If the major process loads (feed rate and feed density) are reasonably constant and the only corrections required are for variations in heat losses or tube fouling, feedback control will suffice. If steam flow varies because of demands elsewhere in the plant, a cascade configuration will probably be the proper choice. The selective control may add to the cascade control configuration the feature of protection against running out of steam. If, however, the major load variables change rapidly and frequently, it is strongly suggested that feed forward in conjunction with feedback be considered.

5.4.3 Feedback Control of Evaporator

A typical feedback control system is shown in Fig. 5.30. The product concentration is measured with a density meter (DT) and the amount of steam to the first effect is controlled through a three-mode controller (DC). The internal material balance is maintained by level control (LC) on each effect. Levels in each effect is measured by level transmitter (LT). This control system is found to be very effective and stable for reasonably constant process loads.

5.4.4 Cascade Control of Evaporator

A typical cascade system is illustrated in Fig. 5.31. This control system measures the product density as in feedback control and adjusts the heat input through a flow loop controller (FC) that is being set in cascade from the final density controller. This cascade control arrangement is particularly effective when steam flow variations (outside of the evaporator) are frequent. It should be noted that with this arrangement the valve positioner is not required and can actually degrade the performance of the flow control loop.

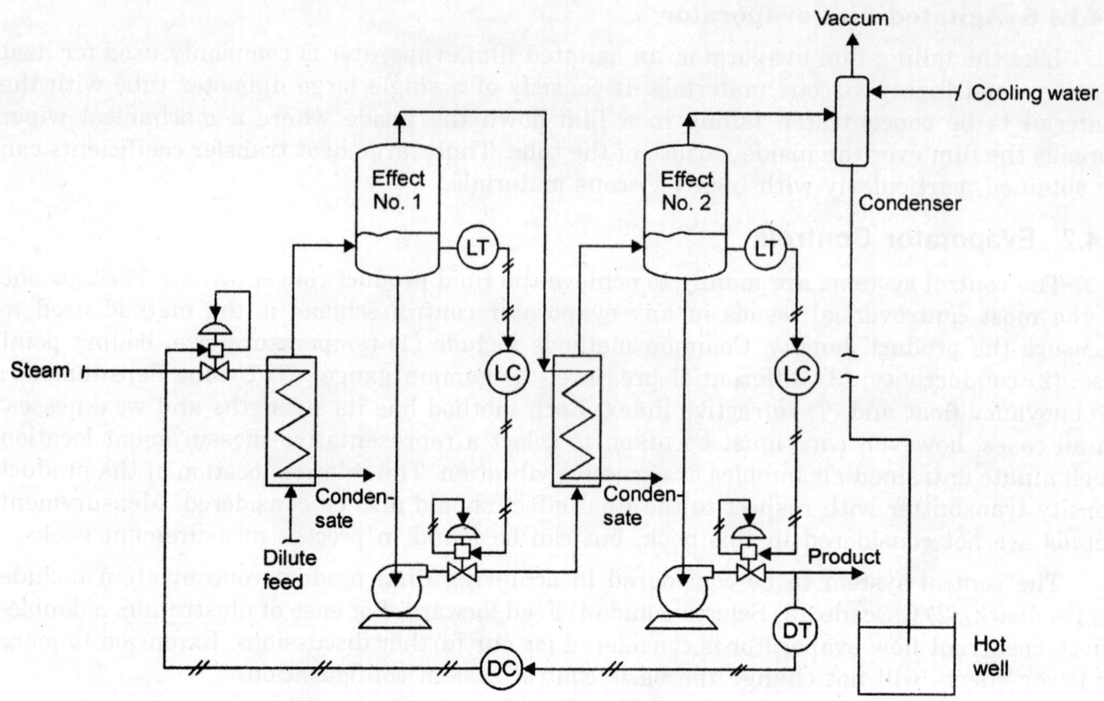

Fig. 5.30 Feedback control system of evaporator.

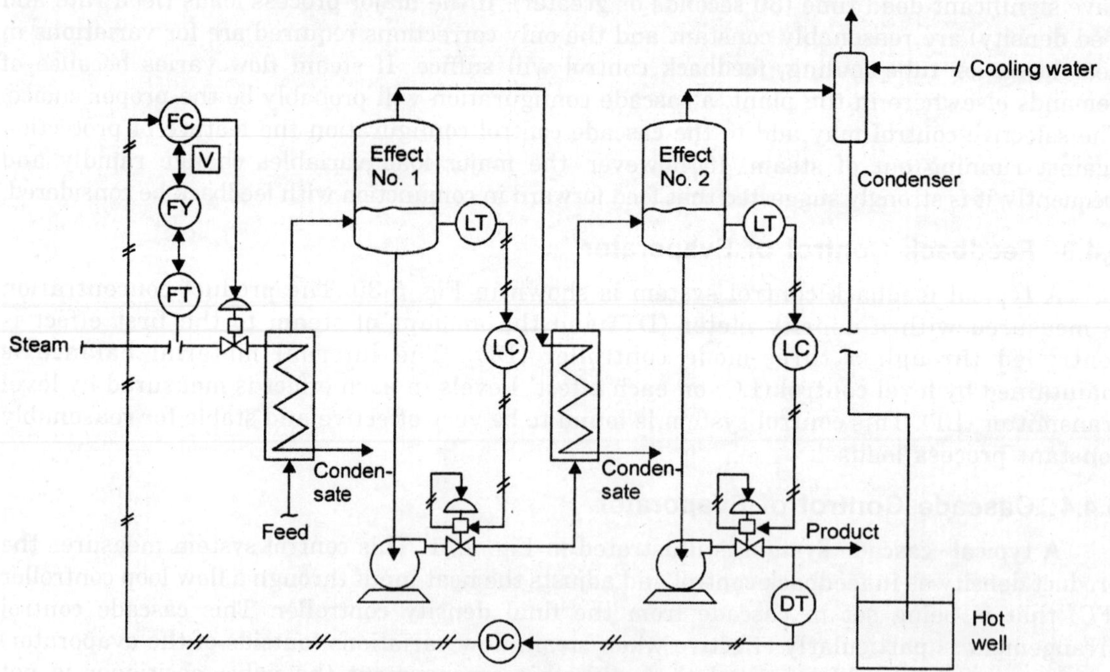

Fig. 5.31 Cascade control system of evaporator

5.4.5 Selective Control of Evaporator

The selective control scheme shown in Fig. 5.32 adds to the cascade configuration the feature of protection against running out of steam. This feature is provided by the valve position controller (VPC), whose measurement is the steam valve opening and whose set point is about 90%. Therefore as the feed gets too dilute and therefore requires more steam to concentrate, the steam valve opens. When it reaches 90% lift, the VPC output signal drops below that of the feed controller (FC), and therefore the low signal selector (FY) transfers the control of the feed valve to the VPC. During the period while the feed is dilute due to some upset, the VPC sets the feed flow rate to a value which corresponds to the allowable maximum opening of the steam valve. The VPC is a PI controller with the integral mode dominating (as in the case in most valve position controllers). Both controllers (FC and VPC) are provided by external feedback taken from the output of the low selector (FY). This way the controller which is in control will have its output signal and its feedback signal at the same values and therefore act as a normal PI controller. The controller which is not selected will operate as a proportional-only controller with a bias, where the bias is the external feedback signal. This guarantees that at the time of switch over the two outputs will be identical, and therefore the switchover will occur bumplessly. At the time of switch over the error in the idle controller has just reached zero, and therefore its output equals its external feedback.

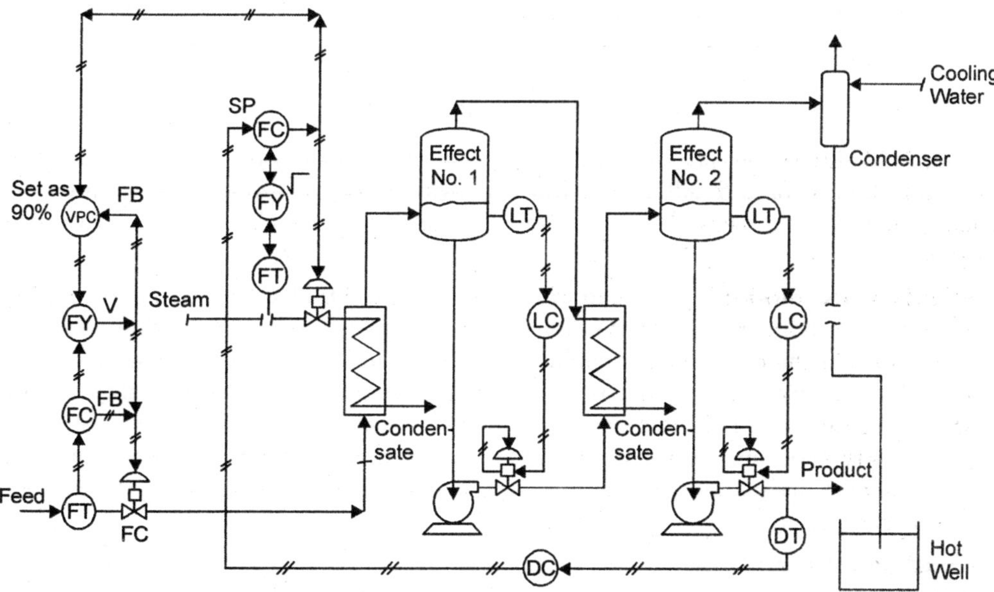

Fig. 5.32 Valve position controller (VPC) cuts back the feedflow and thereby protects the system from running out of steam. The selective control scheme is protected from reset windup by external feedback.

5.4.6 Feedforward Evaporator Control

In most evaporator applications the control of product density is constantly affected by variations in feed rate and feed density to the evaporator. In order to counter these load variations, the manipulated variable (steam flow) must attain a new operating level. In the pure feedback or cascade arrangements this new level was achieved by trial and error as performed by the feedback (final density) controller. A control system able to react to these load variations when they occur (feed rate and feed density) rather than wait for them to pass through the process before initiating a corrective action would be ideal. This technique is called 'feedforward control' which has been discussed in chapter-3 already.

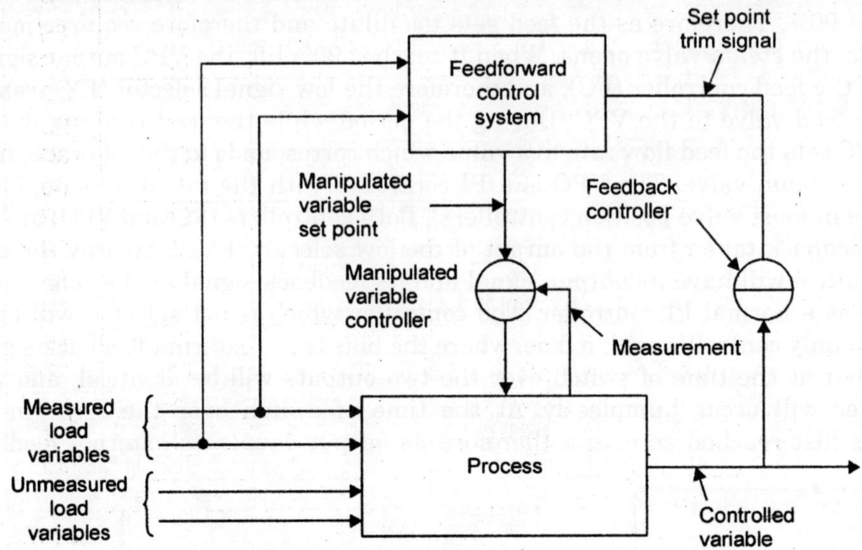

Fig. 5.33 Feedforward system,

The Fig. 5.33 illustrates in block diagram form the features of a feedforward system. The three ingredients of a feedforward control system are (1) the steady state model, (2) feedback trim and (3) process dynamics. We will briefly discuss about these with reference to evaporators.

1. **Steady-state model :** The load variables are classified as either major or minor . A relationship incorporating the major load variables (measurable), the manipulated variables and the controlled variable can be developed. Such a relationship is termed as the steady-state model of the process. Development of the steady-state model for an evaporator involves material and energy balances. A relationship between the feed density, feed rate (major measurable local variables), steam flow rate (manipulated variable) and product density (controlled variable) is to be established and incorporated in the control system.

2. **Feedback trim :** Minor load variables are very hard to measure. In evaporators minor variables might be heat losses and tube fouling. Such load variables are easily handled by a feedback loop. The purpose of the feedback loop is to trim the forward calculation to compensate for the minor and unmeasured load variations. Without this feature the controlled variable could be off the set point. As a general rule, feedback trim is incorporated into the control system at the point at which the setpoint of the controlled variable appears.

3. **Process dynamics :** A change in one of the major loads to the process also modifies the operating level of the manipulated variable. In a co-current flow evaporator , an increase in feed rate will call for an increase in steam flow. The increased federate will rapidly appear at the other end of the train while the increased steam flow is

still overcoming the thermal inertia of the process. This sequence results in a transient decrease of the controlled variable (density) and the load variable passes through the process faster than the manipulated variable. The reverse will occur in case of counter current evaporator operation, that is, manipulated variable will pass thorough the process faster than the load variable. Both the behaviours are shown in Fig. 5.34 (*a*) and (*b*). This dynamic imbalance is normally corrected by inserting a dynamic element (lag, lead-lag, or a combination thereof) in at least one of the load measurements to the feed forward control system.

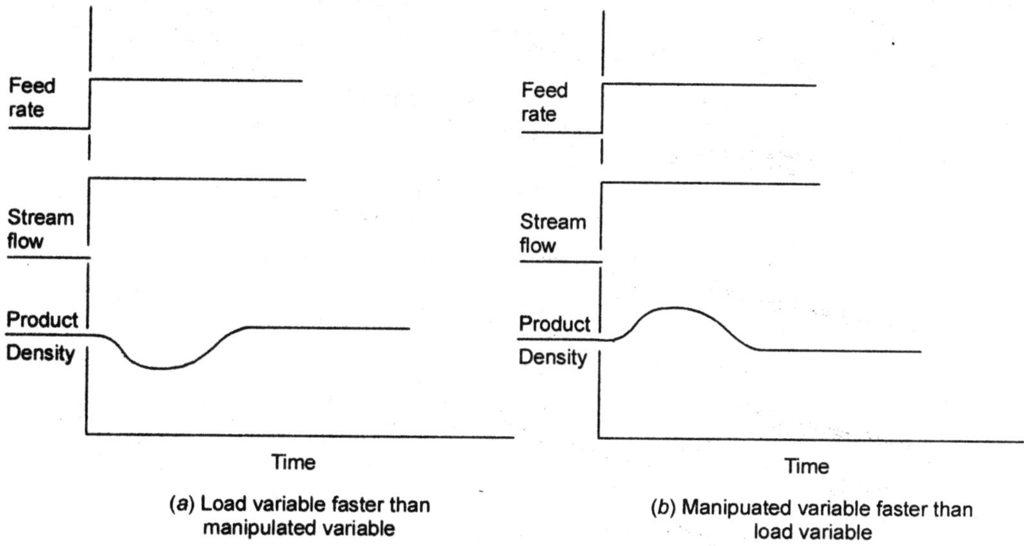

Fig. 5.34 Process dynamics.

A typical control system taking the above three ingredients into consideration is illustrated in Fig. 5.35.

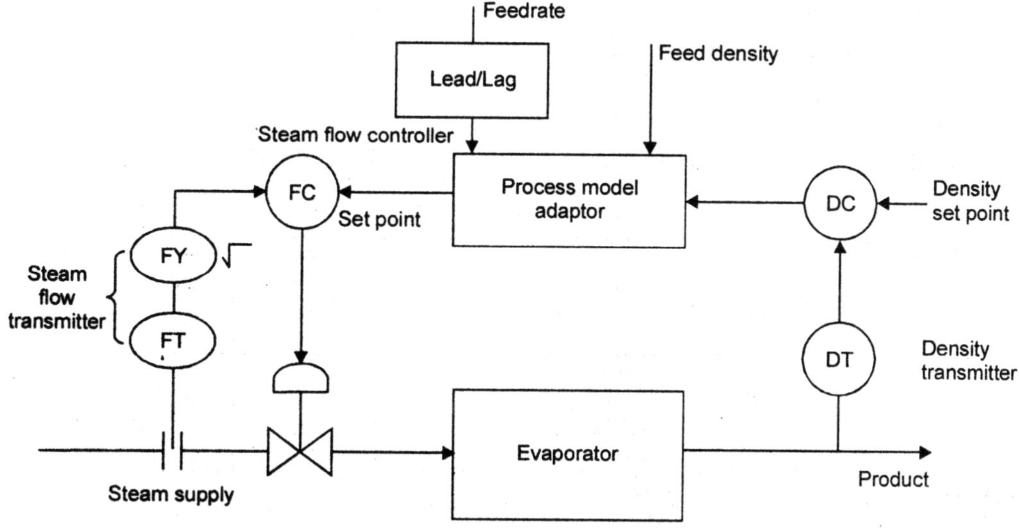

Fig. 5.35 Feedforward control with feedback trimming.

5.5 DRYER

5.5.1 Dryer Operation

Drying refers to an unit operation in which moisture of the substance is removed by thermal means. During drying operation, mass and heat transfer occurs simultaneously. Heat is transferred from the bulk of the gas phase (drying media) to the solid phase and mass is transferred from the solid phase to the gas phase in the form of liquid and vapour through various resistances. In other words, the drying process involves the removal of liquids, such as water or other solvents, by adding heat to vapourise them. Drying is frequently the last operation in the manufacturing processes and is usually carried out after evaporation, filtration or crystallization. Drying operations are mostly encountered in food, chemical , agricultural, pharmaceutical and textile industries.

Based on mode of operation dryers are classified into two types :

1. **Batch dryers :** In the case of batch dryers definite size of batch of wet feed is charged to the dryer and drying is carried over a given period of time.
2. **Continuous dryers :** In the case of continuous dryers, the materials flow in and out continuously and drying is carried out under steady state conditions. Continuous dryers can be further sub divided as given below : (Refer Fig. 5.36)

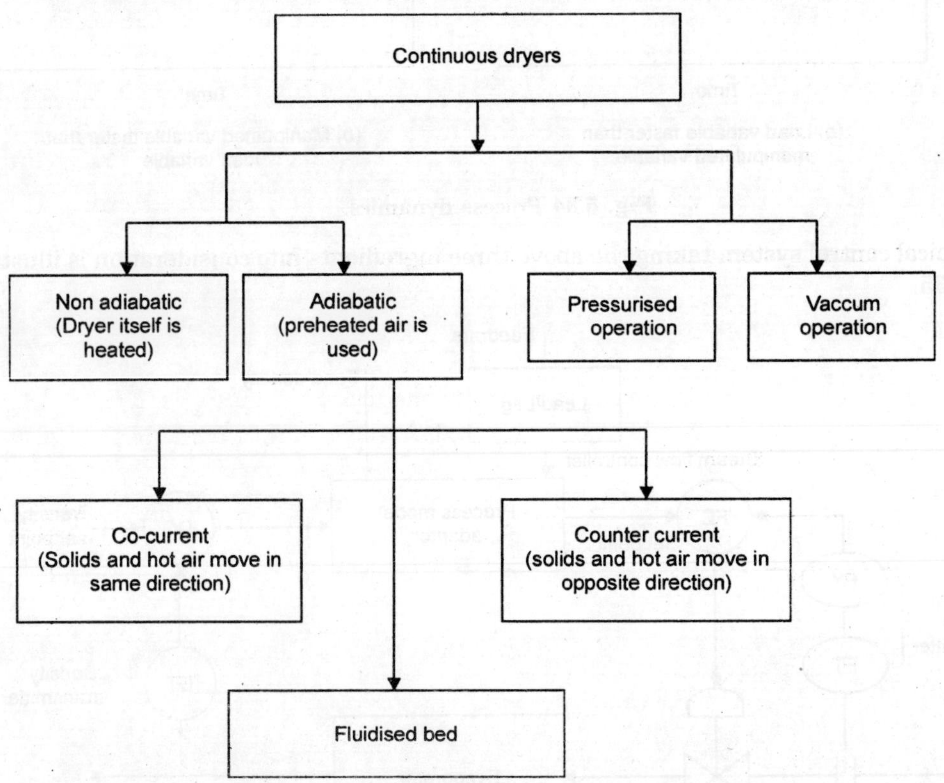

Fig. 5.36 Subdivision of continuous dryers.

5.5.1.1 *Air as a Drying Medium*

Adiabatic dryers obtain their heat from the hot air which enters them and use the sensitive heat to vapourise the moisture from the process solids. In this process the air is cooled, but its energy content (enthalpy) does not change, because the sensitive heat loss is compensated by an equal rise in enthalpy due to increased moisture contents (adiabatic process). Air enters at a high temperature and at a low relative humidity (RH) and leaves at a low temperature and high relative humidity.

The drying curve of a typical material is shown in Fig. 5.37. It consists of four zones.

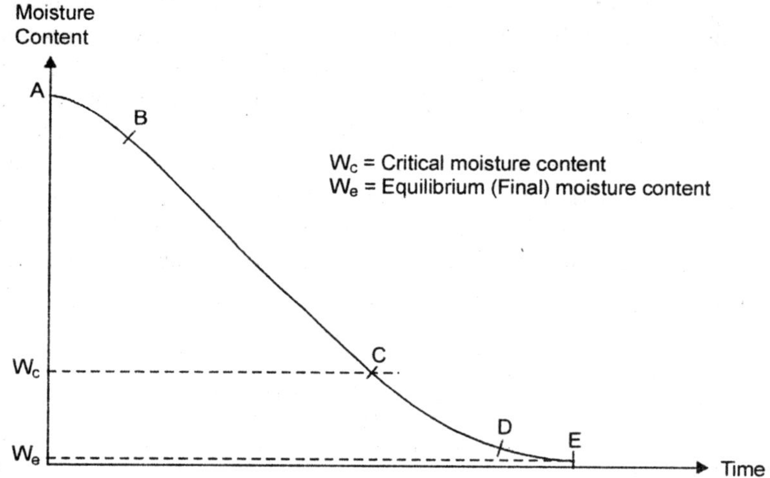

Fig. 5.37 Typical drying curve.

1. **A-B:** represents the period of product entry into the dryer. Since some heat is necessary to bring the material to the initial drying temperature, evaporation during this phase is very slow. It is known as 'preheat zone'.
2. **B-C:** represents the period of evaporation of surface moisture or moisture that migrates readily and is known as 'constant rate zone'.
3. **C-D :** represents the period after surface evaporation. The rate of evaporation drops off after the surface moisture has evaporated. This section is known as 'falling rate zone'.

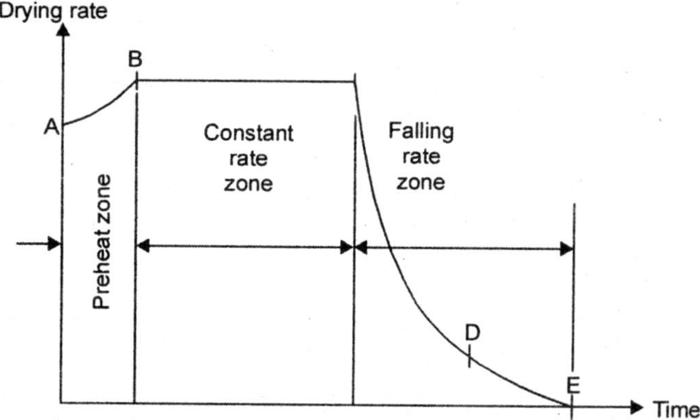

Fig. 5.38 Drying rate curve.

4. **D-E :** represents final phase of evaporation whose rate of evaporation is still slower. This zone forms a part of 'falling rate zone' or sometimes called as 'final zone'.

Drying rate curve shown in Fig. 5.38 represents the derivative of the drying curve shown in Fig. 5.37, that is the variation in rate of drying as a function of time.

5.5.1.2 The Temperature of Air and Solids

Batch dryer

Fig. 5.39 depicts the temperature levels of solid as well as hot air in a batch dryer. The amount of heat transferred from the air to the solid is increased if the temperature difference between them is increased. If the solids are at constant temperature (constant rate zone), the enthalpy of the air will also be constant and the air will only exchange its sensitive heat for the latent heat of the water which joins it. Therefore in the zone B-C, the wet-bulb temperature is also constant. The heating air enters a dryer at some high temperature (T_i) and leaves at an outlet temperature (T_o) which is low at the beginning of a batch drying cycle and rises as the solids heat up and dry. The dry bulb temperature of the outlet air must always be higher than its wet bulb temperature (T_{wb}), because at 100% relative humidity (T_{wb} = *dew* point temperature) the air would no longer be able to pick up moisture from the solids.

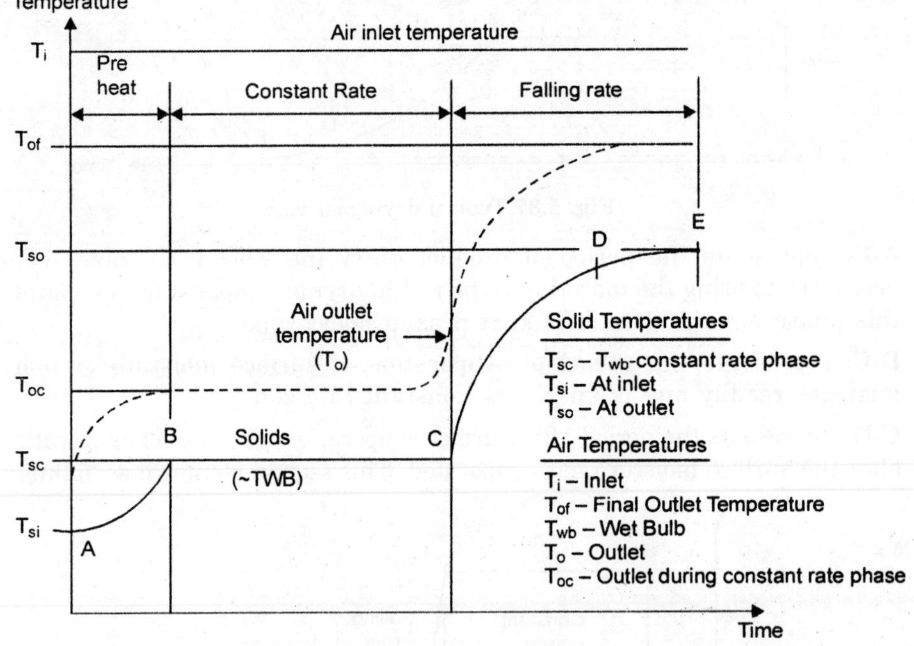

Fig. 5.39 Product temperature rises in a batch dryer from T_{si} to T_{so} as it is heated by air, which enters at T_i and leaves at T_{of}.

Continuous dryer : (Refer Fig.5.40)

The majority of the industrial continuous dryers are co-current longitudinal (adiabatic) units which can be used on thermally sensitive products. Air enters at a high temperature (T_i), which speeds up the drying process but is usually acceptable even with heat-sensitive products, because the solids temperature is limited to the wet-bulb temperature (T_{wb}) as long as the particle surface is wet.

Counter-current longitudinal dryers are used on thermally stable solids (cement, carbon black etc), which must be dried to very low moisture levels.

In continuous adiabatic fluidized bed dryers the average moisture contents of the solids in the bed is almost the same as the moisture content of the product. Most fluid bed dryers are generating products with less than the critical moisture content (W_c in Fig. 5.37) and are operated in the falling drying-rate zone. Under these conditions the air outlet temperature (T_o) determines product dryness, if feed flow and moisture content are constant.

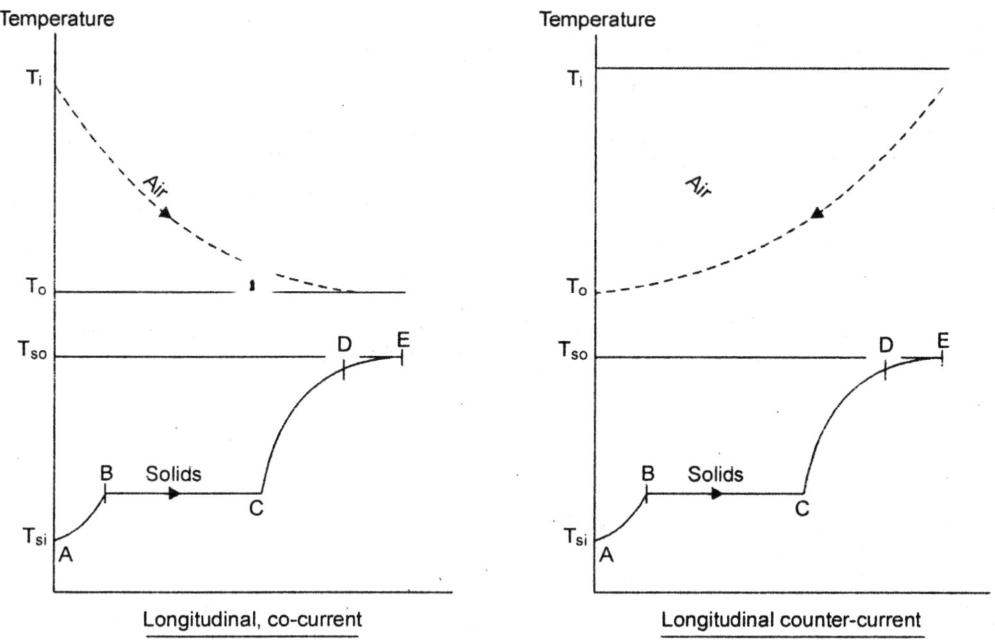

Fig. 5.40 Temperature profiles in longitudinal adiabatic dryers.

5.5.1.3 Dryer Characteristics

Each material to be dried will have a unique drying characteristic, depending upon its substance, the solvent, the affinity of one of these for the other, and the surroundings characteristic of the particular dryer. The theoretical drying rate depends upon feedrate and driving force. The latter is combination of temperature difference between the product and the moisture condition prevailing in the dryer atmosphere. Other factors of equal importance are the intimacy of material surface contact with the heating medium and the degree of agitation or surface renewal. The three essential operating factors necessary to any drying operation are a source of heat, a means of removing the solvent from the environment of the product, and a mechanism to provide agitation or surface renewable. Based on the above three factors, a few representative batch and continuous dryer types are listed in Table 5.1.

5.5.2 Dryers' Controls

The specific property of interest in dryer operation is moisture content of the final product which is unfortunately difficult to measure directly, particularly for continuous systems. In batch systems, one might take samples for moisture analysis at various periods and establish appropriate drying conditions. In case of continuous dryers, empirical runs are made to establish dryness of a product with given feed and dryer characteristics so that operating curves can be developed. The control scheme is designed to maintain these conditions with occasional feedback and with corrections from grab samples. Analysers are also available for moisture analysis that can be adapted for automatic, closed-loop dryer control. Even if a fully reliable solids moisture analyzer is available,some designers will prefer to control the dryer on the basis of a temperature based inferential system and use the analyzer only to update the inferential model. This is because most dryer processes are dominated by deadtime due to the slow transportation of solids, and information on load changes can be obtained sooner by controlling the dryer on the basis of the rapidly changing air temperatures. The more advanced dryer controls use the analysers for feedback trimming of the faster (and partially feed forward) inferential temperature models.

Table 5.1 Dryer Characteristics

Type	Examples	Heat Source	Method of Moisutre Removal	Method of Agitation	Typical Feed
		Batch Dryers			
Atmospheric	Tray	Hot air	Air flow	Manual	Granular or powder
Vacuum	Blender tumbler	Hot surface	Vacuum	Tumbling	Granular or powder
	Tray	Hot surface	Vacuum	None	Solid or liquid
	Rotating blade	Hot surface	Vacuum	Mixing blade	Solid or liquid
Specialty	Fluid bed	Hot air	Air flow	Air	Granular or powder
		Continuous Dryers			
Heated cylinder	Double-drum	Hot surface	Air flow	None	Liquid or slurry
Tumbling	Rotary	Hot air	Air flow	Tumbling	Granular
	Turbo	Hot air	Air flow	Wipe to successive shelves	Granular
Air stream	Spray	Hot air or combustion gas	Air flow	Not required	Liquid or sluurry
	Flash	Hot air	Air flow	Air	Granular or powder
	Fluid bed	Hot air	Air flow	Air	Granular

Dryer controllers usually include proportional and reset modes only, because dryer dynamics do not generally warrant the use of the rate response. Most dryer control systems are still relatively unsophisticated. Most moisture specifications are somewhat liberal in recognition of the difficulties present in moist and dry material handling. We will be discussing here the following sample control schemes both in batch dryers and in continuous dryers.

(A) Batch Dryers

1. Atmospheric dryer controls.
2. Vacuum dryer controls.
3. Fluid-bed dryer controls (Batch type).

(B) Continuous Dryers

1. Heated-cylinder type dryer controls (Double-drum dryer).
2. Rotary dryer controls.
3. Turbo dryer controls.
4. Spray dryer controls.
5. Fluid-bed dryer controls (continuous type).

5.5.3 Batch Dryer Controls

Batch dryers are particularly well adapted for drying relatively small quantities ; especially in cases in which batch identity might be of value.

5.5.3.1 Atmospheric Dryer Controls

Atmospheric dryer operates at close to atmospheric pressure and obtains the heat for drying from the hot air supply. The two most common forms are the tray dryers, in which trays covered with material to be dried are loaded into racks, and the truck dryers, which are similar except that the racks of trays are mounted on tracks.

The available control parameters are the air velocity and distribution, temperature, and humidity. The velocity and distribution of the air are usually set manually by dampers and are not changed. Humidity is also a manual setting by virtue of adjusting the damper in the recirculation line. If recirculation is not used and if product requirements warrant, dehumidification is used on the inlet air. Both of these provisions are shown in Fig. 5.41.

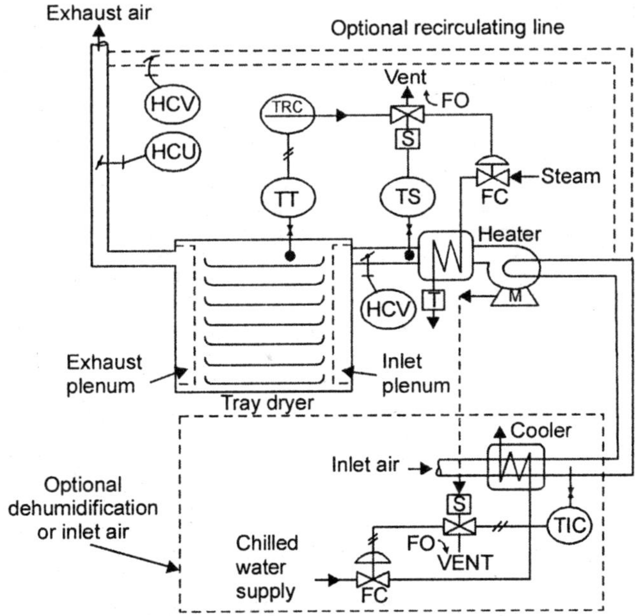

Fig. 5.41 Atmospheric tray dryer.

Control of the temperature within the dryer is accomplished by regulation of the steam to the heating coil. The temperature sensor can be placed in the dryer as shown or in the inlet air or in outlet air. The sizing of the steam valve is determined by the dryer airflow and temperature requirements, and by the pressure of steam used. It is necessary during the latter portion of the drying cycle to reduce steam flow to a level sufficient only to heat the air to the dryer temperature. A temperature switch (T_s) is included in the heater discharge to limit this maximum inlet temperature.

A more advanced batch dryer control scheme is one in which the dew point of the incoming air is controlled automatically and the termination of the drying is not based on time but rather on the measurement of outlet temperature.

5.5.3.2 Vacuum Dryer Controls

The operation of a vacuum dryer relies on the principle of heat transfer by conduction to the product while it is contained in a vessel under vacuum. As listed in Table 5.1, there are three types of vacuum dryers. The double-cone 'blender dryer' tumbles the product in a jacketed vessel. The tray dryer, in its simplest form, is a tray unit in a vacuum chamber with hollow shelves, through which a heat transfer medium is circulated. Such dryers usually start operation at a very low temperature and are also known as 'freeze dryers'. The shell of the 'rotary dryer' is fixed and jacketed with material agitation being supplied by a hollow mixing blade for the passage of heat-transfer fluid.

Parameters that can be controlled in batch vacuum dryers are absolute pressure (vacuum), rate of tumbling, or agitation and the temperature of the heat transfer medium. The general trend is to control vacuum and rate of tumbling manually and concentrate on the transfer fluid temperature for automatic control. A typical scheme for a blender-dryer is shown in Fig. 5.42.

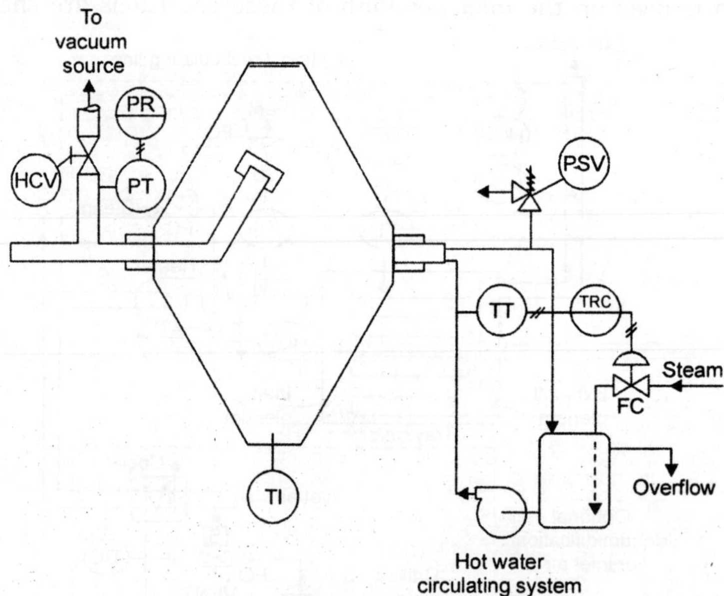

Fig. 5.42 Vacuum blender-dryer

5.5.3.3 Fluid-Bed Dryer Controls : (Batch type)

In this dryer, the product is held in a portable cart with a perforated bottom plate. Heated air, blown (pressure) or sucked (vacuum) through the plate, fluidizes the product to cause drying. A typical control scheme with cascade mode is illustrated in Fig. 5.43. Air flow is manually adjusted to get proper bed fluidization and the control is restricted to temperature control of the inlet air. The bed is sufficiently well mixed so that a high inlet temperature can be used at the beginning of the cycle without danger of damaging the product as long as the temperature is reduced during the latter stages of drying (during falling-rate phase). This is accomplished by the use of a controller on the outlet air (Master) that adjusts the set point of the inlet air controller (slave). As the bed temperature rises during drying, the set point of the inlet air temperature controller is reduced.

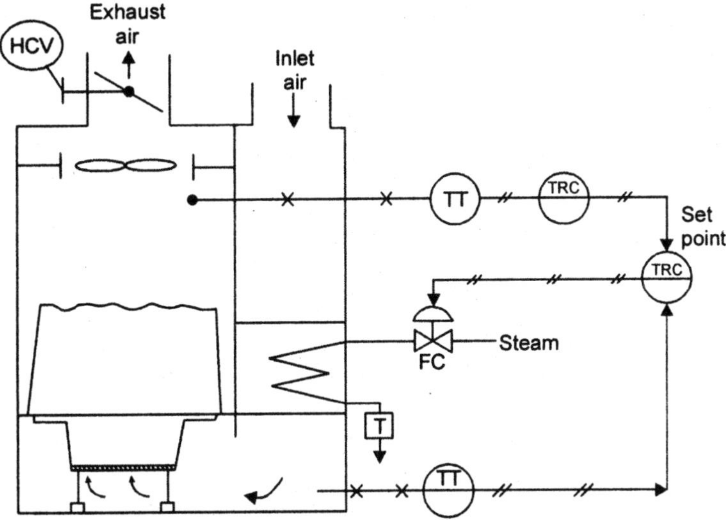

Fig. 5.43 Batch fluid-bed dryer with cascade control.

5.5.4 Continuous Dryer Controls

Continuous dryers are used in large-scale industrial processes producing dry solid materials. The control of a continuous dryer is similar to that of other process equipment with a few important exceptions given below.

1. As the measurement of actual dryness (moisture content) is difficult to measure, inferential model of secondary variable such as temperature is used for control. Analysers are used to trim or update the inferential model.
2. The majority of continuous dryers require control of the flow and temperature of an air system. Though temperature detection of air stream is inherently sluggish, it does not affect the control as majority of the dryers do not require fast response (except for flash and spray dryers).
3. Controllers for dryers generally use two-mode units with proportional and reset modes.

5.5.4.1 Heated-Cylinder Type Dryer Controls

A double-drum heated-cylinder type dryer with its common controls is depicted in Fig. 5.44. Feed liquid is fed into the 'valley' between the heated cylinders. The drums, rotating downward at the center, receive a coating of the liquid with the thickness depending upon the spacing between the rolls. The material must be dry by the time it rotates to the doctor knife, where it is cut off the roll.

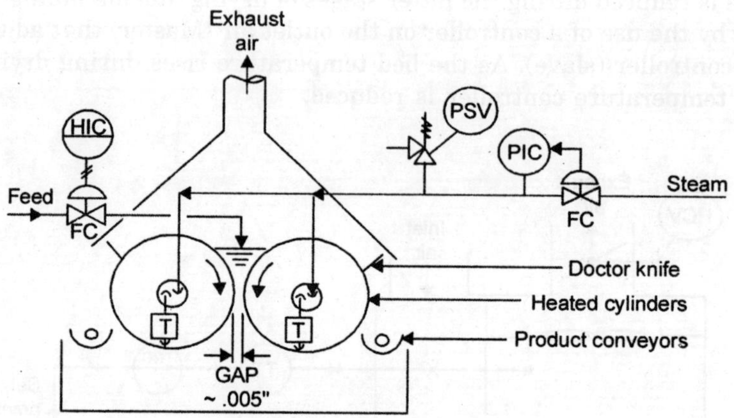

Fig. 5.44 Double-drum dryer

The variables available for control are the speed of the cylinders, the spacing between them, the liquid level in the valley, and the steam pressure in the cylinders. The speed and spacing are usually adjusted manually. The liquid level is maintained by throttling the feed stream by a manual loading station (HIC). Steam pressure is controlled automatically. The overall heat transfer is a function of the surface temperature (steam pressure) and of the moisture content (Thermal conductivity) of the process material. Because the heat transfer area is fixed, the best method to respond to load variations is to change the steam pressure, which inturn will change the dryer surface temperature. There can be a ratio control working with steam flow and pressure signals to take care of load variations.

5.5.4.2 Rotary Dryer Controls

Rotary dryers are 'tumbling dryers' in which material is tumbled, or mechanically turned over, during the drying process. Fig. 5.45 shows one form of rotary dryer. It consists of hollow cylindrical shell set with its axis at a slight angle to the horizontal so that material is consequently advanced through the dryer from one end to another end. As the shell is rotated, the material is lifted by the flights and then dropped through an air stream. The speed of rotation, the angle of elevation, and the air velocity determine the material hold up time. Variable – speed drives are sometimes incorporated to change the rotation rate, but they are usually manually controlled. A typical control scheme for a rotary, counter flow dryer is shown in Fig. 5.46. Primary controls maintain the airflow and inlet air temperature. Secondary considerations are pressure control on the outlet air to maintain pressure within the dryer and temperature and level alarms at the dry product outlet. A temperature switch is also provided at the air outlet to prevent dryer overheating when feed is stopped. Direct setting of the air inlet temperature is based on the supposition that the product temperature approaches it before the solids reach the discharge end. An improved control system can be obtained if some of the heat content of the exhaust air is reused and if the load variations are automatically controlled.

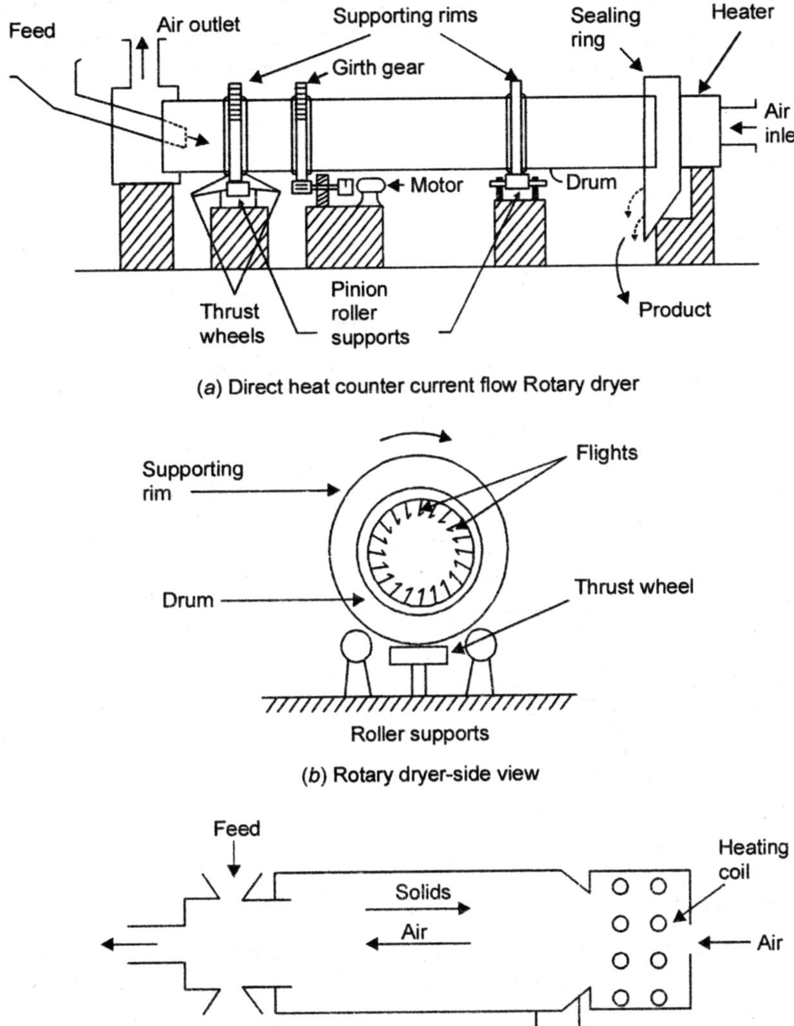

(a) Direct heat counter current flow Rotary dryer

(b) Rotary dryer-side view

(c) Schematic view of rotary dryer

Fig. 5.45 Rotary dryer.

5.5.4.3 Turbo Dryer Controls

In a turbo dryer the material is dried on rotating horizontal shelves (Refer Fig. 5.47). They are arranged in a vertical stack, and the product is wiped from each shelf through a slot after a little less than one revolution. A leveler bar then spreads it evenly on the shelf below. In addition to the general counter current of hot air, fan blades on a central shaft impart a horizontal velocity pattern. Since the internal fan provides consistent air circulation, it is feasible to control the air throughout. Motorized dampers are provided for the lower air input and for the combined flow to the upper sections. The division between the upper inlets is adjusted by manual dampers. The inlet air temperature is controlled by throttling of the steam valve, and the motorized dampers are adjusted by the corresponding temperatures. Circulation rates of the center fan and the shelves and the feed rate of material are manually controlled.

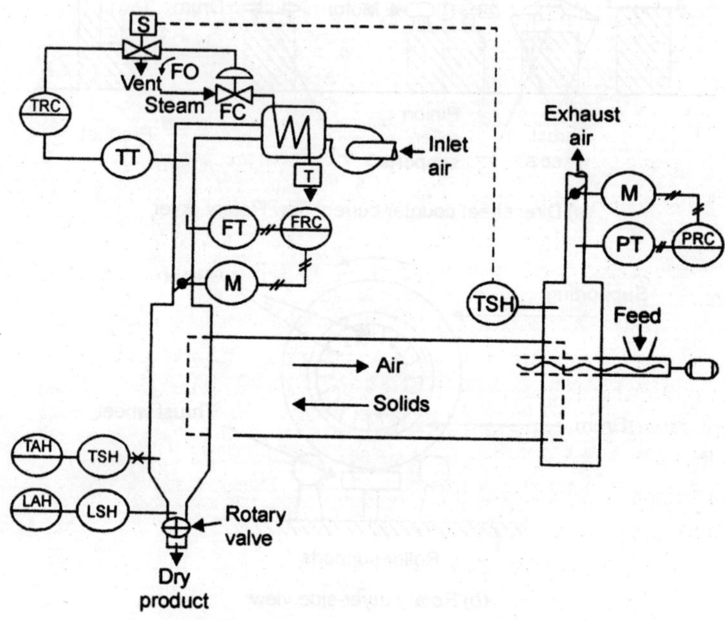

Fig. 5.46 Rotary dryer controls.

5.5.4.4 Spray Dryer Controls

The holdup time in spray dryer is in the order of tenths of a second. A liquid or thin slurry feed is atomized in a chamber in which there is a large flow of hot air (Refer Figure 5.48). The inlet air temperature must be quite high, and therefore direct -fired heaters are used whenever possible. The feed is introduced by a high pressure, manually controlled pump. Air flow is also manually regulated and balanced. Process conditions are maintained by temperature control near the outlet end. The temperature controller regulates the firing rate of the fuel-air mixture through a fuel control unit. A temperature switch is provided to shut off both the feed and the fuel in case of fire or other abnormally high temperature conditions.

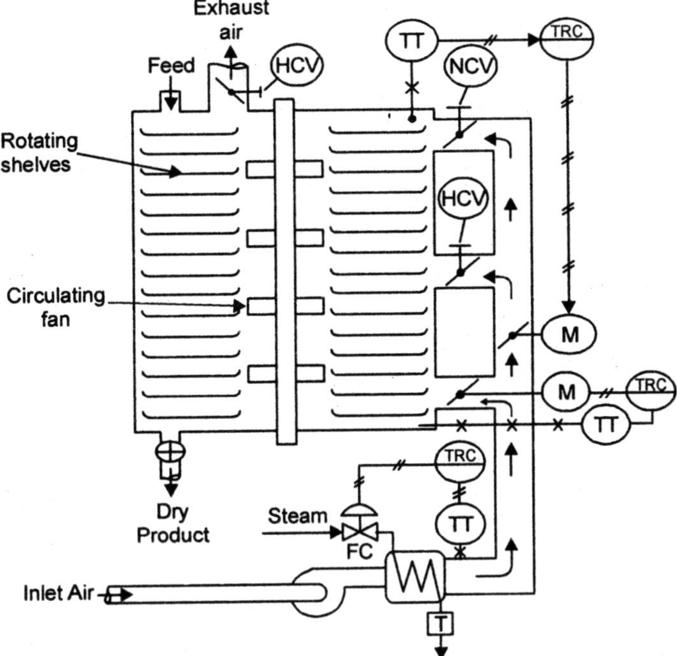

Fig. 5.47 Turbo dryer.

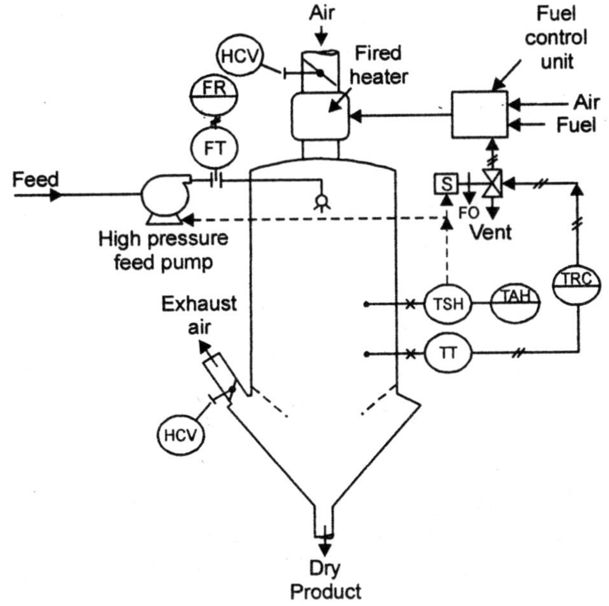

Fig. 5.48 Spray dryer.

5.5.4.5 Fluid-Bed Dryer Controls : (Continuous Type)

The fluid-bed dryer has a fluidized bed of material maintained by an air flow upward through a perforated plate. (Refer Fig. 5.49). Feed is controlled by a variable-speed screw, and discharge is by overflow of the bed through a side arm. The bed is maintained at the desired product moisture level by a temperature controller with its bulb within the fluidized product or in the air space above. Since the system is sensitive to outlet temperature but not to the absolute humidity of the air stream, an increase in humidity of the entering air can cause a reduction in inlet air temperature instead of increasing it. This is because the effect of increased humidity is to reduce the drying rate, which represents less heat loss from the air and therefore raises the outlet temperature. The controller TRC compensates for this rise in temperature by lowering the inlet temperature which further reduces the drying capacity of the air. This effect can be avoided by adapting the system in which air flow rate and inlet temperature are directly controlled (as in Fig. 5.46).

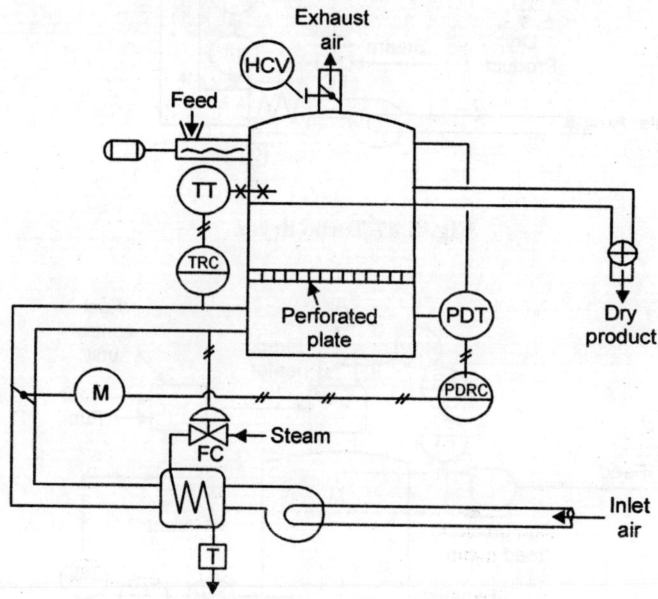

Fig. 5.49 Fluid-bed dryer

Humidity-Corrected Feedforward Control of Dluid Bed Dryer

Yet another way of avoiding humidity effect is by using feedforward control as shown in Fig. 5.50. In a fluidized -bed adiabatic dryer both bed temperature and moisture content can be considered uniform and close to the temperature and dryness of the product. Therefore, the solids temperature profile in a fluidized-bed dryer is not like the curves in Fig. 5.40; rather, it is more like a horizontal line. When the load increases, both inlet and outlet air temperatures (Ti and To) must rise. For a condition in which the product moisture content required is in the 'falling rate region', the product moisture can be expressed as

$$Xp = \text{Constant} \ln \left(\frac{Ti - Tw}{To - Tw}\right) \tag{5.3}$$

where X*p* is product wetness and T*i*, T*o*, T*w* respectively are air inlet, outlet, and wet-bulb temperatures. Note the absence of the variables : feedrate, air rate, feed moisture, air humidity. Thus, any one of these variables can be changed without affecting product dryness if the relationship of these three temperatures is maintained. For a given dryer and a given product moisture content,

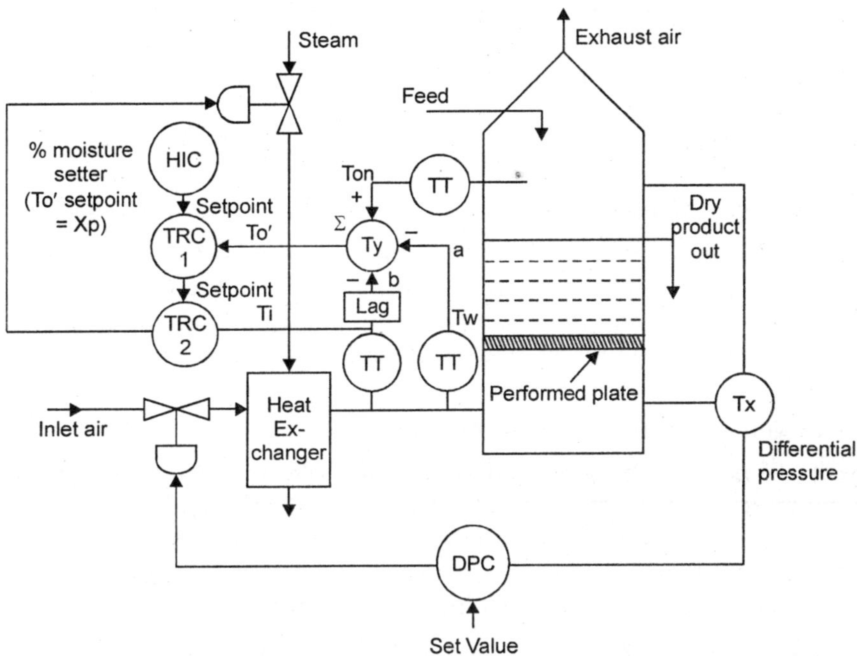

Fig. 5.50 Humidity-corrected Feedforward control of Fluid-bed dryer.

$$\left(\frac{Ti - Tw}{To - Tw}\right) = K \tag{5.4}$$

is constant. With this definition of *K*, the outlet air temperature required to compensate for load changes (T*o*′) is given by

$$\frac{(Ti - Tin) - (Tw - Twn)}{(To' - Ton) - (Tw - Twn)} = K \tag{5.5}$$

from which

$$To' = Ton - \overbrace{K_1(Ti - Tin)}^{b} - \overbrace{(K_2)(Tw - Twn)}^{a} \tag{5.6}$$

$$To' = Ton - b - a$$

In this equation the second subscript, *n*, denotes normal load conditions. As shown in Fig. 5.50, the correcting terms *a* and *b* in the equation (5.6) are subtracted from the detected value of T*o* in TY, and the resulting T*o*′ (corrected) signal becomes the measurement to TRC-1. The set point of TRC-1 is manually set via HIC to allow adjustment for new product dryness requirements, and therefore TRC-1 in effect is a moisture controller.

If feed rate or feed moisture content increases, this reduces T*o*, and therefore T*o*′ measurement drops. This causes an increase in TRC-1 output, which raises the set point of TRC-2. If air humidity is increased, this increases T*w*, which also reduces T*o*′ measurement and therefore raises the setpoint of TRC-2. To stabilize the system, the 'lag' is introduced.

To control moisture, temperatures at both ends of the dryer must change, and the difference between T*i* and T*o* is a measure of dryer load. Inlet air quantity is controlled with the help of differential pressure measured and the DPC controller as shown in Fig. 5.50.

5.6 HEAT EXCHANGER

5.6.1 Heat Exchanger Operations

The transfer of heat is one of the most basic unit operations of the processing industries. A heat exchanger exchanges heat between two streams, heating one and cooling the other. Heat can be transferred between the same phases (liquid to liquid, gas to gas etc) or phase change can occur on either the process side (condenser, evaporator, reboiler etc) or the utility side (steam heater) of the heat exchanger.

Heat exchangers are classified into three groups:

1. **Direct contact exchangers :** The hot and cold streams are brought into direct contact (mixed) and heat is transferred. These are particularly common when one heat is solid or entrained with a solid (air dryers etc) or for vapour-liquid systems where only the liquid product is of value (spray dryers, cooling towers etc). Use for liquid-liquid systems is limited to immiscible pairs.
2. **Regenerating exchanger :** A regenerating exchanger is one where the heat is transferred in steps: first from the hot phase to a storage medium, then from storage to the cold fluid. A sand tank or a rotary slab might be the storage phase.
3. **Recuperating exchanger :** Recuperators are most widely used heat exchangers in industries. In this arrangement, the hot and cold fluids are separated by a wall, and the heat transferred to and through the wall. This class includes double pipe (hair pin), shell and tube, and plate and frame heat exchangers.

5.6.2 Heat Exchanger Controls

The controls in heat exchanger will mostly involve measurement of temperature and control of flowrate. The controls for cooling purpose and heating purpose may differ little bit as far as the location of sensors and control valves are concerned. Three way valves of two different designs, for diverting and for mixing, are very often used. Liquid level controls (condensers) and pressure control loops are also used. Feedback, feedforward, cascade and adaptive control techniques are used depending on the process requirement. We will be discussing mainly the controls used in the following heat exchanger types.

1. Liquid-to-liquid heat exchangers
2. Steam heaters
3. Condensers
4. Reboilers
5. Multipurpose systems

5.6.3 Liquid-to-liquid heat exchanger controls

These exchangers are used for both cooling the hot process fluid using coolant and heating the cold process fluid using hot liquid. Fig. 5.51 shows simple feedback control systems used for the above purposes. The figure illustrates the cooler and heater installations with the control valve mounted on the exchanger inlet and outlet, respectively. From a control quality point of view, it makes little difference whether the control valve is upstream or down stream to the heater. The inlet side is usually preferred, because this allows the exchanger to operate at lower pressure than that of the return header.

Components Selection

Control valve : Whenever the process load drops, the same amount of valve adjustment by TIC will have more effect, because the process fluid spends more time in the exchanger. As load is reduced, the exchanger becomes relatively oversized and therefore faster and more effective. As a drop in load tends to increase the process gain-that is, tends to make the loop excessively sensitive and prone to cycling- an increase in load does the opposite. As load rises, the exchanger becomes less and less effective, more and more undersized; therefore, it takes more time for the TIC to make a correction. An increase in load thus will make the loop sluggish, as the residence and dead times are reduced and the process gain is lowered. The use of equal percentage valve is recommended to minimize the above effect and compensate for the variable process gain. The process gain drops with increased load whereas the equal percentage valves's gain rises with load (or as it opens more and more) and hence the total loop gain can be held relatively constant.

Controller : In majority of the installations, a three-mode controller would be used for heat exchanger service. The derivative (rate action) becomes essential in long time-lag systems or when sudden changes in heat exchanger throughput are expected. Proportional band setting must be wide to maintain stability (usually between 10% and 100%). The integral (reset) control mode is required to correct the temperature offsets caused by process load changes.

Thermal element : The selection and location of the thermal element (temperature sensor) is also important. This element must be placed in a representative location, without increasing measurement time lag. It must be located far enough from the heat exchanger for adequate mixing of the process fluid, but close enough so that the introduced delay will not be substantial.

Three-way valves (By pass control) : The limits within which process temperature can be controlled are a function of the nature of load changes expected and of the speed of response for the process. In many installations, the process timelag in the heat exchanger is too great to allow for effective control during load changes. In such cases it is possible to circumvent the dynamic characteristics of the exchanger by partially bypassing it and blending the warm process liquid with the cooled process fluid as shown in Fig. 5.52. The resulting increased system speed of response together with some cost savings are the main motivations for considering three-way valves in such services. The three-way valve used in Fig. 5.52 is called 'diverter valve'. A three-way valve can also be used as 'mixing valve' as shown in Fig. 5.53 to control the cooler.

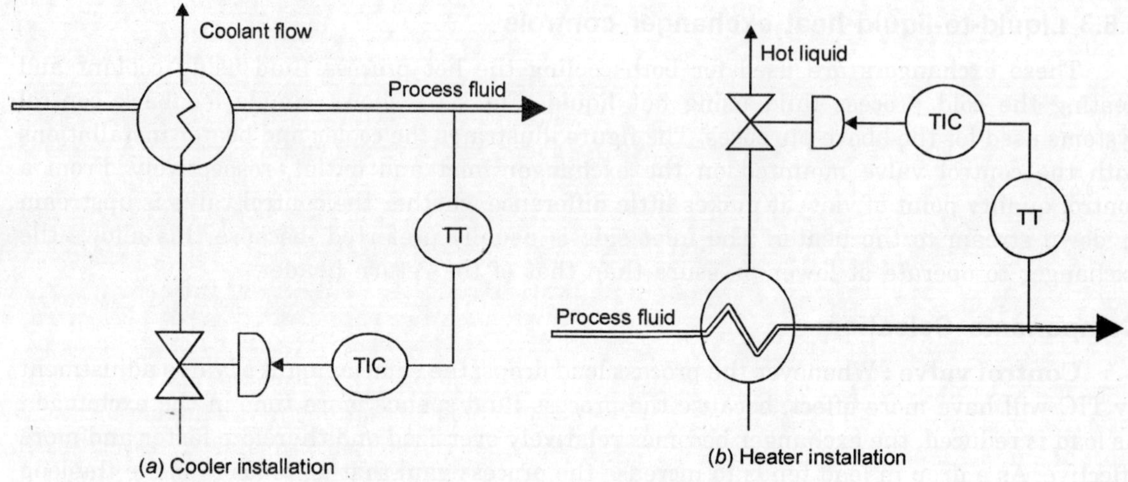

Fig. 5.51 Liquid-to-liquid heat exchanger

Stable operation of the diverter and mixing valves is achieved by flow tending to open the plugs in both the cases. Refer the expanded view of the valve plugs shown in respective Figs. 5.52 and 5.53. If a mixing valve is used for diverting service or if a diverting valve is used for mixing service, the operation becomes unstable because if flow direction is reversed, the fluid itself will try to push the valve plugs closed (bath tub effect).

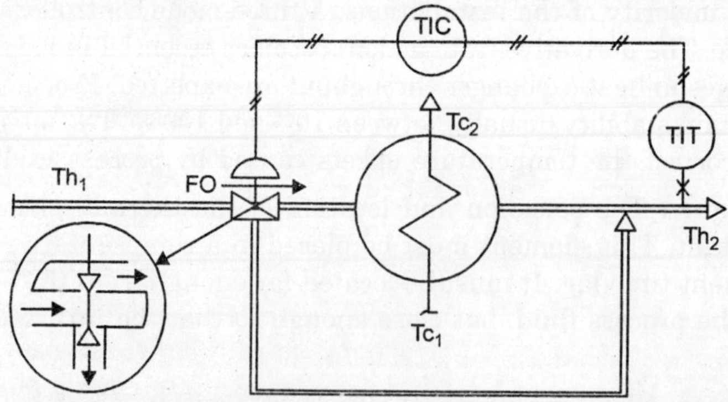

Fig. 5.52 When a diverter valve is used to control, it does not eliminate non linearity. But it does speed up the response and minimize fouling.

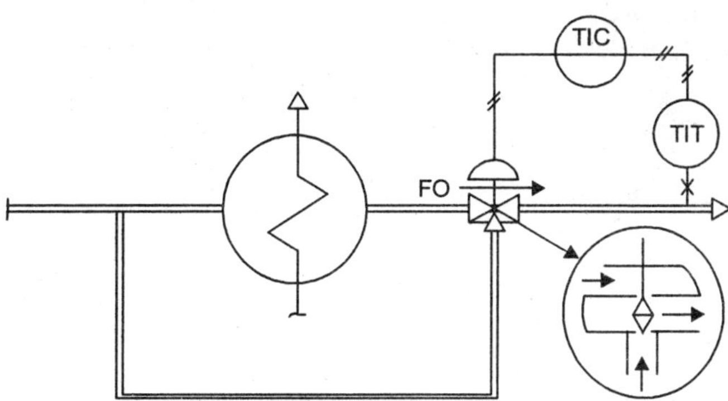

Fig. 5.53 A mixing valve can be used to control the cooler. It has a flow to open inner valve for stability.

The three-way valve does not change the non-linear nature of the heat exchanger (as equal percentage valve does), but the dynamic response is improved, because the by pass will shorten the time delay between a change in valve position and the response at the temperature sensor. Three way valves are unbalanced designs and are normally provided with linear ports. The unbalanced nature places a limitation on allowable shut off pressure difference across the valve, and the linear ports eliminate the potential of compensating for the variable gain of the process. For balancing the three way valve, it is recommended that a manual balancing valve be installed in the exchanger by-pass as shown in Fig. 5.54. This valve is so adjusted that its resistance to flow equals that of the exchanger. The resistance to flow in such installations will be maximum when one of the paths is closed and the other is fully open, whereas minimum resistance will be experienced when the flow is divided equally between the two paths.

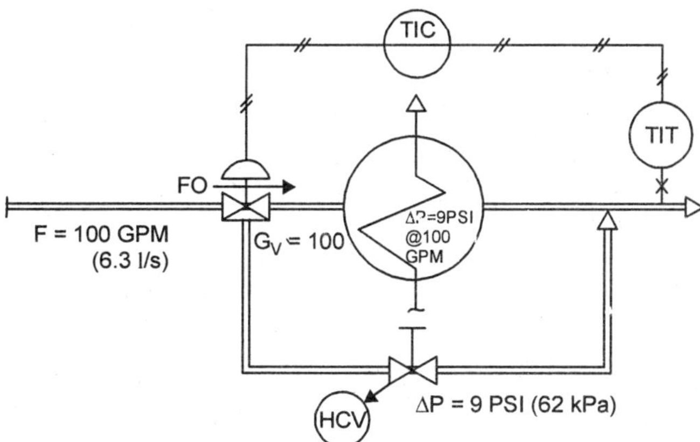

Fig. 5.54 It is desirable to install a balancing valve in the exchanger bypass.

Two Two-Way Valves

Sometimes , it is desirable to improve the system response speed by the use of exchanger by pass control in situations where, for reasons of temperature or other considerations, three way valves cannot be used. In such situations, the installation of two two-way valves is the logical solution. As illustrated in Fig. 5.55, the two valves should have opposite failure positions. Therefore when one is open the other is closed and at 50% of the controller output signal (9 psi or 12 mA) both are half way open. In order for these valves to give the same control as a three way valve would, it is necessary to provide them with linear plugs.

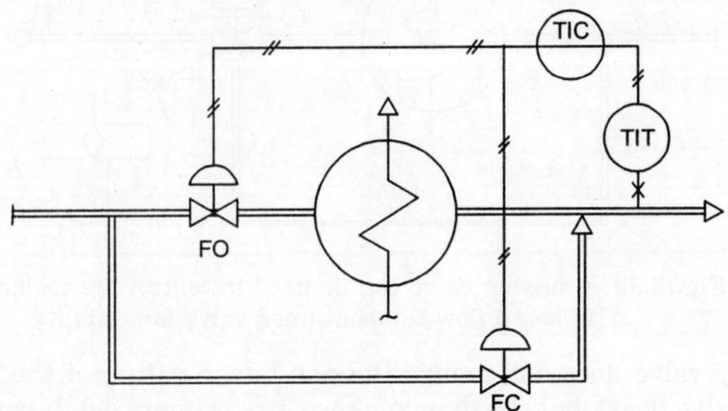

Fig. 5.55 Exchanger bypass control can be achieved using two two-way valves.

To summarise, by pass control is applied to circumvent the dynamic characteristics of heat exchangers, thus improving their controllability. By pass control can be achieved by the use of either one three-way valve or two two-way valves. Three-way valves are not normally recommended for high-temperature or high-pressure differential services.

5.6.4 Steam Heater Controls

A feedback control shown in Fig. 5.56 is frequently used on steam-heated exchangers. The steam-heated exchanger is non linear. The steady state gain is the derivative of outlet temperature (T_2) with respect to the steam flow (Fs).

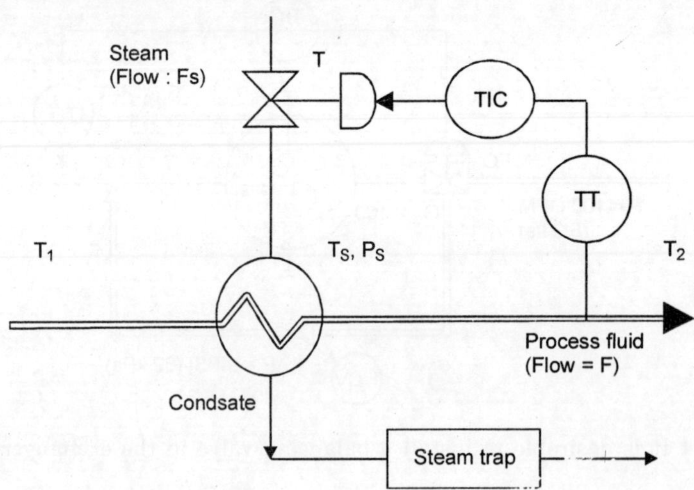

Fig. 5.56 Steam heater-feedback control

$$K_P = \frac{dT_2}{dFs} = \frac{\Delta Hs}{FC_P} \tag{5.7}$$

where T_2 = Outlet temperature of the process fluid

Fs = Steam flow

Hs = Latent heat of steam

ΔHs = Latent heat of vapourisation

F = Process fluid flow

C_P = Specific heat of process fluid.

Therefore the process gain varies inversely with flow. In a step response test, the outlet temperature (T_2) reaches 63.2 % of its final value after a time equivalent of its residence time, which is tube volume divided by flow. Therefore, the time constant, dead time, residence time, period of oscillation and process gain all vary with flow. As flow (load) drops to 50%, the process gain is doubled; therefore, the TIC loops have a tendency to become unstable at low loads and to become sluggish at high roads. In order to eliminate cycling, these loops are usually tuned at minimum load, resulting in sluggish response at higher loads. One way to compensate for the drop in process gain as the flow increases is to use an equal-percentage valve whose gain.increases with load. This is sufficient if ($T_2 - T_1$) is fairly constant. Because of the high rangeability requirement, the desirability of equal-percentage valve is more pronounced. Even the use of the above valve may not be sufficient sometimes. One solution is to use a large and a small valve in parallel.

Minimum condensing pressure : The condensing pressure is a function of load when the temperature is controlled by throttling the steam inlet as shown in Fig. 5.56. The minimum condensing pressure should always be higher than the total of trap back pressure and trap differential pressure. Otherwise, the condensate will start accumulating in the exchanger instead of getting discharged. This causes coverage of more and more of heat transfer area resulting in increase of condensing pressure. When this pressure rises sufficiently to discharge the trap, the condensate is suddenly blown out and the effective heat transfer area increases instantaneously. This can result in cycling as the exchanger surface is covered and uncovered. This may cause noise and hammering as the steam bubbles collapse on contact with the accumulated cooler condensate.

Condensate throttling : Mounting the control valve in the condensate line is sometimes proposed as a solution to minimum condensate pressure problem (Refer Fig. 5.57).

The throttling of the valve causes variations only in the condensate level inside the partially flooded heater and has no effect on the pressure which stays constant. Therefore there is no problem in condensate removal.

In this case, when the load is decreasing, the valve is likely to close completely before the condensate builds up to a higher enough level to match the new lower load with a reduced heat transfer area. In this direction, the process is slow, because steam has to condense before the level can be affected. When the load increases, the process is fast, because just a small change in control valve opening is sufficient to drain off enough condensate to expose an increased heat transfer surface. When these 'non symmetrical' dynamics are present, control is bound to be poor. If controller is tuned for increasing load direction, sluggish performance in decreasing load will occur, and if it is done the other way, cycling can occur when the load rises. Hence this arrangement is not a totally suitable solution.

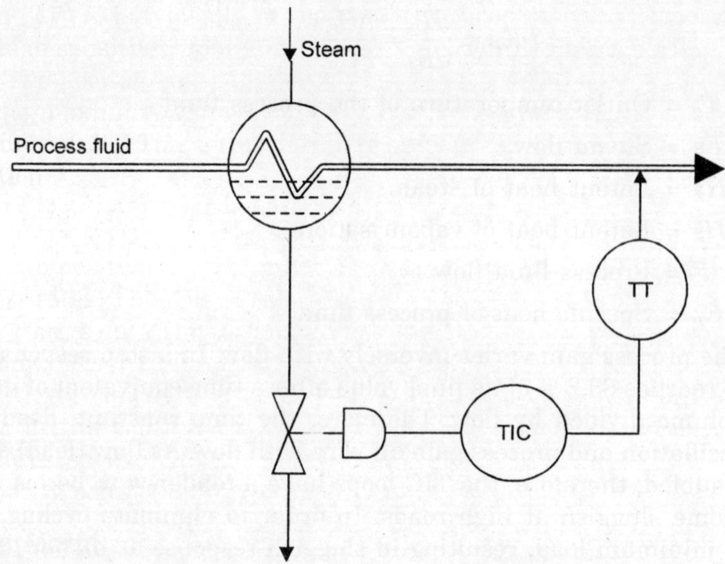

Fig. 5.57 Control valve in the condensate line.

Level controller : The control system shown in Fig. 5.56 provided quick response but was unable to discharge its condensate at low loads. The system shown in Fig. 5.57 eliminated the condensate problem, but at the price of worsening control response. Because the low condensing pressure situation is a result of the combination of low load and high heat transfer surface area, it is possible to prevent it from developing by reducing the heat transfer area. This can be done by replacing steam trap by a level control loop. As the required level is a function of the load, the set point to the level controller can be adjusted automatically. Such an arrangement is illustrated in Fig. 5.58.

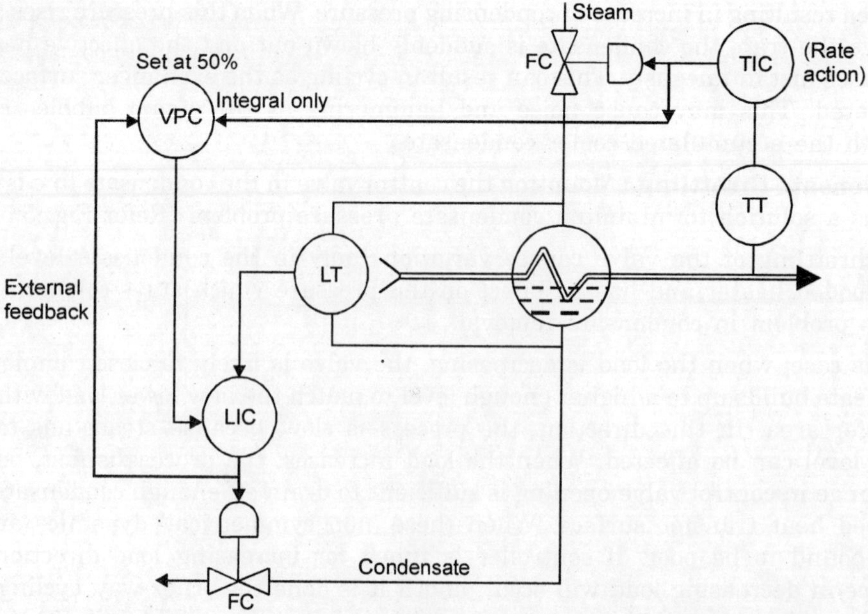

Fig. 5.58 Steam heater with condensate level controller

Load is detected by the valve position controller (VPC) having a setpoint of, say, 50%. When the load rises, the steam valve opens beyond 50% and the VPC increases the active heat transfer area in the heat exchanger by lowering the setpoint of the condensate level controller (LIC). The VPC is an integral only controller and therefore will not respond to measurement noise or valve cycling. The integral time is set to give slow floating action that is fast enough to respond anticipated load changes. The external feedback protects from reset windup in the VPC when LIC is switched to local setpoint.

Bypass control : The concept used in liquid to liquid cooler with the help of three-way valves to circumvent the transient characteristics of coolers can be applied to steam heater also. One additional advantage here is that the bypass created an additional degree of freedom; therefore, steam can now be throttled as a function of some other property. The logical decision is to adjust the steam feed so that it maintains the condensing pressure constant. This then eliminates problems associated with condensate removal. Fig. 5.59 illustrates a bypass control discussed above.

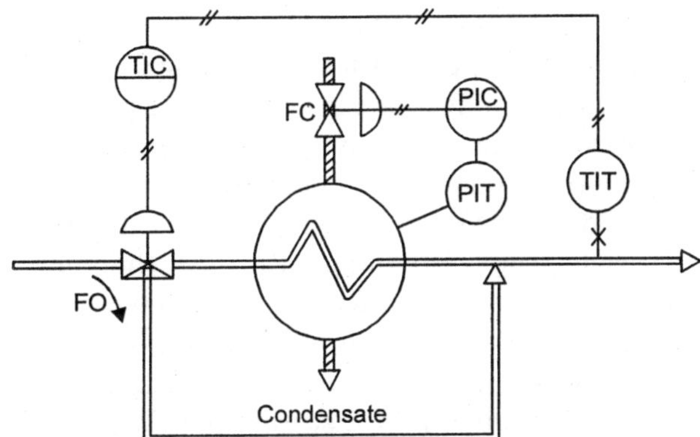

Fig. 5.59 Bypass control of steam heater provides the added degree of freedom needed for independent control of condensing pressure

Interaction between Parallel Steam Heaters : (Refer Fig. 5.60)

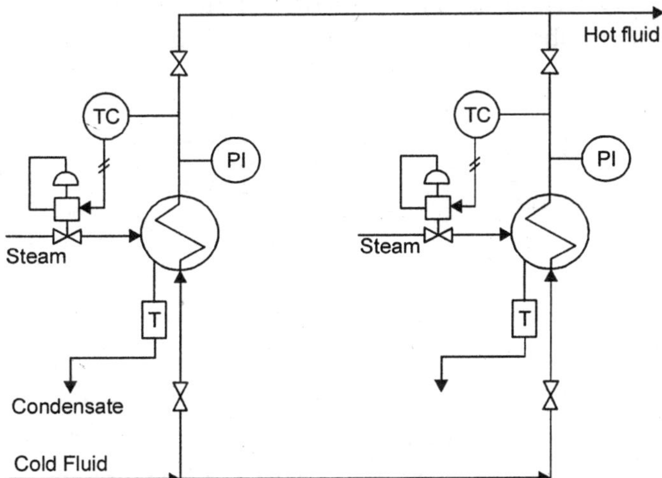

Fig. 5.60 Parallel heaters operating close to the boiling point of the process fluid can experience serious interaction problems

If the process fluid is heated to a temperature approaching its boiling point, serious interaction can occur between parallel heaters (Refer Fig. 5.60). The mechanism of developing this oscillation is as follows : a sudden drop in flow causes overheating and vapourisation in one of the heaters. The vapour formation increases the back pressure and further reduces the flow, eventually forcing all flow through the other cold exchanger, while the hot exchanger discharges slugs of liquid and vapour. After a period of noise and vibration, when the hot exchanger has discharged all of its liquid, the back pressure drops and flow is resumed, drawing feed from the cold one. This causes the cold exchanger to overheat, and the cycle is repeated with the roles of the exchangers reversed. This type of interaction can be eliminated by providing distribution controls. The control system used will make sure that the load is equally distributed between exchangers.

5.6.4.1 Cascade control of a Steam Heater

Cascade loops are invariably installed to prevent outside disturbances from entering the process. In a steam heater installation, the steam header pressure variations can be a disturbance. The conventional feedback control system shown in Fig. 5.56 cannot respond to a change in steam pressure until its effect is felt by the process temperature sensor. In other words, an error in the detected temperature has to develop before corrective action can be taken. The cascade loop, in contrast, responds immediately, correcting the process temperature. Instead of pressure, steam flow can also be considered as a disturbance. Fig. 5.61 shows a temperature -pressure cascade loop with temperature as the controlled variable (Master control) and the pressure as the manipulated variable (slave control). Fig. 5.62 shows a temperature-flow cascade loop with flow as the manipulated variable. In both the cases, the primary variable (temperature) is slow, and the secondary or manipulated variable (pressure or flow) is capable of responding quickly to disturbances. Therefore, if disturbances occur upsetting the manipulated variable (steam pressure or steam flow), these disturbances will be sensed immediately and corrective action will be taken by the secondary controller so that the primary variable (process temperature) will not be affected. In order for a cascade loop to be successful, the slave must be faster than the master. A rule of thumb is that the time constant of the primary should be ten times that of the secondary, or that the period of oscillation of the primary should be three times that of the secondary.

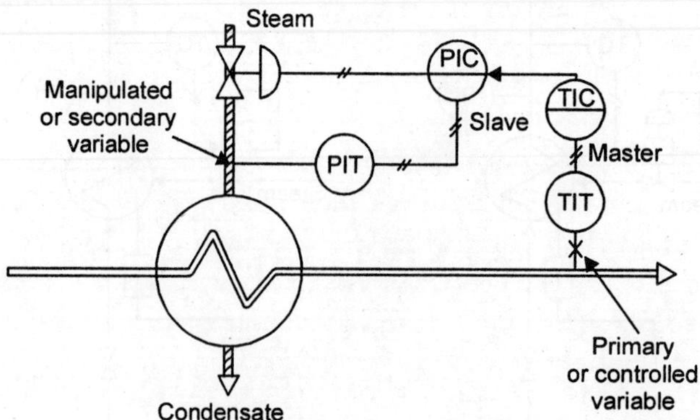

Fig. 5.61 A temperature pressure cascade loop on a steam heater increases the speed of response.

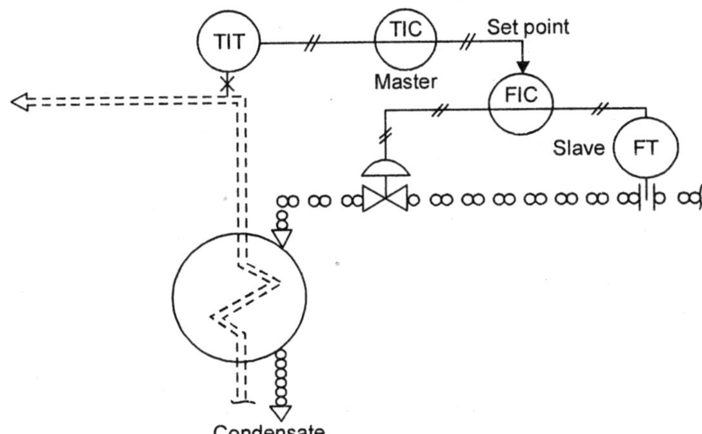

Fig. 5.62 Temperature-flow cascade loops are commonly used on steam reboilers.

5.6.4.2 Feedforward Optimization of a Steam Heater

Recollecting our earlier discussions on feedback and feedforward control systems, we may make the following statements. Feedback mode of control necessitates that the disturbance variable affect the controlled variable itself before correction can take place. Feedforward mode of control responds to a disturbance such that it instantaneously compensates for an error that the disturbance would otherwise have caused in the controlled variable later. Fig. 5.63 illustrates a steam heater under feedforward control.

This control system consists of two main segments. The feedforward portion of the loop detects the major load variables (the flow F and temperature T_1 of the entering process fluid) and calculates the required steam flow (F_{SP}) as a function of these variables. When the process flow increases, it should be matched with an equal increase in steam flow instantaneously. Because instantaneous response is not possible, the next best thing is to add more steam than needed as soon as possible. This is the dynamic correction function served by the lead-lag element in the feedforward loop.

The feedback portion of the loop (TIC) has to do much less work in this configuration, as it only has to correct for minor load variables, such as heat losses to the atmosphere, steam enthalpy variations, and sensor errors. The feedback and feedforward portions of the loop complement each other perfectly. Feedforward is responsive, fast, and sophisticated, but inaccurate; feedback is capable of regulating in response to unknown or poorly understood load variations, and although it is slow, it is accurate.

Adaptive gain : To have stable loop behaviour in a process where its gain is inversely varying with manipulated variable, things like equal percentage valve or multiplier in a circuit to compensate the inversion action is required. The gain achieved thus is called 'adaptive gain'.

The heat transfer process in a steam heater is a variable gain process. As we have already seen, the steady state gain of a steam heater process is as follows :

$$\frac{dT_2}{dFs} = \frac{\Delta Hs}{FC_P} \tag{5.7}$$

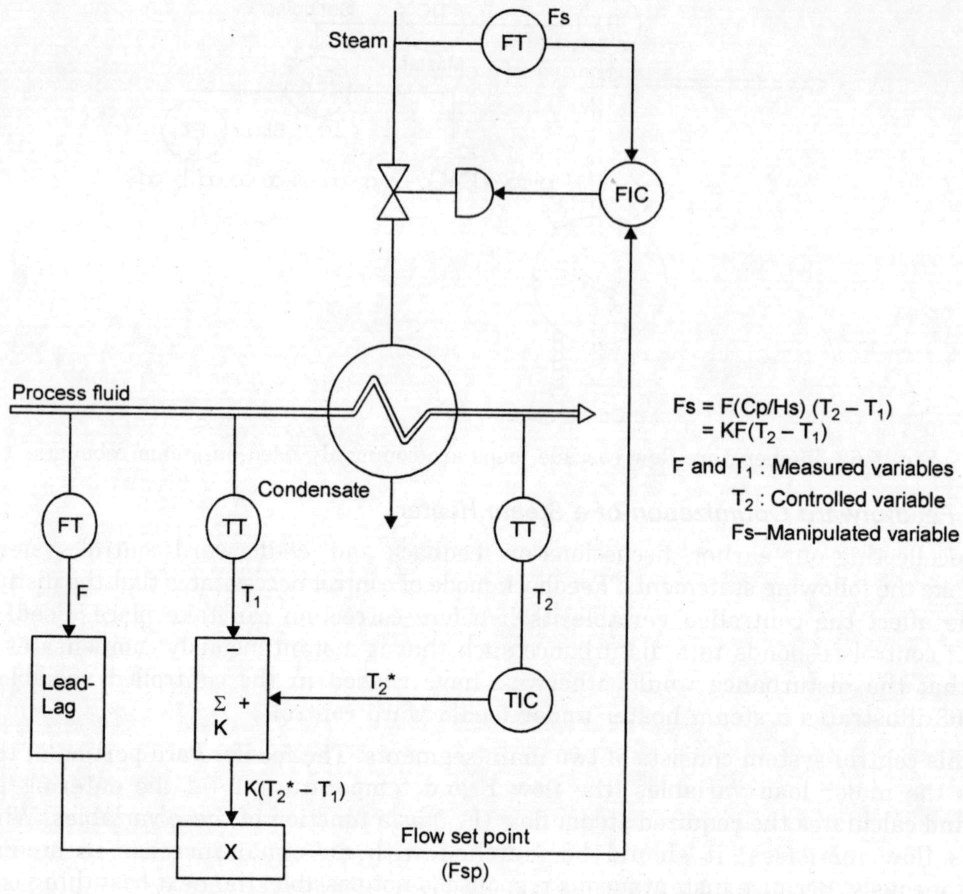

Fig. 5.63 Feedforward optimization of steam heater

Therefore, the process gain varies inversely with flow (F). If the temperature rise ($T_2 - T_1$) is also a variable, even the use of an equal-percentage valve cannot correct for this non linearity in the process. In that case, the only way to keep the process gain constant is to use the feedforward system shown in Fig. 5.63. There, a reduction in process flow causes a reduction in the gain of the multiplier which exactly cancels the increase in process gain. Thus, the feedforward loop provides 'gain adaptation' as a side benefit, because as the process gain varies inversely with flow, it causes the controller gain to vary directly with flow. The result is constant total loop gain and therefore stable loop behaviour.

The feedforward as well as cascade control concepts described are not limited to steam heaters but can be used on all types of heat exchangers. The only needed modifications are the ones that are required to correctly reflect the heat balance equation of the particular heat transfer unit.

5.6.5 Condenser Controls

Depending on whether the control of condensate temperature or of condensing pressure is of interest, the systems shown in Fig. 5.64 and Fig. 5.65 can be considered. Both of these throttle the cooling water flow through the condenser (manipulated variable is cooling water flow). When it is not desirable to throttle the cooling water, the system illustrated in Fig. 5.66 can be considered. Here the exposed condenser surface is varied to control the rate of condensation. Where non condensables are present, a constant purge may serve to remove the inerts. The non symmetrical response of such controls have already been described in connection with Fig. 5.57.

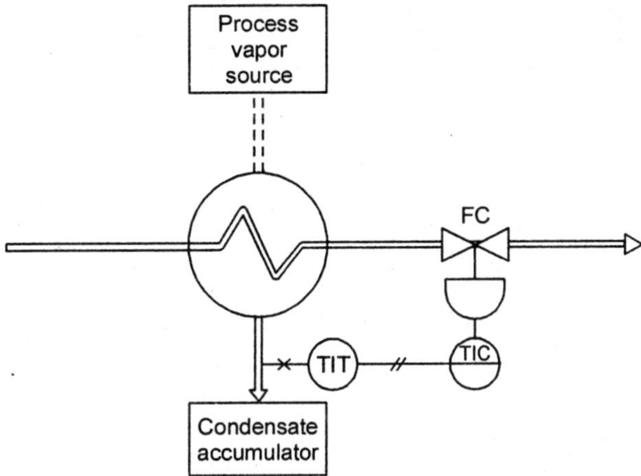

Fig. 5.64 Condensate temperature control throttles the cooling water flow through the condenser.

An important point to understand is that heat transfer efficiency is the highest when both the coolant flow and the heat transfer area are at their maximums. Therefore, the goal of optimization in Fig. 5.65 is to open the coolant valve fully; in Fig. 5.66, the goal of optimization is to eliminate flooding. These goals are achieved by slowly lowering the PIC setpoint until the cooling capacity is fully utilized and none of it is wasted through throttling. The preferred method of condenser control is to vary the heat transfer area through partial flooding. The controller on a partially flooded condenser is usually tuned as a level controller with derivative action added.

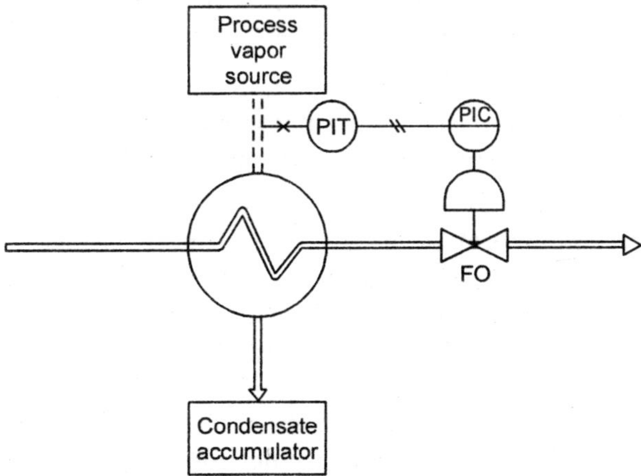

Fig. 5.65 The cooling water flow through the condenser can be throttled to control pressure.

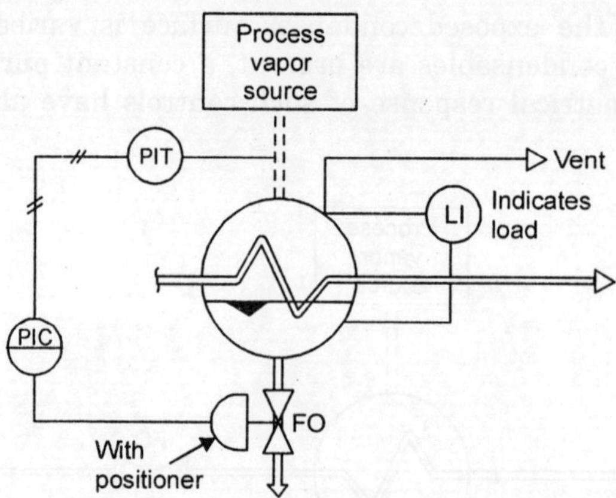

Fig. 5.66 Condenser control can be achieved by changing the wetted surface area.

Distillation Condensers

A variety of condenser designs is available for use in connection with distillation towers. The air cooled designs are sensitive to ambient variations. These units are throttled by manipulating their inlet louvers or by adjusting the blade pitch of the fans. If multispeed or variable speed fans are available, the condenser rangeability can be increased while energy is saved.

When the condenser is water-cooled and its inlet pressure is controlled by partial flooding as discussed earlier (Fig. 5.66), the main problem is response speed. A change in valve position does not immediately affect the heat transfer area. The response speed is slower when the level needs to be increased and faster when it is to be decreased. This 'nonsymmetricity' has already been discussed in connection with Fig. 5.57. In controlling distillation towers, the elimination of this non symmetry is one of the important goals of control. The simplest method of handling this problem is to bleed off a fixed amount of gases and vapours to a lower-pressure unit, such as to an absorption tower, if such is present in the system. Otherwise it is possible to install a vent condenser to recover the condensate vapours from this purge stream.

Lowering the condenser below the accumulator not only reduces installation cost and makes maintenance more convenient but also eliminates the non-symmetricity from the process.

When the condensing temperature of the process fluid is low, water is no longer an acceptable cooling medium. One standard technique of controlling a refrigerated condenser is illustrated in Fig. 5.67. Here the heat transfer area is set by the level control loop and the operating temperature is maintained by the pressure controller. When process load changes, it affects the rate of refrigerant vapourisation, which is compensated for by level-controlled make up. Usually the pressure and level settings are made manually, but can be automatically adjusted as a function of load as explained in connection with Fig. 5.58.

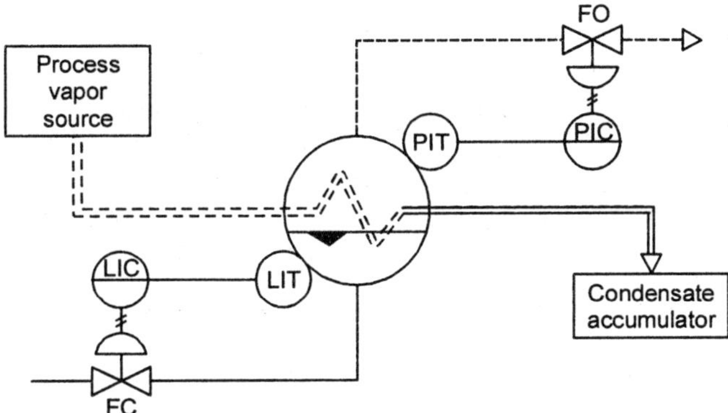

Fig. 5.67 Condensers can be controlled using refrigerant coolants

For some liquid mixture, the temperature required to vapourise the feed would need to be so high that decomposition would result. To avoid this, it is necessary to operate the column at pressures below atmospheric pressure (vacuum).

Though we have discussed briefly about condensers and their controls used in distillation column here, we will be dealing elaborately about the distillation condensers while studying distillation column controls.

5.6.6 Reboiler Controls

When a steam heated reboiler is used, only one degree of freedom is available; therefore, only one controller can be installed without overdefining the system. This one controller is usually applied to adjust the rate of steam addition. With regard to minimum condensing pressure considerations, the same thing applies as has been discussed earlier in connection with steam heaters. Fig. 5.68 and Fig. 5.69 show the two basic alternatives for controlling the reboiler : either to generate vapours at a controlled superheat temperature or to generate saturated vapours at a constant rate set by the rate of heat input. Naturally, there are other, more sophisticated reboiler control strategies, using compostion, temperature difference, or derived variables as the means of control, but in their effects they are all similar to the systems in Figs. 5.68 and 5.69.

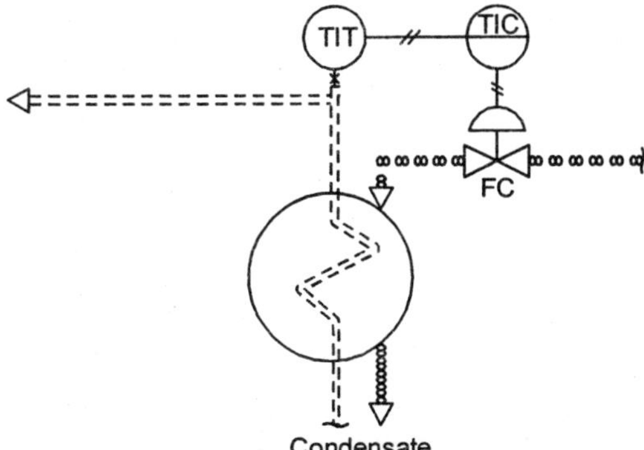

Fig. 5.68 Reboilers can be controlled by generating vapours at controlled temperatures.

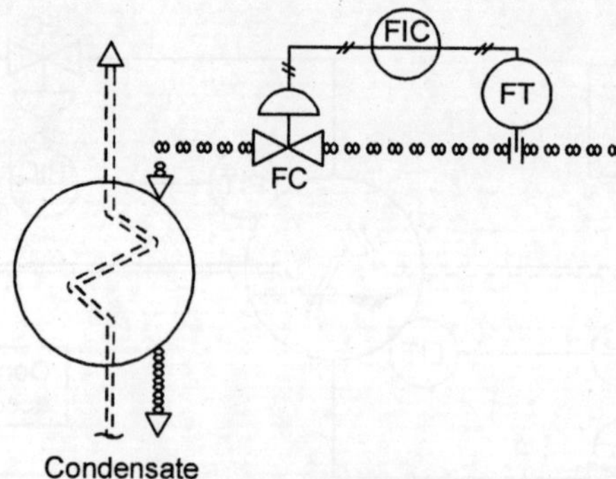

Fig. 5.69 Heat input control sets the steam flow to reboilers.

Fired Reboilers

When heat duties are great and distillation tower bottom temperatures are high, fired reboilers are used. The fired reboiler heats and vapourises the distillation tower bottoms as this liquid circulates by natural convection through the heater tubes. A common control scheme is shown in Fig. 5.70. The reboiler return temperature, an important variable, is controlled by TRC-1 by throttling the fuel gas control valve TV-1. The high temperature alarm TAH-1 is provided to warn the operator that the process fluid has suddenly reached an excessive temperature, indicating that manual adjustments are needed to cut back on the firing. LSL-1, the low level switch, and PSL-1, the low fuel pressure switch are hooked up with solenoid 'S' to avoid overheating of tubes and dangerous air-fuel mixture respectively. Whenever the solenoid operates, the diaphragm of the control valve(TV-1) is vented to close the fuel valve. All others are indicating instruments.

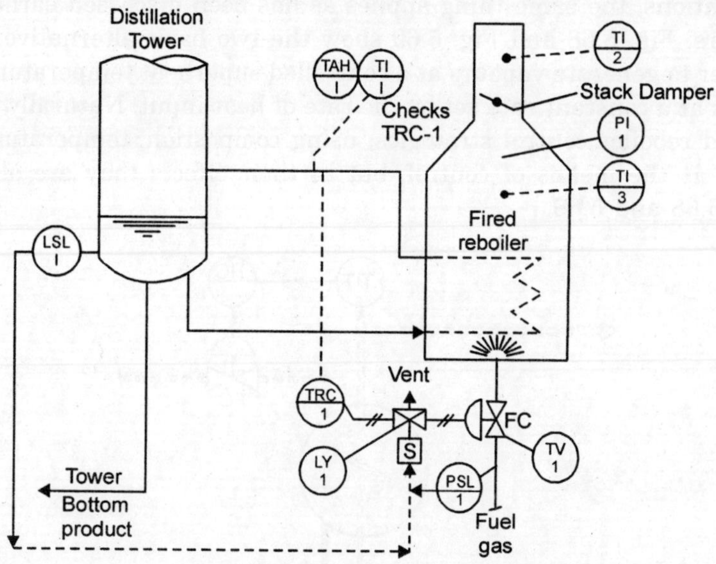

Fig. 5.70 Fired reboiler controls

Fired Heaters and Vaporizers

The unit feed heater of a crude oil refinery is representative of the class of furnaces that includes fired heaters and vaporizers. Crude oil, prior to distillation in the 'crude tower' into the various petroleum fractions must be heated and partially vapourised. The heating and vapourisation is done in the crude heater furnace, which consists of a firebox with preheating coils and vapourising coils. The prime variables in this process are listed below :

1. Flow control of feed to the unit
2. Proper splitting of flow in the parallel paths through the furnace
3. Correct amount of heat supplied to the crude tower.

Fig. 5.71 shows the typical process controls for this type of furnace. The crude feed rate to the unit is set by the flow controller (FRC-2). The flow is split through the parallel paths

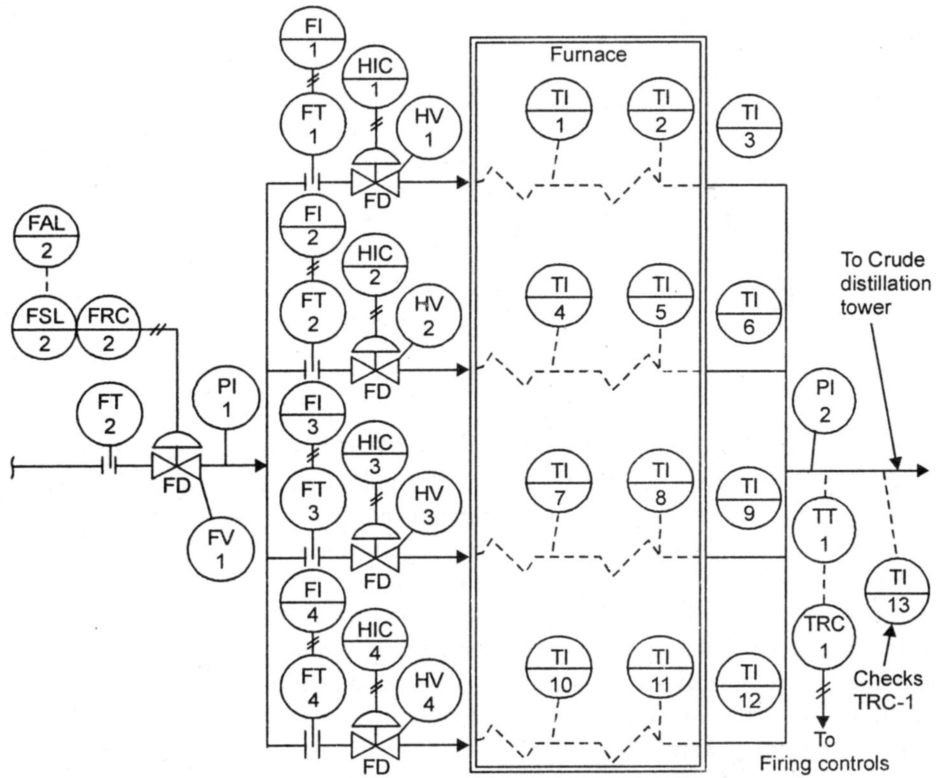

Fig. 5.71 Crude heater-vapourizer process controls.

of the furnace by remote manual adjustment of the control valves HV-1 through HV-4 via the manual stations HIC-1 through HIC-4. FI-1 through FI-4 are flow indications and TI-1 through TI-12 are temperature indications for observation. The heat supplied to the tower furnace is controlled via TRC-1, whose setpoint is decided by the unit operator. The operator depends on experience and on the results in the fractionator to determine the proper temperature setting. The firing controls are standard ones which are not shown in Fig. 5.71.

5.6.7 Multipurpose Heat Transfer System

So far we have discussed the control of isolated heat transfer units. In the majority of critical installations, the purpose of such systems is not limited to the addition or removal of heat, but involves making use of both heating and cooling in order to maintain the process temperature constant. Such a task necessitates the application of multipurpose systems, incorporating many of the features that have been discussed individually earlier.

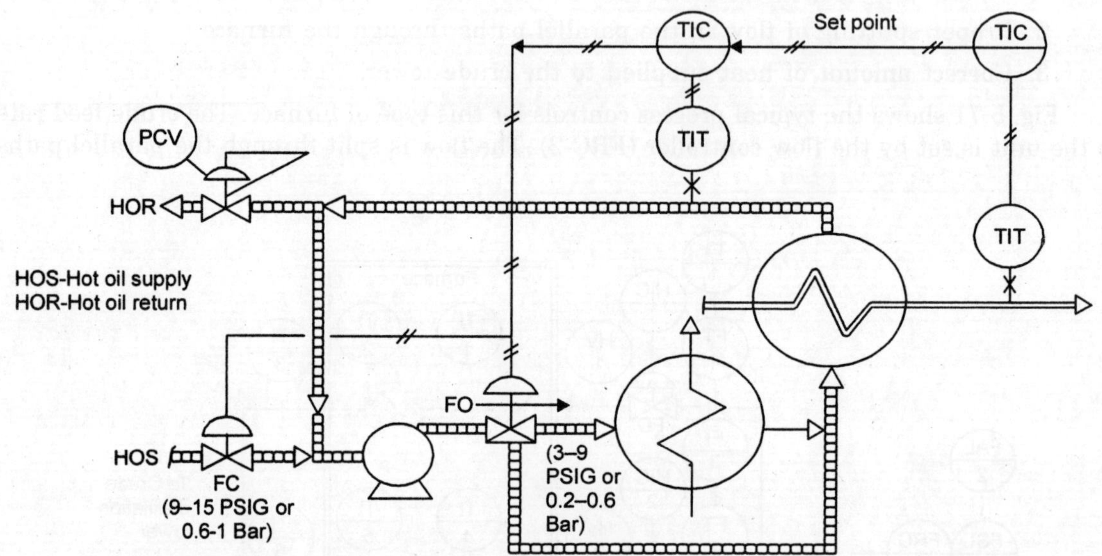

Fig. 5.72 This recirculating multipurpose heat transfer system uses hot oil as its heat source and water as its coolant.

Fig. 5.72 depicts a design that uses hot oil as its heat source and water as the means of cooling, arranged in a recirculating system. The points made earlier in connection with three-way valves, cascade systems and split-range control apply here. When the process temperature is above the desired set point, the output signal to the valves will be reduced. When the value of the signal is between 9 and 15 psi (0.6 and 1 kg/cm^2), the three-way valve is fully open to exchange bypass and the two-way valve is partially open to supply hot oil. If this reduction is not sufficient to bring the temperature down to set point, the signal will further decrease, thereby fully closing the two-way valve at 9 psi (0.6 kg/cm^2) and beginning to open the flow path through the cooler from the three-way valve. At a 3 psi (0.2 kg/cm^2) signal, the total cooling capacity of the system is applied to the recirculating oil stream, which in that case flows through the cooler without bypass. The most important feature of this design is that it operates on a 'split-range signal'. The cascade loop is used to overcome the non symmetrical nature of the process dynamics (lags and responses are different for the cooling and heating phases.)

Fig. 5.73 illustrates a design in which low cost and rapid response to load changes are main considerations. These characteristics are provided by using the minimum hardware and by circumventing the transient characteristics of the exchangers.

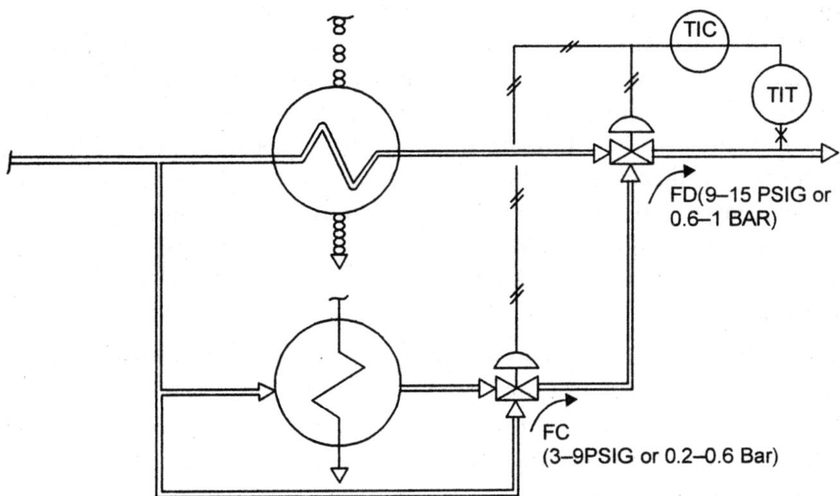

Fig. 5.73 This multipurpose temperature control system blends process streams at different temperatures

5.7 DISTILLATION PROCESS (COLUMN)

5.7.1 Operation

Distillation is defined as a process in which a liquid or vapour mixture of two or more substances is separated into its component fractions of desired purity, by the application and removal of heat. Distillation is the most common class of separation processes and probably one of the best-understood unit operations. Distillation is based on the fact that the vapour of a boiling mixture will be richer in the components that have lower boiling points. Therefore when this vapour is cooled and condensed, the condensate will contain more volatile components. At the same time, the original mixture after vapourisation will contain more of the less volatile material. Distillation columns are designed to achieve this separation efficiently. It consumes enormous amount of energy, both in terms of cooling and heating requirements. It can contribute to more than 50% of plant operating costs. The best way to reduce operating costs of distillation units is to improve their operation efficiency via process optimization and control.

Distillation equipment : There are some basic variations to the distillation process. One way of classifying distillation column types is to look at how they are operated. Thus we have batch and continuous columns. In batch operation, the feed to the column is introduced batch-wise. That is, the column is charged with a 'batch' and then the distillation process is carried out. When the desired task is achieved , a next batch of feed is introduced. In batch distillation the feed concentration is rich in light components at the beginning and lean in light components at the end.

Continuous columns process a continuous feed stream. They are capable of handling high throughputs and are the most common of the two types. Here the feed concentration is relatively constant. In this section, the emphasis is on the continuous processes. Although we often speak of controlling a distillation tower, many more measurements and controls are actually associated with equipment other than the tower. Hence it might be useful to review the equipment used in distillation.

The column : The primary piece of distillation equipment is the main tower. It is also called as 'column' or 'fractionator'. All the three terms are used interchangeably. The tower, column or fractionator has two purposes: first, it separates a feed into a vapour portion that ascends the column and a liquid portion that descends; second, it achieves intimate mixing between the two counter-current flowing phases. The purpose of mixing to get an effective transfer of the more volatile components in the ascending vapour and a corresponding transfer of the less volatile components into the descending liquid. In continuous distillation, the feed is introduced continuously into the side of the distillation column. If the feed is all liquid, the temperature at which it first starts to boil is called the 'bubble point'. If the feed is all vapour, the temperature at which it first starts to condense is called the 'dew point'.

Continuous columns can be further classified according to the nature of the feed that they are processing. In a 'Binary Column' the feed contains only two components. In a 'multi-components column' the feed contains more than two components. In a multi product column, the column has more than two product streams.

The separation of phases is accomplished by differences in vapour pressure, with the lighter vapour rising to the top of the column and the heavier liquid flowing to the bottom. The position of the column above the feed is called the 'rectifying section' and the portion below the feed is called the 'stripping section'. The intimate mixing is obtained by one or more of several methods. A simple method is to fill the column with lumps of an inert material, or packing, that will provide surface for contacting the vapour and liquid. Another effective way is to use a number of horizontal plates, or trays, which cause the ascending vapour to be bubbled through the descending liquid. Fig. 5.74 shows one such tray design arrangement. Tray designs are also numerous and varied. Tray designs include, bubble cap plate units, sieve plates, turbogrid trays, v-grid trays, and other speciality-type units, though bubble caps and sieve trays are the most common in distillation applications.

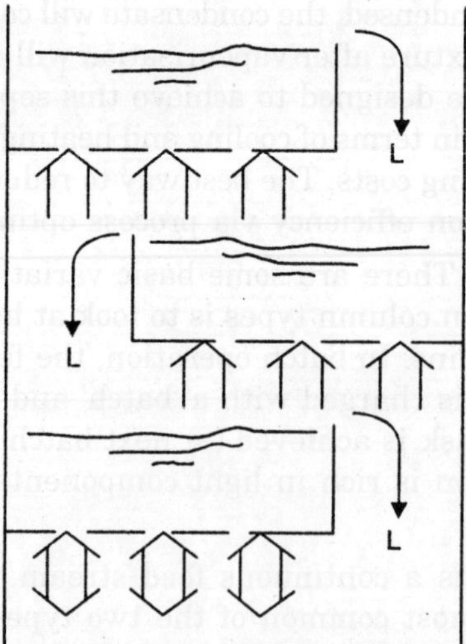

Fig. 5.74 Intimate contact and therefore equilibrium is obtained as the vapour bubbles ascend through the liquid held up on each tray, as the liquids descends down the column.

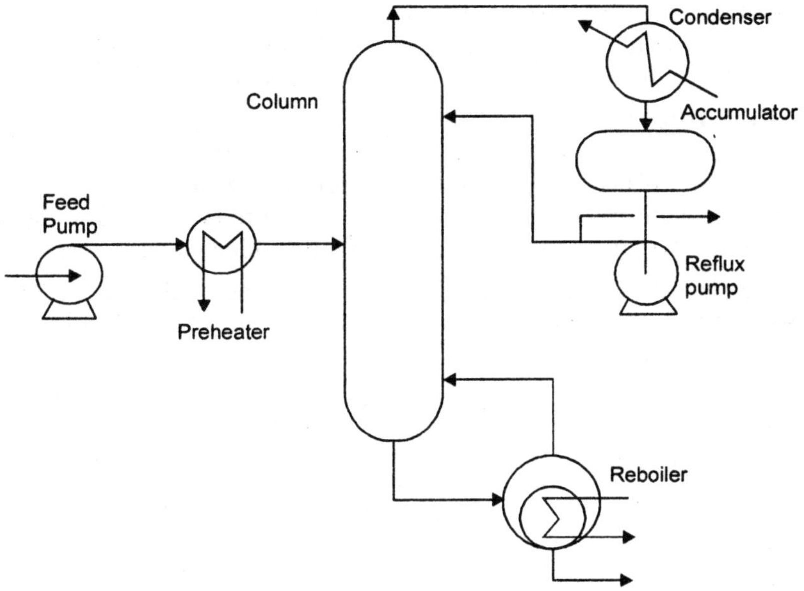

Fig. 5.75 Distillation equipment

The other equipment associated with the column are shown schematically in Fig. 5.75. They include :

1. Condenser
2. Reboiler
3. Inter heaters/Inter coolers
4. Feed preheater

1. Condenser and accumulator : The overhead vapour leaving the column is sent to a cooler, or condenser, and is collected as a liquid in a receiver, or accumulator. A part of the accumulated liquid is returned to the column as 'reflux'. A reflux pump is used for this purpose as well as for, distillate which is withdrawn as 'Over-head product'. In many cases, complete condensation is not accomplished. Such condensers are called partial condensers. Common condensers include bin-fans and water coolers. However, for efficiency of heat recovery, heat exchange with another process stream is often performed. Propane is the most common refrigerant used. The condenser and accumulator are the key pieces of equipment with respect to controlling the pressure in the column.

2. Reboilers : The liquid leaving the column bottom is heated in a reboiler. A reboiler is a special type of heat exchanger used to provide the heat necessary for distillation. Part of the liquid is vapourised and returned into the column as boil-up. The remaining liquid is withdrawn as a bottom product, or residence. Reboilers are available in widely varying designs. They may be internal, but most are external to the column. They also may be of natural circulation or forced circulation. The kettle reboiler is the most common external forced circulation design. Vertical and horizontal thermosyphon reboilers operate by natural circulation. In these, flow is induced by the hydrostatic pressure imbalance between the liquid inside the tower and the two-phase mixture in the reboiler tubes. In the forced circulation reboilers, a pump is used to ensure circulation of the liquid past the heat transfer surface.

Reboiler design variations are illustrated in Fig. 5.76. Reboilers may be designed so that boiling occurs inside vertical tubes, inside horizontal tubes, or on the shell side. Common heat sources include hot oil, steam or fuel gas (fired reboilers).

3. Interheaters/Intercoolers : In some cases, additional vapour or liquid is withdrawn from the column at points above or below the point at which the feed enters. All or a portion of this side stream can be used as intermediate product. Sometimes, economical column design dictates that the side stream be cooled and returned to the column to furnish localized reflux. The equipment that does this is called an 'intercooler' or side stream cooler.

Sometimes localized heat is required. In this case, some of the liquid in the column is removed and passes through an 'inter heater' or side stream reboiler before being returned to the column.

4. Feed Preheater : Often the feed is preheated before entering the column. Common preheat mediums include the bottoms product or low-pressure steam. Preheating is often convenient method to recover heat that would otherwise be wasted.

Feed pump, reflux pump etc are the auxiliary equipment associated with column operation.

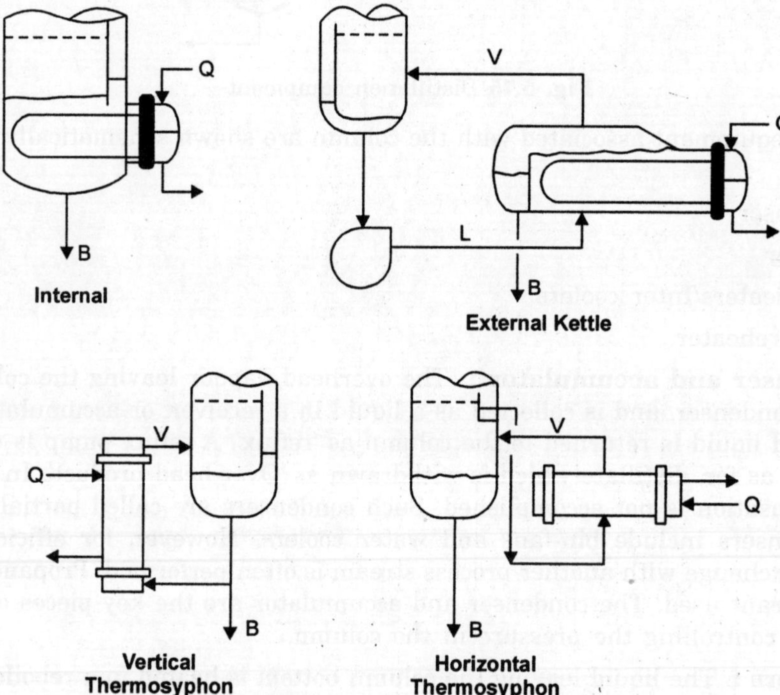

Fig. 5.76 Reboiler design variations. External kettle reboilers often use forced circulation (pump), while the thermosyphon designs depend on natural circulation. The horizontal thermosyphon reboiler takes its liquid from the bottom tray, while the others take it from the column bottoms.

5.7.2 Distillation Controls

Out of batch and continuous distillation columns and of two products and multiproducts columns,we will restrict ourselves for further discussions to continuous Binary distillation columns. The control of continuous binary distillation columns to give products of constant composition is one of the most difficult problems in process control. To begin with, it is very difficult to obtain accurate, continuous measurements of the composition of nearly pure streams. Temperature measurement can often be used to infer composition. Since distillation separates materials according to their difference in vapour pressures and since vapour pressure is a temperature controlled function, temperature measurement has historically been used to indicate composition.

The large hold up of liquid in the column, reboiler and reflux drum tends to make distillation control systems sluggish. The effective time delay can be made small by locating the control point at or close to the end of the column.

The dynamic analysis of distillation columns is difficult because of the counter flow of vapour and liquid, which introduces a special type of composition interaction between the stages, and because of non linear equilibrium relationship, which makes gain of each stage different. In addition to composition lags , there are also lags in the flow of vapour and liquid through a column, and the lags in liquid flow have large effect on the control system performance.

Before taking any decision on control strategies for distillation column, one should clearly understand the operating objectives of the column. The important such objectives are listed below :

1. Products (top and bottom) compositions should be constant and as required.
2. To increase the throughput.
3. Enhancing the column stability.
4. Operating against equipment constraints.
5. Economy in energy consumption and overall economic benefits.
6. Safety of men and machineries.

Evaluation of control strategies : The first decision involves configuration of the top and bottom control loops, which directly determine product compositions. Once these strategies are tentatively determined, strategies for the remaining variables like column pressure, accumulator level, bottom level etc become much easier to select.

Pairings of controlled and manipulated variables are normally made according to Single-Input/Single-Output (SISO) method. Sometimes Multiple-Input/Multiple-Output (MIMO) variables are paired. Some SISO variables combination is tabulated in Table 5.2.

Table 5.2 Pairing of controlled and manipulated variables-Binary distillation column

S.No.	Controlled variables	Manipulated variables
1	Overhead composition	Reflux flow rate
2	Bottoms composition	Reboiler heating medium flow rate
3	Accumulator level	Distillate flow rate
4	Bottoms level	Bottoms flow rate
5	Column pressure	Condenser flowrate/vapour bypass rate.

The pairings can follow three general control structures :

1. **Energy balance control** uses reflux and reboiler heatant flow to control compositions, thus fixing the energy inputs.
2. **Material balance control** uses the distillate and bottoms product flows to control compositions, thus fixing the overall material balance.
3. **Ratio control** utilizes a ratio of any two flow rates at each end of the column. A common example is the control of reflux-to-distillate ratio.

Control loop interactions and 'to be avoided' combinations are to be analysed and decided out of experience before evaluating control strategies.

5.7.3 Control Configurations

The first step in the design of a control system must be the development of a steady-state process model. The model defines the process with equations developed from the material and energy balances of the unit. The model is kept simple by the use of one basic rule: The degrees of freedom for control limit the controlled variables (product compositions in our case) specified in the equations. Some of the variables that can be manipulated to control a ideal binary distillation column are shown in Fig. 5.77. For example, for a given feed rate only one degree of freedom exists for material balance control. If overhead product (distillate) is a manipulated variable (controlled directly to maintain composition), then the bottom product cannot be independent but must be manipulated to close the overall material balance according to the following equations :

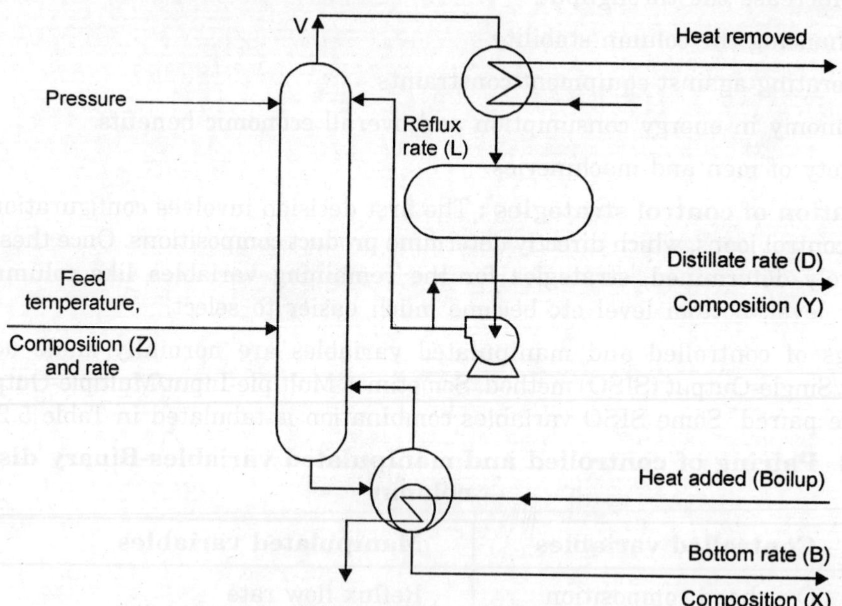

Fig. 5.77 Variables that fix the distillation operation

$$F = D + B \tag{5.9}$$

where F = Feed rate (the inflow).

D = Overhead distillate rate (an outflow).

B = Bottoms rate (an outflow).

Accumulation = Inflow – Outflow

$$A = F - (D + B) \tag{5.10}$$

Since accumulation is zero at steady state, then B is dependent upon F and D,

$$B = F + D \tag{5.11}$$

Or if the bottoms product is the manipulated variable :

$$D = F - B \tag{5.12}$$

If the compositions of the feed, distillate product, and bottoms product are known, then the component material balance is also possible. For a given feed composition and desired product compositions, only one bottoms-to-feed ratio, B/F (product split), will satisfy the overall and component material balances. Therefore, by fixing the bottoms flow, the distillate flow will be fixed. However, fixing the product split does not fix either the distillate or bottoms composition since many combinations are possible with the same value of B/F.

Energy balance : The energy balance and separation obtained are closed related. Conceptually, product composition control can be thought of as a problem of the rate of heat addition Q_B at the bottom of the fractionator and the rate of heat removal Q_T at the top of the column. A series of energy balances produces lot of additional equations. Here we will concentrate on the final important ones.

The internal vapour boil-up rate V_B equals the heat added by the reboiler divided by the heat of vapourisation (ΔH) of the bottoms product.

$$V_B = Q_B/\Delta H \tag{5.13}$$

The vapour rate V above the feed tray equals the vapour boil-up rate plus the vapour entering with the feed

$$V = V_B + F.V_F \tag{5.14}$$

where F = Feed rate

V_F = Vapour fraction in the feed F.

The liquid rate below the feed tray L_f, equals the internal reflux rate plus the liquid in the feed :

$$L_f = L_i + (1 - V_F)\, F \tag{5.15}$$

The distillate rate, D, equals the vapour rate, V, above the feed tray minus the internal reflux :

$$D = V - L_i \tag{5.16}$$

The bottoms rate, B, equals the liquid rate, L_f, minus the boil-up, V_B :

$$B = L_f - VB \tag{5.17}$$

The criterion for separation is the ratio of external reflux (L) to distillate (D) versus the ratio of boil-up (V_B) to bottoms (B). Manipulating reflux affects separation equally as well as manipulating boil-up, albeit in opposite directions. Consequently, only one degree of freedom exists to control separation. Thus, for a two-product tower, two equations define the process. One is an equation describing separation and the other is an equation for material balance.

Dynamic model : Since the tower does not operate always at steady state, it is essential to also account for the dynamics of the process. This extends the steady-state internal flow model and requires additional considerations. Time lags in the reflux flow through trays, changes in boil-up rates, liquid inventory at the bottom of the tower, degree of separation and orientation of separation etc are to be taken into account for developing dynamic model.

Separation equations : The control of product compositions for a fractionator is primarily a matter of control of the internal flows. In considering product separation, the degree of separation and the orientation of separation are important.

$$\text{Degree of separation} = \text{Log}_e \frac{(\%LK_D \times \%HK_B)}{(\%HK_D \times \%LK_B)} \tag{5.18}$$

$$\text{Orientation of separation} = \frac{\%HK_D}{\%LK_B} \tag{5.19}$$

(for a given degree of separation)

where

$\% \, LK_B = x =$ mole fraction of the light key component in the bottoms

$\% \, LK_D = y =$ mole fraction of the light key component in the distillate

$\% \, HK_B =$ mole fraction of the heavy key component in the bottoms

$\% \, HK_D =$ mole fraction of the heavy key component in the distillate

The relationship between the light key component x and the energy balance was developed as a function of separation 'S' as :

$$S = \frac{y(1-x)}{x(1-y)} \tag{5.20}$$

The relationship between separation (S) and the ratio of boil-up to feed (V/F) over a reasonable operating range is :

$$V/F = a + bS \tag{5.21}$$

where 'a' and 'b' are functions of the relative volatility, the number of trays, the feed composition and the minimum V/F. The control system therefore computes V based on the equation

$$V = F \left\{ a + b \, \frac{y(1-x)}{x(1-y)} \right\} \tag{5.22}$$

Since 'y' is held constant, the bottom composition controller adjusts the value of the parenthetical expression if an error should appear in x.

Let $V = \dfrac{y(1-x)}{x(1-y)}$, and the control equation becomes

$$V = F\,(a + b\,[V/F]) \tag{5.23}$$

where $[V/F]$ = the desired ratio of boil up to feed.

Fig. 5.78 illustrates four of the most common control configurations of a binary column, where it is assumed that feed flow and tower pressure are kept constant. The above control equations for controlling internal product flow rates are dependent upon the configurations of the control system used. Comparing case-1 with case-2 , we find the bottoms product rate is controlled with bottom level controller as a master and 'B' rate controller as slave in a cascade fashion in case-1 where as in case-2 the bottom level is controlled with reboiler heat energy Q_B as slave controller in a cascade mode. Comparing case-1 with case-3, we find the accumulator level is controlled with distillate (D) output controller as slave in case-1 where as in Case-3 the reflux rate (L) controller as slave in cascade mode. Comparing case-3 and case-4, we find the accumulator level control as a master works in conjuction with reboiler heat input controller as slave.

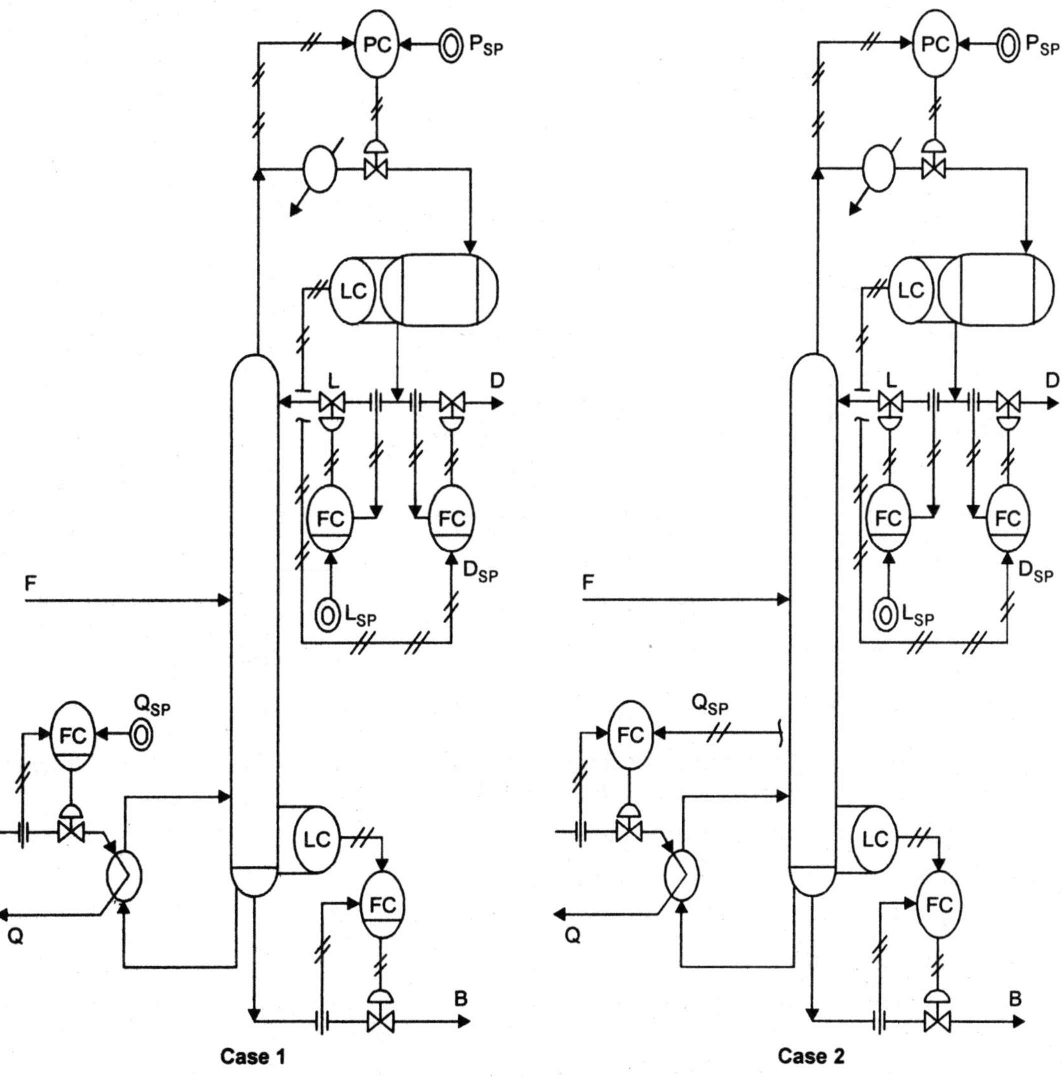

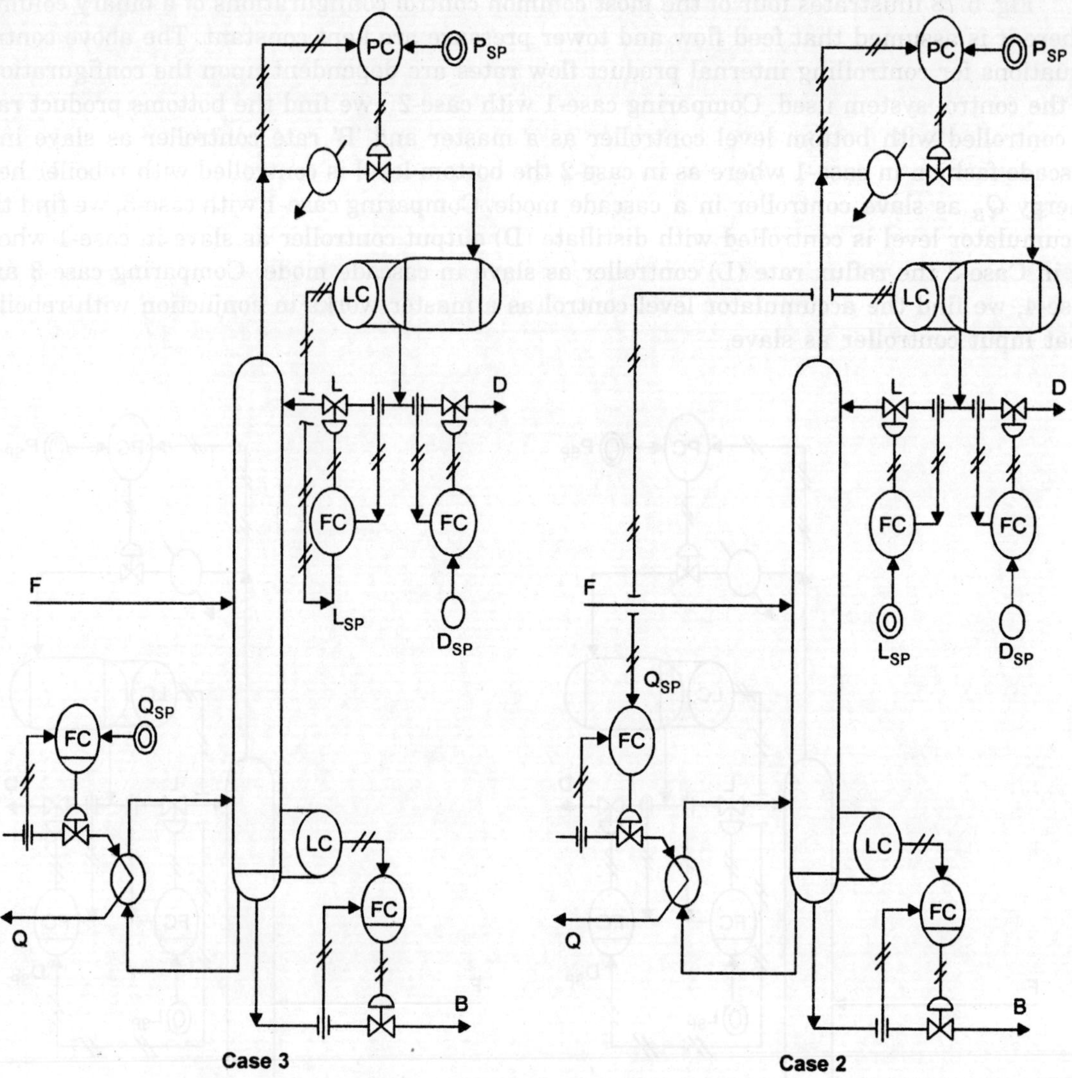

Fig. 5.78 Four cases of conventional distillation control configurations.

Fig. 5.79 shows another configuration with four feedback loops that satisfy the following objectives of maintaining the four variables at desired values :

1. Composition of the distillate stream, x_D
2. Composition of the bottoms stream, x_B
3. Liquid hold up in the accumulator or reflux drum, M_{RD}.
4. Liquid hold up in the base of the column, M_B.

With the above introduction to the operation of binary distillation column and control strategies, the following important controls will be elaborated further :

1. Composition control (product quality control).

2. Feed forward control techniques.
3. Column pressure control.
4. Feed flow rate and temperature control.
5. Benzene column control - An example .

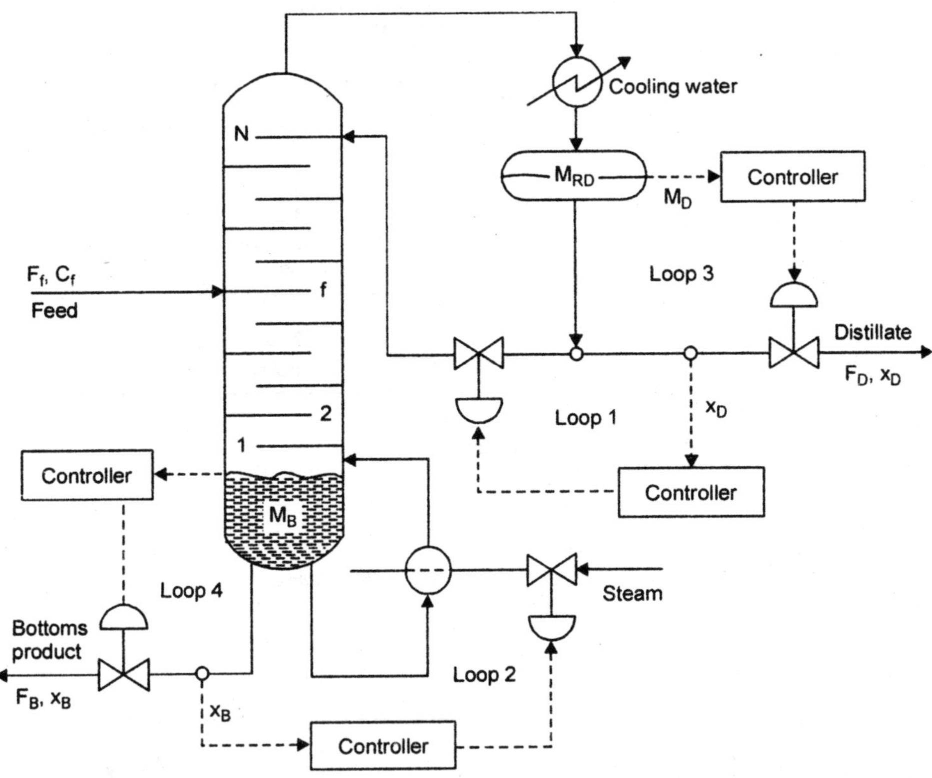

Fig. 5.79 Feedback loops of a binary distillation column.

5.7.4 Composition Control (Product Quality Control)

Conceptually, product quality control or composition control is a problem of making precise adjustments to the rate of heat addition and the rate of heat removal from the tower. Heat removal determines the internal reflux flow rate, and the internal reflux as measured on the top tray is a direct reflection of the composition of the distillate. Heat added determines the internal vapour rate. These internal vapour and liquid flow rates determine the circulation rate, which inturn determines the degree of separation between two key components. Once interaction of various variable pairings has been established and the column's operating objectives and disturbance variables are considered, the primary composition control loops of the column can be selected. Measurement of these control variables can be either direct or inferred. Three different systems for the distillate composition control namely (*a*) feedback (*b*) feed forward, and (*c*) inferential is shown in Fig. 5.80 which is self explanatory. The feedback system uses the composition analyzer (normally considered to be costly and unreliable) to measure the distillate composition and manipulate the reflux ratio. The feedforward system measures the composition of feed and manipulates the reflux ratio. The third one, the inferential control system uses temperature measurements (secondary measurements) to estimate or infer the composition in the distillate and manipulate the reflux ratio. In the following paragraphs we will see similar control systems, but with little more details.

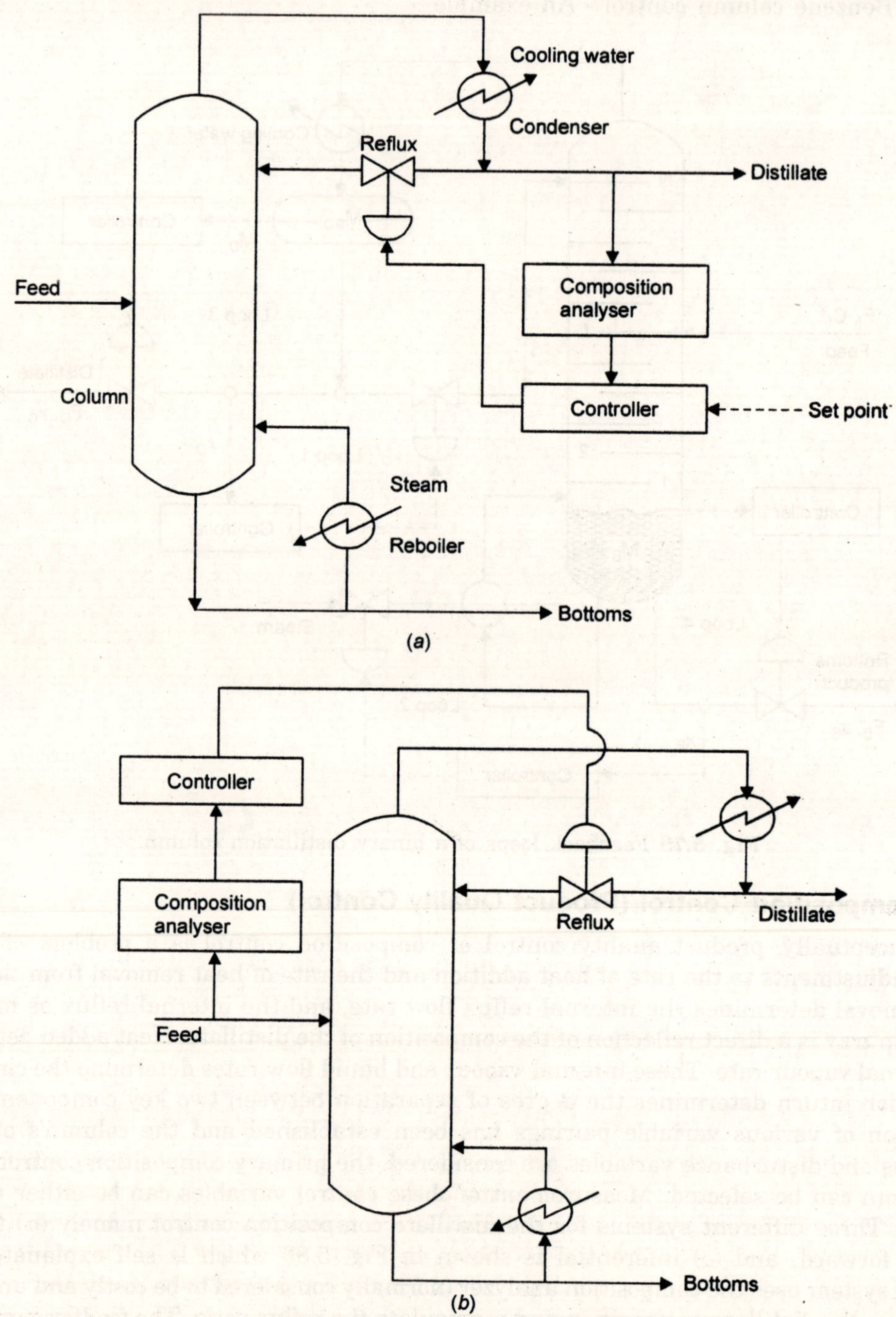

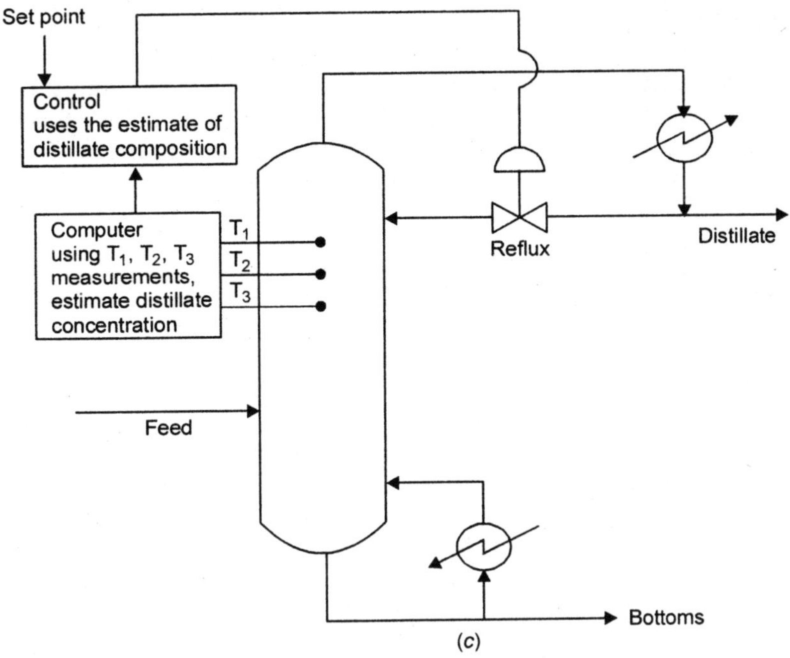

(c)

(*a*) Feedback (*b*) Feedforward (*c*) Inferential

Fig. 5.80 Three different systems for the distillate composition control of a simple distillation column

5.7.4.1 Inferring Composition from Temperature (Inferential Control)

Temperature measurement can often be used to infer composition. Since distillation separates materials according to their difference in vapour pressure and since vapour pressure is a temperature-controlled function, temperature measurement has historically been used to indicate composition. This presumes that the column pressure remains constant (or that the pressure variations are compensated for in the temperature measurement) and that feed composition is constant. Then any change in composition within a column will result in a temperature change.

The best location for the temperature sensor has to be suitably selected. We should measure the temperature on a tray that strongly reflects changes in composition. When composition of the bottom product is important, it is desirable to maintain a constant temperature in the lower section. This can be done by letting the temperature measurement manipulate the reboiler steam supply by resetting the steam flow controller set point as shown in Fig. 5.81. It is a cascade control system with TRC as master controller and FRC as slave controller. Bottoms product output line is provided with a control valve which will be operated with the level signal. That means the output of bottoms product is regulated maintaining the bottom level constant.

When the composition of the distillate product is important, it is desirable to maintain a constant temperature in the upper section of the column as shown in Fig. 5.82. This again utilizes the cascade control principle with temperature controller TRC as master and the reflux flow controller FRC as slave. The reflux flow is the manipulated variable. The output of distillate is controlled through an another cascade loop with accumulator level controller as master one and the distillate flow controller as slave one. Distillation temperature is an indication of composition only when column pressure remains constant and hence the column pressure has to be maintained constant with a help of a feedback controller (PRC). Manipulated variable here is the coolant flow to the condenser. We will discuss in detail the different ways of column pressure controls in the next section. Normally the point of column pressure control is near the temperature control point. This arrangement helps to fix the relation between temperature and composition at this particular point.

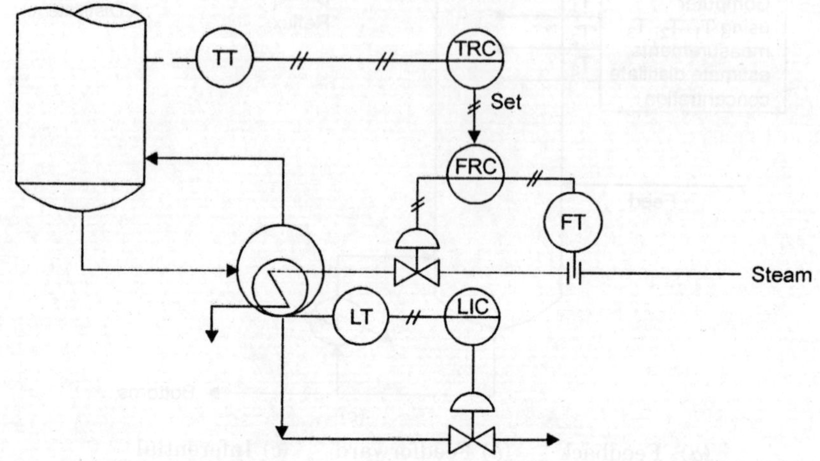

Fig. 5.81 Temperature cascaded heat addition to the reboiler.

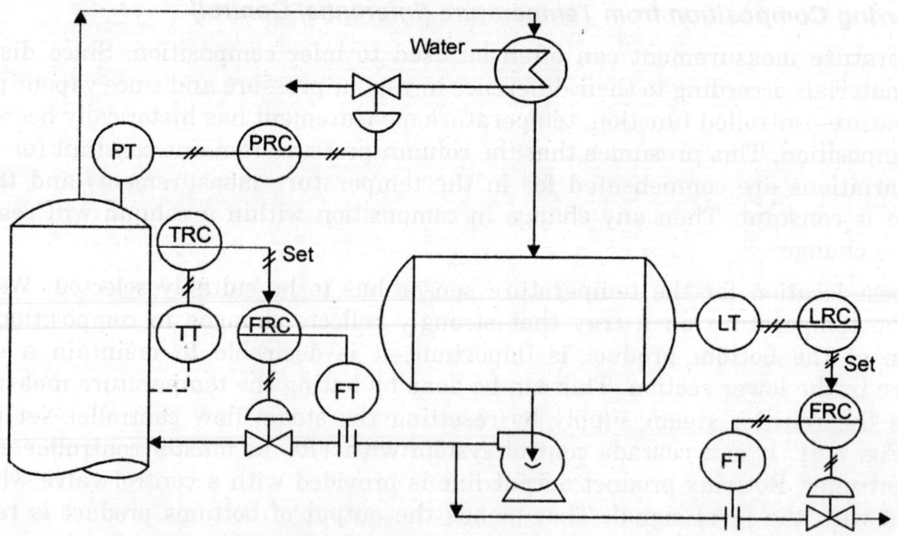

Fig. 5.82 Temperature cascaded reflux flow for improved overhead composition control

5.7.4.2 Composition Control Using Analysers

Although additional investment is needed for installing an analyzer for composition measurement and involves rigorous maintenance, a savings from improved operation usually results. Out of several types of analysers for composition, the chromotograph is the most versatile.

A simple feedback control using chromotograph for distillate withdrawal is shown in Fig. 5.83. Reflux flow is independently controlled. The chromatograph continuously analyses a sample of vapour from one of the intermediate trays. The chromatograph output is used by a controller to modulate the product draw off valve. Such a direct control is often used when the dead time of each analysis update is less than the response time of the process. Since control of product specification is often the objective of the fractionator, direct composition control would seemingly be better than control by temperature, which is an inferred property. The composition controller provides feedback correction in response to feed composition changes, pressure variations and variations in tower efficiencies.

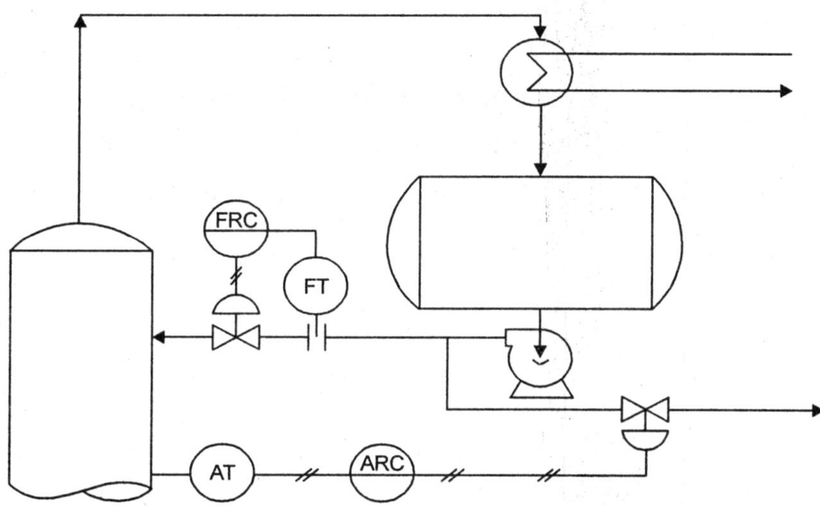

Fig. 5.83 Distillate withdrawal controlled by chromotograph

Figure 5.84 shows a configuration where the chromatograph continuously analyses a liquid sample from the condenser rundown line. The sample probe and sampling system provide a representative vapour sample to the chromatograph. The chromatograph measurement is used by a master controller (ARC) to manipulate the reflux flow by setting the set point to the reflux flow controller (FRC), the slave controller of the total cascade loop.

Often the analyzer is so slow that it degrades the controllability of the process. In that case, some type of dead-time compensation is used as discussed in Chapter-3 (3.9.1). A Smith-predictor compensator or similar compensator is used to model the process to predict what the analyzer measurement should be between analysis updates. The model is then compared to the actual measurement, and the input to the controller is biased by the difference. Fig. 5.85 shows the same configuration as Figure 5.84 except that the analyzer controller is equipped with a first-order Smith -predictor dead-time compensator. The multiplier, lag and dead-time calculation blocks represent the predicted analysis. The lag block represents first order process. This predicted response is subtracted from the actual measurement to give a differential of the actual process from its model. This delta (Δ) is added to the model without deadtime to give a pseudo-measurement to the analyzer controller. Thus, the analyzer measurement, which has a significant dead time due to sampling and cycle time, provides a trim to the predicted measurement of the model.

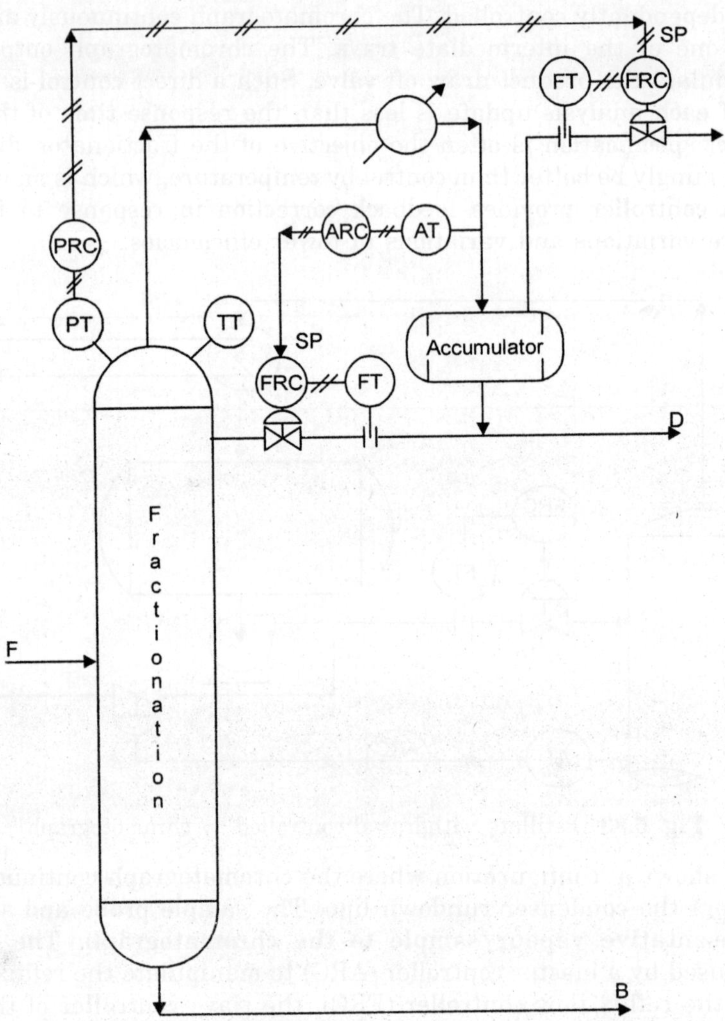

Fig. 5.84 Analyser controller cascade to reflux flow controller.

In yet another configuration, analyzer control cascaded to temperature control is used when stable temperature on a particular tray is desired and the tower operates at a constant, maintainable, and controllable pressure.

An example is cascading the analyzer controller to the overhead temperature of the tower, which in turn is cascaded to the reflux flow rate. Since temperature is an indicator of composition at this pressure, the analyzer controller only serves as a trim correcting for variations in feed composition. Fig. 5.86 shows this triple cascade of analyzer to temperature to reflux flow controller.

Feed composition analyzer, boiling point analyzer and viscosity analyzer are some more analytical instruments which can be used for effective control. Proper sampling of material in a column is necessary if analysers are to control effectively. A poor sampling system often is responsible for the unsatisfactory performance of plant analysers.

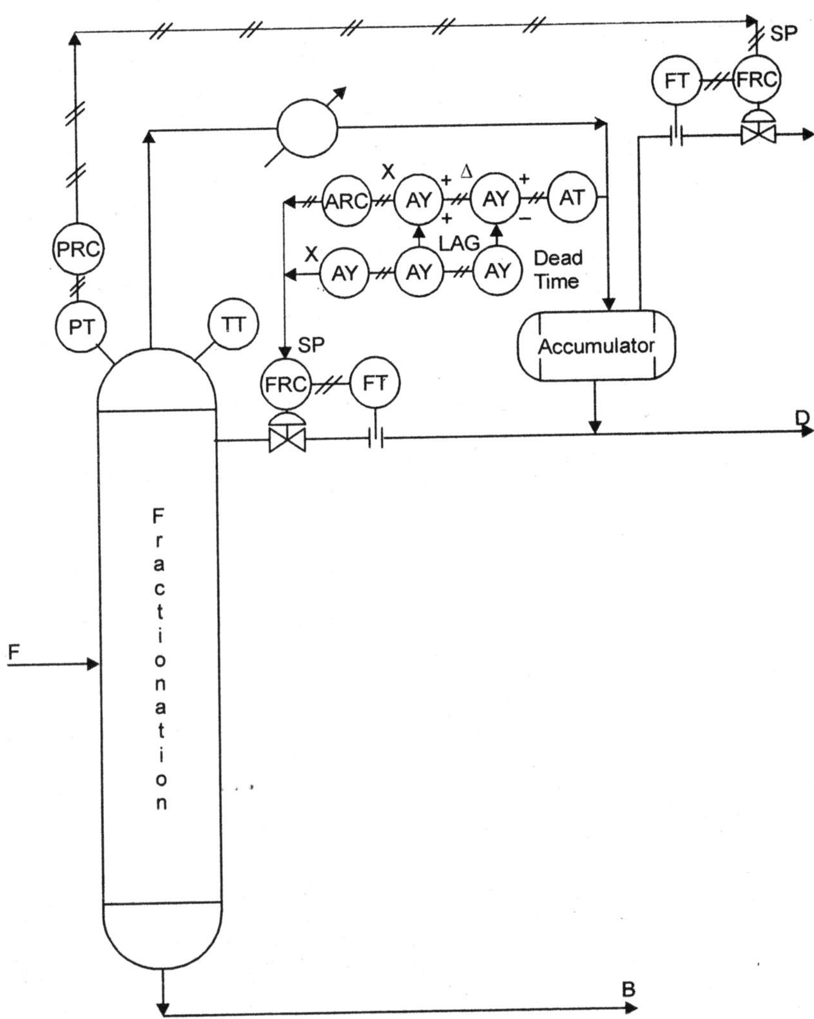

Fig. 5.85 Analyser controller with dead-time compensation cascaded to reflux flow control.

5.7.5 Feed Forward Control Techniques

Feedforward control techniques react to variations in disturbance variables, predict the disturbances effects, and take corrective action before the tower is significantly affected. Feedback control attempts to maintain the set point of a controlled variable by measuring its value at the outlet of the tower. In most cases a combination of feedforward and feedback techniques can correct process deviations in the shortest time. This correction is accomplished by considering process dynamics (dead time and time lags), the nonlinearities between separation efficiency and column loading, loops interactions and process measurements. The type of disturbances which feedforward control is most often used to compensate for include:

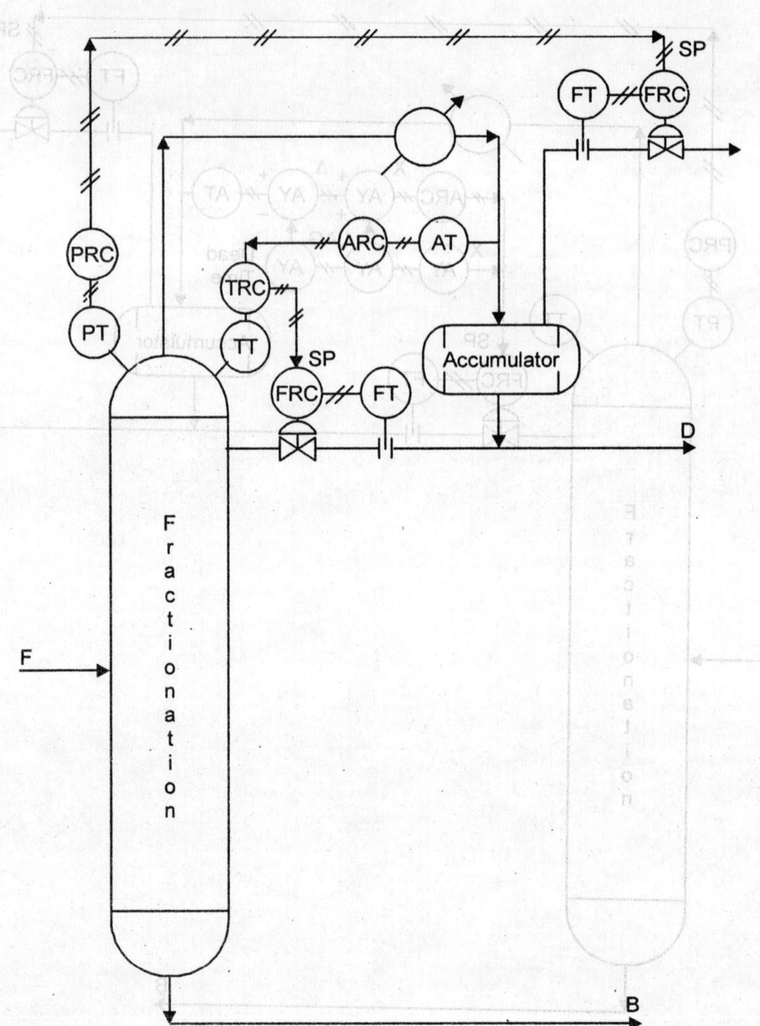

Fig. 5.86 Analyser control cascaded to temperature control further cascaded to reflux flow control.

1. Feed flowrate.
2. Top and reflux temperatures (due to change in ambient temperatures).
3. Reflux flow rate.
4. Tower pressure.
5. Feed composition.
6. Feed temperature or enthalpy.
7. Reboiler heat.

The application of feed forward techniques involves the use of the models and equations already discussed under control configurations (5.7.3), but dynamically tuned to approximate the response of the distillation tower.

Flow control of distillate : The column interactions that otherwise might necessitate the use of an internal reflux control system can be eliminated in some cases when the flow of distillate product draw off is controlled and reflux is put under accumulator level control. The steady-state material balance around the accumulator (Refer Fig. 5.87) is expressed by :

$$V = L + D \tag{5.24}$$

where V = boil-up (vapour rate)

L = Reflux rate

D = Distillate rate

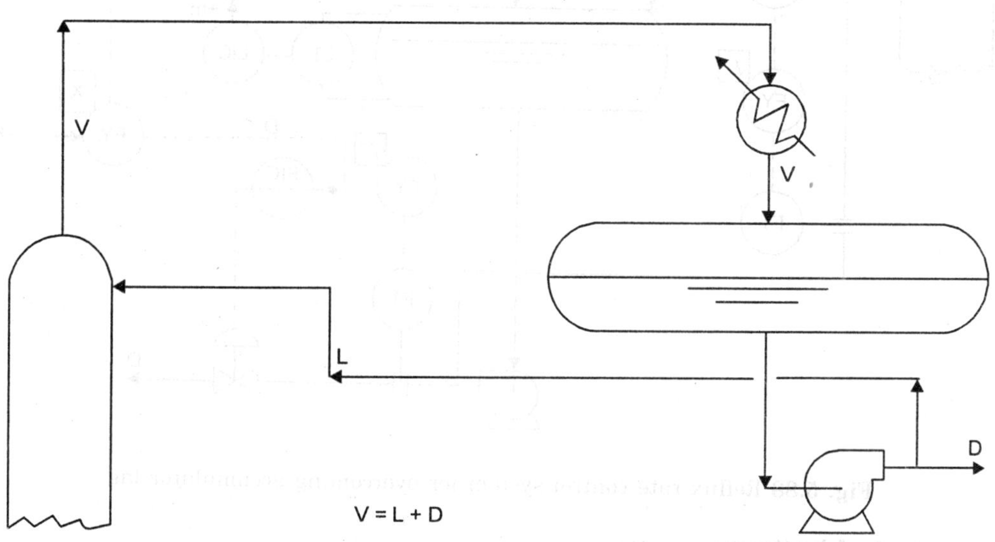

Fig. 5.87 Reflux accumulator material balance

To overcome the accumulator lag, the reflux rate, L, must be manipulated in direct response to a change in distillate rate, D, rather than by waiting for the response of a level controller. If V is constant (K), equation (5.24) is solved for L, which is the manipulated variable in this part of the system.

$$L = K - D \tag{5.25}$$

For this equation to be satisfied, L must be decreased one unit for every unit D is increased, and vice versa.

If V is indeed constant and both the computations and the flow manipulations are perfectly accurate, no level controller is needed. If these conditions cannot be met, a trimming function is introduced. The system equation becomes

$$L = m - K D \tag{5.26}$$

where m = The output of the level controller

K = An adjustable coefficient.

The resulting control system is shown in Fig. 5.88.

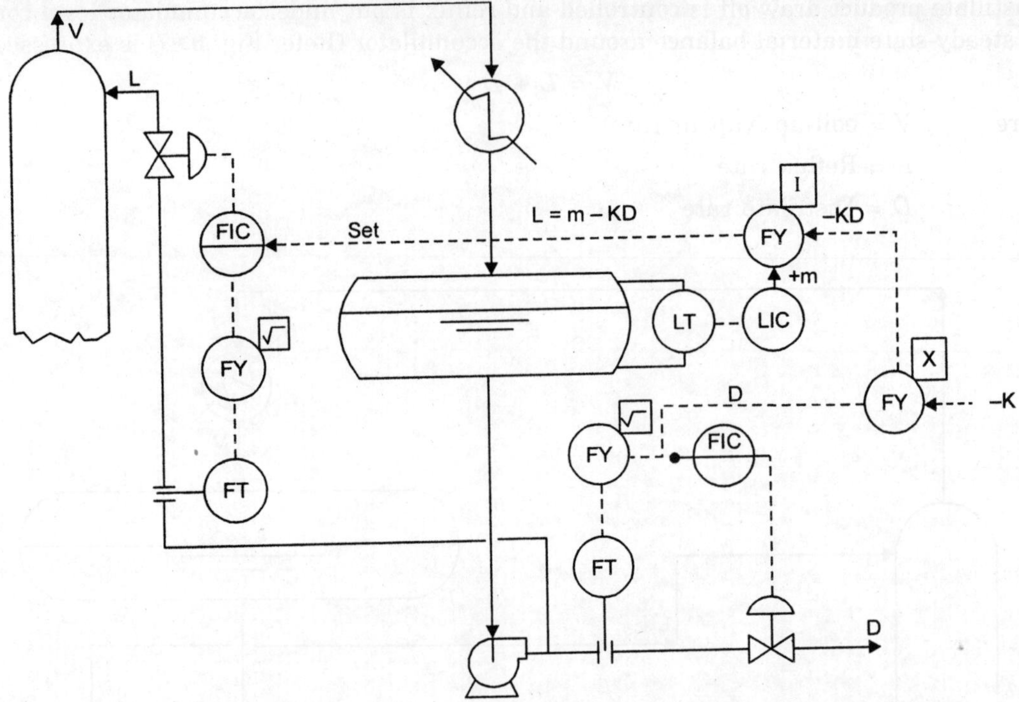

Fig. 5.88 Reflux rate control system for overcoming accumulator lag.

Flow control of bottoms

A similar system can be used to column bottoms, where the bottoms product is flow controlled and the bottom's level is maintained by the manipulation of the heat input or boilup (V). The equation for the system is

$$V = m - K\,B \tag{5.27}$$

where

m = The output of the bottom's level controller.

B = Bottoms product flow.

K = Adjustable coefficient.

V = Boilup.

Since these models are only approximations of the real process, inaccuracies do exist. In the majority of the feedforward applications, their purpose is not to replace feedback but to minimize the amount of work that the feedback part of the loop has to do. That means the final control system must be able to measure and quantify the disturbance, then react before the fractionator separation can be upset in the first place.

Feedforward distillation control system with constant separation

A distillation column operating under common separation conditions has one fewer degree of freedom than others, because its energy-to-feed ratio is constant. At a given separation, for each concentration of the key component in the distillate, a corresponding concentration exists in the bottoms. In other words, for a constant feed composition, holding the concentration of a component constant in one product stream fixes it in the other. Fig. 5.89 shows an example of a constant separation feedforward system in which distillate is the manipulated variable. A material balance on the light key component gives.

$$Fz = Dy + Bx = Dy + (F - D)x \tag{5.28}$$

$$D = F\left(\frac{z-x}{y-x}\right) = F\left(\frac{D}{F}\right) \tag{5.29}$$

If the flow measurements are of the differential pressure type, then

$$D^2 = F^2\left(\frac{z-x}{y-x}\right)^2 = F^2\left(\frac{D}{F}\right)^2 \tag{5.30}$$

Because boil-up must change in proportion to feedrate, a second forward loop is obtained for setting heat input :

$$Q = F\left(\frac{Q}{F}\right) \text{ or } Q^2 = F^2\left(\frac{Q}{F}\right)^2 \tag{5.31}$$

where z, y, x = mole fraction of the key light component in feed, overheads, and bottoms respectively.

D/F = required distillate-to-feed ratio.

Q/F = required energy-to-feed ratio.

The block labeled 'dynamics' in Fig. 5.89 is a special module designed to influence the transient response. This is because the time response of the distillate to a feed rate change must be dynamically matched. The dynamic block is generally a dead-time module and a lead-lag module in series. In the steady state its output equals its input.

Similar feedforward control systems can be developed for :

1. Maximum recovery of one worthy product than other.
2. Composition control of two products (constant separation technique discussed above takes care of one product only).
3. Two products control with interactions.
4. With feed composition compensation.

5.7.6 Column Pressure Control

Most distillation columns are operated with constant pressure control. Where the temperature inferred composition control is practiced, constant pressure control is necessary, otherwise temperature measurements are to be compensated for pressure variations. But several advantages can be achieved through floating-pressure operation in most columns. The primary advantage of floating-pressure control is the ability to operate at minimum column pressure within the constraints of the system. Lower pressure reduces the volatility of distillation components, thereby reducing the heat input required to effect a given separation. Other advantages include increased reboiler capacity and reduced reboiler fouling due to lower tower temperatures. A constant pressure control is a normal feed back control. In this section we will be dealing with the following pressure control strategies in connection with distillation column.

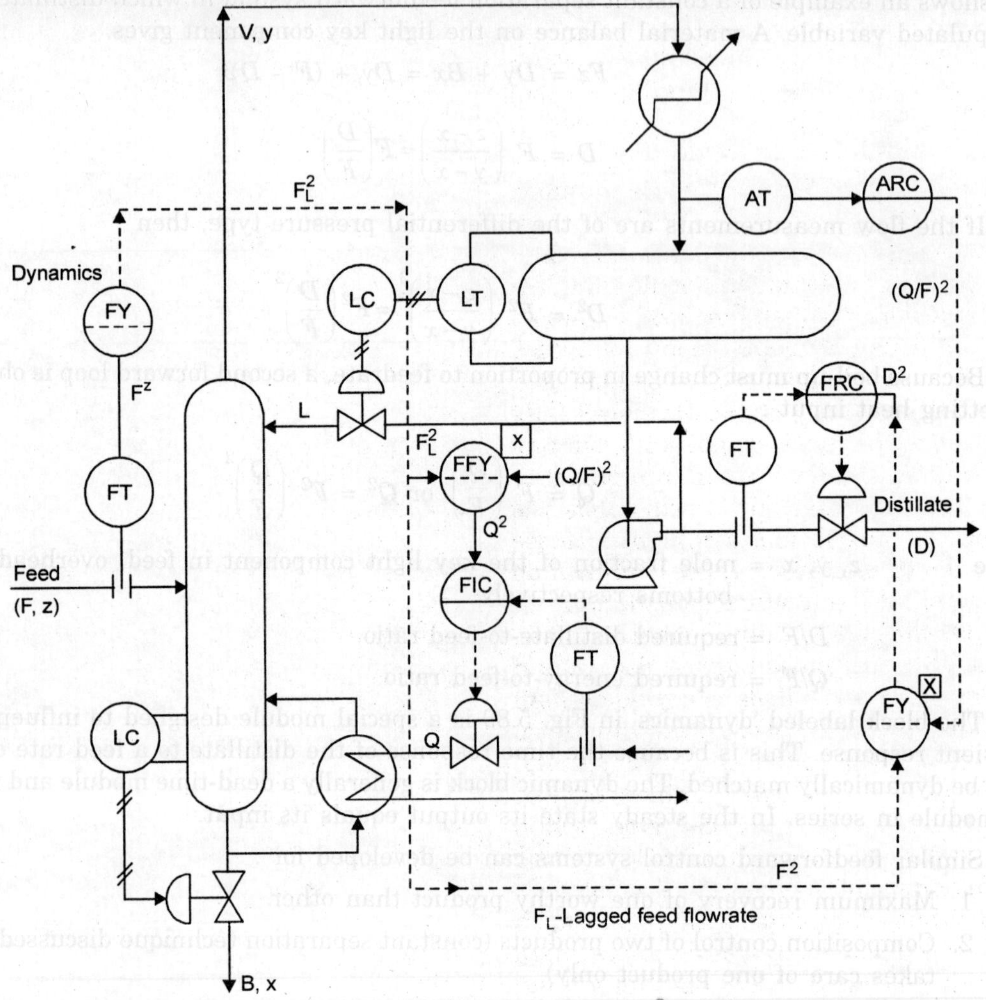

Fig. 5.89 Feedforward distillation control system with constant separation

1. Floating-pressure control strategies for the following conditions.
   (*a*) Liquid distillate withdrawn when non condensables (Inerts) are present.
   (*b*) Vapour distillate withdrawn when non condensables are present.
   (*c*) Liquid distillate withdrawn with negligible noncondensables.
2. Vacuum column pressure control.
3. Vapour recompression pressure control.

5.7.6.1 Floating Pressure Control Strategies

(*a*) Liquid distillate with inerts : In some separation processes the problem of pressure control is complicated by the presence of large percentage of inert gases. The noncondensables must be removed to avoid accumulation and blanketing off the condensing surface, thereby causing loss of column pressure control. The simplest method of handling this problem is to bleed off a fixed amount of gases and vapours either to a low pressure unit available in the system or through a vent condenser. A typical column pressure control with inerts present is depicted in Fig. 5.90. As the condensables build up in the condenser, the pressure controller (PRC-1) will tend to open the control valve (PCV-1) to maintain the proper rate of condensation. The signal to this valve could also be used to operate a purge control valve (PCV-2), as the opening of PCV-1 passses a certain operating point (a form of split range control). This could be done either by means of a calibrated valve positioner or by use of second pressure controller (PIC-2).

(b) Vapour distillate with inerts : In this case, the overhead product is removed from the system as a vapour, and consequently the pressure controller can be used to modulate this flow as shown in Fig. 5.91. The system will quickly respond to changes in this flow. A level controller is installed on the overhead receiver to regulate the cooling water to the condenser. It will condense only enough condensate to provide the column with reflux. The condenser must have a short residence time for the water to minimise the level control time lag. If the condenser is undersized, the cooling water flow should be maintained at a constant rate. Then the level controller can regulate a stream of condensate through a small vapouriser and mix it with the vapour from the pressure control valve as shown in Fig. 5.92. If the cooling water has bad fouling tendencies, it would be preferable to use a control system as shown in Fig. 5.93, using the pressure controller to regulate a vapour by pass around the condenser.

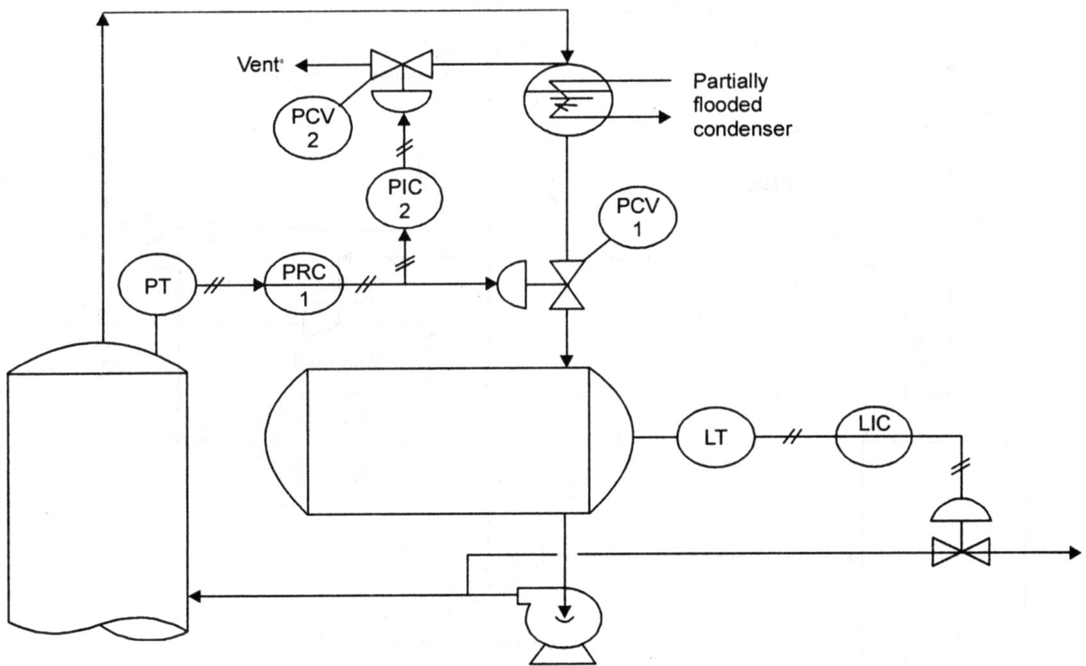

Fig. 5.90 Column pressure control with inerts present.

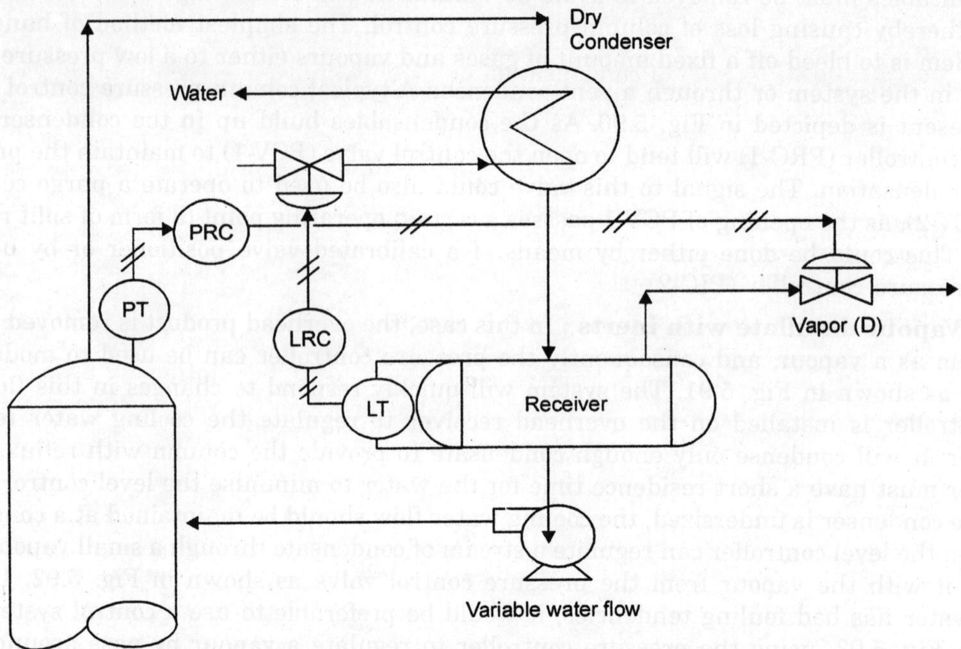

Fig. 5.91 Column pressure with vapour distillate and inerts present.

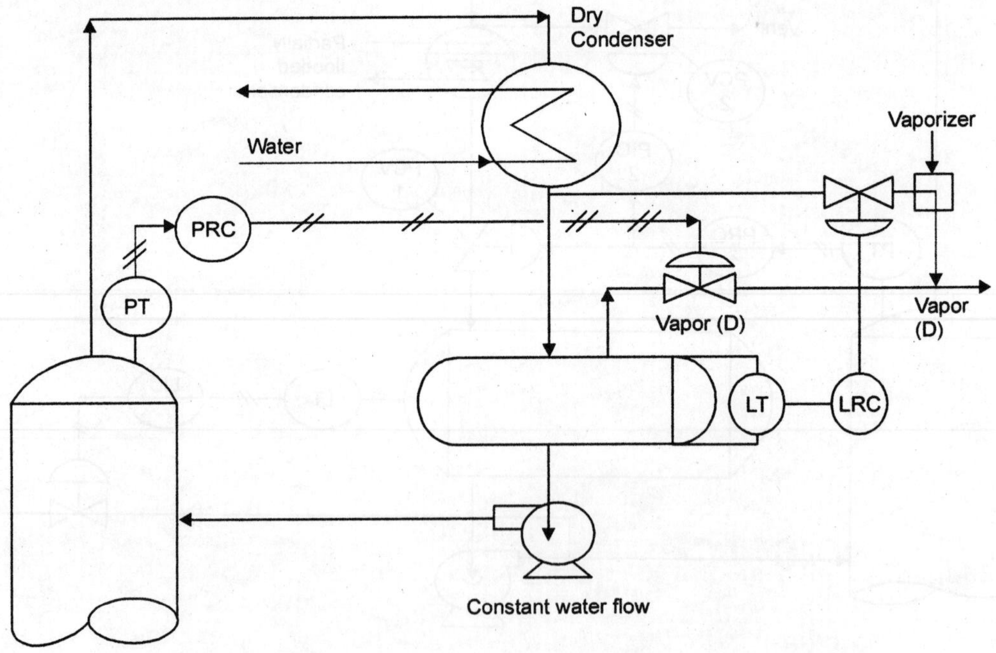

Fig. 5.92 Column pressure with vapour distillate and inerts present.

(*c*) **Liquid distillate with negligible inerts :** This situation is the most common, and pressure is controlled usually by adjustment of the rate of condensation in the condenser. The method of controlling the rate of condensation depends upon the mechanical construction of the condensing equipment. We will be discussing three different methods of controlling.

(1) **Cooling water control :** The control valve is placed on the cooling water line from the condenser as shown in Fig. 5.93. This system is recommended only when the cooling water contains chemicals, to prevent fouling of the tubes in the event of high temperature rise in the condenser tubes. With a properly designed condenser, the pressure controller needs only proportional control, because a narrow throttling range is sufficient.

(2) **Partially flooded condenser :** To avoid timelags, the water rate is to be kept constant and the condensing rate is to be controlled by regulating the area of surface exposed to the vapours. This is done by placing a control valve in the condensable line and modulating the flow of condensable from the condenser. When the pressure is dropping, the valve reduces the condensate flow, causing it to flood more tube surface and consequently reducing the condensing surface exposed to the vapors. The condensing rate is thereby reduced, and hence the pressure rises.

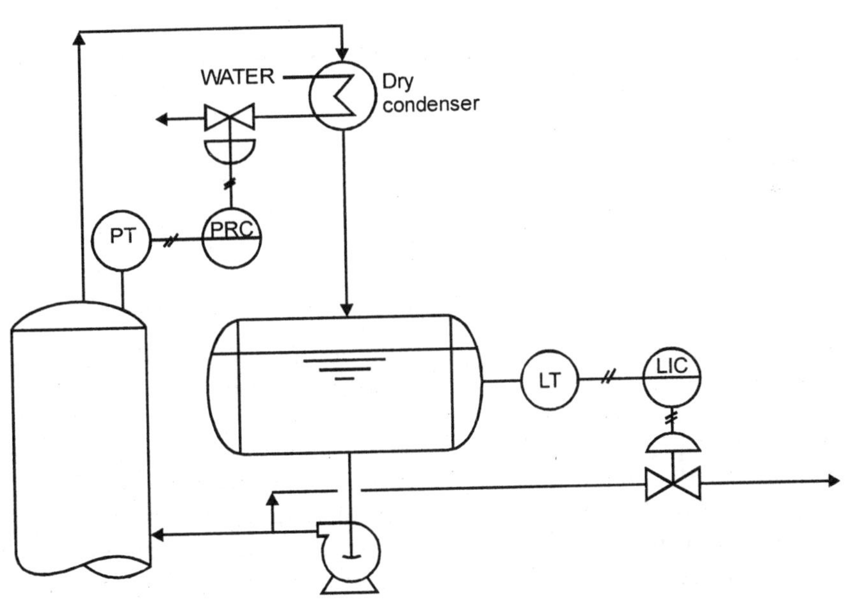

Fig. 5.93 Column pressure control by throttling condenser water.

(3) **Hot vapour by pass control :** A third possible installation for use with liquid distillation with inerts is used when the condenser is located below the receiver. The control valve is placed in a bypass from the vapour line to the accumulator as shown in Fig. 5.94. When this valve is open, it equalizes the pressure between the vapour line and the receiver. This causes the condensing surface to become flooded with condensate because of the 3 to 5 metres of head [it is usual practice to elevate the bottom of the accumulator 3 to 5 metres above the suction of the pump in order to

provide a positive suction head on the pump] in the condensate line from the condenser to the receiver. The flooding of the condensing surface causes the pressure to build up because of the decrease in the condensing rate. Under normal operating conditions the sub cooling that the condensate receives in the condenser is sufficient to reduce the vapour pressure in the receiver. The difference in pressure permits the condensate to flow up the 3m to 5m of pipe between the condenser and the accumulator.

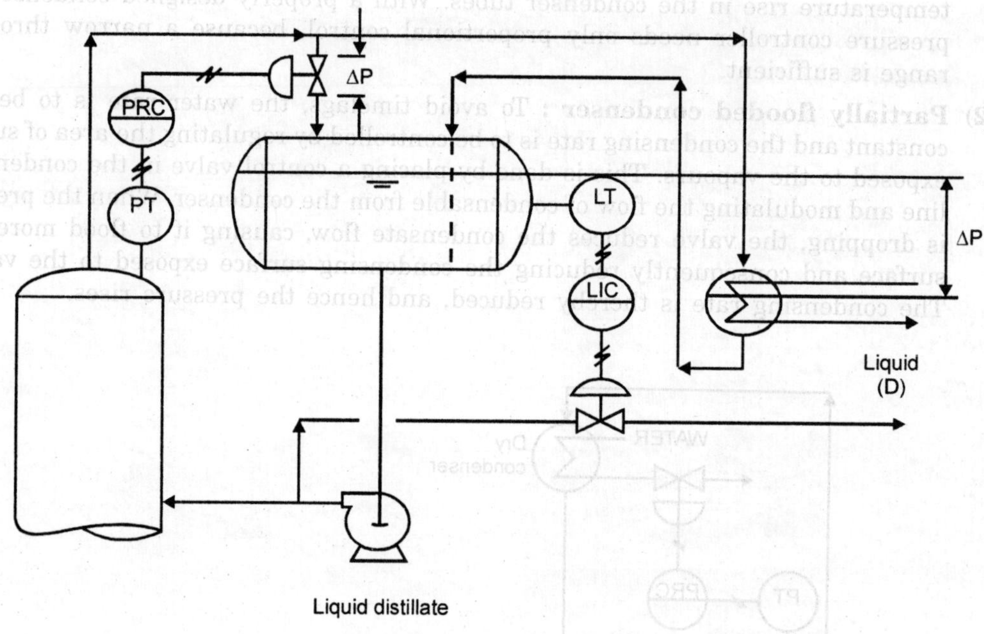

Fig. 5.94 Column pressure controlled by hot gas bypass.

Another system controls the pressure in the accumulator by throttling the condenser bypass flow as shown in Fig. 5.95. The column pressure is maintained by throttling the flow of vapour through the condenser. The operation of the system is as follows : If column pressure rises, PRC-1 opens PCV-1. This increases the vapour pressure in the condenser, which pushes some of the condensate out of it and increases the condensing surface area available for the vapours. The rate of condensation is increased, thereby and the column pressure is lowered back to the set point of PRC-1. At this higher rate of condensation the pressure drop (ΔP) across PCV-2 is also reduced (the valve opens). If the column pressure drops, the opposite sequence occurs : PCV-1 closes and the flooding of the condenser increases, reducing the rate of condensation and increasing the pressure drop (ΔP) across PCV-2 by slightly closing it. The setting of PRC-1 must always be above that of PRC-2.

The most common pressure control is shown in Fig. 5.96. This type of control features the pressure controller throttling the hot vapour bypass as shown previously. A second pressure controller on the accumulator is generally set at 5 psi below the tower pressure and is used to vent the inerts that may build up in the system.

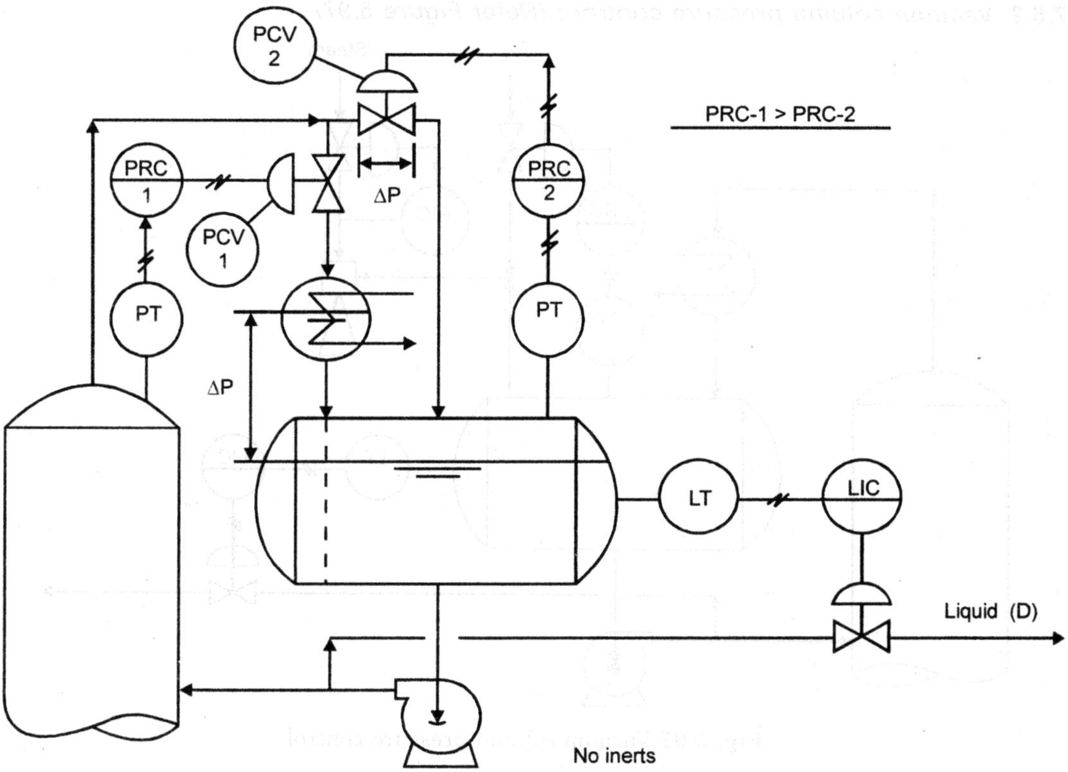

PCV
2
PRC-1 > PRC-2
PRC
1
ΔP
PRC
2
PCV
1
PT
PT
ΔP
LT
LIC
Liquid (D)
No inerts

Fig. 5.95 High-speed column pressure control-no inerts.

Partially
flooded
condenser
PRC
PT
PRC
Inerts
PT
LT
LRC
Liquid
(D)
With inerts

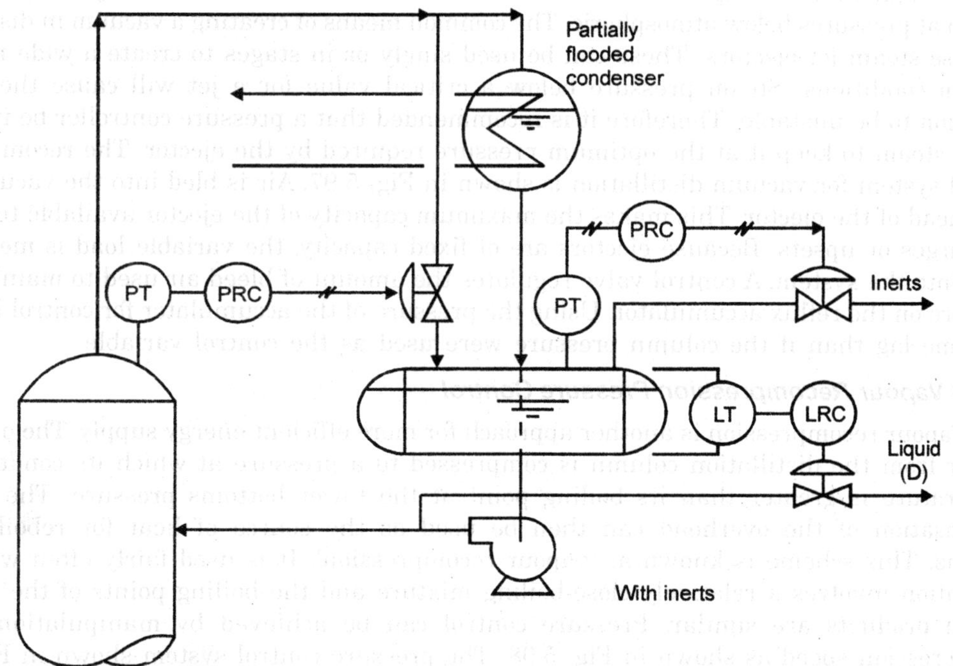

Fig. 5.96 High-speed column pressure control – with inerts.

5.7.6.2 Vacuum column pressure control : (Refer Figure 5.97)

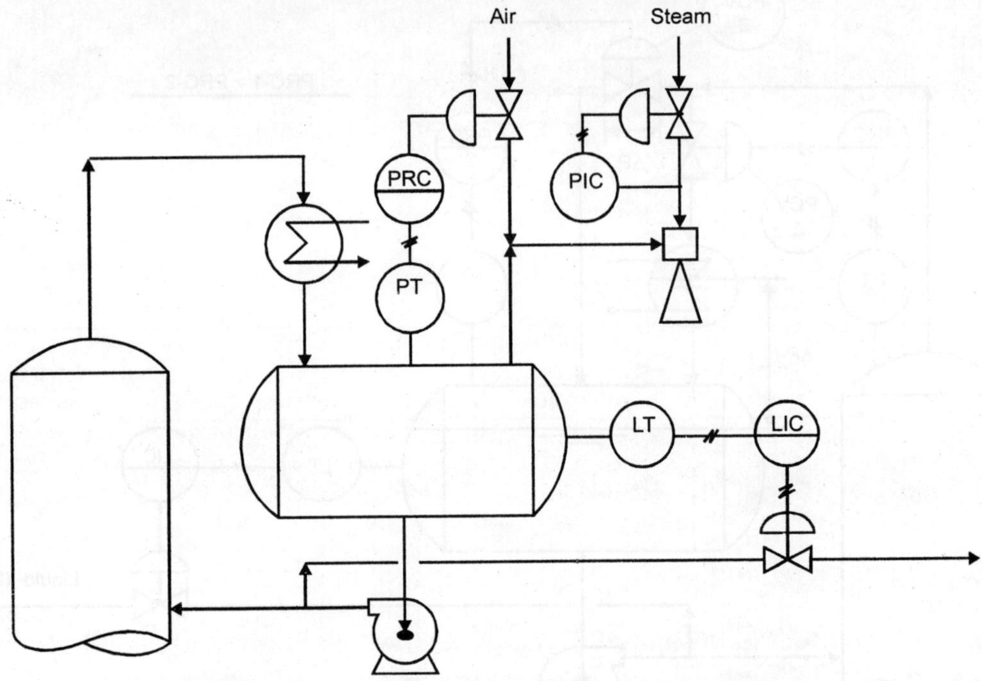

Fig. 5.97 Vacuum column pressure control

For some liquid mixtures, the temperature required to vapourise the feed would need to be so high that decomposition would result. To avoid this, it is necessary to operate the column at pressures below atmospheric. The common means of creating a vacuum in distillation is to use steam jet ejectors. These can be used singly or in stages to create a wide range of vacuum conditions. Steam pressure below a critical value for a jet will cause the ejector operation to be unstable. Therefore it is recommended that a pressure controller be installed on the steam to keep it at the optimum pressure required by the ejector. The recommended control system for vacuum distillation is shown in Fig. 5.97. Air is bled into the vacuum line just ahead of the ejector. This makes the maximum capacity of the ejector available to handle any surges or upsets. Because ejectors are of fixed capacity, the variable load is met by air bleed into the system. A control valve regulates the amount of bleed air used to maintain the pressure on the reflux accumulator. Using the pressure of the accumulator for control involves less time lag than if the column pressure were used as the control variable.

5.7.6.3 Vapour Recompression Pressure Control

Vapour recompression is another approach for more efficient energy supply. The overhead vapour from the distillation column is compressed to a pressure at which its condensation temperature is greater than its boiling point at the tower bottoms pressure. The heat of condensation of the overhead can then be used as the source of heat for reboiling the bottoms. This scheme is known as 'vapour recompression'. It is used fairly often when the distillation involves a relatively close-boiling mixture and the boiling points of the top and bottom products are similar. Pressure control can be achieved by manipulation of the recompression speed as shown in Fig. 5.98. The pressure control system shown in Fig. 5.98

is used often in cryogenic demethanizers. Propylene fractionators make use of vapour recompression via heat pumps. Fig. 5.99 shows the pressure controls for this particular tower.

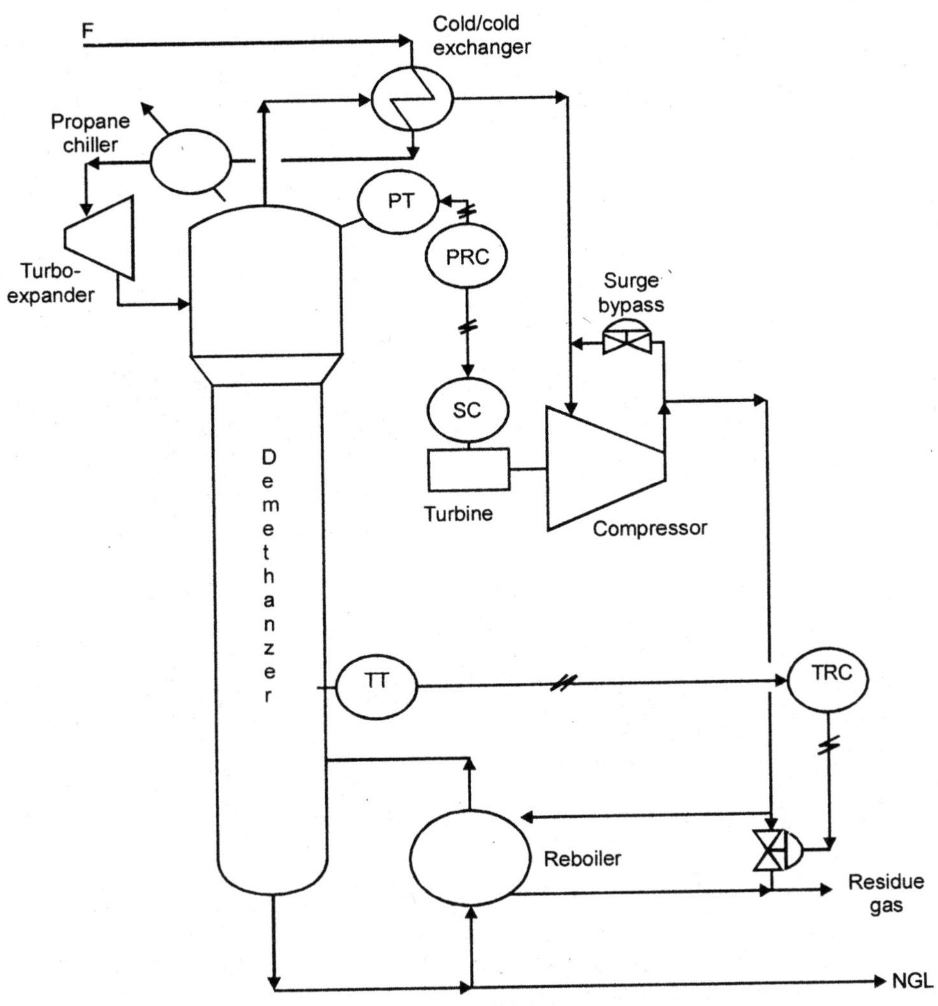

Fig. 5.98 Vapour recompression pressure control.

5.7.7 Feed Flow Rate and Feed Temperature Control

One of the best means of stabilizing operation in almost any continuous flow unit, including distillation, is by holding the flow rates and temperatures constant.

5.7.7.1 Feed Flow Rate Control

A flow controller in the feed line can maintain a constant flow rate. The feed rate to a distillation column is often determined by the flow from upstream processing equipment such as reactors, extraction columns, or other distillation towers. Let us take a case for our discussions where the distillate product of one column is fed to a second column. Any changes that occur to the first column are reflected in the quantity of feed to the second. The flow of products from the previous column is normally controlled by accumulator liquid level controller. An additional flow controller in cascade mode can be introduced as shown in Fig. 5.100 to iron out temporary variations caused by liquid level changes. The flow set point is continuously adjusted by the level controller in a cascade arrangement. The control algorithm for the level controller is usually selected to be non linear to allow the level to float in the surge tank without changing the FRC set point, which would otherwise upset the feed to the column. Therefore, the non linear controller is so configured that as long as the level in the surge tank is between 25% and 75% (middle 50%), the set point to the FRC remains constant. This will allow the surge tank to fulfil its purpose and smooth out the load variations between the related processes. If the level drops below 25% or rises above 75%, the FRC set point is reduced or increased respectively to protect from draining or flooding the tank.

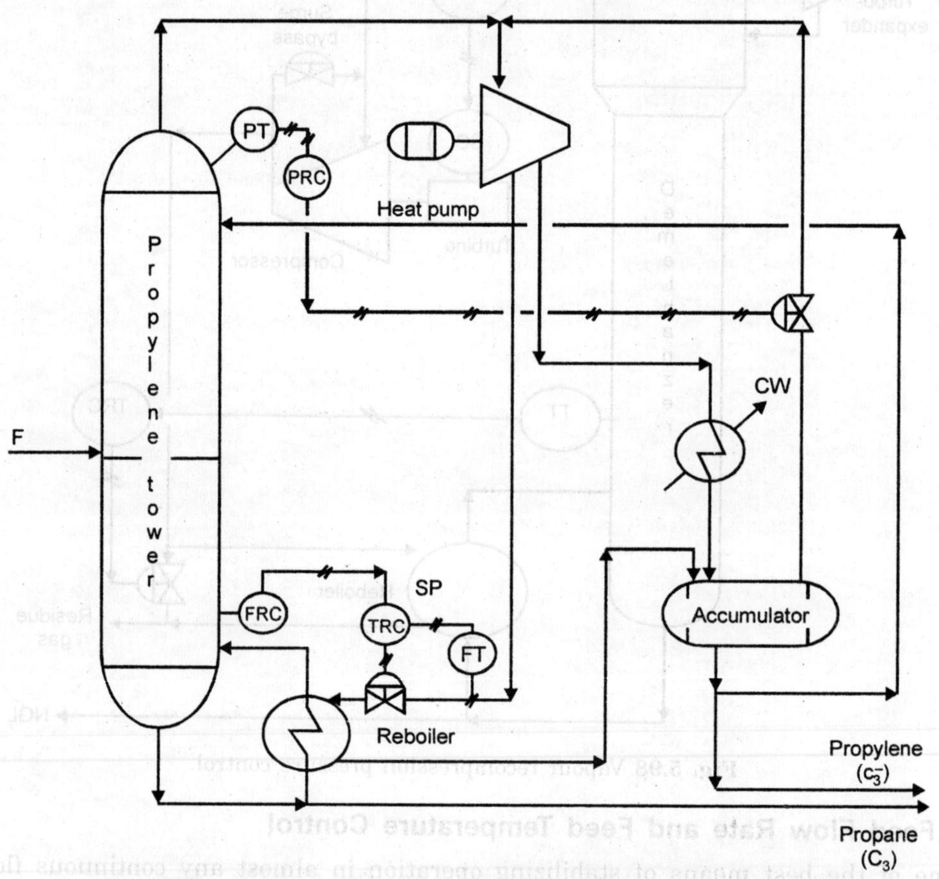

Fig. 5.99 Propylene tower vapour recompression pressure control.

If feed-rate disturbances must be accepted by the column, a feedforward control system as shown in Fig. 5.101 can be used to minimize the impact of the disturbances. The ratio, '*m*', is selected by material balance around the column. The use of adaptive tuning or other non linear level control techniques on the level controller can minimize feed-rate disturbances. The key is to allow the accumulator to utilize its capacity to accommodate transient material balance accumulations and act as a surge drum to minimize feed flow changes to the next unit.

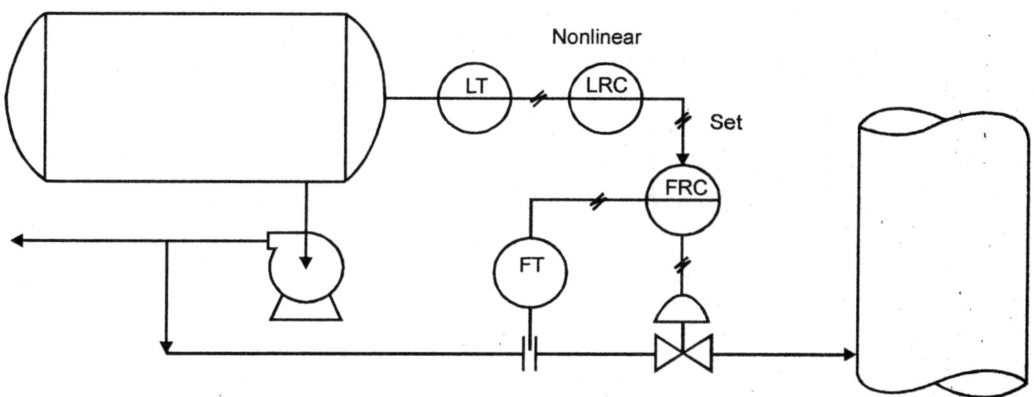

Fig. 5.100 Cascade control of feed to second column.

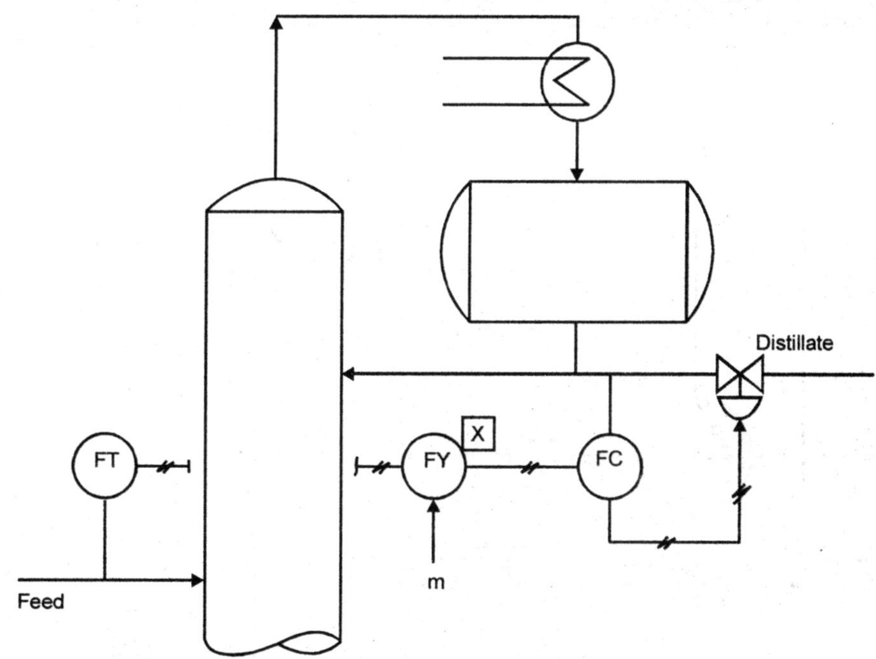

Fig. 5.101 Feedforward control minimizes feed rate disturbances

5.7.7.2 Feed Temperature Control

Automatic control of the feed temperature is a refinement that makes operation of column a little steadier. The temperature of the feed determines how much additional heat must be added to the column by the reboiler. For efficient separation, it is desirable to have the feed at its bubble point when it enters the column. Steam may be used as the source to heat the feed. The use of cascade control loop as shown in Fig. 5.102 can provide superior temperature control. To maintain feed temperature, a three-mode controller (TIC) is generally used.

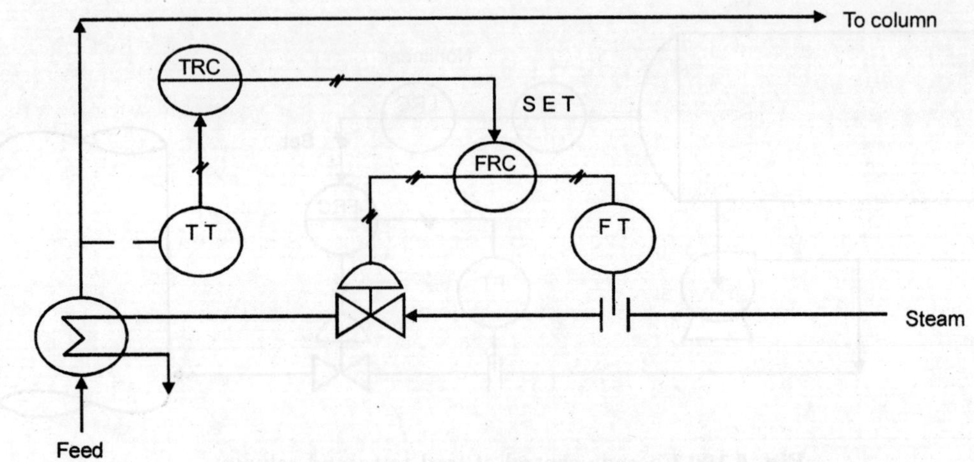

Fig. 5.102 Improved column feed temperature control.

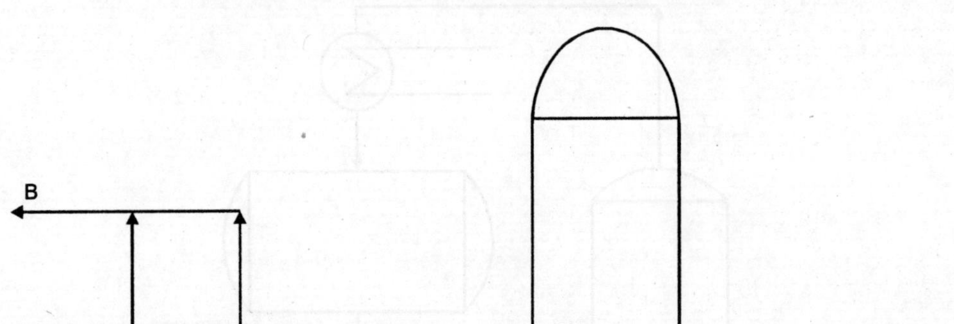

Fig. 5.103 Feed temperature control with economizer.

Another common feed preheating configuration is to use an economizer on the feed stream. An economizer is a heat exchanger designed to take advantage of waste heat in order to preheat the feed. Often, if the bottoms product is to be stored, it must be cooled. Exchanging with the feed stream accomplishes both objectives of feed preheating and product cooling. The temperature control is achieved by manipulation of a bypass valve around the economizer as shown in Fig. 5.103. If heat from the product stream is not sufficient, a second exchanger using steam can be used with the controls as shown in Fig. 5.102.

5.7.8 Benzene Column Control — An Example

The material balance control system is designed for a single purpose: to adjust the plant's production rate at the level desired by the management and keep it at this level. In other words, the material balance control is not designed to maintain the operating variables in certain units at constant values (set points) against changes in various disturbances. This is accomplished by additional control systems called product quality control system. We will see in this section the development of a product quality control system for a benzene column.

Benzene Column Control

Consider the feed as a pseudo-binary mixture composed of benzene (≡ A) and toluene + diphenyl (≡ B). We can specify the following four controlled variables for a binary distillation column for benzene.

1. Benzene product quality.
2. Distillate rate (fractional recovery of benzene in the overhead product).
3. Liquid level in the overhead accumulator.
4. Liquid level at the bottom of the column.

The following four manipulated variables are also available :

1. Distillate flow rate (D)
2. Bottoms flow rate (B)
3. Reflux rate (R)
4. Steam flow rate (S).

The basic disturbances are feed flow rate and composition, and the cooling water temperature in the overhead condenser. Table 5.3 shows all possible control loop configurations for the benzene distillation column.

The alternatives can be screened and the best selected.

Argument-1 : Trying to control the liquid level at the bottom of the column with reflux flow or distillate flow rate involves very long time responses because the action of the manipulated variable must travel the whole length of the distillation column before it is felt by the controlled variable. Therefore the control loop configurations 4, 5, 10, 12, 14, 16, 17, 18, 20, 21, 23 and 24 are ruled out.

Argument-2 : A long time response is also involved when we try to control the level in the overhead accumulator by manipulating the bottoms flow rate for the same reason above. This rules out configurations 1 and 8.

Argument-3 : A long time response, not quite as long as above, is involved when we try to control the accumualtor's level with the steam flow rate. This rules out configurations 2 and 7.

Argument-4 : It is quite complicated to control the distillate composition or flow rate with the bottoms flow rate. It also involves long time responses. Consequently, configurations 6, 11, 13 and 15 are ruled out.

At this point, we realize that only configurations 3, 9, 19 and 22 are left out for further consideration. Which one of the these four will be the best and how to select it ?

Table 5.3 Possible control loop configurations for the benzene distillation column

Configuration number	Benzene product composition	Distillate rate	Overhead accumulator level	Column bottom level	Method for composition control[b]	Comment[a]
1	D	R	B	S	Direct M.B.	(2)
2	D	R	S	B	Direct M.B.	(2)
3	D	S	R	B	Direct M.B.	
4	D	S	B	R	Direct M.B.	(1)
5	D	B	S	R	Mixed	(1)
6	D	B	R	S	Mixed	(3)
7	R	D	S	B	V/F	(2)
8	R	D	B	S	V/F	(2)
9	R	S	D	B	Indirect M.B.	
10	R	S	B	D	Indirect M.B.	(1)
11	R	B	D	S	V/F	(3)
12	R	B	S	D	V/F	(1)
13	B	D	R	S	Mixed	(3)
14	B	D	S	R	Mixed	(1)
15	B	R	D	S	Direct M.B.	(4)
16	B	R	S	D	Direct M.B.	(1)
17	B	S	R	D	Direct M.B.	(1)
18	B	S	D	R	Direct M.B.	(1)
19	S	D	R	B	V/F	(3)
20	S	D	B	R	V/F	(1)
21	S	R	B	D	Indirect M.B.	(1)
22	S	R	D	B	Indirect M.B.	
23	S	B	D	R	V/f	(1)
24	S	B	R	D	V/F	(1)

(*a*) (1) Long time response

(2) Long time response and increasing complexity

(3) V/F control is not desirable in general

(4) Column bottom level control by steam flow is desirable

(*b*) Direct M.B – Direct material balance control

Indirect M.B – Indirect material balance control

V/F – Vapour to feed ratio control

We must note that the foregoing discussion and elimination of 20 out of 24 control structures applies to any distillation column. Which one we will select among the remaining four depends on the characteristics of the particular column we consider. Thus for the benzene column we select configuration 3 as the best, for the following reasons.

1 On a cold day or during a rainstorm the temperature of the cooling water in the overhead condenser drops and the overhead vapour passing through the condenser produces subcooled liquid. When the subcooled liquid returns back (through the reflux flow) to the top tray of the column, it causes less vapour to go overhead. Less vapour overhead causes the liquid level in the accumulator to drop. If the accumulator level is controlled by reflux flow, the latter will decrease. Thus the disturbance caused by the cooling water temperature drop does not propagate down the column in terms of increased liquid overflows. A new equilibrium takes place quickly on the top tray, thus isolating the effect of the disturbance from the rest of the column. Configurations 3 and 19 control the level of the overhead accumulator by the reflux flow rate and thus are retained while 9 and 22 are rejected.

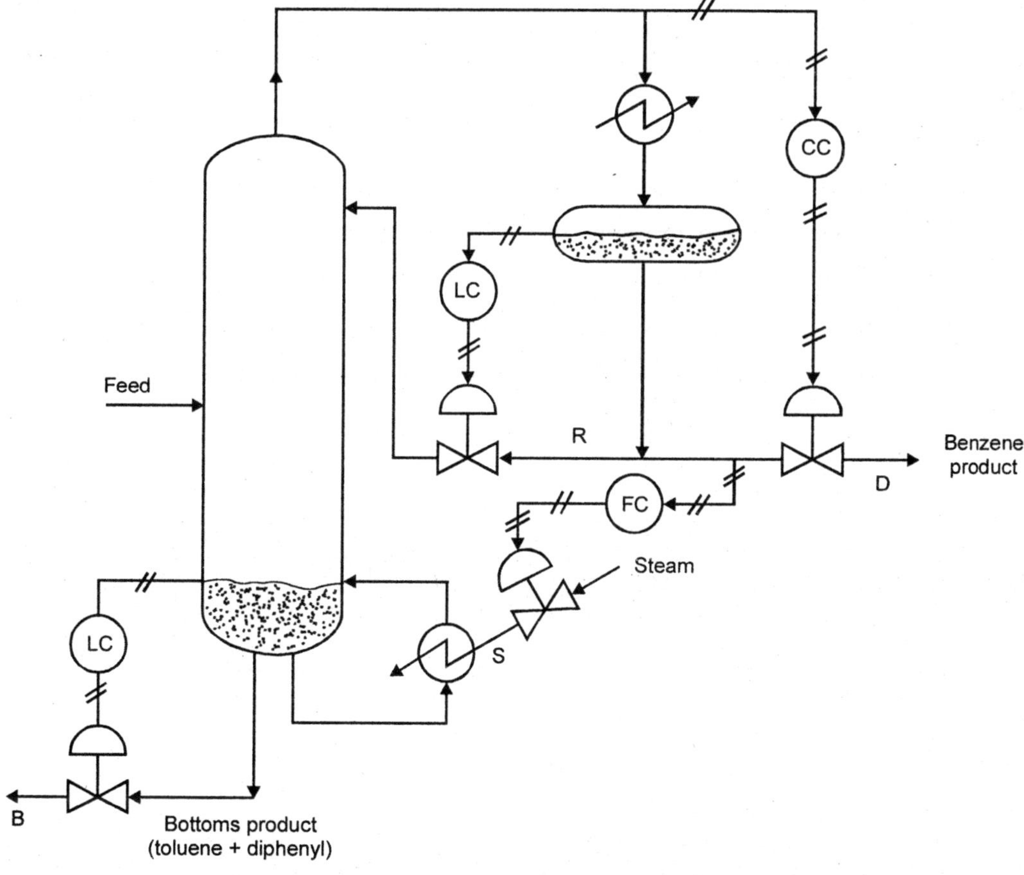

Fig. 5.104 Control loops around a distillation column.

2. In configuration 3, the benzene product composition is controlled directly by the distillate flow, whereas in configuration 19 it is controlled indirectly by the steam rate. Therefore, the configuration 3 is more responsive and is retained as better.

Figure 5.104 shows the four control loops for the benzene column.

In this chapter we have broadly discussed about operation and controls of various industrial units : Boilers, Reactors, Mixers, Dryers, Evaporators, Heat exchangers and Distillation columns. The units considered put together may form only a tip of an ice berg. They are many more industrial, especially chemical industrial, units with complicated and sophisticated controls possible. The process control engineer should keep the following requirements in mind while designing a control system : 1. Safety of men and machine, 2. Product quality, 3. Environmental regulations, 4. Operational constraints, and the last but not the least, 5. economy.

5.8 PROBLEMS AND SOLUTIONS

Problem 5.1

Design a ratio control system for a steam atomized oil burner system for a boiler. Assume the oil pressure range as 0–10 kg/cm^2 and steam pressure range as 0–15 kg/cm^2. The oil to steam pressure ratio requirement is between 1:1 to 1 : 1.5. Assume the atomization control is independent of fuel controller.

Solution :

Given : Range of oil pressure transmitter : 0–10 kg/cm^2

Range of steam pressure transmitter : 0–15 kg/cm^2

Ratio requirement : 1:1 to 1:5

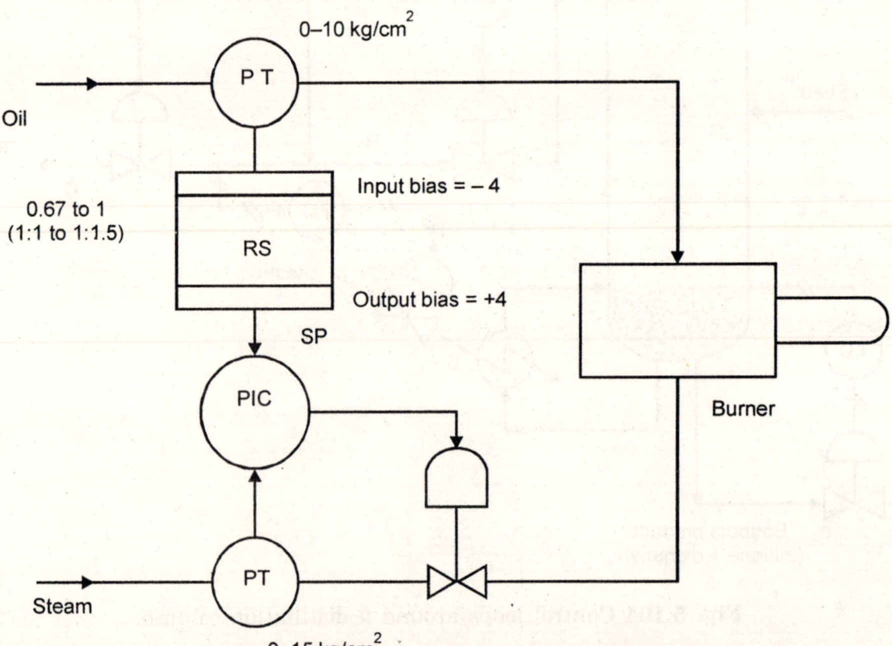

Fig. 5.105 Atomising steam ratio control.

The electrical output of both the transmitters will be 4–20 mA (standard signals)

∴ The ratio station setting for 1:1 = 0.67

for 1:1.5 = 1

[∵ When 1: 1.5 ratio is required (*i.e.*, 10:15), the electrical current output from both the transmitters will be equal (20 mA at the maximum range) and hence the electrical ratio is 1. When the ratio requirement is 1:1, the electrical current output ratio will be 1: 0.67 (output of oil pressure transmitter = 20 – 4/10 = 1.6 mA/Kg/cm^2 and steam pressure transmitter = 20 – 4/15 = 1.07 mA/Kg/cm^2, 1.07/1.6 = 0.67)]

The current ratio of 0.67 to 1 can be comfortably selected. The resulting ratio control system can be drawn as shown in Fig. 5.105.

Let us check the suitability of the circuit for different oil pressures at different ratio set as below :

Case-1 : Ratio is set at 1:1 (Electrical ratio 1:0.67)

Oil pressure kg/cm^2	2.5	5	7.5	10
Transmitter output mA	8	12	16	20
After input bias mA	4	8	12	16
After ratio station mA	4 × 0.67 = 2.68	5.36	8.04	10.72
After output bias mA	6.68	9.36	12.04	14.72
Steam pressure after balance	2.5	5.0	7.5	10
Ratio achieved ≅	1:1	1:1	1:1	1:1

Case-2 : Ratio is set at 1: 1.25 (Electrical ratio 1:0.83)

Oil pressure kg/cm^2	2.5	5	7.5	10
Transmitter output mA	8	12	16	20
After input bias mA	4	8	12	16
After ratio station mA	3.32	6.64	9.96	13.28
After output bias mA	7.32	10.64	13.96	17.28
Steam pressure after balance	3.11	6.23	9.34	12.46
Ratio achieved ≅	1.25	1.25	1.25	1.25

Case-3 : Ratio is set at 1:1.5 (Electrical ratio 1:1)

Oil pressure kg/cm^2	2.5	5	7.5	10
Transmitter output mA	8	12	16	20
After input bias mA	4	8	12	16
After ratio station mA	4	8	12	16
After output bias mA	8	12	16	20
Steam pressure after balance	3.75	7.5	11.25	15
Ratio achieved ≅	1.5	1.5	1.5	1.5

Oil pressure transmitter output becomes set point to PIC after passing through input bias, ratio station and output bias. Assuming PIC controls the oil pressure as per set point, the ratio achieved is tabulated.

Problem 5.2

Design a parallely operated fuel gas-air control system for a boiler or reboiler incorporating safety measures.

Solution :

Let us first analyse the meaning of safety measures here. In a temperature controlled combustion chamber or furnace the following conditions are to be met as safety measures.

1. Whenever the controller demands more heat energy, air should increase first followed by gas to avoid explosions which can occur when gas increases first and air is not able to follow due to some reasons.
2. Whenever the controller requires to cut down heat energy, gas should decrease first followed by air for the same reason as well as for better temperature control.

A schematic to satisfy the above condition is shown below :

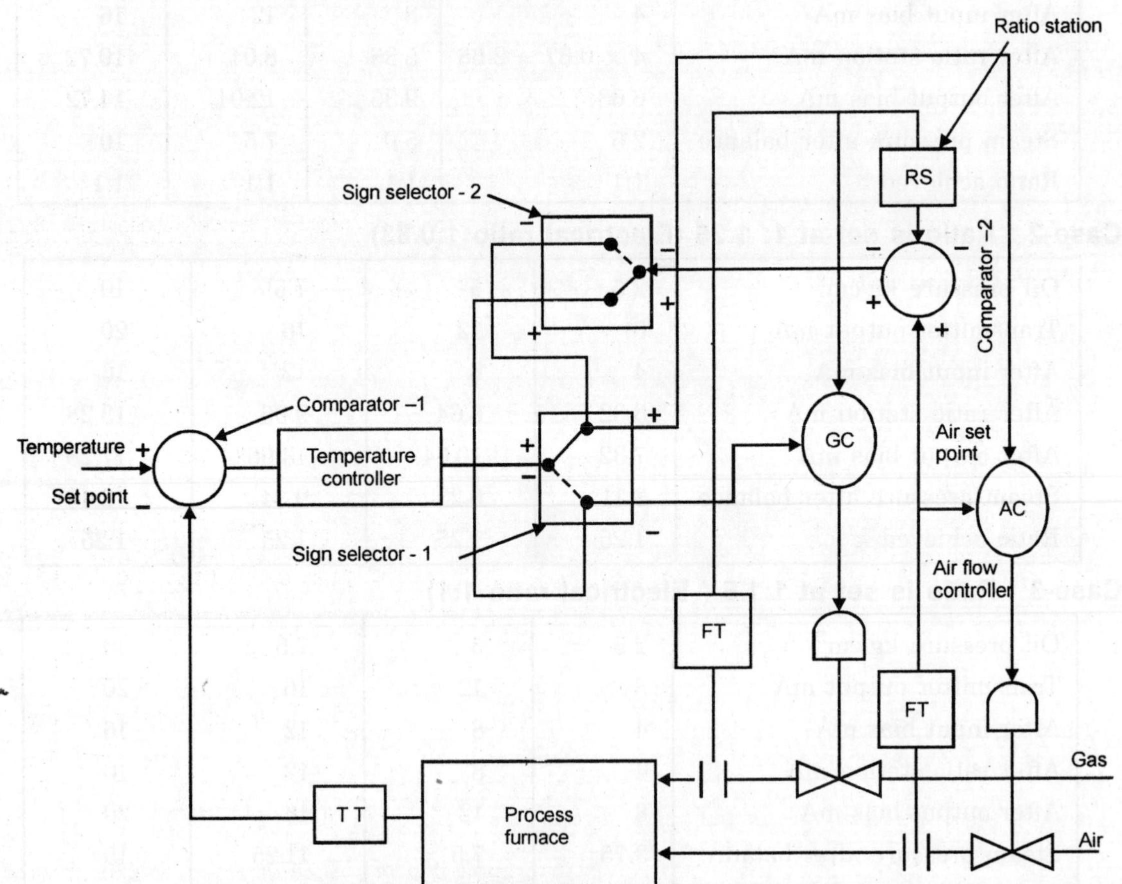

Fig. 5.106 Cross connected fuel-air control system.

The working of the above control system can be explained as follows :

Case-1 : Furnace temperature is less than the set point: It requires more heat energy. As per our condition the air should increase first and then gas to satisfy the heat requirement. As set point is more than furnace temperature, comparator-1 and hence temperature controller gives positive signal. The sign selector-1 selects the signal and sends it to air flow controller to increase the air flow. The increase in air flow makes the comparator-2 to give positive signal to sign selector-2 to connect the signal to gas controller to increase gas flow. The ratio between air and gas is thus maintained with the help of ratio station.

Case-2: Furnace temperature is more than the set point: It requires less heat energy. The reverse happens now to decrease gas first and then air.

At balance the fuel gas/air ratio is maintained for proper combustion as required.

Problem 5.3

Design an inferential distillate composition control system for a binary distillation column separating a binary mixture of pentane and hexane into two product streams of pentane (distillate) and hexane (bottoms). Control objective is to maintain the production of the distillate stream with 95 mole % pentane in the presence of changes in the feed composition.

Solution :

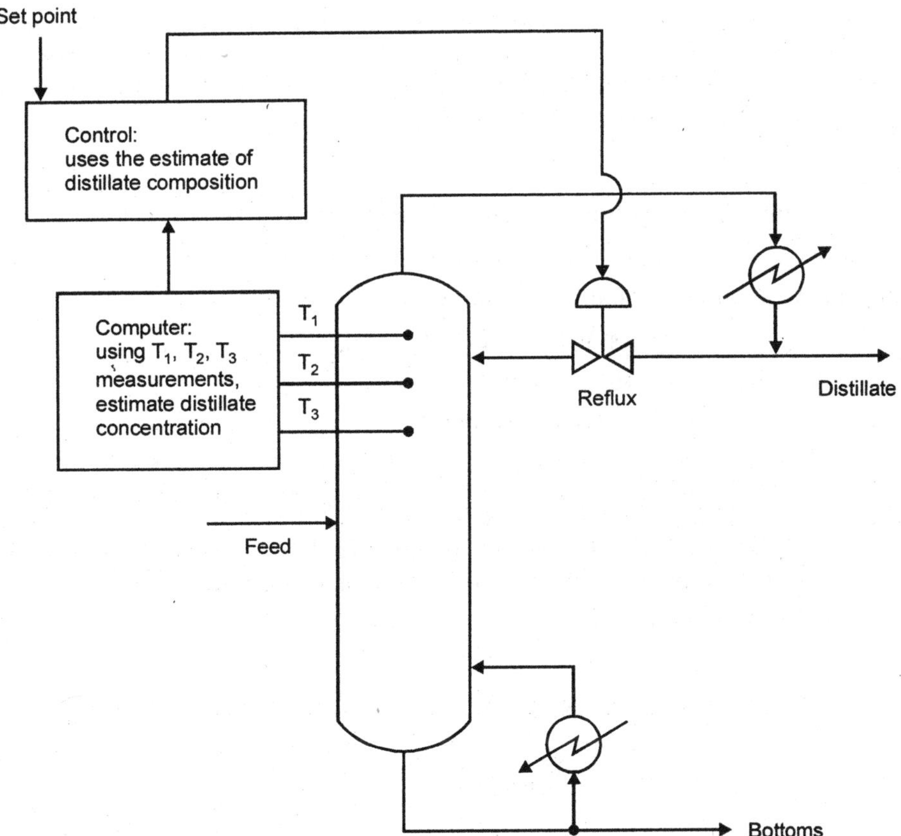

Fig. 5.107 Inferential-Distillation composition control of a simple distillation column.

It is clear that the reflux flow rate into the column has to be the manipulated variable. It is possible to measure the composition straight away and use a feedback control loop to adjust the reflux flow ratio to maintain the composition in the distillate. A feed forward control can also be used by measuring composition in the feed and control reflux ratio to maintain composition in the distillate. Both the systems depend on the composition analysers. Such analysers are normally very costly, low reliable and maintenance intensive. Hence , many a times it is preferred to go for inferential control system.

We know that the measurement of temperature of the liquid at various trays along the column will help to infer the composition. The temperature measurements are usually simple and reliable. Using material and energy balances around the trays of the column and the thermodynamic equilibrium relationships between liquid and vapour streams, we can develop a mathematical relationship that gives us the composition of the distillate if the temperature of some selected trays are known.

Figure 5.107 shows such a control scheme that uses temperature measurements (secondary measurements) to infer the composition of pentane in the distillate.

It is possible to use analyzer for verification and updation of the estimate in the inferential block or computer.

Problem 5.4

Develop a suitable control schematic for a mixing station where Gas 'A' and Gas 'B' are to be mixed to get Gas 'C'. The requirement is to get variable volume of gas 'C' but with constant calorific value of 3000 Kcals/Nm^3. [Assume calorific value of Gas 'A' and Gas 'B' as 4000 Kcals/Nm3 and 1000 Kcals/Nm^3 respectively.]

Solution :

Let V_A, V_B and V_C are volume of gases *A*, *B* and *C*.

and C_A, C_B and C_C are calorific values of gases *A*, *B* and *C*.

At any point of time the following equation is valid.

$$V_A C_A + V_B C_B = V_C C_C \tag{5.32}$$

As calorific values are already known,

$$4000\ V_A + 1000\ V_B = 3000\ V_C$$

or

$$4V_A + V_B = 3\ V_C \tag{5.33}$$

Controlled variable is V_C and there are two manipulated variables possible (V_A and V_B). We may keep one fixed and the other varied. What are the limitations?

The other finding equation is

$$V_A + V_B = V_C \tag{5.34}$$

Solving equations 5.33 and 5.34 for V_A and V_B in terms of V_C we get,

$$V_A = \frac{2}{3} V_C$$

$$V_B = \frac{1}{3} V_C$$

Hence we may develop a simple ratio controller loop to satisfy our requirement as below :

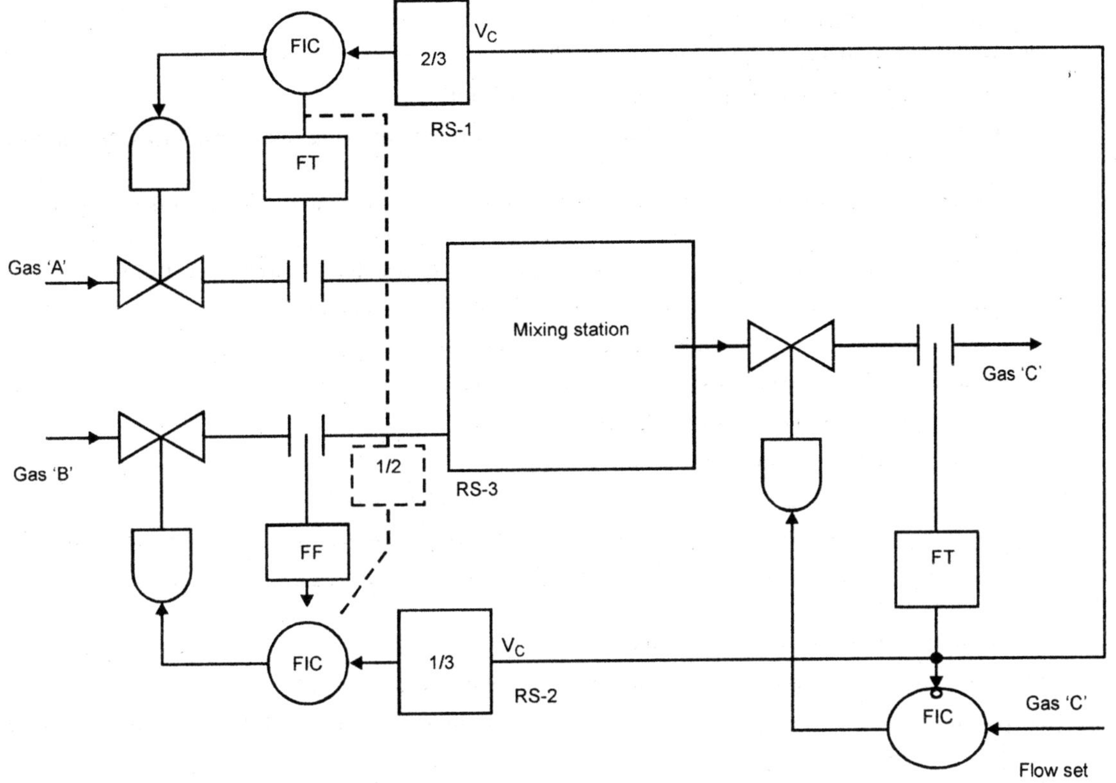

Fig. 5.108 Mixing station control - problem 5.4

Alternatively,

$$V_B = \frac{1}{3} V_C = \frac{1}{3}\left(\frac{3}{2}\right) V_A.$$

$$V_B = \frac{1}{2} V_A.$$

Instead of RS-2, we may connect RS-3 with a ratio ½ taking V_A signal (shown in dotted lines).

The circuit needs no more elaboration.

5.9 PROBLEMS AND QUESTIONS

1. Name the analytical instruments used for the composition analysis in distillation column.
2. Explain the term 'capacity' of an evaporator.
3. What is meant by counter-current operation of a multi effect evaporator ?
4. With neat sketches, explain the cascade control strategy used for the control of temperature in an exothermic chemical reactor. What are the limitations of cascade control?

5. Discuss in detail the feed forward control strategies used in distillation column control.
6. With neat sketches, discuss the control of 'temperature' and 'pressure' in condensers.
7. Name some of the continuous dryers.
8. What is vapouriser ? Explain.
9. Name four main equipment in a distillation unit.
10. Explain the control scheme for product quality control by inferring composition from temperature in a distillation process.
11. Explain the temperature control system using three way valve in a liquid to liquid heat exchanger.
12. Explain with diagram the application of cascade control for a batch type fluid-bed dryer.
13. Name two disadvantages of 'Once-through' cooling in a reactor.
14. Define 'bubble point' and 'dew point' in connection with distillation process.
15. Differentiate 'co-current' and 'counter-current' dryers.
16. Name the four zones in a drying process.
17. Name atleast four important heat exchangers.
18. Name the three ingredients of a feed forward control system.
19. Explain the humidity corrected feed forward control of a continuous fluid bed dryer.
20. Explain the application of selective control in an otherwise cascade control for a two-effect evaporators.
21. Name six different types of evaporators.
22. Draw the graph showing different zones of drying process.
23. What is the relation between the amount of heat transfer, solid temperature and air temperature ?
24. Write the factors on which the drying characteristic of a material depends on.
25. Why is it preferred to operate the dryers on temperature based inferential systems ?
26. What is meant by non-linearity in process gain ? How to overcome this in a control loop?
27. What is meant by 'adaptive gain' ?
28. Explain 'multipurpose' systems with respect to heat exchanger.
29. Explain the principle of operation and control systems used with respect to vacuum dryer.
30. Explain the control scheme for turbo dryer.
31. How spray dryer is different from other dryers ? Explain with control schemes.
32. Explain 'recompression' process.
33. Why inerts (non condensables) should be vented out ?
34. Explain a control system using multiple coolants for temperature control in a reactor .
35. Describe the column pressure control in cases of liquid and vapour distillates with and with negligible inerts present.
36. Consider the distillation systems shown in Fig. 5.109 (*a*) and (*b*). The columns are thermally uncoupled in the first, whereas in the second they are thermally coupled (*i.e.*, the overhead of the first column is cooled by the bottoms of the second column). For which system is the control system easier ? Elaborate your answer.
37. Identify a control system for the thermally coupled distillation column [Fig. 5.109 (*b*)]. See that the control system covers the start up of the plant.

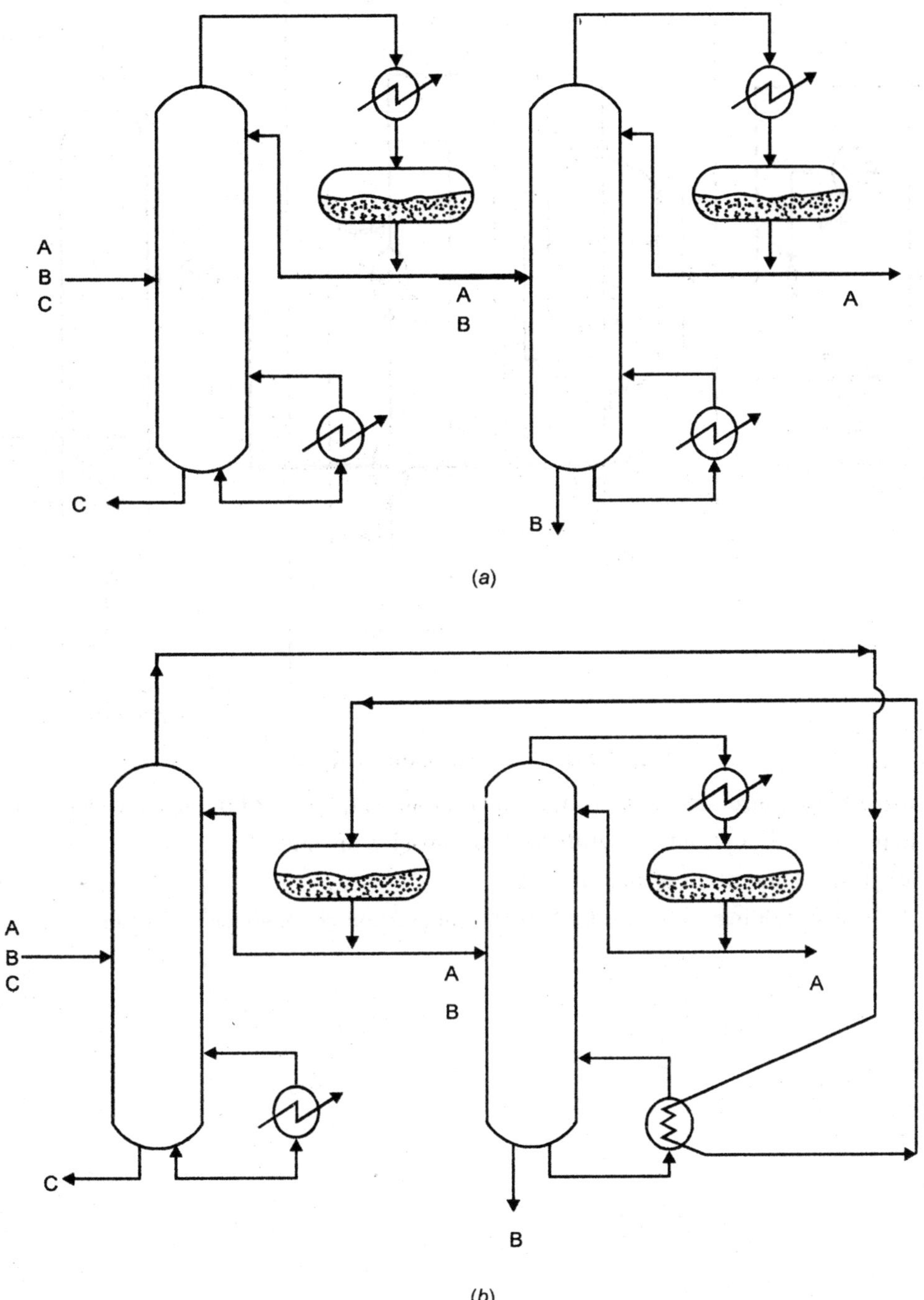

Fig. 5.109 For (Q. 36 and 37)

38. Generate the control loop configuration for a simple chemical process shown in Fig. 5.110.

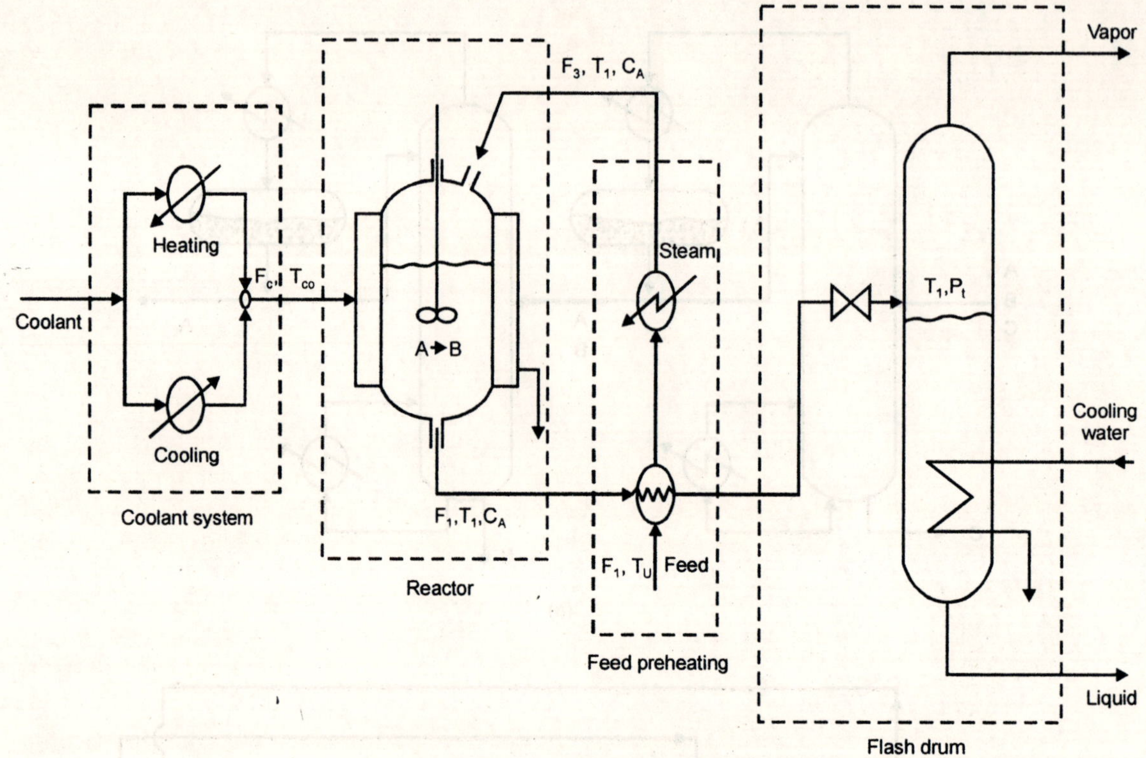

Fig. 5.110 Chemical plant of Q : 38

39. Explain the control scheme for bottom product composition control through reboiler temperature.

40. Explain the role of induced draft fan and forced draft fan :

 (*a*) In the combustion of fuel in a boiler

 (*b*) In maintaining balanced draft in the combustion chamber of the boiler.

6

SIGNAL CONDITIONING

6.1 INTRODUCTION

For a successful process control, the parameter which is to be controlled should be sensed by a suitable sensor, converted into a signal which will truly represent the parameter and presented to the controller for further action. At this juncture one should clearly understand the following terms namely: (1) Sensor, (2) Transducer, and (3) Signal conditioner.

6.1.1 Sensor

A sensor is an element which detects a change. It simply changes its character when an input quantity undergoes a variation. For example, in a strain gauge, a change in strain causes a change in resistance.

A sensor can also be defined as a device that is used to sense or detect any particular change in the quantity that is being measured and makes an appropriate change in its output.

6.1.2 Transducer

A transducer produces an output signal which is a function of the measured quantity. The input signal quantity may be temperature, pressure, flow, speed, vibration etc. The output signal may be a voltage, current or air pressure.

In other words, a transducer senses the physical variable to be measured (i.e. measurand) and converts it to a suitable signal. Transducer can also be defined as a device which converts one form of energy into another form of energy.

6.1.3 Sensor vs Transducer

One should understand clearly the relation between a sensor and a transducer. For example; a strain gauge connected in a Wheatstone bridge converts strain into voltage. Here the strain gauge is called a sensor and the strain gauge along with Wheatstone bridge is called a transducer. A change in strain causes a change in resistance in a strain gauge. In order to convert this change into a voltage variation, a bridge circuit and voltage source are needed. Strain gauge, bridge circuit and the voltage source put-together is a transducer. The transducer is something more than a sensor. Transducer is a bigger set, in which sensor is a part or subset. Sometimes they are one and the same as one set. For example, a thermocouple is a sensor and also a transducer. It might be noted here that some authors use these two words namely sensor and transducer inter-changeably. However, the word sensor is preferred for the initial measurement device because transducer represents a device that converts any signal from one form to another. In other words, all sensors are transducers, but not all transducers are sensors.

6.1.4 Signal Conditioner

The output signal from the transducer normally contains interference requiring further processing by the measurement system. Many a times it becomes necessary to perform certain operations on the signal before it is transmitted further. This process is done by means of signal conditioning. The process of converting the transducer signal to usable format is known as signal conditioning. The signal conditioning of the transducer output signal may include one or more of the following:

1. Amplification
2. Demodulation/Modulation
3. Filtering
4. Impedance matching
5. Linearisation
6. Analog to digital conversion
7. Signal-level and/or bias changing
8. Attenuation
9. Integration /Differentiation
10. Addition/Subtraction

The signal conditioner performs one or more such tasks. Since electrical signals have distinct advantages in this respect, more so with the development of electronics, a signal conditioner is now basically an electronic gadget. For remote measurements, signal transmitters are necessary in addition to signal conditioner.

6.1.5 Functional Block Diagram of a Measurement System

The various functions of a measurement system can be represented by a block diagram as shown in Fig. 6.1.

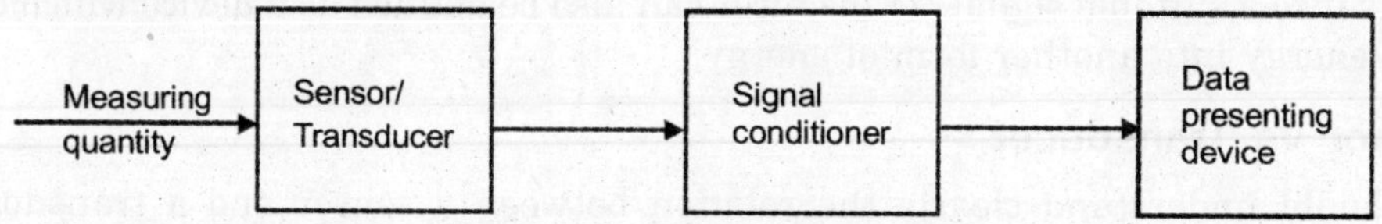

Fig. 6.1 Functional block diagram of a measurement system

As the first two blocks are already discussed, the last block needs some discussions here. The last block namely the data presenting device presents the measured quantity to the observer. The data can be presented by any one of the following ways:

1. Analog indication by means of deflection of a pointer.
2. Digital indication using displays like LCD, LED etc.
3. Recording the variation of signal as a function of time.
4. Totalising/integrating the signal over a period of time.
5. Interfacing with a computer or controller for further processing.

6.2 ANALOG SIGNAL CONDITIONING

As pointed out in Section 6.1, signals are produced by transducers/transmitters. The transducers are of two types—active and passive. Active transducers generate signals themselves and hence no external source of energy is necessary to excite them. Thermocouple belongs to this category. Active transducers work normally on the following principles of operation: (*i*) Thermoelectric, (*ii*) Piezoelectric, (*iii*) Photovoltaic, (*iv*) Electromagnetic and (*v*) Galvanic.

Passive transducers, on the other hand, do not generate any energy. They need be excited by the application of electrical energy from outside. The extracted energy from the measurand produces a change in their electrical state which can be measured. Photo resistor belongs to their category. Passive transducers work normally on the following principles of operation: (*i*) Resistive, (*ii*) Inductive, (*iii*) Capacitive, (*iv*) Magneto resistive, (*v*) Hall effect based, (*vi*) Photoconductive, (*vii*) Thermo resistive and (*viii*) Elasto resistive.

Signal conditioning refers to operations performed on signals from transducers (Active or Passive) to convert them to a form suitable for interfacing with other elements in the process control loop. In this section, we will be dealing only with analog conversions, where the conditioned output is still an analog representation of the variable. Even in applications involving digital processing, some type of analog conditioning is usually required before analog-to-digital conversion is made. Analog signal conditioning involves many processes which can be broadly divided into two categories namely linear processes and non-linear processes as shown in Table 6.1.

Table 6.1 Two categories of signal conditioning processes

Linear processes	Non-linear processes
(*a*) Amplification	(*a*) Modulation
(*b*) Attenuation	(b) Demodulation
(*c*) Integration	(*c*) Sampling
(*d*) Differentiation	(*d*) Filtering
(*e*) Addition	(*e*) Clipping and clamping
(*f*) Subtraction	(*f*) Squaring
	(*g*) Linearisation
	(*h*) Multiplication by another function

6.2.1 Principles of Analog Signal Conditioning

Analog signal conditioning provides the operations necessary to transform a sensor/transducer output into a form necessary to interface with other elements of the process control loop. We often describe the effect of the signal conditioning by the term 'Transfer Function'. By this term we mean the effect of the signal conditioning on the input signal. Thus, a simple voltage amplifier has a transfer function of some constant that, when multiplied by the input voltage, gives the output voltage. Some general types of signal conditioning practically adopted are discussed in this section.

6.2.1.1 Signal Level and Bias Changes

Adjusting the level (magnitude) and bias (zero value) of some voltage representing a process variable is one of the most common types of signal conditioning. For example, some sensor output varies from 0 to 80 mv as process variable changes over a measurement range. However, the receiving equipment like recorders, indicators and controllers may require a voltage that varies from 1*V* to 5 *V* (A standard voltage range accepted internationally) for the same variation of the process variable. Zero shifting or a bias adjustment is required to convert '*O*' input to 1*V* output. Amplification has to be done to increase the 80 mv span input to 4*V* span output. The process is represented in Fig. 6.2.

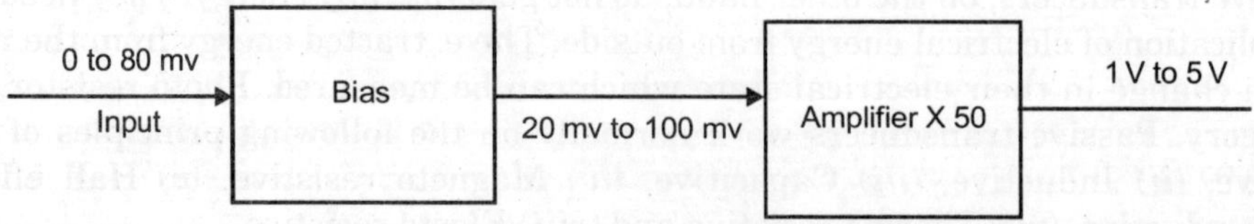

Fig. 6.2 Signal level and bias changes

6.2.1.2 Linearisation

Many a times the characteristics of the sensor/transducer will be non-linear. The non-linearity exists between the process variable (input) and the sensor output. Even if they are approximately linear, they may present problems when precise measurements of the variable and controls are required. The purpose of linearisation is to provide an output that varies linearly with process variable even if the sensor output does not.

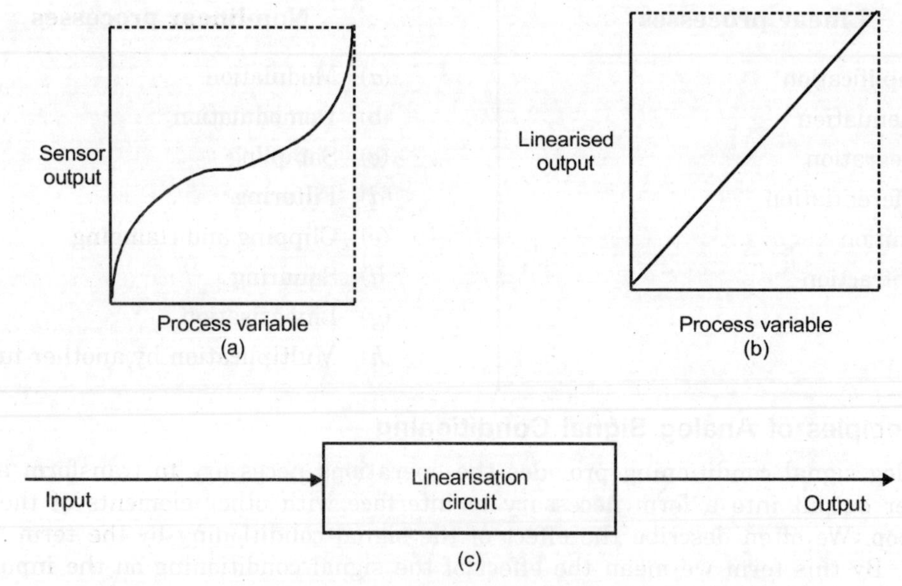

Fig. 6.3 Linearisation

Figure 6.3 indicates symbolically a linearisation circuit. The sensor output, which is non-linear with the process variable as shown in Fig. 6.3(a), is conditioned in the linearisation circuit shown in Fig. 6.3(c) so that a linear output is produced which is linear to process variable [Refer Fig. 6.3(b)]. Such circuits are difficult to design and usually operate only within narrow limits. The modern approach to this problem is to provide the non-linear signal as input to computer and perform the linearisation using software. Virtually any non-linearity can be handled in this manner and, with the speed of modern computers, in nearly real time.

6.2.1.3 Conversions

Signal conditioning is used very often to convert one type of electrical variation into another. It is necessary to provide circuits for a large class of sensors which exhibit changes of resistance with changes in a dynamic variable to convert the resistance changes either to current or voltage signals. Passive circuits like bridge and divider are used generally for this purpose.

Nowadays, the process control standard demands 4–20 mA signals for transmitting and 1–5 *V* for receiving. Thus, voltage-to-current and current-to-voltage converters are often required. Also the use of computers in process control requires conversions from analog data into digital data (ADCs) and vice versa (DACs).

6.2.1.4 Filtering

Spurious signals of considerable strength are always expected in the industrial environment. Line frequency signals, motor start transients, induced signals from nearby power cables etc. are the unwanted signals in the process control loop. In many such cases, it is necessary to use high-pass, low-pass or notch filters to eliminate unwanted signals from the loop. Such filtering can be accomplished by passive filters using only resistors, capacitors, and inductors, or active filters, using gain and feedback.

6.2.1.5 Impedance Matching

Impedance matching is an important element of signal conditioning when transducer internal impedance or line impedance can cause errors in measurement of a dynamic variable. Both active and passive networks are employed to provide such matching.

6.2.1.6 Concept of Loading

The loading of one circuit by another is one of the most important concerns in analog signal conditioning. A display device draws energy from the measuring circuit itself, thus often loading the circuit and lowering the value of the measurand. Also in many applications the transducers do not generate enough power to drive display devices. Hence is the necessity of instrumentation amplifiers which generate enough power to drive display device on one hand and protect the measurand from being loaded on the other. We may discuss the effect of loading with a transducer as input to a signal conditioner, say an amplifier in this case and display devices as shown in Fig. 6.4.

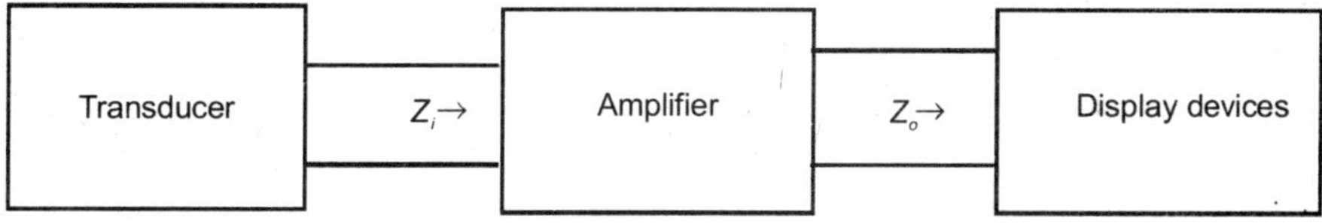

Fig. 6.4 Concept of impedance matching

The desirable properties of the amplifier in this case are:

1. The amplifier should have a high input impedance (ideally infinity) so that the transducer is not loaded. At least it should be more than ten times that of transducer output impedance.

2. The amplifier should exhibit very low output impedance (ideally zero) in order to minimize loading of the amplifier by the subsequent display device.

Quantitatively, we can evaluate loading as follows. According to Thevenin's theorem, the output terminals of any two terminal elements can be defined as a voltage source in series with an output impedance. Assuming the output impedance of the transducer is a resistance R_x, the Thevenin equivalent circuit for the transducer can be shown as in Fig. 6.5 with V_x as voltage source. V_y being the output voltage connected to load R_L. R_L could be the input resistance of the amplifier. A current will flow, and voltage will be dropped across R_x. The loaded output voltage V_y will thus be given by

$$V_y = V_x\left[1 - \frac{R_x}{R_x + R_L}\right] \qquad \text{...(6.1)}$$

The voltage that appears across the load (i.e., input impedance of the amplifier) is reduced by the voltage dropped across R_X (internal resistance of the transducer). It is clear from the equation that the loading influence can be reduced by making $R_L >> R_X$. The same explanation holds good for amplifier output impedance and display device's input impedance.

If the electrical quantity of interest is frequency or a digital signal, then loading is not such a problem. Loading is important mostly when signal amplitudes are important.

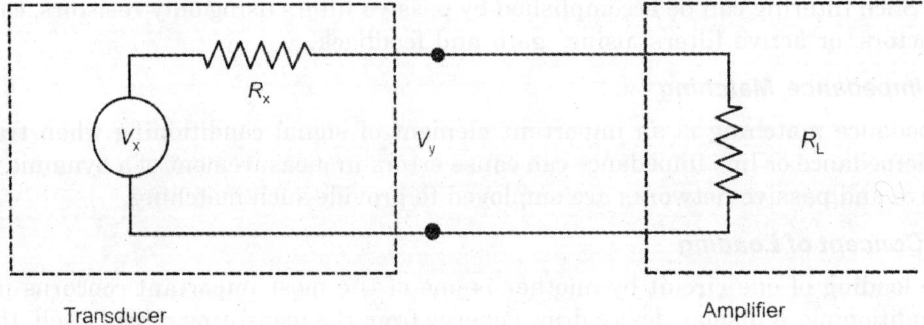

Fig. 6.5 Thevenin equivalent circuit of a transducer

6.2.2 Signal Conditioning Using Passive Circuits

Passive techniques like divider, bridge and filter circuits are extensively used for signal conditioning for many years. Although modern active circuits often replace these techniques, there are still many applications where their particular advantages make them useful.

6.2.2.1 Divider Circuits

The elementary voltage divider shown in Fig. 6.6 can be used to provide conversion of resistance variation into a voltage variation as per the relationship given below.

$$V_D = \frac{V_S}{R_1 + R_2} R_2 \qquad \text{...(6.2)}$$

Where V_S = Supply voltage

R_1 = Divider resistor

R_2 = Sensor resistor (Varies with measured variable)

V_D = Divider output voltage

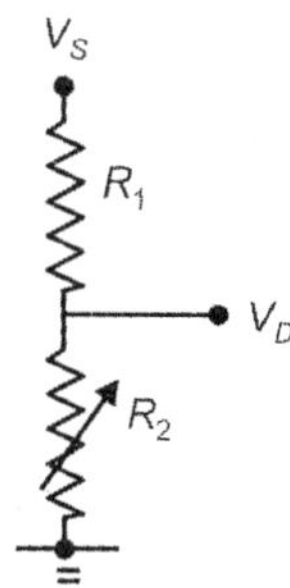

Fig. 6.6 Voltage divider

It is important to consider the following issues when using a divider for conversion of resistance to voltage variation:

1. The divider voltage V_D will not vary linearly even if R_2 varies linearly with measured variable.
2. The effective output impedance (parallel combination of R_1 and R_2) may not necessarily be high. Hence loading effects must be considered.
3. As current flows through both R_1 and R_2, the power rating of both R_1 and sensor (R_2) must be considered.

6.2.2.2 *Bridge Circuits*

To convert impedance variations into voltage variations, bridge circuits are used. For increased sensitivity, amplification can be used to increase the voltage level of the bridge output.

Wheatstone Bridge: Wherever a sensor changes resistance with process variable changes, this simplex and most common bridge is used. Refer Fig. 6.7 for the basic DC Wheatstone bridge. The voltage difference or voltage offset across terminals 'a' and 'b' can be expressed as below assuming the impedance of the detector D as infinity·

$$\Delta V = \frac{VR_3}{R_1 + R_3} - \frac{VR_4}{R_2 + R_4} \text{ or } \Delta V = V\left[\frac{R_3 R_2 - R_1 R_4}{(R_1 + R_3)(R_2 + R_4)}\right] \quad ...(6.3)$$

The bridge is said to be in 'null' condition when $\Delta V = 0$ [The voltage across the detector D is zero]. This occurs when $\frac{R_1}{R_2} = \frac{R_3}{R_4}$ or $R_1 R_4 = R_2 R_3$. Thus the application of Wheatstone bridges to process control applications using high input impedance detectors is possible.

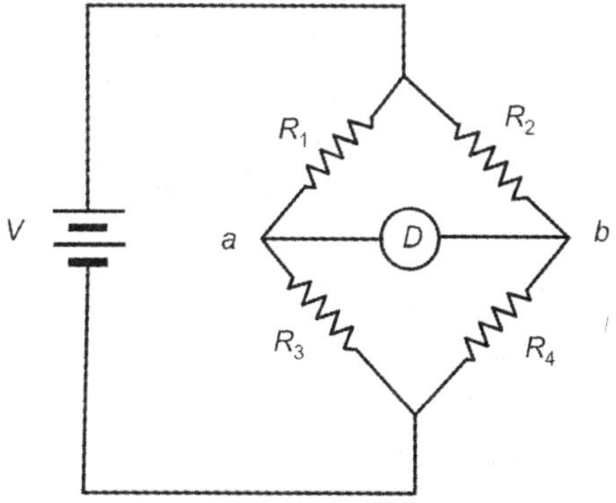

Fig. 6.7 Wheatstone bridge

Galvanometer Detector

When a galvanometer is used for a null detector, it is convenient to use the Thevenin equivalent circuit of the bridge as shown in Fig. 6.8.

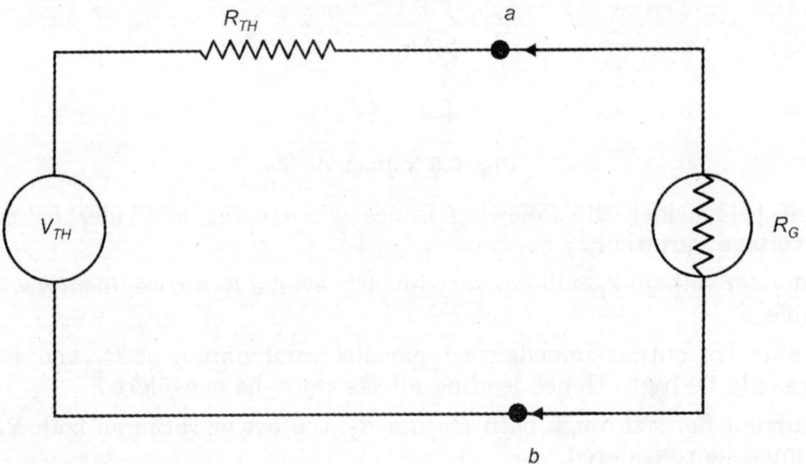

Fig. 6.8 Thevenin equivalent of the bridge with galvanometer as detector

$$V_{TH} = V\left[\frac{R_3R_2 - R_1R_4}{(R_1+R_3)(R_2+R_4)}\right] \quad ...(6.4)$$

$$R_{TH} = \frac{R_1R_3}{R_1+R_3} + \frac{R_2R_4}{R_2+R_4} \quad ...(6.5)$$

The offset current (The current through the galvanometer with internal resistance R_G) is given by:

$$I_G = \frac{V_{TH}}{R_{TH}+R_G} \quad ...(6.6)$$

Bridge resolution is a function of the resolution of the detector used to determine the bridge offset.

Lead Compensation

In many process-control applications, a bridge circuit may be located at considerable distance from the sensor whose resistance changes are to be measured. In such cases the lead resistance changes may cause problems. To avoid such problems 'lead compensation' is used, where any changes in lead resistance are introduced equally into both arms of the bridge circuit, thus causing no effective change in bridge offset. One such compensation circuit is shown in Fig. 6.9.

R_4 being the sensor and lead (3) being power lead, leads (1) and (2) are exposed to the same environment and hence changes are expected to be same. The virtual effect on the offset will get compensated and for all practical purposes it can be assumed as negligible.

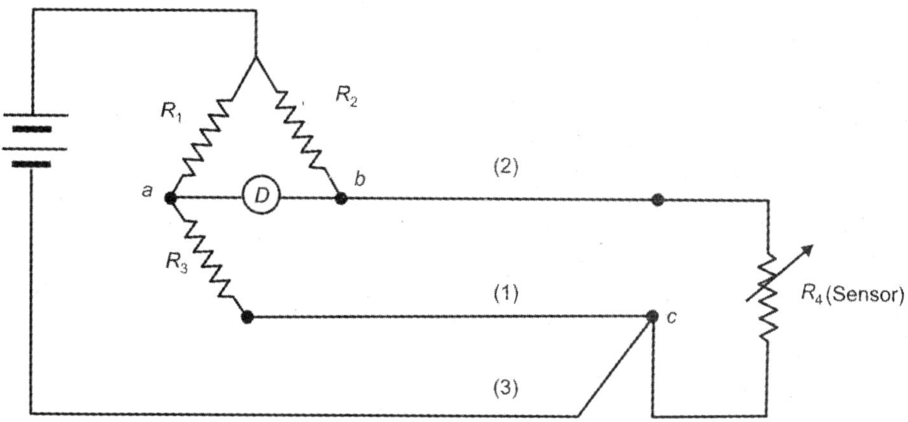

Fig. 6.9 Lead compensation circuit

Current Balance Bridge

In the past, many process control applications used a feedback system in which the bridge offset voltage was amplified and used to drive a motor whose shaft altered a variable resistor to renull the bridge. Such a system does not suit the modern technology of electronic processing. A technique that provides for an electronic nulling of the bridge and that uses only fixed resistors can be used with the bridge. The method uses a 'current' to null the bridge. The standard Wheatstone bridge can be modified for this purpose by modifying one of the arms resistor into two, R_4 and R_5. A current I is fed into the bridge as shown in Fig. 6.10 to reach the null. A closed loop system can even be constructed to provide the bridge with a self-nulling ability.

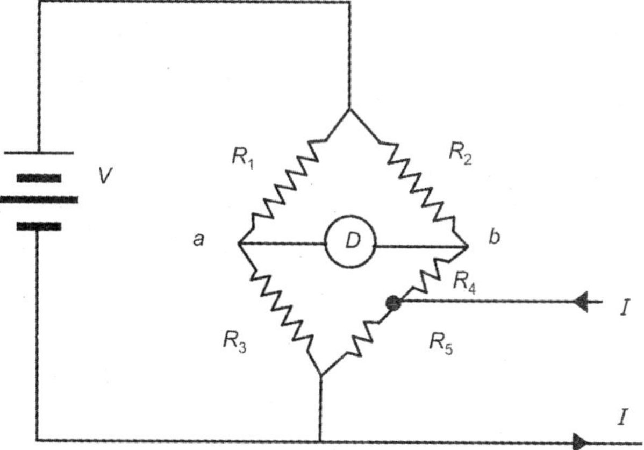

Fig. 6.10 The current balance bridge

Potential Measurements Using Bridges

Bridge circuit can also be used to measure small potentials at a very high impedance. The potential to be measured is to be injected in series with detector, as shown in Fig. 6.11. The voltage appearing across the null detector becomes for basic bridge [Ref. Fig. 6.11(a)],

$$\Delta V = V_x + V_a - V_b$$

$$= V_x + \frac{R_3 V}{R_1 + R_3} - \frac{R_4 V}{R_2 + R_4} \quad \text{...(6.7)}$$

$$= 0 \text{ at null condition.}$$

For current balance bridge the null condition equation is: [Ref. Fig. 6.11(b)]

$$V_x - IR_5 = 0 \quad \text{...(6.8)}$$

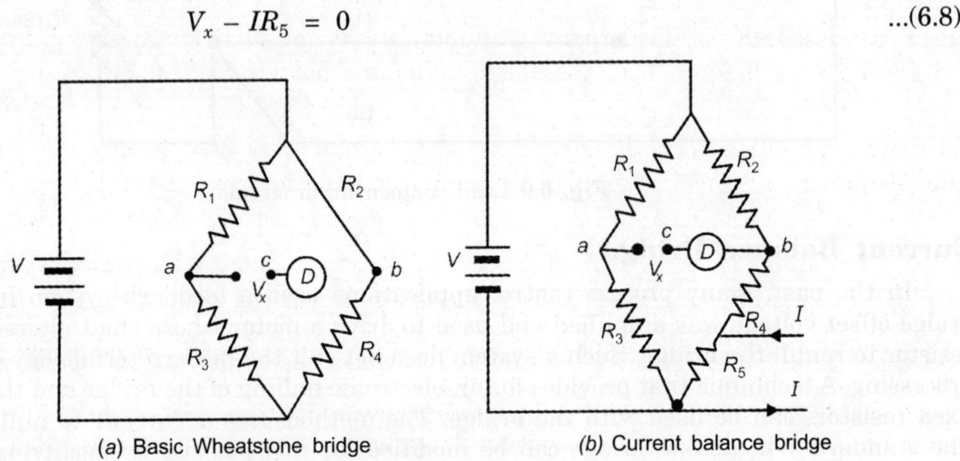

Fig. 6.11 Potential measurement using bridge

AC Bridges

AC bridge behaviour is almost the same as the DC bridge discussed so far. It employs AC excitation instead of DC voltage and for analysis, impedances are considered instead of resistors alone. The general AC bridge circuit is shown in Fig. 6.12.

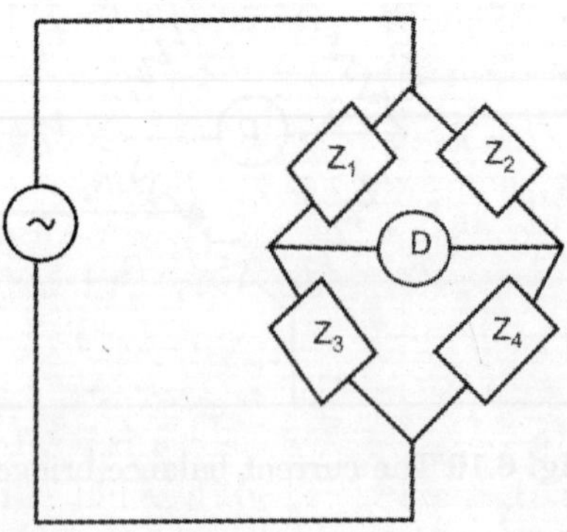

Fig. 6.12 A general AC bridge circuit

The bridge offset voltage is represented as

$$\Delta E = E\left[\frac{Z_3Z_2 - Z_1Z_4}{(Z_1+Z_3)(Z_2+Z_4)}\right] \quad ...(6.9)$$

At null condition, $\Delta E = 0$

and
$$\frac{Z_1}{Z_2} = \frac{Z_3}{Z_4} \quad ...(6.10)$$

Which is analogous to $\frac{R_1}{R_2} = \frac{R_3}{R_4}$ for DC bridge.

Bridge Applications

The primary application of bridge circuits in modern process-control signal conditioning is to convert variations of resistance into variations of voltage. The voltage variation is further conditioned. Non-linearity is present both in DC and AC bridges between offset voltage and the resistor/impedance changes. Even if the sensor has linear relationship between the process variable being measured and the resistance or impedance changes, such linearity is lost when a bridge is used to convert this to a voltage variation. Amplification and linearisation circuits can be employed for further processing.

6.2.2.3 RC Filters

In a measurement system, the signal is more often than not corrupted by unwanted noise signals generated by many factors. The output signal of transducer is fed to the signal conditioning unit. In order to measure the output signal of the transducer originating on account of variation of physical change, it is desirable that the output signal be produced faithfully. For faithful reproduction of signal it becomes necessary to eliminate unwanted signals, noise or spurious ones. The filters are thus designed to pass the signals of wanted frequencies and to reject the signals of unwanted harmonics and noise. In other words, to improve signal-to-noise ratio, it is necessary to use filters.

Basic electrical filters are of two forms namely active and passive filters. Active filter uses active elements like opamp in addition to passive elements like resistance, inductance and capacitance. Passive filter uses passive circuit elements like resistors, inductors and capacitors. In this section we will be discussing about passive filters. In terms of their response to frequencies, filters are ideally classified into four categories:

1. Low-pass filter, 2. High-pass filter, 3. Band-pass filter, and 4. Band-reject filter.

1. Low-pass filter: A low-pass filter transmits all frequencies form zero (DC) to a pre-determined cut-off frequency (ω_c) without loss. For inputs with frequency components $\omega > \omega_c$, it gives a zero output (Refer Fig. 6.13).

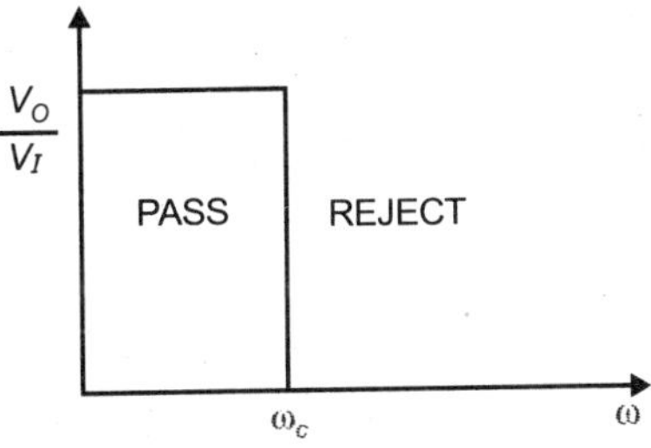

Fig. 6.13 Low-pass filter

2. High-pass filter: A high-pass filter transmits all frequencies above a predetermined cut-off frequency (ω_c) without loss. For inputs having $\omega < \omega_c$, it gives a zero output (Refer Fig. 6.14).

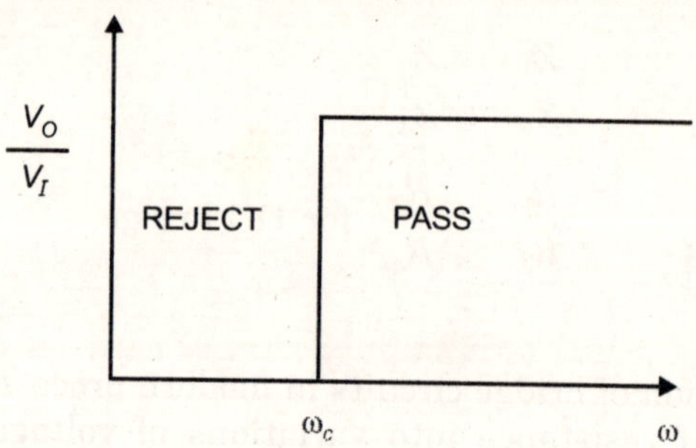

Fig. 6.14 High-pass filter

3. Band-pass filter: A band-pass filter transmits all frequencies within low cut-off frequency ($\omega_{c\ (low)}$) and high cut-off frequency ($\omega_{c\ (high)}$). For inputs having $\omega < (\omega_{c\ (low)})$ and $\omega > (\omega_{c(high)})$, it gives a zero output. (Refer Fig. 6.15).

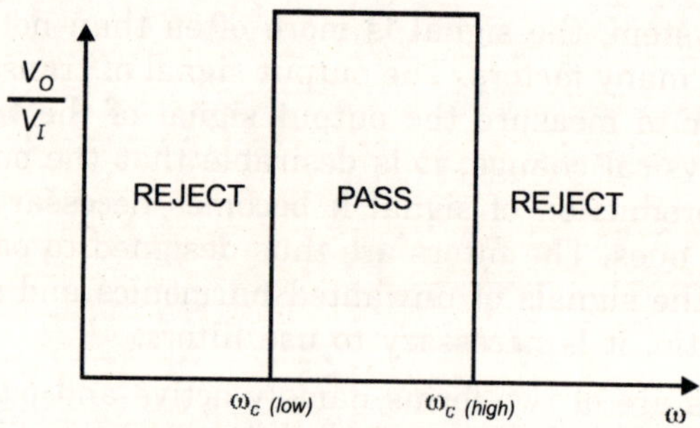

Fig. 6.15 Band-pass filter

4. Band-reject filter: A band-reject filter's action is the reverse of the band-pass filter. It transmits all frequencies below $\omega_{c\ (low)}$ and above $\omega_{c\ (high)}$ and in between it gives a zero output (Refer Fig. 6.16).

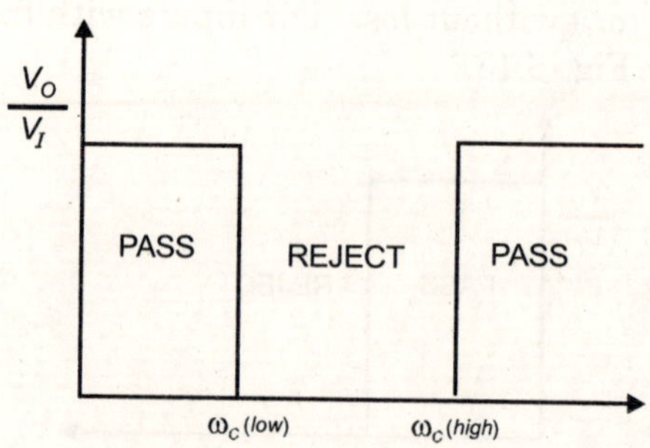

Fig. 6.16 Band-reject filter

The characteristics as depicted in Figs. 6.13 to 6.16 are only ideal ones while practical circuits produce some what different responses as we will be discussing hereon.

Low-Pass RC Filter

The simple circuit shown in Fig. 6.17(a) is called a low-pass RC filter. It blocks high frequencies and passes low frequencies. It will not be possible for the RC filter to exhibit the ideal characteristics as shown in Fig. 6.13, but it can approach that ideal with varying degrees of success. The variation of rejection with frequency is shown in Fig. 6.17(b). In this graph, y-axis is the ratio of output voltage to input voltage without regard to phase. When it is very small or zero, the signal is effectively blocked. The x-axis is normally the logarithm of the ratio of the input signal frequency to a 'critical frequency'. The critical frequency (ω_c) is that frequency for which the ratio of the output to the input voltage is approximately 0.707. In terms of the resistor and capacitor, the critical frequency is given by

$$\omega_c = 2\pi fc = \frac{1}{RC} \quad ...(6.11)$$

The output-to-input voltage ratio for any signal frequency can be computed by

$$\left|\frac{V_O}{V_I}\right| = \frac{1}{\sqrt{1+(\omega/\omega_c)^2}} \quad ...(6.12)$$

A typical filter design is accomplished by finding the critical frequency that will satisfy the design criteria.

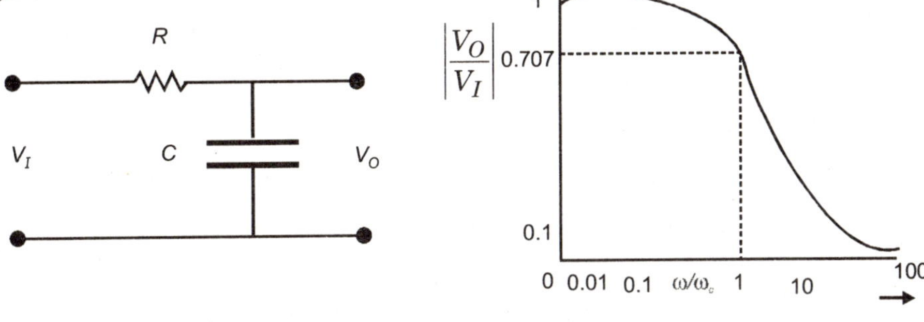

(a) Low-pass RC filter

(b) Response of low-pass RC filter

Fig. 6.17 Low-pass RC filter and its response

High-Pass RC Filter

A high-pass RC filter can be constructed using a resistor and a capacitor as shown in Fig. 6.18(a). Similar to the low-pass RC filter, the rejection is not sharp but distributed over a range around a critical frequency which is defined by the same value as that of low-pass RC filter. The graph of voltage output to input versus logarithm of frequency to critical frequency is shown in Fig. 6.18(b). Here again the magnitude of $V_O/V_I = 0.707$ when the frequency is equal to critical frequency. An equation for the ratio of output voltage to input voltage as a function of the frequency for the high-pass filter is found to be

$$\left|\frac{V_O}{V_I}\right| = \frac{\omega/\omega_c}{\sqrt{1+(\omega/\omega_c)^2}} \quad ...(6.13)$$

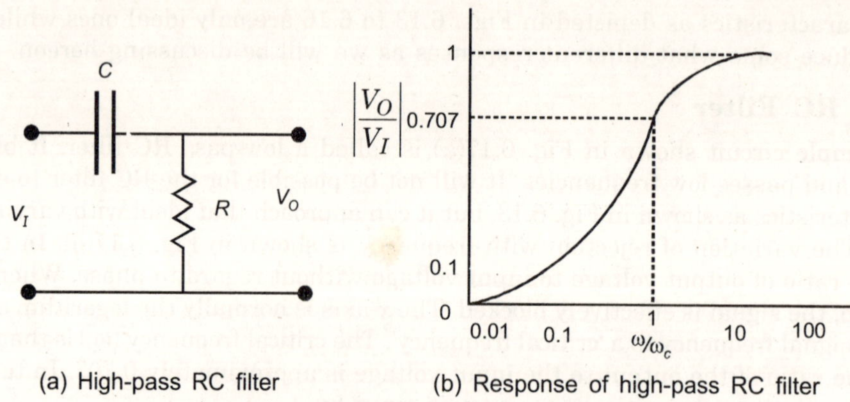

(a) High-pass RC filter (b) Response of high-pass RC filter

Fig. 6.18 High-pass RC filter and its response

Band-Pass RC Filter

Passive band-pass filters can be designed with resistors and capacitors, but more efficient versions use inductors and/or capacitors. The band-pass RC filter is shown in Fig. 6.19(a). It is simply a low-pass filter followed by a high-pass filter. The lower critical frequency is that of the high-pass filter where the higher critical frequency is that of the low-pass filter. The response is shown in Fig. 6.19(b).

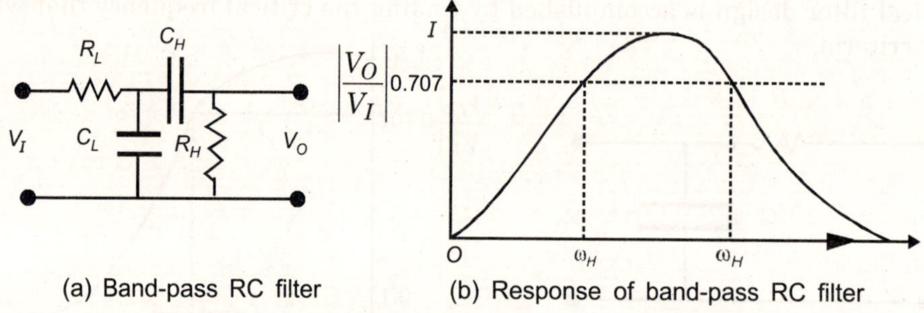

(a) Band-pass RC filter (b) Response of band-pass RC filter

Fig. 6.19 Band-pass RC filter and its response

Band-Reject RC Filter

Band-reject filter is one that blocks a specific range of frequencies. Often such a filter is used to reject a particular frequency or a small range of frequencies that are interfering with a data signal. It is very difficult to realise such filters with passive RC combinations. It is possible of course with inductors and capacitors, but the most success is obtained using active circuits.

One very special band-reject filter, which can be realised with RC combinations, is called a 'notch filter' because it blocks a very narrow range of frequencies. Such a circuit is called as 'twin-T filter' and is shown in Fig. 6.20(a). The characteristics of this filter are determined strongly by the value of the grounding resistor (R_1) and capacitor (C_1). For the particular combination of $R_1 = \pi R/10$ and $C_1 = 10C/\pi$, the filter response is shown in Fig. 6.20(b). The critical 'notch' frequency occurs at a frequency given by

$$\omega_n = 0.785\,\omega_c \text{ where } \omega_c = \frac{1}{RC}$$

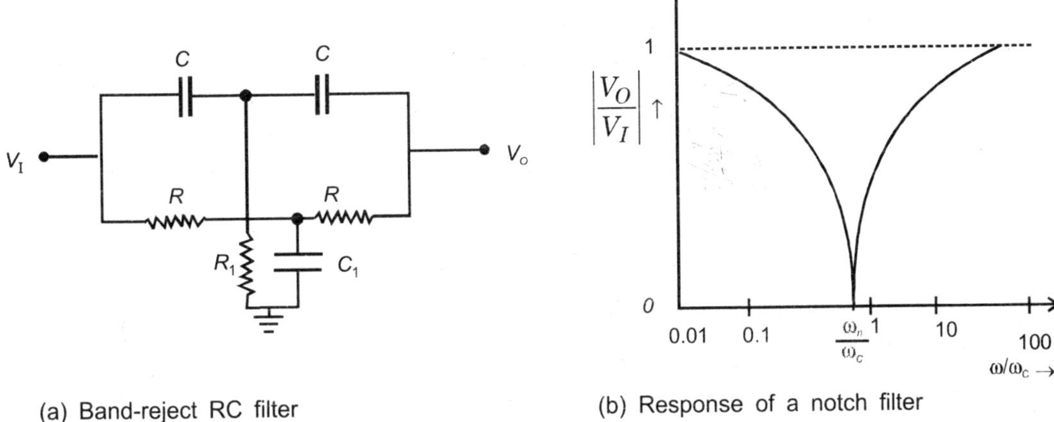

(a) Band-reject RC filter (b) Response of a notch filter

Fig. 6.20 Band-reject RC filter and response

It must be noted that much more improved band-reject and notch filters can be realised using active circuits, particularly using op amps.

6.2.3 Operational Amplifiers (OP amps)

We have already discussed the many diverse requirements for signal conditioning in process control. We considered common, passive circuits that can provide some of the required signal operations, the divider, bridge, and RC filters. Amplification is one process in signal conditioning which can be used to amplify the weak signals from transducers to make them strong enough to be displayed or used in controllers. Normally the instrumentation amplifiers generate enough power to drive display devices on one hand and protect the measurand from being loaded on the other. The type of amplifier that finds wide applications as the building block of signal conditioning applications is called as operational amplifier (op amp in short).

The following fundamental requirements for an Instrumentation amplifier are met by op amp circuits:

1. High input impedance
2. High gain accuracy
3. High CMMR
4. High gain stability with low temperature coefficient
5. Low DC offset
6. Low output impedance.

Op amp is a direct coupled high gain amplifier to which negative feedback is added to control its overall response characteristics. It is used for computing mathematical functions such as addition, subtraction, multiplication, integration etc. It has two input terminals and one output terminal as shown in Fig. 6.21. Terminal with (–) sign is called inverting input terminal and terminal with (+) sign is called non-inverting input terminal. The op amp consists of a number of direct coupled transistors, diodes, resistors, capacitors, etc., but for all practical purpose, the details of the circuitry need not be considered and it can be represented by a triangular symbol. It typically requires connection of bipolar power supplies both + V and $-V$ with respect to ground.

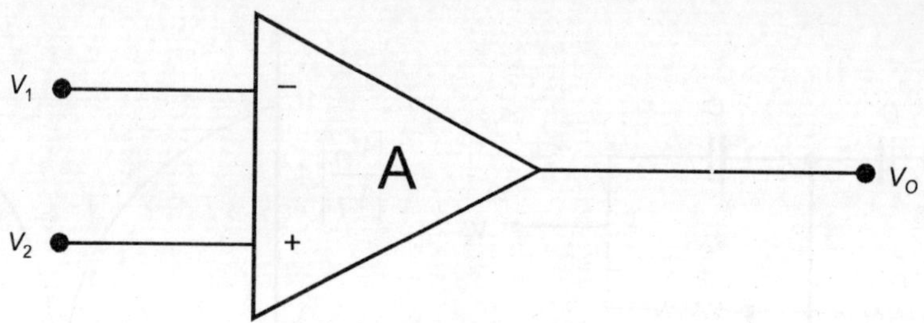

Fig. 6.21 OP amp symbol

It is understood that all voltages are with respect to the ground and therefore the ground line is not shown.

By definition,

$$V_o = A(V_2 - V_1) \quad ...(6.14)$$

where A is the voltage gain of the op amp, which is also referred to as the open-loop d.c gain.

The schematic block diagram of *IC* op amps is shown in Fig. 6.22.

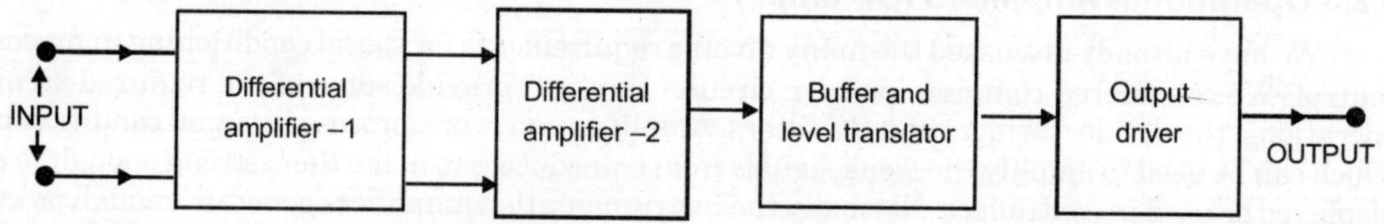

Fig. 6.22 Block diagram of op amp

It consists of four stages as given below:

1. First stage is a differential amplifier with double ended output.
2. Second stage again a differential amplifier with single ended output.
3. Third stage is buffer which is usually an emitter follower whose input impedance is very high so that it prevents loading of high gain stage.
4. Fourth stage is called output driver. The output stage is designed to provide low output impedance as demanded by ideal op amp characteristics.

A large variety of circuits were developed with direct application to process control and instrumentation technology. In general, it is much easier to develop a circuit for a specific service using op amps than discrete components. In this section we will discuss some of the typical practical circuits which are useful for process control as signal conditioning units.

6.2.3.1 *Inverting Amplifier*

This circuit inverts the input signal and may have either attenuation or amplification, depending on the ratio of input resistance R_1, and feedback resistance R_2. The circuit for this amplifier is shown in Fig. 6.23.

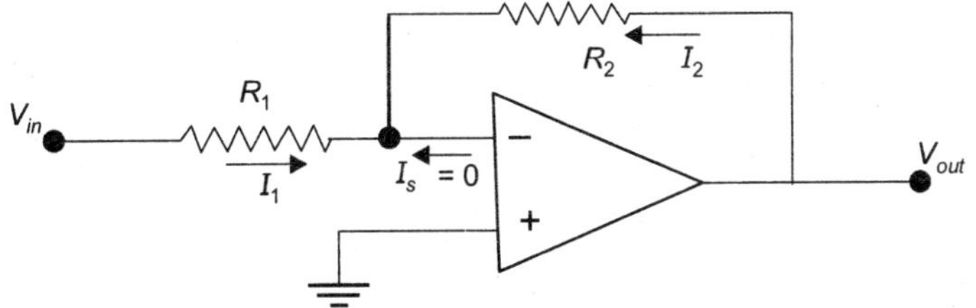

Fig. 6.23 Inverting amplifier

Resistor R_1 connects the input voltage V_{in} to the inverting input of the op amp, and resistor R_2 feedbacks the output to the same point. The common connection is called the summing point. We can see that with feedback and the (+) grounded, $V_{in} > O$ saturates the output negative and $V_{in} < O$ saturates the output positive. With feedback, the output adjusts to a voltage such that,

1. The summing point voltage is equal to the (+) op amp input level, zero in this case.
2. No current flows through the op amp input terminals because of the assumed infinite impedance.

In this case, the sum of currents at the summing point must be zero.

$$I_1 + I_2 = 0$$

where
I_1 = Current through R_1
I_2 = Current through R_2

Because the summing point potential is assumed to be zero, by Ohm's law we have

$$\frac{V_{in}}{R_1} + \frac{V_{out}}{R_2} = 0 \text{ and hence}$$

$$\boxed{V_{out} = \frac{R_2}{R_1} V_{in}} \qquad ...(6.15)$$

Thus, the circuit of Fig. 6.23 is an inverting amplifier with gain R_2/R_1, that is shifted 180° in phase (inverted) from the input. This can be attenuator by virtue of making $R_2 < R_1$.

Normally the following two rules are made for design purposes using op amps:

Rule 1. Assume that no current flows through the op amp input terminals (both inverting and non-inverting).

Rule 2. Assume that there is no voltage difference between the op amp input terminals – that is, $V+ = V-$.

6.2.3.2 Non-Inverting Amplifier

Fig. 6.24 shows an non-inverting amplifier constructed using op amp.

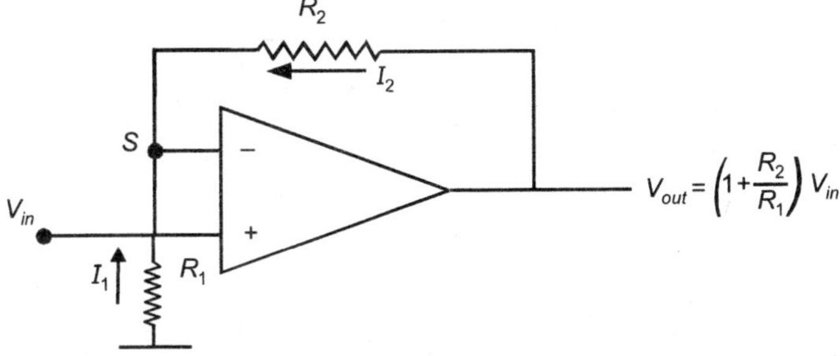

Fig. 6.24 Non-inverting amplifier

As no voltage difference appears across the input terminals, the voltage at summing point '*S*' is assumed as V_{in}.

$$I_1 + I_2 = 0$$

$$\frac{V_{in}}{R_2} + \frac{V_{in} \quad V_{out}}{R_2} = 0$$

$$\therefore \qquad V_{out} = \left[1 + \frac{R_2}{R_1}\right] V_{in} \qquad ...(6.16)$$

From the above equation, we can note that this can never be used for voltage attenuation because the ratio is added to 1. The input impedance is very high and the output impedance is very low.

6.2.3.3 Voltage Follower (Unity Gain Amplifier)

Consider the non inverting amplifier using op amp as discussed in 6.2.3.2 and make the value of $R_2 = 0$ and R_1 open. The output voltage and input voltage relation will become:

$$\frac{V_{out}}{V_{in}} = 1 + \frac{R_2}{R_1} = 1 + \frac{0}{\infty} = 1 \qquad ...(6.17)$$

The circuit will look like as shown in Fig. 6.25

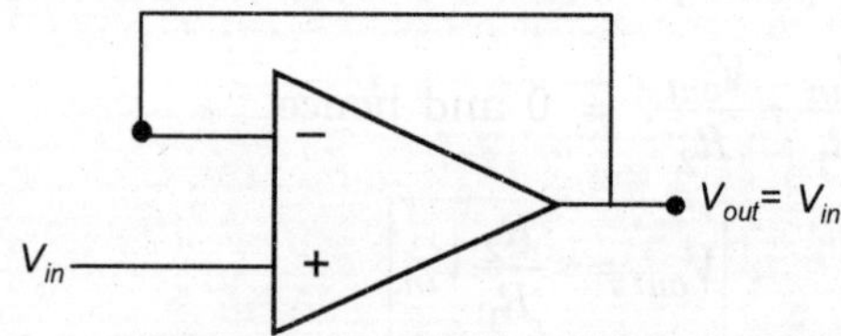

Fig. 6.25 Voltage follower

This circuit will have unity gain and very high input impedance. V_{out} tracks the V_{in} over a range defined by the plus and minus saturation voltage outputs. Current output is limited to the short circuit current of the op amp, and output impedance is very low. The unity gain voltage follower is essentially an impedance transformer in the sense of converting a voltage at high impedance to the same voltage at low impedance.

6.2.3.4 Summing Amplifier (Addition)

The arrangement as in Fig. 6.26 is used to obtain an output which is a linear combination of a number of input signals.

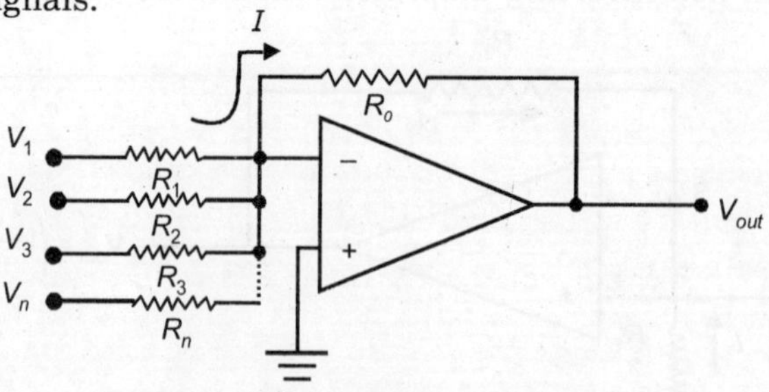

Fig. 6.26 Summing amplifier

Since a virtual ground exists at the op amp input, we have

$$I = \frac{V_1}{R_1} + \frac{V_2}{R_2} + \frac{V_3}{R_3} + \ldots\ldots + \frac{V_n}{R_n} \quad \ldots(6.18)$$

Hence,

$$V_{out} = -R_0 I = -\left[\frac{R_0}{R_1}V_1 + \frac{R_0}{R_2}V_2 + \frac{R_0}{R_3}V_3 + \ldots\ldots + \frac{R_0}{R_n}V_n\right] \quad \ldots(6.19)$$

If $R_1 = R_2 = R_3 \text{ -------- } = R_n = R_0$, then

$$V_0 = -\sum V_n \quad \ldots(6.20)$$

In case non-inverting addition is desired, the resistance ladder may be connected to the non-inverting input with the inverting input grounded.

6.2.3.5 Subtraction

A simple subtraction unit using op amp is shown in Fig. 6.27.

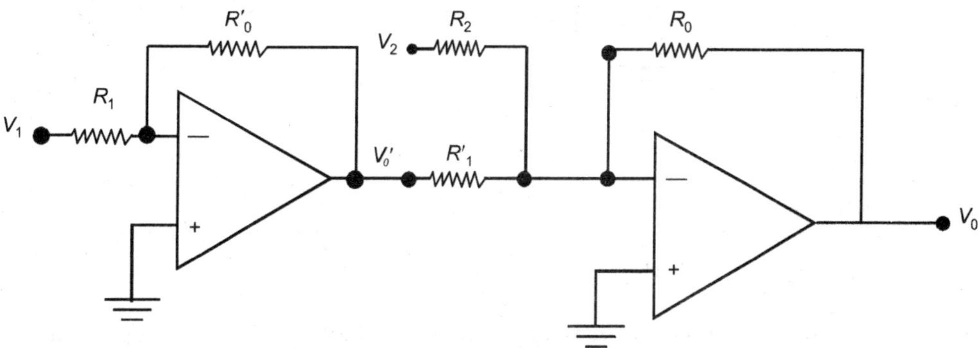

Fig. 6.27 Subtractor

Output from the first stage $= V_0' = -\dfrac{R_0'}{R_1}V_1$

Output after the second stage $= -\left[\dfrac{R_0}{R_2}V_2 + \dfrac{R_0}{R_1'}V_0'\right]$

$$V_0 = -\left[\frac{R_0}{R_2}V_2 - \frac{R_0\,R_0'}{R_1'\,R_1}V_1\right] \quad \ldots(6.21)$$

Thus, if $R_1 = R_2 = R'_1 = R'_0 = R_0$, we get

$$\boxed{V_0 = V_1 - V_2} \quad \ldots(6.22)$$

In addition to summation and subtraction, multiplication and division can be performed by choosing suitable values of R_o and R_i in the basic op amp configuration.

6.2.3.6 *Integrator*

The configuration consisting of an input resistor and feedback capacitor, as shown in Fig. 6.28, works as an integrator. Using the ideal analysis, we can sum the current at the summing point as

$$\frac{V_{in}}{R} + C\frac{dV_{out}}{dt} = 0 \qquad \text{...(6.23)}$$

which can be solved by integrating both terms so that

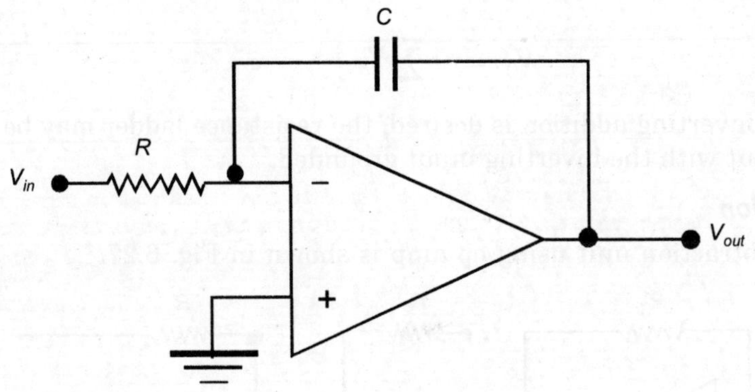

Fig. 6.28 Integrator

the circuit response is

$$V_{out} = -\frac{1}{RC}\int V_{in}\,dt \qquad \text{...(6.24)}$$

This result shows that the output voltage varies as an integral of the input voltage with a scale factor of (–) $1/RC$. This circuit is employed in many cases where integration of a transducer output is desired.

A linear ramp can be generated by keeping the input voltage V_i constant at K. Then,

$V_{out} = -\frac{K}{RC}t$, which is a linear ramp with the negative slope of K/RC.

6.2.3.7 *Differentiator*

Op amp can also be used to construct a circuit with an output proportional to the derivative of the input voltage. It can be realized with only a single capacitor and a single resistor as shown in Fig. 6.29. Using ideal analysis to sum currents at the summing point gives the equation

$$C\frac{dV_{in}}{dt} + \frac{V_{out}}{R} = 0 \qquad \text{...(6.25)}$$

Solving for the output voltage shows that the circuit response is

$$V_{out} = -RC\frac{dV_{in}}{dt} \qquad \text{...(6.26)}$$

Therefore, the output voltage varies as the derivative of the input voltage.

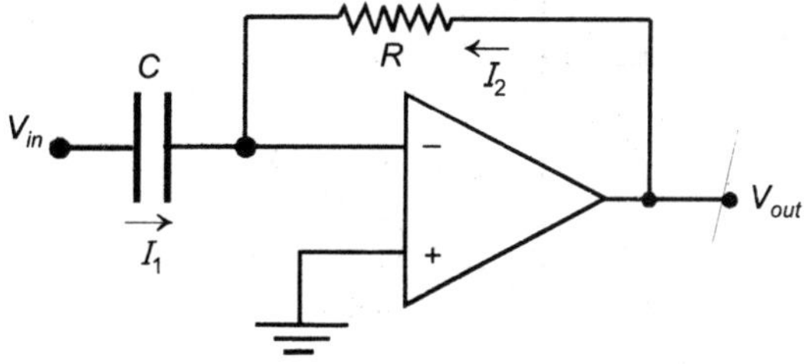

Fig. 6.29 Differentiator

6.2.3.8 *Voltage-to-Current Converter*

Since 4–20 mA is considered a standard signal in process control, it is often necessary to employ a linear voltage-to-current converter. Such a circuit must be able to drive number of different loads without changing the voltage-to-current transfer characteristic (load independent circuit). One such voltage-to-current converter is shown in Fig. 6.30. An analysis of this circuit shows that the relationship between current and voltage is given by

$$I = \frac{-R_2}{R_1 R_3} V_{in} \qquad \text{...(6.27)}$$

provided $\qquad R_1(R_3 + R_5) = R_2 R_4.$

The circuit can deliver current in either direction as required by a particular application.

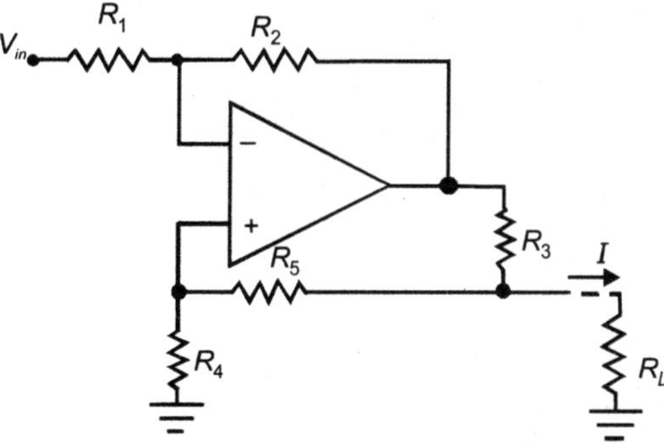

Fig. 6.30 Voltage-to-current converter

In this circuit, the maximum load resistance is always less than the saturation voltage divided by the maximum current. The minimum load resistance is zero.

6.2.3.9 *Current-to-Voltage Converter*

We also need current-to-voltage conversion in the process control signal transmission system. This can be achieved using op amp as shown in Fig. 6.31.

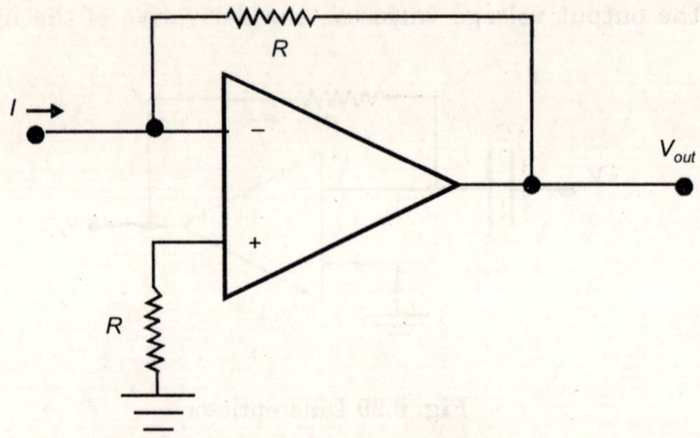

Fig. 6.31 Current-to-voltage converter

The relation between output voltage and input current is given by the equation

$$V_{out} = -IR \qquad ...(6.28)$$

The resistor, R, in the non-inverting terminal is employed to provide temperature stability to the configuration.

6.2.3.10 Linearisation

Linearisation of signals can also be achieved with the help of op amp, by placing a non-linear element in the feedback loop of the op amp, as shown in Fig. 6.32. The summation of currents provides

$$\frac{V_{in}}{R} + I(V_{out}) = 0$$

Where V_{in} = Input voltage

R = Input resistance

$I(V_{out})$ = Non-linear variation of current with voltage

Solving the equation yields

$$V_{out} = G\left(\frac{V_{in}}{R}\right) \qquad ...(6.29)$$

Where V_{out} = Output voltage

$G\left(\frac{V_{in}}{R}\right)$ = a non-linear function of the input voltage [actually the inverse function of $I(V_{out})$]

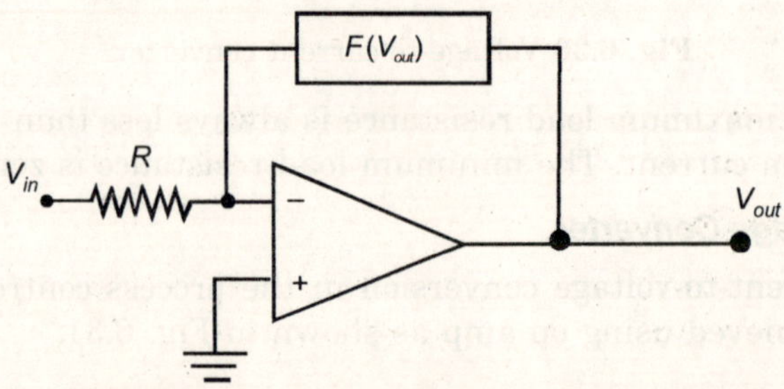

Fig. 6.32 Linearisation with non-linear feedback

Thus, as an example, if a diode is placed in the feedback as shown in Fig. 6.33, the function $I(V_{out})$ is an exponential.

$$I(V_{out}) = I_o \exp(\alpha V_{out}) \quad ...(6.30)$$

Where I_O = Amplitude constant,

α = Exponential constant.

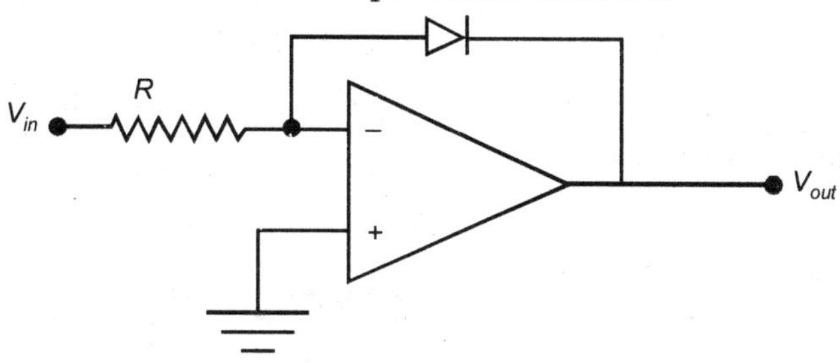

Fig. 6.33 Logarithmic amplifier

The inverse of the above equation is a logarithm, $V_{out} = \frac{1}{\alpha} \log_e(V_{in}) - \frac{1}{\alpha} \log_e(I_o R)$ and hence a logarithmic amplifier.

6.2.3.11 Differential Amplifier

The use of an op amp as a differential amplifier is very useful in instrumentation. Wheatstone bridges, strain gauges, thermocouples and hot-wire anemometers generate small difference signal which usually must be amplified. The differential instrumentation amplifiers provide an output that is a precise multiple of the difference between two input signals. An ideal differential amplifier provides an output voltage with respect to ground that is some gain times the difference between two input voltages:

$$V_{out} = A_D (V_1 - V_2) \quad ...(6.31)$$

Where A_D is the differential gain and both V_1 and V_2 are input voltages with respect to ground.

Common Mode Rejection

The common-mode input voltage is the average of the voltage applied to the two input terminals,

$$V_{cm} = \frac{V_1 + V_2}{2} \quad ...(6.32)$$

An ideal differential amplifier will not have any output that depends on the value of the common-mode voltage; that is, the circuit gain for common-mode voltage, A_{cm} will be zero. The common-mode rejection ratio (CMRR) of a differential amplifier is defined as the ratio of the differential gain to the common-mode gain. The common-mode rejection (CMR) is the CMRR expressed in dB.

$$CMRR = \frac{A_D}{A_{cm}} \quad ...(6.33)$$

$$CMR = 20 \log_{10} (CMRR) \quad ...(6.34)$$

Clearly, the larger these numbers, the better the differential amplifier. Typical values of CMR range from 60 to 100 dB.

The most common differential amplifier circuit using op amp is shown in Fig. 6.34. The circuit uses two pairs of matched resistors, R_1 and R_2. When the matching is perfect and the op amp is ideal the output voltage is given by the equation,

$$V_o = \frac{R_2}{R_1}(V_2 - V_1) \quad \text{...(6.35)}$$

Voltage followers are used on the input to provide high input impedance. Such an amplifier is called 'Instrumentation Amplifier'

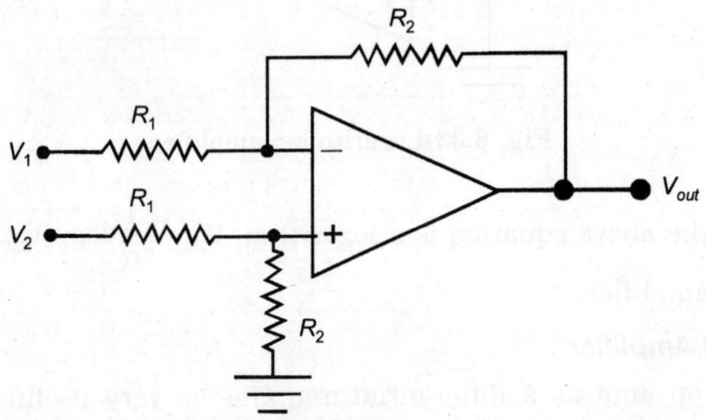

Fig. 6.34 Common differential amplifier

Instrumentation Amplifier

Differential amplifiers with high input impedance and low output impedance are given the special name of 'Instrumentation amplifier'. They find a host of applications in process-measurement systems, principally as the initial stage of amplification for bridge circuits. Fig. 6.35 shows one such instrumentation amplifier in common use.

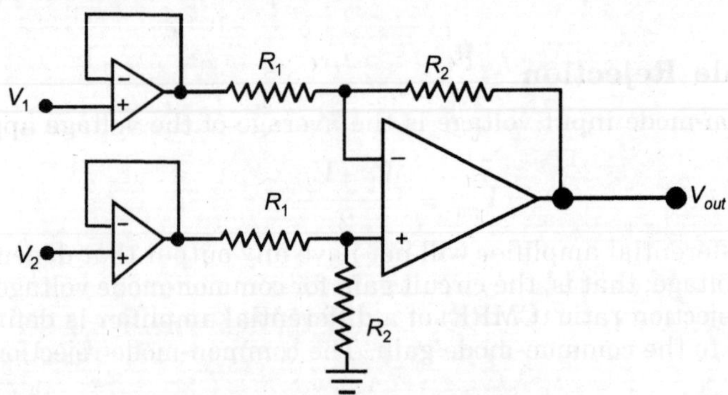

Fig. 6.35 Instrumentation amplifier

In the circuit discussed, the changing gain becomes difficult and requires changing two resistors, that too with matched ones. The circuit shown in Fig. 6.36 allows for selection of gain, within certain limits, by adjustment of a single resistor, R_G.

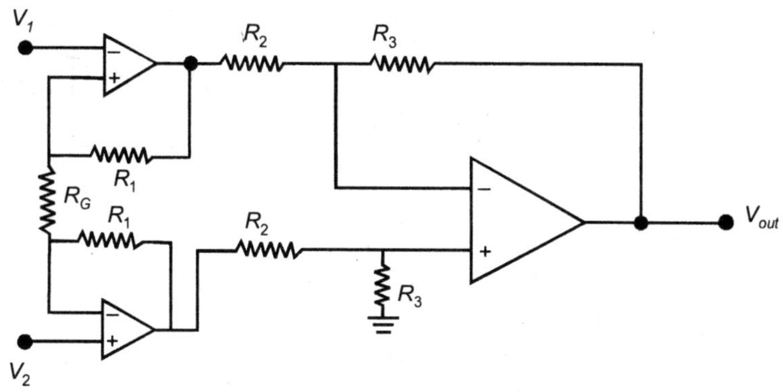

Fig. 6.36 Instrumentation amplifier with gain adjustment

6.2.4 DC Signal Conditioning

With this background of bridges, amplifiers, filters etc., we can now set up a generalised DC signal conditioning system as shown in Fig. 6.37.

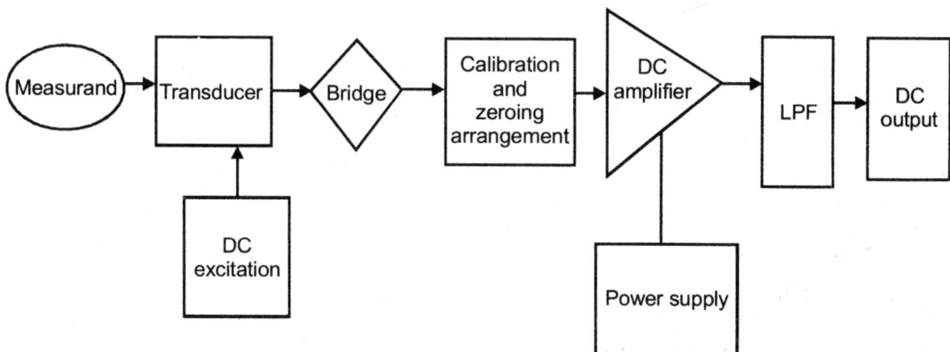

Fig. 6.37 DC signal conditioning system

In such a system the bridge is generally of the wheatstone variety, the transducer constituting one arm or more. Calibration and zeroing arrangement may be a simple manually adjustable potentiometer or an elaborate automatic compensating device. The DC amplifier need to possess a high CMRR, but it must have a good thermal stability.

The advantages of DC signal conditioners are that they are cheaper, can be calibrated easily and overload recovery is better. The chief disadvantage is drifting.

DC signal conditioning systems are generally used for common resistance transducers such as potentiometers and resistance strain gauges.

6.2.5 AC Signal Conditioning System

AC signal conditioning system contains similar blocks as that of DC signal conditioning system but for a carrier oscillator which provides excitation to the AC Bridge and a phase sensitive detector as shown in Fig. 6.38.

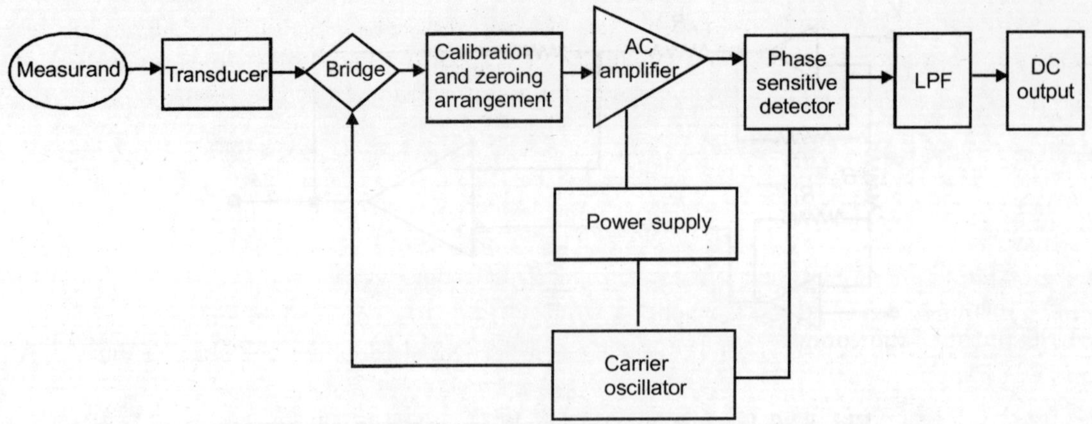

Fig. 6.38 AC signal conditioning system

Transducers, which constitute part of bridges, are generally excited by AC sources of frequency 50 Hz–200 kHz. One has to choose this carrier frequency carefully so that it is atleast 10 times that of signal frequency. A phase-sensitive detection is necessary to ensure polarity of the DC output. Freedom from drift and very high signal-to-noise ratio are some of the advantages.

AC signal conditioning systems are generally used for variable reactance transducers and for systems where signals are to be transmitted via cables to connect transducers to the signal conditioning unit.

6.2.6 Design Considerations

The main issues which are to be considered while designing an analog signal conditioning system are briefly given here.

Step 1: Definition of measurement objective: The parameter to be measured (measured variable like pressure, temperature etc.), range of the measurement, accuracy and linearity required, and noise level in the environment are to be defined.

Step 2: Selection of the sensor: For the defined parameter the proper sensor has to be selected taking into consideration the sensor output (resistance change, mv or ma etc.,), transfer function, response time, range, power requirement etc.

Step 3: Design of analog signal conditioning: Desired output, range, linearity, input impedance, output impedance etc. are to be considered here.

Step 4: Other important points: Loading effects of both sensor and signal conditioner, relation between output and input and the final device to which the signal conditioner output goes etc. are to be kept in mind.

If enough information is not available to address the above steps, the designer must exercise good technical judgment in accounting for the design.

6.3 DIGITAL SIGNAL CONDITIONING

Digital computers started playing a major role in process control applications. Computers, which were familiar in applications like accounting, finance, banking, insurance, data-storing, statistics etc. have become very handy in industrial application in the name of process computers. They are used for controlling individual equipment and for that matter the complete process of a factory from a central place. Distributed digital control systems (DCS), programmable logic controllers (PLC) and data acquisition system (DAS) have come to stay in process industries. In addition there are lots of direct applications of digital electronics in process lines. All these digital electronic systems require digital data to be presented to them in a digital format. In other words, the data are to be digitally conditioned before presenting to them.

As process computers are mostly digital electronic devices (at the start of the computer era, analog computers mostly assembled with op amps were tried for process control) nowadays, all the information they work with has to be digitally formatted. For controlling temperature, the temperature signal has to be represented digitally. Hence the need for digital signal conditioning–to condition process-control signals to be in an appropriate digital format.

Digital signal conditioning in process control means finding a way to represent analog process information (as most of the transducers give analog output by nature) in a digital format. Though greater accuracy is not expected in using digital techniques to represent data, there are many advantages. Digital data are much more immune from spurious influences that would cause subsequent inaccuracy, such as noise, amplifier gain changes, power supply drifts, and so on.

The use of digital computers in process control systems is more valuable because of the following reasons:

1. More useful to control various control loops from a centralised control room. Even here the functionwise distribution of control is possible.
2. Mathematical operation like linearization, scale and bias changes, unit conversions etc. can be achieved at ease.
3. Complicated control equations can be solved easily. Modeling of processes (static as well as dynamic) can be done. All with must faster rate is possible.
4. Large industrial complex can be integrated by networking of computers. Even units at different far away locations can be linked to the corporate computer to have access to some of the critical parameters by CEOs.

In the following sections, emphasis will be given to digital-to-analog conversion (DAC) and analog-to-digital conversion (ADC) techniques and data acquisition systems (DAS). It is assumed that the reader has sufficient background on digital fundamentals like binary, octal and hexadecimal numbers, fractional binary numbers, Boolean algebra, digital logic circuits, programmable logic controllers (PLCs) and computer interfaces.

6.3.1 Converters

Most of the sensors/transducers used to measure process variables produce analog signals. To interface this signal with a computer or digital logic circuit, it is necessary first to perform an analog-to-digital (*A*/*D*) conversion. A digital-to-analog (*D*/*A*) conversion becomes essential when the digital signal is required to drive an analog device.

6.3.1.1 Comparators

Comparator is the simple and most wanted device in digital signal conditioning.

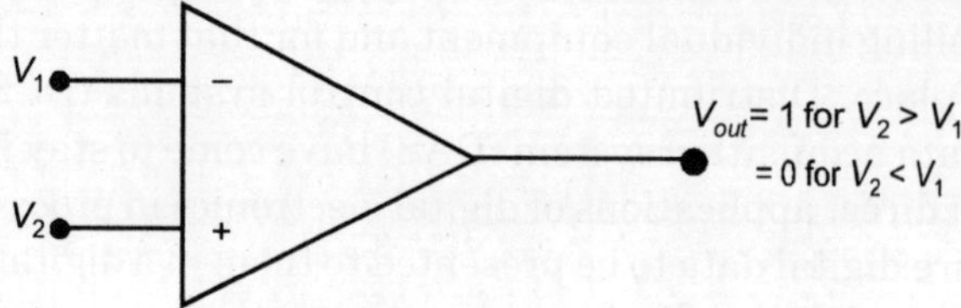

Fig. 6.39 Basic comparator

This normally uses an op amp for the purpose as shown in Fig. 6.39. This device simply compares the two analog voltages on its input terminals. Depending on which voltage is larger, it gives either 'high or 1' output or 'low or 0' output. The comparator is extensively used for alarm signals to computers or digital processing systems. Many a times it becomes integral part of the A/D and D/A converters. Normally one of the input voltages, V_1 or V_2 , will be the variable one and the other a fixed value called a trip, trigger, or reference voltage. The reference voltage may be provided from a divider using available power supplies.

Open-Collector Comparators

In open-collector comparators, the output terminal of the comparator is connected internally to the collector of a transistor in the comparator. An external resistor or relay or load is connected from the output to an appropriate power supply. The output terminal will show either a 'O' (OV) if the internal transistor is ON or 1 (V_S) if the internal transistor is OFF. One such open-collector comparator is shown in Fig. 6.40.

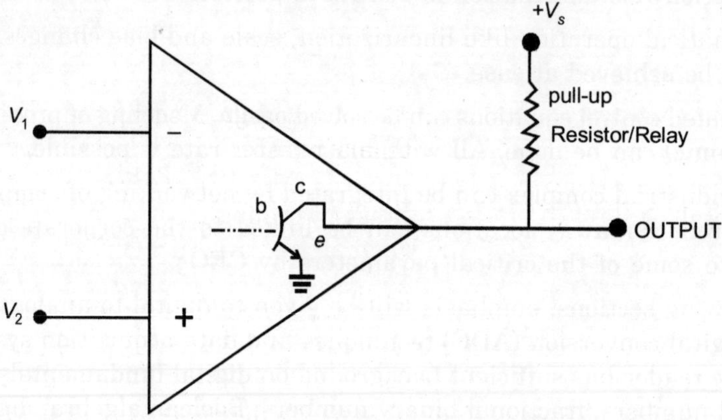

Fig. 6.40 Comparator with open collector output

Hysteresis Comparator

If the signal voltage has noise or approaches the reference value too slowly, the comparator output may fluctuate between 'O' and '1' state. Such fluctuation of output may cause problems with the equipment designed to interpret the comparator output signal.

The above problem can be solved by providing a dead band or hysteresis window to the reference level about which output changes occur. Once the comparator has been triggered high, the reference level is automatically reduced so that the signal must fall to some value below the old reference before the comparator goes to the low state. One common technique to achieve the above is illustrated in Fig. 6.41(a). R_f and R are added to perform the hysteresis function.

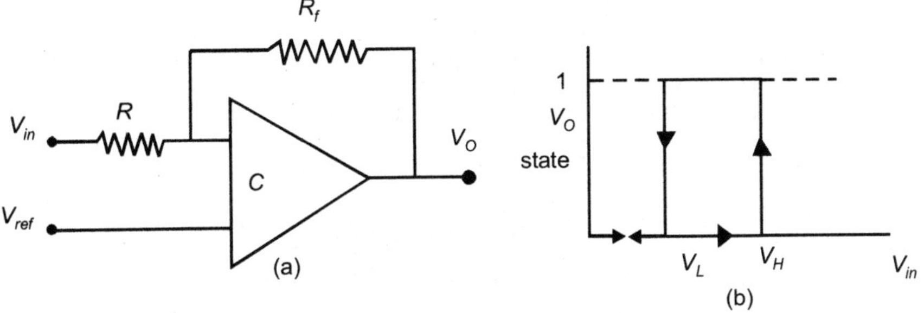

Fig. 6.41 Hysteresis comparator

Under the condition that $R_f \rangle\rangle R$, the response of the comparator is shown in Fig. 6.41 (b). The condition for which the output will go high ('1' state) is defined by the condition.

$$V_{in} \geq V_{ref} \text{ [i.e.,} V_H = V_{ref}]$$

Once having been driven high, the condition for the output to drop back to the low ('O' state) is given by the relation

$$V_{in} \leq V_{ref} - \frac{R}{R_f} V_o = V_L$$

The dead band or hysteresis is given by $\frac{R}{R_f} V_o$ and is thus selectable by the choice of resistors.

6.3.1.2 Digital-to-Analog Converters (DACs)

Digital-to-Analog converters are used to convert digital signals to analog signals. The digital information may be in the form of a binary number with some fixed number of digits. The binary number is called a binary word and the digits are called bits of the word. A unipolar DAC converts a digital word into an analog voltage by scaling the analog output to be zero when all bits are zero and some maximum value when all bits are one. This can be mathematically represented by treating the binary number that the word represents as a fractional number. The output of DAC can be defined as a scaling of some reference voltage as shown below.

$$V_{out} = V_R \left[b_1 2^{-1} + b_2 2^{-2} + \ldots\ldots\ldots + b_n 2^{-n} \right] \quad \ldots(6.36)$$

The minimum V_{out} is zero, and the maximum is determined by the size of the binary word.

Thus, a 4-bit word has a maximum of

$$V_{max} = V_R \left[2^{-1} + 2^{-2} + 2^{-3} + 2^{-4} \right]$$
$$= 0.9375\ V_R$$

and an 8-bit word has a maximum of

$$V_{max} = V_R \left[2^{-1} + 2^{-2} + \ldots\ldots + 2^{-7} + 2^{-8} \right]$$
$$= 0.9961\ V_R$$

An alternate equation given below is easier to use.

$$V_{out} = \frac{N}{2^n} V_R \qquad ...(6.37)$$

where N = base 10 whole-number of DAC input.

Suppose for a 8-bit converter with a 5.0 V reference,

if (1) Input: 00000000, output = 0

(2) Input: 11111111, output = 4.9805 V as per eqn. (6.36)

= 5 V as per eqn. (6.37).

(3) Input: 10110111, output $= \left[2^{-1}+0+2^{-3}+2^{-4}+0+2^{-6}+2^{-7}+2^{-8}\right] V_R$

$= [0.5+0+0.125+0.0625+0.0156+0.0078+0.0039] V_R$

$= 0.7148 \times 5 = 3.574$ as per eqn. (6.36)

or

$$\text{output} = \frac{N}{2^n} V_R = \frac{183}{256} \times 5 = 3.574 \text{ as per eqn. (6.37)}$$

It is clear that equation (6.37) is easier to use.

Bipolar DAC

Depending up on the application demand, DACs can be designed to give output in plus and minus ranges. A simple offset-binary is frequently used, where in the output is simply biased by half the reference voltage of equation (6.37). The bipolar DAC relationship is then given by

$$V_{out} = \frac{N}{2^n} V_R - \frac{1}{2} V_R \qquad ...(6.38)$$

When N is zero, $V_{out} = -\frac{V_R}{2}$ which is the minimum output value. When N is at maximum value (2^n-1), the output will be $V_{out} = \left(\frac{2^n - 1}{2n}\right) V_R - \frac{1}{2} V_R = \frac{V_R}{2} - \frac{V_R}{2^n}$ which is the maximum output value.

Conversion Resolution

Conversion resolution is a function of the reference voltage and the number of bits in the word. The more bits, the smaller the change in analog output for a 1-bit change in binary word, and hence the better resolution. The smallest possible change is simply given by

$$\Delta V_{out} = V_R 2^{-n} \qquad ...(6.39)$$

Where ΔV_{out} = Smallest output change

V_R = Reference voltage

n = Number of bits in the word.

Basic Structure of a DAC

The basic schematic of DAC is shown in Fig. 6.42. The input is n-bit binary word and is combined with reference voltage to give an analog output signal.

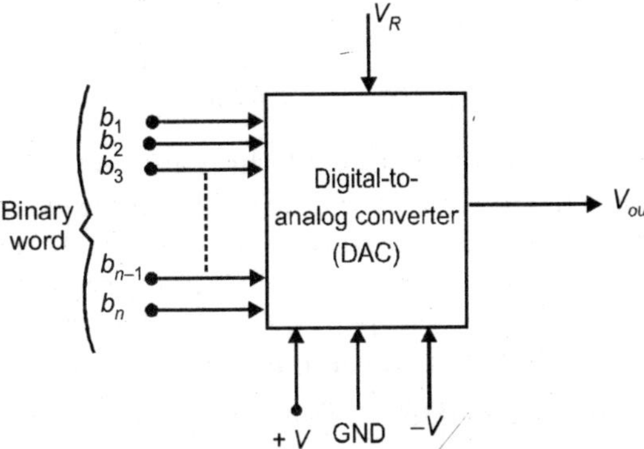

Fig. 6.42 The Basic schematic of DAC

In general, a DAC can be considered as a black box, and no knowledge of internal working is required. For better understanding of the working of DAC we discuss below two types of DACs which are very common.

1. Simple DAC: A simple 4-bit DAC arrangement is shown in Fig. 6.43. It consists of a voltage reference V_R, a network of precision resistors R, 2R, 4R and 8R, and a set of digitally controlled switches s_0, s_1, s_2 and s_3. The op amp acts as a current-to-voltage converter. The switches will either connect or disconnect a resistor to the voltage source according to the input data.

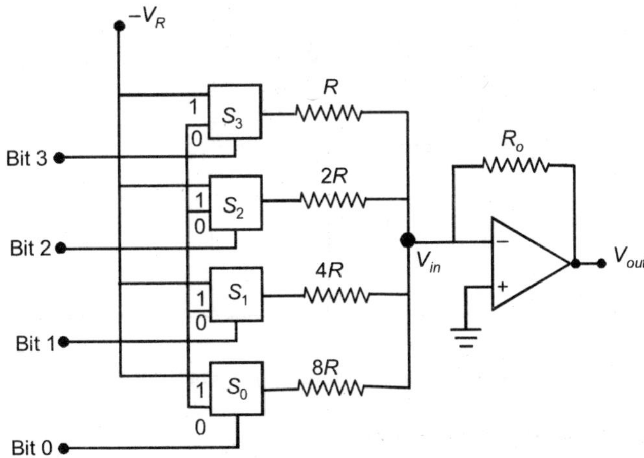

Fig. 6.43 A simple DAC

Case 1: If all bits are zero (0000), no current passes through the resistors and hence $V_{in} = 0 = V_{out}$.

Case 2: If bit-3 is 1 and the rest Os, the current passes only through the resistor R and its value is $-V_R/R$. Then the output voltage equals to $\frac{R_o}{R} V_R$.

Case 3: If all the 4-bits are 1s, the output is

$$V_{out} = \left[1+\frac{1}{2}+\frac{1}{4}+\frac{1}{8}\right]\frac{V_R R_o}{R}$$

or

$$V_{out} = [8+4+2+1]\frac{V_R R_o}{8R} \quad ...(6.40)$$

The analysis of the three cases above and the equation (6.40) clearly show that the analog output is proportional to the decimal value of the digital input. The reliability of this kind of DAC mainly depends on the accuracy and thermal stability of the value of the resistors. It is very difficult to obtain thermally stable and precision resistors specially when the values are high which is likely to happen when number of input bits is high. To avoid such difficulties, the ladder-type DAC is recommended for DACs.

2. Resistive Ladder Network DAC : The circuit arrangement for a resistive ladder type DAC is shown in Fig. 6.44. The ladder basically splits current according to the need. With the *R*-2*R* choice of resistors, it can be shown through network analysis that the output voltage is given by equations (6.36) or (6.37).

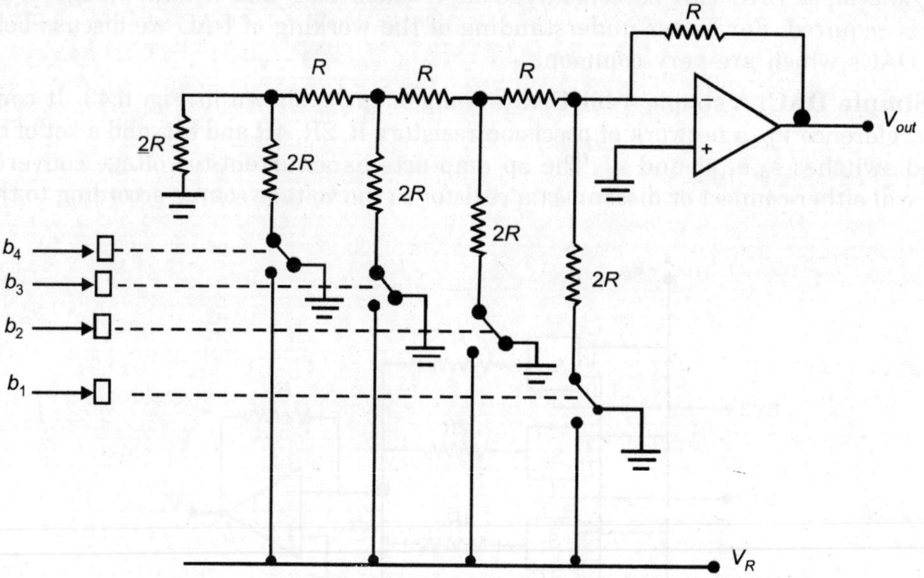

Fig. 6.44 Resistive ladder network DAC

6.3.1.3 *Analog-to-Digital Converters (ADCs)*

With the growing use of digital logic and computers in process control, it is necessary to employ an ADC to provide a digitally encoded signal for the computer. The input-output relation of ADC can be expressed in a similar way to that of the DAC as given in equation (6.36). In this case, however, the interpretation is reversed. The ADC will find a fractional binary number that gives the closest approximation to the fraction formed by the input voltage and reference.

$$b_1 2^{-1} + b_2 2^{-2} + b_3 2^{-3} + \ldots\ldots\ldots\ldots + b_n 2^{-n} \leq \frac{V_{in}}{V_R} \quad \ldots(6.41)$$

Where $b_1, b_2 \ldots\ldots\ldots b_n$ = n-bit digital output

V_{in} = Analog input voltage

V_R = Analog reference voltage

The inequality sign is used because the fraction on the right can change continuously over all values, but the fraction derived from the binary number on the left can change only in fixed increments of $\Delta N = 2^{-n}$. In other words, the only way the left side can change is if the LSB changes from 1 to 0 or from 0 to 1. In either case, the fraction changes only by 2^{-n} and nothing in between.

Therefore, there is an inherent uncertainty in the input voltage producing a given ADC output, and that uncertainity is given by

$$\Delta V = V_R 2^{-n} \quad \ldots(6.42)$$

Bipolar Operation

A bipolar ADC is one that accepts bipolar input voltage for conversion into an appropriate digital output. The most common bipolar ADCs provide an output called offset-binary. This simply means that the normal output is shifted by half the scale so that all zeros corresponds to the negative maximum input voltage instead of zero.

Basic Structure of an ADC

Fig. 6.45 shows the basic schematic of an ADC.

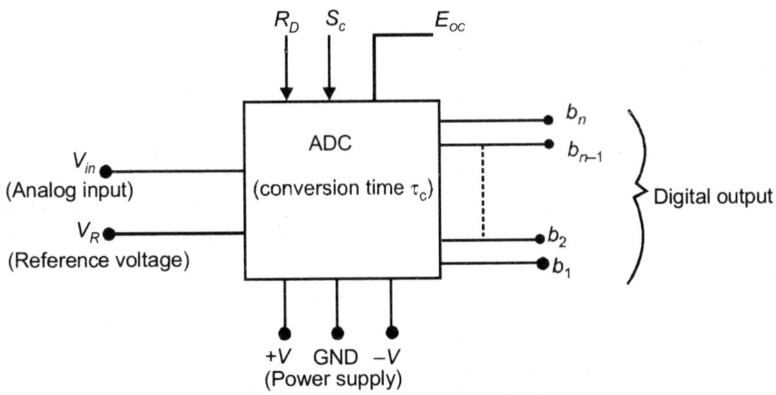

Fig. 6.45 The basic schematic of ADC

In addition to analog input, Reference Voltage, Power supply and Digital outputs, ADC will have common control lines, namely S_C (start-convert), E_{OC} (End-of-conversion) and R_D (Read). These control lines are single-bit digital inputs and outputs designed to control operation of the ADC and allow for interface to a computer.

Conversion Time

It is a very important characteristic of ADCs. It is the processing time taken for the ADC to convert the analog input into digital output. The read operation may occur at any time after the end-of-conversion has been issued by the ADC. The next start-convert (S_C) demand should come after E_{OC}.

There are four ADC techniques generally used. They are:

1. Successive approximation or parallel feedback ADC.
2. Voltage-to-time conversion method.
3. Dual-slope integration method.
4. Voltage-to-frequency conversion method.

1. Successive Approximation Method: As the name implies, in this method a reference voltage is repeatedly divided by 2 and the result is compared with the analog signal at each step. If the result is higher, the bit is set to a '1' starting from MSB (most significant bit). Or else, the bit is set to a 'O'. The another name for such an ADC is parallel-feedback ADC.

A block diagram of the ADC is shown in Fig. 6.46

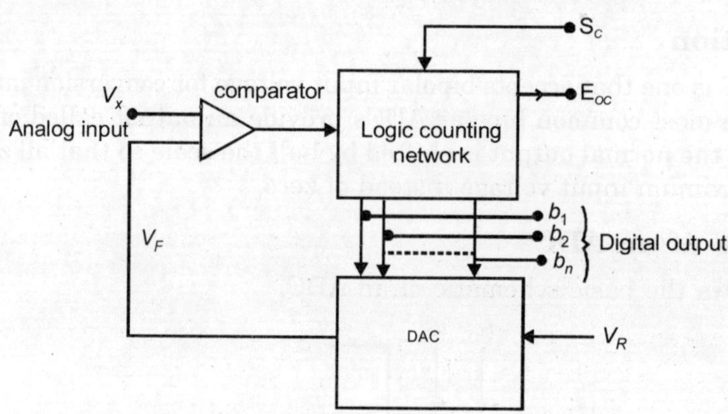

Fig. 6.46 Successive approximation type ADC

The logic circuit is such that it successively sets and tests each bit, starting with the most significant bit of the word. We start with all bits zero. Thus, the first operation will be to set $b_1 = 1$ and test $V_F = V_R 2^{-1}$ against V_X through the comparator. If V_X is greater than V_F, then b_1 will be left at '1' and b_2 is set to '1', and a comparison is made of V_X versus $V_F = V_R (2^{-1} + 2^{-2})$, and so on.

If V_X is less than $V_R\, 2^{-1}$, then b_1 is reset to zero; b_2 is set to 1, and a comparison is made for V_X versus $V_R\, 2^{-2}$. The process is repeated to the least significant bit (LSB) of the word. The operation can be explained best through an example.

Example

What will be the output of a 8-bit successive approximation type ADC if the input is 2.795 *V* and the reference is 5 *V?*

Solution:

1. Set $b_1 = 1$, $V_F = 5 \times 2^{-1} = 2.5$ V, $V_X \rangle\ 2.5$ V, leave $b_1 = 1$

2. Set $b_2 = 1$, $V_F = 2.5 + 5 \times 2^{-2} = 3.75$ V, $V_X < 3.75$, reset $b_2 = 0$
3. Set $b_3 = 1$, $V_F = 2.5 + 5 \times 2^{-3} = 3.125$ V, $V_X < 3.125$,reset $b_3 = 0$
4. Set $b_4 = 1$, $V_F = 2.5 + 5 \times 2^{-4} = 2.8125$, $V_X < 2.8125$, reset $b_4 = 0$
5. Set $b_5 = 1$, $V_F = 2.5 + 5 \times 2^{-5} = 2.6563$, $V_X > 2.6563$, leave $b_5 = 1$
6. Set $b_6 = 1$, $V_F = 2.6563 + 5 \times 2^{-6} = 2.7344$, $V_X > 2.7344$, leave $b_6 = 1$
7. Set $b_7 = 1$, $V_F = 2.7344 + 5 \times 2^{-7} = 2.7735$, $V_X > 2.7735$, leave $b_7 = 1$
8. Set $b_8 = 1$, $V_F = 2.7735 + 5 \times 2^{-8} = 2.7930$, $V_X > 2.7930$, leave $b_8 = 1$

Thus the output of the ADC is 10001111 which denotes the fraction of the reference voltage and thus equaling to $\left(\frac{1}{2}+0+0+0+\frac{1}{32}+\frac{1}{64}+\frac{1}{128}+\frac{1}{256}\right) \times 5 = 2.793$, which is in error by 0.002 V, but less than $\frac{1}{2}$ LSB.

2. Voltage-to-time Conversion Method: A staircase wave form is generated and compared with the input analog signal at each step. A digital counter measures the time period required to match the levels of the input and staircase. This time period is calibrated with voltage. The arrangement is shown schematically in Fig. 6.47 (a). The clock provides pluses at regular intervals. The pulse count increases linearly with time and, therefore the DAC output V_F generates a staircase waveform [Fig. 6.47]. Suppose, initially $V_F < V_X$, then the positive comparator output causes the counter to count UP. But, V_F goes on increasing with every pulse from the clock until it exceeds V_X when the UP-DOWN control line changes and the counter starts counting DOWN. No sooner is V_F lowered by 1 step than the control changes state for counting UP and the count increases by 1 LSB. The process keeps repeating with the result that the digital output reads ± 1 LSB around the correct value.

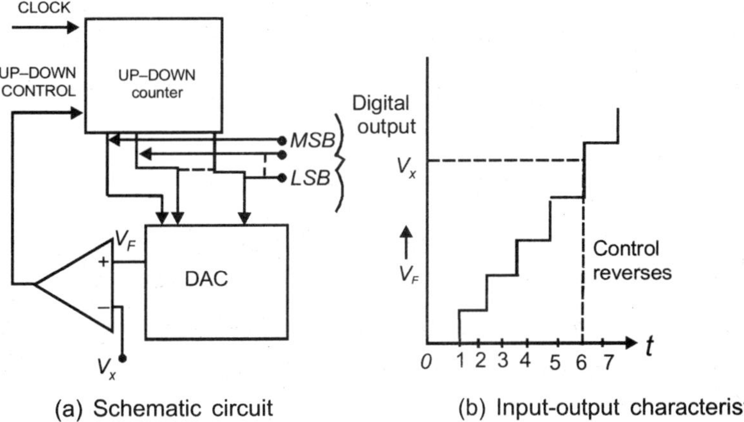

(a) Schematic circuit (b) Input-output characteristics

Fig. 6.47 Voltage-to time conversion type ADC

3. Dual-slope Integration Method: In this method an integrator is used to integrate first an input signal voltage (V_X) for a fixed period of time and then an accurate voltage reference (V_R) with the reverse slope, and the time required to return to starting voltage is measured. The input signal voltage is obtained from the ratio of the time periods and the value of the reference voltage (V_R). A widely used schematic is shown in Fig. 6.48.

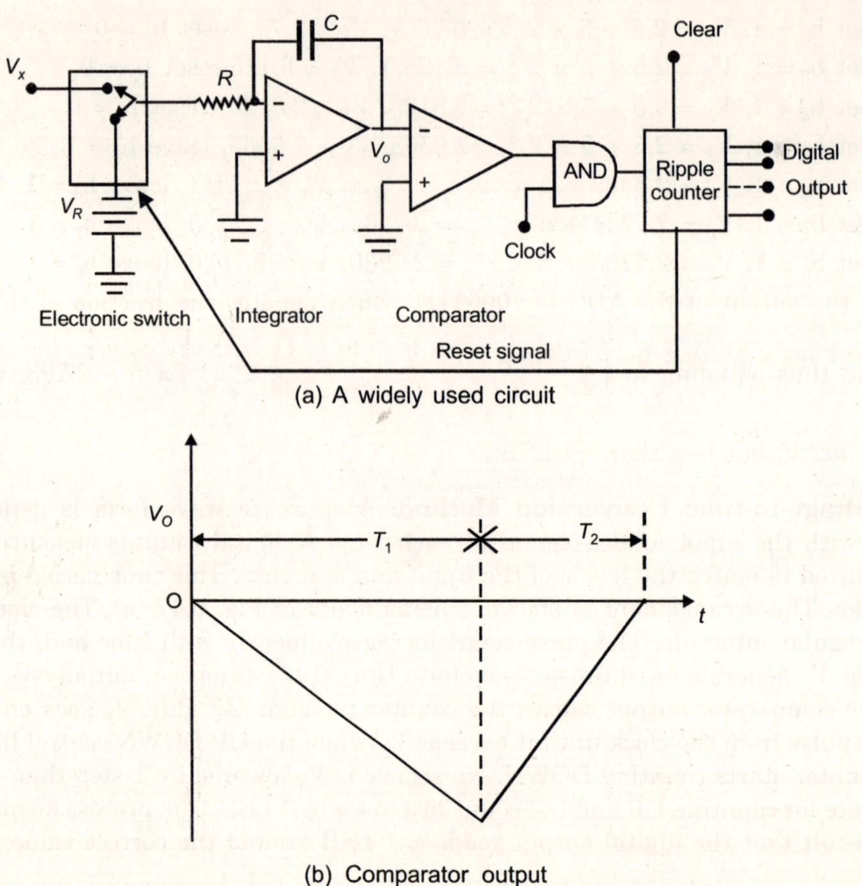

Fig. 6.48 Dual-slope integration type ADC

If T_1 is the fixed time period for which the integrator integrated the input signal voltage V_X, then

$$V_{OT_1} = -\frac{V_X T_1}{RC} \qquad ...(6.43)$$

assuming that the capacitor was completely discharged at the beginning. The negative sign indicates that the input voltage (V_X) is positive. After T_1 seconds, the switch is thrown to the reference voltage V_R. Then the integrator will begin to ramp towards zero at a rate of $\frac{V_R}{RC}$, assuming that V_R is of opposite polarity as V_X. Here, since the integrator starts from V_{OT_1}, we can write

$$V_{OT_2} = V_{OT_1} + \frac{V_R T_2}{RC} = 0 \qquad ....(6.44)$$

where T_2 is the time required to reach zero voltage level. From equations (6.43) and (6.44), it is easy to find

$$V_X = \frac{T_2}{T_1}|V_R| \qquad ...(6.45)$$

The following points may be noted:

1. The equation (6.45) does not include R or C of the integrator.

2. It includes the two time periods in the form of a ratio. Hence a stable rather than an accurate clock is needed.

3. Because the input is integrated over a period of time, the variation of the input signal is averaged and therefore, no sample and hold device at the input is necessary.

The actual operation of the system shown in Fig. 6.48 is explained below:

Initially, the ripple counter is cleared and the switch is connected to V_X. The analog signal is integrated for a fixed number of clock pulses (say 'n_1'). If the clock period is T, after $T_1 = n_1 T$, the ripple counter reads zero. This change of state sends a control signal to the electronic switch and V_R gets connected to the input of the integrator. V_R is negative, but $|V_R|$ is greater than V_X. Hence the integration time T_2 is less than T_1. As long as V_o is negative, the output of the comparator remains high, the AND gate remains enabled and the counter keeps counting pulses. Say, after n_2 counts, i.e., $T_2 = n_2$T, the voltage falls to the level of V_x. At that moment V_o falls to zero, the AND gate is disabled and the counter stops counting pulses.

Equation (6.45) can now be rewritten as

$$V_X = \frac{n_2 T}{n_1 T}|V_R| = n_2 \frac{|V_R|}{n_1} \quad ...(6.46)$$

Since $|V_R|$ and n_1 are constants, $V_x \alpha\, n_2$ which is the count displayed at ripple counter.

4. Voltage-to-Frequency Conversion Method: The voltage-to-frequency conversion method is based upon converting the analog voltage into a variable frequency and then using frequency as input to a counter for a fixed interval of time. The output of the counter is then a measure of the frequency and thus the input voltage.

Figure 6.49 shows a voltage-to- frequency converter. It consists of an integrator that feeds a comparator which, in turn, drives a monostable multivibrator (one-shot). One output of the one-shot feeds a counter and another actuates an electronic switch which discharges the integrator via a current source.

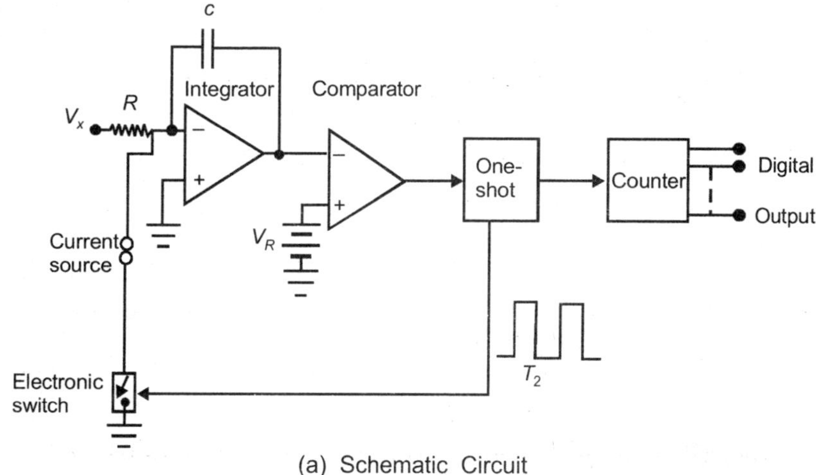

(a) Schematic Circuit

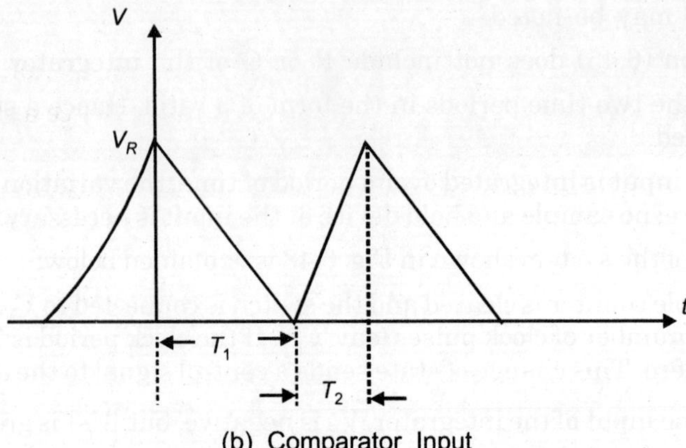

(b) Comparator Input

Fig. 6.49 Voltage-to-frequency conversion ADC

The input voltage V_X to the integrator outputs a ramp with negative slope. The comparator compares the output with reference voltage V_R. No sooner the integrator outputs a zero than the one-shot produces an output pulse. Let T_1 be the time required to reach zero voltage. Then,

$$\frac{V_X}{RC} T_1 = V_R \quad \text{or} \quad T_1 = \frac{V_R}{V_X} RC \qquad \text{...(6.47)}$$

The output pulse of the one-shot actuates the electronic switch to drain current and for the duration of the pulse T_2 the output of the integrator ramps with a positive slope till the voltage level reaches V_R. In fact during this period the capacitor discharges to the desired level of V_R. Hence,

$$V_R = \frac{1}{C}\left[I \quad \frac{V_X}{R}\right] T_2 \quad \text{or} \quad T_2 = \frac{V_R C}{I \quad \frac{V_X}{R}} \qquad \text{...(6.48)}$$

Where I = Current source output. Simplifying equations (6.47) and (6.48), we get

$$T = T_1 + T_2 = \frac{RC}{V_X}\left[\frac{1 \quad \frac{V_X}{R}}{C}\right] T_2 + T_2$$

$$T = \frac{IR}{V_X} T_2 \qquad \text{...(6.49)}$$

Hence, the frequency 'f' of the output signal, which the counter counts, is

$$f = \frac{1}{T} = \frac{V_X}{IRT_2} \qquad \text{...(6.50)}$$

It is clear from the equation (6.50) that $F \alpha V_X$, other factors being constants of the circuitry.

6.3.2 Sample-and-Hold Circuit (SHC)

The finite conversion time of the ADC has serious consequences on the rate of change of signals presented for conversion. An ADC performs the conversion process by referring back to the input signal while conversion is taking place. Obviously, if the input is changing while this process is taking place, errors will occur. Consequently, the ADC output will be in error if the magnitude of the input voltage changes by more than one LSB voltage. To solve this problem, what is needed is simply that the signal not change during the conversion process. Therefore, the answer is to hold the value constant during that process. This is accomplished with a sample-and-hold circuit. A sample-and-hold circuit sample an input signal and holds on to its last sampled value until the input is sampled again.

The basic concept of the sample-and-hold circuit is shown in Fig. 6.50, where the SHC is connected to the input of an ADC. When the electronic switch is closed, the capacitor voltage will track the input voltage, $V_C(t) = V_{in}(t)$. At some time, t_s, when a conversion of the input voltage is desired, the electronic switch is opened, isolating the capacitor from the input. Thus, the capacitor will hold (stay charged) to the voltage when the switch opened. $V_C = V_{in}(t_s)$.

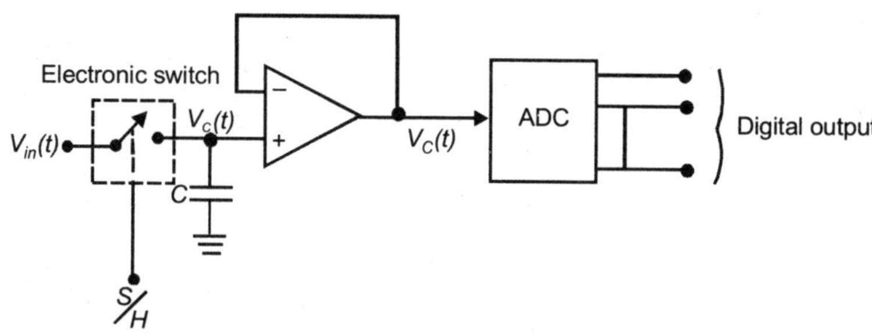

Fig. 6.50 Sample-and-hold circuit-basic concept

The voltage follower allows this voltage to be impressed upon the ADC input, but the capacitor does not discharge because of the very high input impedance of the follower. The start-convert is then issued, and the conversion proceeds with the input voltage remaining constant. When the conversion is complete, the electronic switch is reclosed and tracking continues until another conversion is needed. FET can be used as electronic switch in S/H circuits as shown in Fig. 6.51.

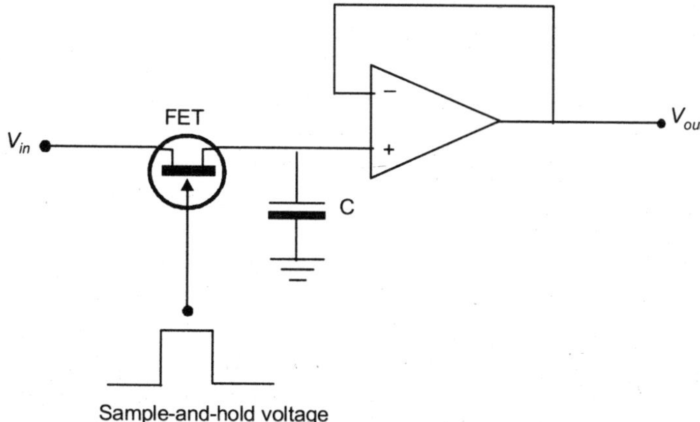

Fig. 6.51 FET as an electronic switch for S/H unit

6.3.3 Microprocessor-Compatible Converters: (Computer Interface)

Figure 6.52 shows a simple model of a computer system. The processor is connected to external equipment via three parallel sets of digital lines. The 'data lines' carry data to and from the processor. The 'address lines' allow the computer to select internal location for input and output. The 'control lines' carry information to and from the computer related to operations, such as reading, writing, interrupts, and so on. This collection of lines is called the 'bus' of the computer. The term 'interface' refers to the hardware connections and software operations necessary to input and output data using connections to the bus.

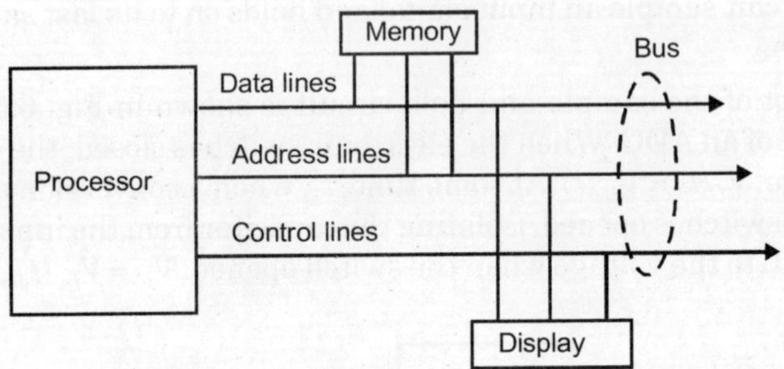

Fig. 6.52 Generic model of a computer bus system

A whole line of ADCs have been developed that interface easily with microprocessor-based computers. The ADCs have built-in tri-state outputs so that they can be connected directly to the data-bus of the computer. Data from ADC are placed on the data bus lines only when the computer issues an appropriate enable command (often called a READ). Fig. 6.53 shows how the ADC appears when connected to the environment of the microprocessor-based computer. The ADC appears much the same as memory (or display). The decoding circuitry is necessary to provide the start-convert command, to input the convert-complete response from the ADC, and to issue the tri-state enable back to ADC.

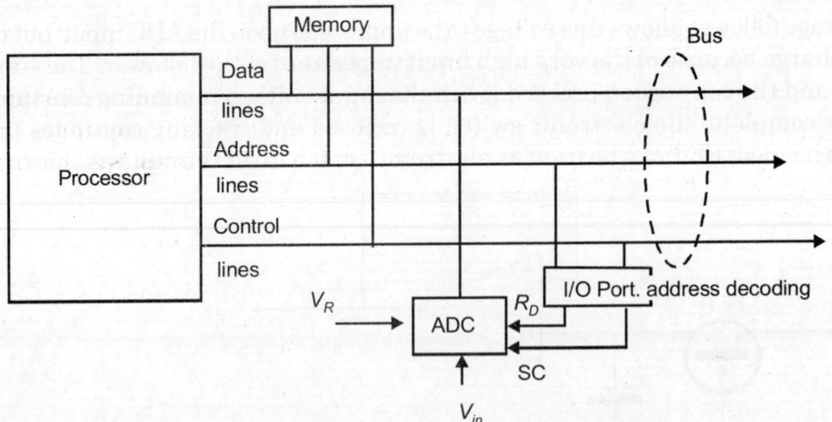

Fig. 6.53 Computer compatible ADC

6.3.4 Data-Acquisition Systems

To implement direct digital control in the process industries, microprocessor-based personal computers are used extensively. They are designed using a bus that consists of the data lines, address lines, and control lines as already shown in Fig. 6.53. Essential equipment like RAM, ROM, disk, and CD_ROM are all communicated via these bus lines with the processor. The computer also connects the bus lines to a number of printed circuit board (PCB) sockets, using an industry standard configuration of how the bus lines are connected to the socket. These sockets are referred to as expansion slots. Special PCBs called data-acquisition systems (DASs) have been developed for the purpose of providing for input and output of analog data. These are used when the personal computer is to be used in a control system. We will be seeing in short about the hardware and software of data-acquisition system in this section.

6.3.4.1 DAS Hardware

The hardware equipment include analog multiplexer, ADC with sample and hold circuit, Address decoder/command processor and DAC with latch as shown in Fig. 6.54. DAS will have number of analog input channels. The analog multiplexer (MUX) allows the DAS to select data from a number of analog inputs. The multiplexer acts like a multiple set of switches arranged in such a fashion that any one of the input channels can be selected to provide its voltage to the ADC through sample and hold circuit. In some cases, the DAS can be programmed to take channel samples sequentially.

Whenever the DAS is requested to obtain a data sample, the S/H is automatically incorporated into the process. The ADC conversion time constitutes the major part of the data sample acquisition time, but the S/H acquisition time must also be considered to establish maximum throughput with the help of address decoder/command processor. The computer can select to input a sample from a given channel by sending an appropriate selection on the address lines and control lines of the computer bus. These are decoded to initiate the proper sequence of commands to the MUX, ADC, and S/H. Another common feature is the ability to program the DAS to take a number of samples from a channel with a specified time between samples. In this case, the computer is notified by interrupt when a sample is ready for input.

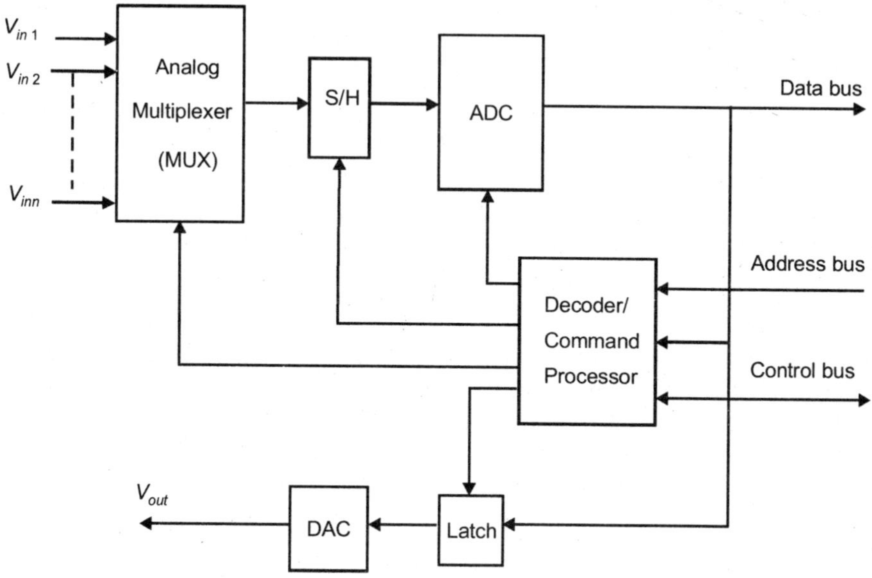

Fig. 6.54 Layout of a DAS board

For output purposes, the DAS often includes a latch and DAC. The address decoder/command processor is used to latch data written to the DAS, which is then converted to an appropriate analog signal by the DAC.

6.3.4.2 DAS Software

The process of selecting a channel and initiating a data input from that channel involves some interface between the computer and the DAS. This interface is facilitated by software that the computer executes. The software is often provided by the DAS manufacturer in the form of programs on disk. The software can also be written by the user.

The flow chart of the basic sequence of operations that must occur when a sample is required from the DAS is shown in Fig. 6.55.

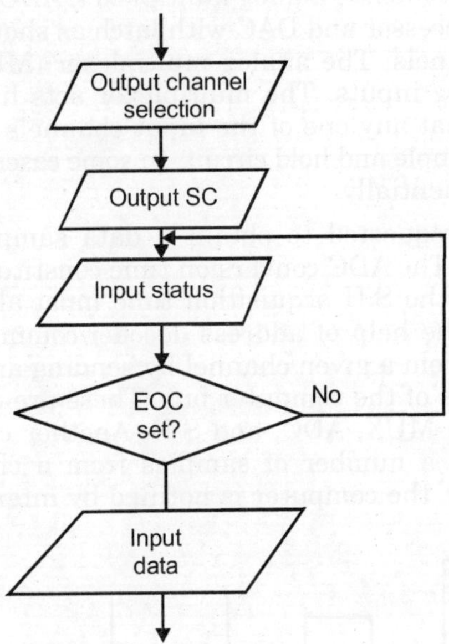

Fig. 6.55 DAS software-flow chart

The sequence starts with selection of a channel for input. This is accomplished by a write to the DAS decoder that identifies the required channel. The MUX then places that channel input voltage at the S/H unit. The software then issues a start-convert (SC) command according to the specifications of the DAS. This is often accomplished by a write to some base + offset address. The DAS internally activates the hold mode of the S/H and starts the converter.

The end-of-convert (EOC) is provided in a status register in the DAS. The contents of this status register can be read by the processor by a port input of a base + offset address. The appropriate bit is then tested by the software to deduce whether the EOC has been issued. Once the EOC has been issued, the software can input the data itself by a read of an appropriate address, again a base + offset, which enable tri-states, placing the ADC output on the data bus.

6.4 PROBLEMS AND SOLUTIONS

Problem 6.1

A temperature sensor, thermocouple, gives a voltage proportional to temperature with a sensitivity 0.5 mv / °C. The sensor has an output resistance of 2.0 k Ω. If the sensor is connected to an amplifier with input resistance of 5 kΩ and gain of 200, find the amplifier output for a temperature of 100°C,

(*a*) *Neglecting the loading effects.*

(*b*) *Taking loading effects into account.*

Solution:

(*a*) Neglecting the loading effects: The unloaded output of the sensor

$$V_T = 0.05\text{mv/°C} \times 100\text{°C} = 5 \text{ mv}$$

The output of the amplifier = 5 mv × 200 = 1000 mv = 1 V

(*b*) Taking loading effects into account:

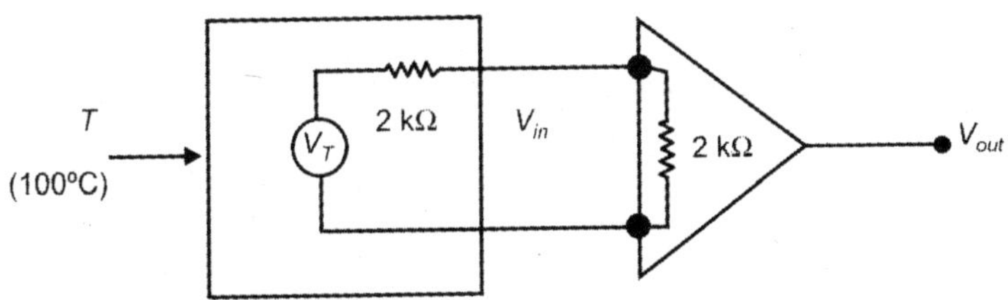

Fig. 6.56 Equivalent circuit for problem 6.1

$$V_{in} = V_T\left[1 - \frac{2.0\text{k}\Omega}{2\text{k}\Omega + 5\text{k}\Omega}\right]$$

Where $V_T = 5$ mv

so that

$$V_{in} = 5\,[1 - 0.2857]$$
$$= 3.5714$$
$$V_{out} = 3.5714 \times 200 = 71428 \text{ mv}$$
$$= 0.7143 \text{ V}$$

Problem 6.2

The resistance of a sensor varies from 5 kΩ to 20 kΩ for a dynamic variable over a range. If the sensor is connected in a divider circuit with constant resistance of 15 kΩ and supply voltage 10 V, then find

(*a*) *The minimum and maximum of output voltage,*

(*b*) *The range of output impedance,*

(*c*) *The range of power dissipated by the sensor.*

Solution:

The sensor and the divider circuit can be represented as in Fig. 6.57.

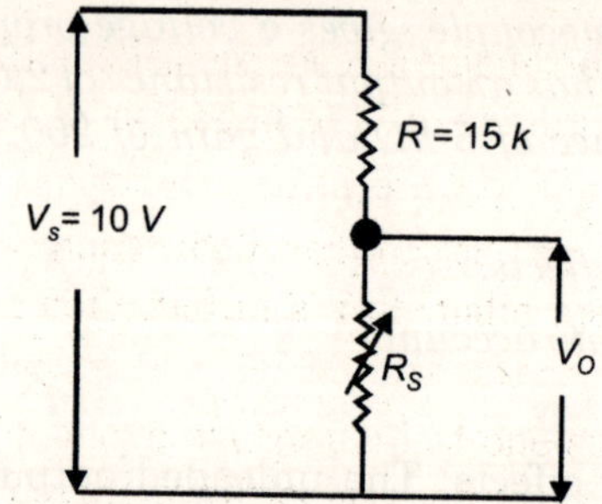

Fig. 6.57 Divider circuit (Problem 6.2)

(*a*) The minimum output voltage $= V_{o\,min} = \dfrac{R_{S\,min} \times V_S}{R + R_{S\,min}}$

$$= \frac{5\ \text{k}\Omega \times 10\ V}{(15\ \text{k}\Omega + 5\ \text{k}\Omega)} = \frac{50}{20} = 2.5\ \text{V}$$

The maximum output voltage $= V_{Omax} = \dfrac{R_{S\,max} \times V_s}{R + R_{Smax}}$

$$= \frac{20\ \text{k}\Omega \times 10V}{15\ \text{k}\Omega + 20\text{k}\Omega} = \frac{200}{35} = 5.7\ \text{V}$$

(*b*) The minimum output impedance = Parallel combinition of R_s min & R

$$= \frac{R_{S\,min} \times R}{R_{S\,min} + R} = \frac{5 \times 15}{5 + 15} = \frac{75}{20}$$

$$= 3.75\text{k}\Omega$$

Similarly the maximum output impedance $= \dfrac{20 \times 15}{20 + 15} = \dfrac{300}{35} = 8.57\,\text{k}\Omega$

(*c*) The range of power dissipated $= \dfrac{(V_{omin})^2}{R_{Smin}}$ *to* $\dfrac{(V_{omax})^2}{R_{S\,max}}$

= 1.25 mw to 3.79 mw.

Problem 6.3

A Wheatstone bridge circuit has resistance of $R_1 = R_2 = R_3 = 5.00\ k\Omega$ and $R_4 = 5.5\ k\Omega$ and a 10.00 V power supply. If a galvanometer with a 100 Ω internal resistance is used for a detector, find the offset current.

Solution:

Refer Fig. 6.7 and Fig. 6.8. From equation (6.4), the offset voltage is V_{TH}.

$$V_{TH} = V\left[\frac{R_3R_2 - R_1R_4}{(R_1+R_3)(R_2+R_4)}\right]$$

$$= 10\left[\frac{5\times5 - 5\times5.5}{(5+5)(5+5.5)}\right] = \frac{-2.5}{10\times10.5} = -23.8 \text{ mv}$$

$$V_{TH} = -23.8 \text{ mV}$$

From equation (6.5),

$$R_{TH} = \frac{R_1R_3}{R_1+R_3} + \frac{R_2R_4}{R_2+R_4} = \frac{5\times5}{5+5} + \frac{5\times5.5}{10.5}$$

$$= \frac{25}{10} + \frac{27.5}{10.5} = 2.5 + 2.62 = 5.12$$

$$R_{TH} = 5.12 \text{ k}\Omega$$

From equation (6.6),

$$I_G = \frac{V_{TH}}{R_{TH}+R_G} = \frac{-23.8}{5.12+0.1} = \frac{-23.8 \text{ mv}}{5.22 \text{ k}\Omega}$$

$$= -4.56\ \mu\text{A}$$

The offset current of 4.56 μA flows from '*b*' to '*a*'

Problem 6.4

A current balance bridge, as shown in Fig. 6.10, has resistors $R_1 = R_2 = 15\ k\Omega$, $R_3 = 2\ k\Omega$, $R_4 = 1.5\ k\Omega$ and $R_S = 0.5\ k\Omega$. Assuming the impedance of the null detector very high, find the current required to null the bridge if R_3 changes by $5\ \Omega$. The supply voltage is 10V.

Solution:

To verify that the bridge is at null with $I = 0$ for the nominal resistance values, we may calculate V_a and V_b.

$$V_a = \frac{(10\text{ V})(2\text{ k}\Omega)}{(15+2)\text{k}\Omega} = 1.1765\text{V} \qquad \left[\because V_a = \frac{V\times R_3}{R_1+R_3}\right]$$

$$V_b = \frac{(10\text{V})(1.5+0.5)\text{k}\Omega}{(15+1.5+0.5)\text{k}\Omega} = 1.1765\text{V} \left[\because V_b = \frac{V\times(R_4+R_5)}{R_2+R_4+R_5}\right]$$

Since $V_a = V_b = 1.1765$ V when $I = 0$, we are sure that the bridge is at a null.
When R_3 changes by 5 Ω to 2.005 kΩ

$$V_a = \frac{10\times2.005}{15+2.005} = 1.1791$$

Hence the voltage at b must increase by (1.1791 – 1.1765) V or 2.6 mv to renull the bridge.

That means, the current I should flow through R_5 to satisfy

$$\Delta V = IR_5$$

$$2.6\ mv = I \times 0.5\ k\Omega$$

$$\therefore \quad I = \frac{2.6\ mv}{0.5 k\Omega} = 5.2\ \mu A.$$

Problem 6.5

A current balance bridge, as shown in Fig. 6.10 has resistors $R_1 = R_2 = 5\ k\Omega$, $R_4 = 1800\Omega$, $R_3 = 2\ k\Omega$, $R_5 = 200\Omega$ and a high impedance null detector. Find the current required to null the bridge if R_3 changes by 5Ω. The supply voltage is 10V.

Solution:

$$V_a = \frac{VR_3}{R_1 + R_3} = \frac{10\ V \times 2\ k\Omega}{(5+2)\ k\Omega} = \frac{20}{7} = 2.857\text{V}$$

Assuming $I = 0$,

$$V_b = \frac{V(R_4 + R_5)}{(R_2 + R_4 + R_5)} = \frac{10 \times (1800 + 200)\Omega}{5\text{k}\Omega + 1800\ \Omega + 200\ \Omega}$$

$$= \frac{10 \times 2\ \text{k}\Omega}{7\text{k}\Omega} = \frac{20}{7} = 2.857\ \text{V}$$

∴ For the nominal resistance value given, the bridge is at a null with $I = 0$.

When R_3 increases by 5 Ω to 2005 Ω,

$$V_a = \frac{10 \times 2005}{5000 + 2005} = 2.862\text{V}$$

which shows that the voltage at 'b' must increase by (2.862 – 2.857) V or 5 mv to renull the bridge. This can be provided by a current I satisfying the equation

$$IRS = \Delta V$$

$$\therefore \quad I \times 200 = 5\ \text{V}.$$

$$\therefore \quad \text{I} = \frac{5 \times 1000}{200} \mu\text{A} = 25\ \mu\text{A}$$

Problem 6.6

Consider a bridge circuit shown in Fig. 6.11 for potential measurement. The values of R_1 and R_2 are 2 kΩ and that of R_3 = 1 kΩ. The power supply is 15 V.

(*a*) *Assuming R_4 being a variable resistance and the bridge is nulled at $R_4 = 1.2\ k\Omega$, find the unknown potential.*

(*b*) *If the bridge is of current balancing type with $R_4 = 0.8\ k\Omega$ and $R_5 = 0.2\ \Omega$, find the current necessary to null the bridge for the same potential.*

Solution:

(*a*) Using equation (6.7) for null condition

$$\text{unknown potential} = V_X = \frac{R_4 V}{R_2 + R_4} - \frac{R_3 V}{R_1 + R_3}$$

$$= \frac{1.2 \times 15}{2 + 1.2} - \frac{1 \times 15}{2 + 1}$$

$$= 5.625 - 5 = 0.625\text{V}$$

$$V_X = 625\ \text{mv}.$$

(*b*) Since $R_1 = R_2$ and $R_3 = R_4 + R_5$, the bridge will be nulled with $V_X = 0$ and $I = 0$. When V_X is present, I has to be supplied to null the bridge as per equation (6.8).

$$V_X - IR_5 = 0$$

$$\therefore \qquad I = \frac{V_X}{R_5} = \frac{625\ mv}{0.2\ k\Omega} = 3.125\,\text{mA}$$

Problem 6.7

A band pass filter consists of two RC networks connected in cascade. The low pass filter consists of a resistor $R_L = 20\ k\Omega$ and $C_L = 100\ pf$ and the high pass filter consists of $R_H = 2\ M\Omega$ and $C_H = 0.01 \mu f$. Find the lower and upper cut off frequencies, the pass band and the pass band gain.

Solution: Refer Fig. 6.19.

$$\text{Lower cut off frequency } f_L = \frac{1}{2\pi R_H C_H}$$

$$= \frac{1}{2\pi \times 2 \times 10^6 \times 0.01 \times 10^{-6}}$$

$$= 7.96 \text{ Hz}$$

$$\text{Upper cut off frequency } f_H = \frac{1}{2\pi R_C C_L}$$

$$= \frac{1}{2\pi 20 \times 10^3 \times 100 \times 10^{-12}}$$

$$= \frac{10^6}{4\pi} = 79.6\,\text{kHz}$$

$$\text{Pass band} = f_H - f_L = 79.6 \text{ kHz} - 7.96 \text{ Hz}$$

$$= 79.59 \text{ kHz}$$

$$\text{Pass band gain} = A = \frac{R_H}{R_L + R_H} = \frac{2 \times 10^6}{20 \times 10^3 + 2 \times 10^6}$$

$$= 0.99$$

Problem 6.8

A type – J thermocouple (Fe-K) is used to measure a furnace temperature for a range of 0°C to 600°C. Develop a signal conditioning circuit to give an output of 1 to 5 V for the range. The circuit must have very high input impedance.

Solution:

This problem has to be approached in three steps.

Step 1: Find out the mv-output generated by the thermocouple for 0°C and 600°C

Step 2: Develop an equation for the output of the signal conditioning circuit and the input which is the output of thermocouple.

Step 3: Design a suitable circuit to satisfy the above equation and with high input impedance.

Step-1: The mv-output for an Iron-constanton (Fe-K) thermocouple can be taken from standard 'temp.–mv' table.

$$0°C \rightarrow 0\ mv$$
$$600°C \rightarrow 33.096\ mv \simeq 0.0331V$$

Step-2 : We can write a straight line equation relating input and output satisfying the conditions.

$V_{out} = m\ V_{in} + V_o$ where 'm' is the slope of the line and represents gain ($m > 1$) or attenuation ($m < 1$) required, and V_o is the bias or intercept, that is, the value V_{out} would be V_o if V_{in} = O. Thus we form two equations to solve for m and V_o for the conditions of the problem

$$1 = m\ (0) + V_o$$
$$5 = m\ (0.0331) + V_o$$

Wet get $V_o = 1$ and $m = 120.00$ and hence the equation becomes

$$V_{out} = 120\ V_{in} + 1$$
$$= 120\ (V_{in} + 0.008)$$

Step-3: A differential amplifier with a gain of 120 and a fixed input of –0.008 volts to the inverting side will do the function. One such arrangement is shown below. Voltage divider and zener diode are used to get a constant bias voltage of –0.008 V.

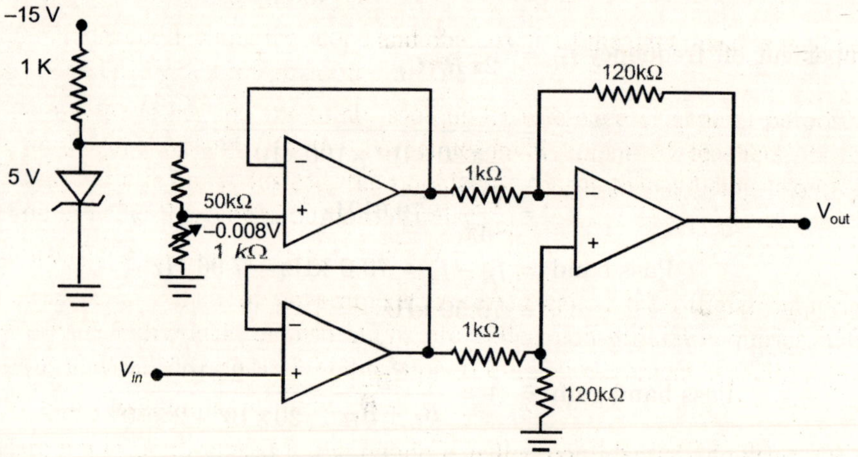

Fig. 6.58 Circuit for problem 6.8

Problem 6.9

A Pt 100 RTD is used to measure the temperature of a bearing in turbo generator. The range of operation is from 100°C to 400°C. Develop an analog signal conditioner to provide 4 to 20 mA standard output signal for this temperature range.

Solution:

Step 1: To find the resistance values of Pt 100 RTD for 100°C and 400°C, refer the standard tables:

$$100°C \rightarrow 138.50\ \Omega$$
$$400°C \rightarrow 247.06\ \Omega$$

Step 2: The resistance variation of 138.50 Ω to 247.06 Ω can be conditioned to get 1 to 5 V (a standard output voltage) first.

$$V_{out} = m\,R_S + V_o$$

Two equations can be formed with available information.

$$1 = 138.5 \text{ m} + V_o \qquad \text{...(1)}$$

$$5 = 247.06 \text{ m} + V_o \qquad \text{...(2)}$$

$$2-1 \text{ gives } 4 = 108.56 \text{ m}$$

$$\therefore \quad \text{m} = 0.0368$$

Solving for V_o from 1

$$V_o = 1 - 138.5 \times 0.0368 = -4.0968$$

$$\therefore \quad V_{out} = 0.0368\, R_s - 4.0968$$

Step 3: Voltage to current converter has to follow suitably to convert 1 to 5 V into a 4 to 20 mA signal. Fig. 6.30 may be considered for this purpose. It we make $R_1 = R_2 = 1$ kΩ in Fig. 6.30, then the equation (6.27) becomes $I = \dfrac{V_{in}}{R_3}$.

for $$V_{in} = 1\text{V},\ 4 \text{ mA} = \frac{1}{R_3}$$

$$\therefore \quad R_3 = \frac{1}{4}\text{k}\Omega = 250\Omega$$

This is also true when $V = 5\text{V}$ to get 20 mA

$$I = \frac{5\text{ V}}{250\ \Omega} = 20 \text{ mA}$$

We may assume $R_5 = 0\ \Omega$ and select

$$R_3 = R_4 = 250\ \Omega.$$

This completes the voltage to current converter

Now the total circuitry may be drawn as shown in Fig. 6.59.

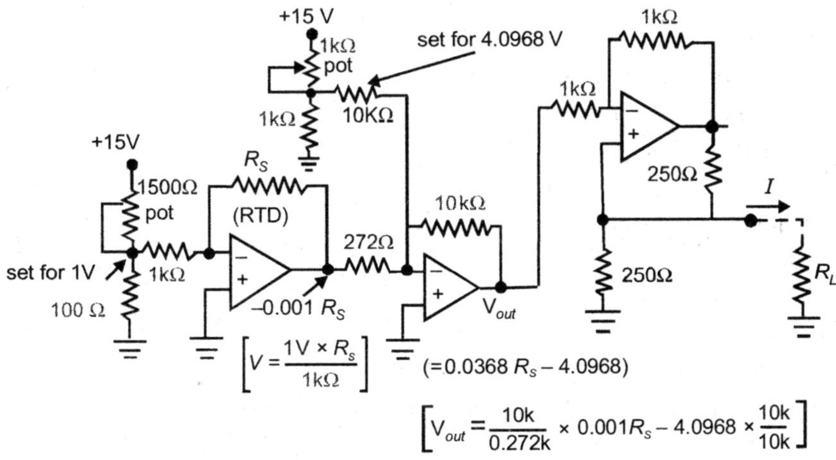

Fig. 6.59 Circuit for problem 6.9

Problem 6.10

A control valve has a linear variation of opening as the input voltage varies from 1 to 5 V. A micro computer outputs an 8-bit word to control the valve opening using an 8-bit DAC to generate the valve voltage.

(*a*) *Find the reference voltage required to obtain a full open valve (5 V).*

(*b*) *Find the percentage to valve opening for a 1-bit change in the input word.*

Solution:

(*a*) The full open-valve condition occurs with a 5 V input. If a 5 V reference is used, a full digital word 11111111_2 will not quite give 5 V, so we use a larger reference. Thus, we have using equation (6.36)

$$V_{out} = V_R\,[b_1 2^{-1} + b_2 2^{-2} + \text{---------------} + b_n 2^{-n}]$$

$$5 = V_R\left[\frac{1}{2}+\frac{1}{4}+\frac{1}{8}+\frac{1}{16}+\frac{1}{32}+\frac{1}{64}+\frac{1}{128}+\frac{1}{256}\right]$$

$$V_R = \frac{5}{0.9961} = 5.02\,\text{V}$$

(b) The percentage of valve change per step is found first from

$$\Delta V_{out} = V_R 2^{-8}$$

$$= 5.02 \times \frac{1}{256}$$

$$= 0.0196$$

Thus,

$$\text{Percent} = \frac{0.0196}{(5-1)} \times 100 = 0.49\%$$

Problem 6.11

Temperature is measured by a sensor with an output of 0.05 mv / °C and further amplified by a factor of 100. Determine the required ADC reference and word size to measure 0°C to 400°C with 0.4°C resolution.

Solution:

At the maximum temperature of 400°C the voltage output from the sensor is

$$(0.05\ \text{mv/°C} \times 400\text{°C}) = 20\ \text{mv}$$

After amplification, it becomes

$$20\ \text{mv} \times 100 = 2000\ \text{mv} = 2\ \text{V}$$

So a 2 V reference is used.

A change of 0.4°C results in a voltage change of

$$0.4 \times 0.05 \times 100 = 2.0\ \text{mv}$$

So we need a word size where

$$0.002\ \text{V} = 2\,(2^{-y})$$

Choose a size 'n' that is the integer part of y plus 1. Thus, solving with logarithms, we find

$$y = \frac{\log(2) - \log(0.002)}{\log(2)}$$

$$= 9.996 \simeq 10$$

so, a 10-bit word is required for this resolution. A 10-bit word has a resolution of

$$V = (2)(2^{-10}) = 0.00195 \text{ V} = 1.95 \text{ mv}$$

which is better than the minimum required resolution of 2 mv.

Problem 6.12

A measurement of temperature using a transducer that outputs 5.0 mv / °C must measure to 120°C. A 6-bit ADC with a 10 V reference is used.

(*a*) *Develop a circuit to interface the sensor and ADC.*

(*b*) *Find the temperature resolution.*

Solution:

The transducer output at 120°C will be

$$(5.0 \text{ mv}/°\text{C}) \times 120°\text{C} = 0.6 \text{ V}$$

(*a*) The interface circuit must provide a gain so that at 120°C the ADC output is 111111. The input voltage that will provide this output is found from

$$V_X = V_R\left(2^{-1} + 2^{-2} + 2^{-3} + 2^{-4} + 2^{-5} + 2^{-6}\right)$$

$$= 10\left(\frac{1}{2} + \frac{1}{4} + \frac{1}{8} + \frac{1}{16} + \frac{1}{32} + \frac{1}{64}\right)$$

$$= 9.84375 \text{ V}$$

Thus, the required gain must provide this voltage when the temperature is 120°C

$$\text{gain} = \frac{9.84375}{0.6} = 16.41$$

The op amp circuit of Fig. 6.60 will provide this gain.

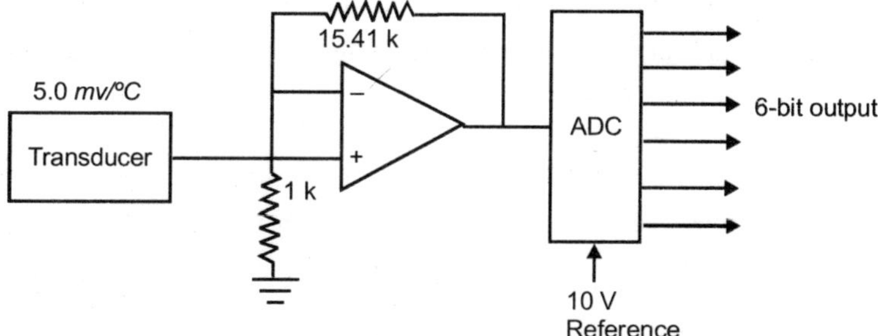

Fig. 6.60 Circuit for problem 6.12

(*b*) The temperature resolution can be found by working backward from the least significant bit (LSB) voltage change of the ADC

$$\Delta V = V_R 2^{-n}$$

$$= 10(2^{-6}) = 0.15625$$

This corresponds to a transducer change of

$$\Delta V_T = \frac{0.15625}{16.41} = 0.00952$$

or a temperature of

$$\Delta_T \frac{0.00952}{0.005\,V/°C} = 1.9°C$$

6.5 PROBLEMS AND QUESTIONS

1. What is meant by signal conditioning?
2. Differentiate sensor, transducer and signal conditioner.
3. A sensor with a nominal resistance of 50 Ω is used in a bridge with $R_1 = R_2 = 100\ \Omega$, $V = 10.0\,V$, and $R_3 = 100\ \Omega$ potentiometer. It is necessary to resolve 0.1 Ω changes of the sensor resistance.

   (*a*) At what value of R_3 will the bridge null?

   (*b*) What voltage resolution must the null detector possess?
4. Draw the circuit diagram of a non-inverting amplifier for a gain of 10 using an op amp. Use minimum resistance value as 10 kΩ.
5. What are the characteristics of an ideal op amp?
6. What is an instrumentation amplifier? What are the important features of an instrumentation amplifier?
7. Define CMRR of an op amp.
8. What are the different types of filters used in instrumentation systems?
9. Draw a typical circuit diagram of an instrumentation amplifier.
10. How can you make a unity gain amplifier with an op amp?
11. An ac Wheatstone bridge with all arms as capacitors nulls when $C_1 = 0.4\ \mu F$, $C_2 = 0.3\ \mu F$ and $C_3 = 0.27 \mu F$. Find C_4
12. Develop a low-pass RC filter to attenuate 0.5 MHz noise by 97%. Specify the critical frequency, values of R and C, and the attenuation of a 400 Hz input signal.
13. Signal – Conditioning analysis shows that the following equation must relate output voltage to input voltage.

    $$V_{out} = 3.35\ V_{in} - 2.5$$

    Design circuits to do this using (*a*) a summing amplifier and (*b*) a differential amplifier.
14. A control system needs the average of temperature from three locations. Sensors make the temperature information available as voltages V_1, V_2, and V_3. Develop an op amp circuit that outputs the average of these voltages.
15. A sensor varies from 1 to 3 kΩ. Use this in an op amp circuit to provide a voltage varying from 1 *V* to 5 *V* as the resistance changes.
16. A pressure sensor outputs a voltage varying as 100 mv/kg/cm^2 and has a 2.5 kΩ output impedance. Develop signal conditioning to provide 0 to 5 V as the pressure varies from 50 to 150 kg/cm^2.
17. What overall accuracy can one expect from the construction of a 16-bit ADC?
18. What do you mean by the resolution of a digital to analog converter (DAC)?
19. Why do you need a sample and hold circuit?
20. What are the methods used for DACs?
21. What are the methods used for ADCs?

22. What is the resolution of an ADC?
23. A sensor provides temperature data as 400 μV/°C. Develop a comparator circuit that goes high when the temperature reaches 500°C.
24. An 8-bit DAC has an input of 11001011_2 and uses a 10 V reference.
    (*a*) Find the output voltage produced.
    (*b*) Specify the conversion resolution.
25. An 8-bit ADC has 15 V reference. Find the output for inputs of 3.5V, 7.8 V and 12 V.

7

DIGITAL CONTROLLERS—COMPUTER-BASED CONTROL

7.1 INTRODUCTION

Modern control system implementations are carried out using computers which are digital. The measurement, final control operations, the strategy for control and the modes of controller action are still the same as what we have already seen under analog controllers. But the functions of the controller have been taken over by a computer. The computer inputs measurement data, determines the error, solves the controller mode equations to determine the feedback, and transmits this feedback to the final control element.

Technology of networks and network communication has become handy to exchange information between computers over networks. The widely used networks are local-area networks (LAN), wide-area networks (WAN) and world wide network (WWW). This concept has been carried over to control systems in the name of 'field buses'. This has led to the development of distributed control system (DCS) wherein sensors, computers, and final control operation signals are exchanged over a common network called the 'field bus'.

There are still situations in control systems wherein basic digital electronics can provide the needed action. Single variable and multivariable alarms, two-position control etc. are some such examples. In this chapter we may not discuss all those systems, but briefly see the computer based process control system.

7.2 COMPUTER-BASED CONTROLLER

Computer-based controller is the most important element in a modern process control systems. Measurement data from the plant process like temperature, pressure, flow, level, concentration, density etc. are inputs to the computer. Computer performs all the calculations for controller modes. The output from the computer is converted into a suitable signal to operate the final control elements in the field. Such an application is historically referred to as direct digital control (DDC).

The DDC directly interfaces to the process for data acquisition and control purpose. It has necessary hardware for directly interfacing (opto-isolator, signal conditioner, ADC. etc) and reading the data from process. It should also have memory and arithmetic capability to execute required P, PI and PID control strategy. At the same time, the interface to control valve (final control element) should also be part of DDC. Fig. 7.1 shows the various functional blocks of a direct digital control system.

The functional blocks in combination with microprocessor are shown. The multiplexer acts like a switch under microprocessor control. It switches and presents at its output the analog signal from a sensor/transmitter. The analog-to-digital converter converts the analog signal to digital value. The microprocessor performs the following tasks:

1. It 'reads' the various process variables from different transmitters through multiplexer and ADC.
2. It 'determines' the error for each control loop and executes control strategy for each loop.
3. It 'output's the correction value to control valve through DAC.

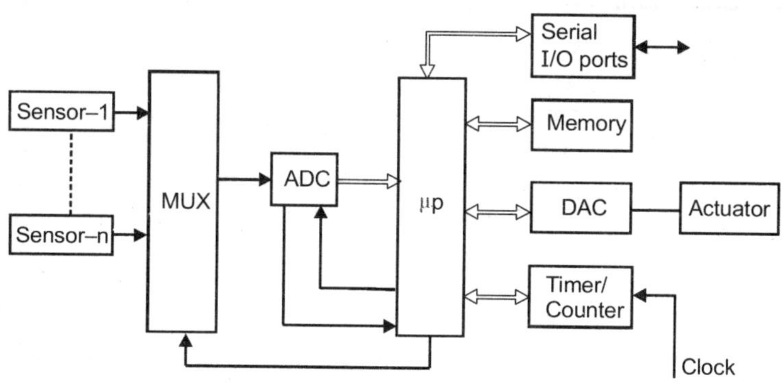

Fig. 7.1 Direct digital control

There are many hardware configurations of the above system. The controlling computer could be similar to a microprocessor-based personal computer (PC) mounted in a rack or even on a desktop. It could be a microprocessor-based computer on a small printed circuit board mounted inside measurement equipment or even a large computer. In this section we will explore the hardware and software configuration of typical process-control computer.

7.2.1 Hardware Configurations

A simplified microprocessor-based computer with a standard array of devices is shown in Fig. 7.2. As all the devices are available as small integrated circuits they can be mounted on a relatively small printed circuit board (PCB). The ROM (Read only memory) a non-volatile memory, holds the programs that the processor executes. The RAM (Random Access Memory or Read-write Memory) is used to hold the transient results of calculations and other results of data processing. The data I/O typically consists of ADCs and DACs as well as digital I/O channels. The network interface card provides for serial communication of the computer over a serial field bus or LAN.

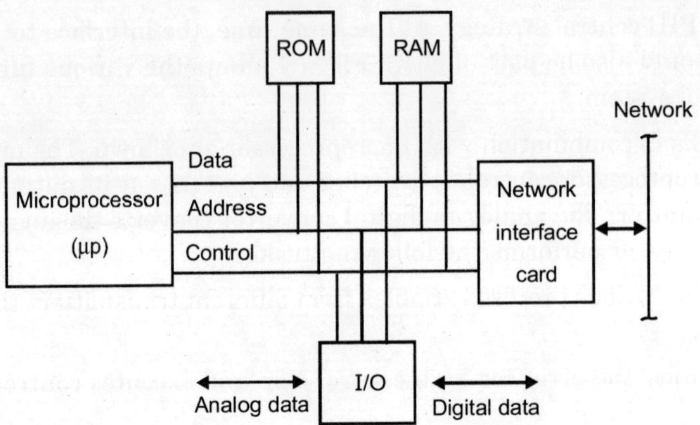

Fig. 7.2 Basic microprocessor-based computer

7.2.1.1 Single Loop Controller

Single loop controller basically controls single variable. It measures the error occurring in one plant variable and attempts to correct it by applying a change to the operating level of one and only one plant control variable. Here a number of single loop controllers are required to control a process. These controllers can be located in the centralized room and receive analog inputs from the sensors and output the analog output signal to the final control elements in the field. Controller simply does the comparison, solves the control mode requirements and supplies the necessary output for the parameter concerned.

Smart Sensors

It is also possible to embed the controller computer directly into a sensor to make it smart sensor. Fig. 7.3 shows one possible implementation of such a system for controlling flow rate through a process pipe line. The sensor and computer are housed directly at the site of the measurement. The feedback signal is delivered to the valve via the standard 4-20 mA current transmission. Operation of the flow control loop is monitored via the serial interface, which is also used to update the set point, controller mode gains, and the operating parameters. It is also possible to eliminate the 4-20 mA connection if the signal-conditioning system of the control valve actuator contains a network interface circuit so that it can be connected to the serial bus. In this case the smart sensor sends feedback information to the valve via the serial bus.

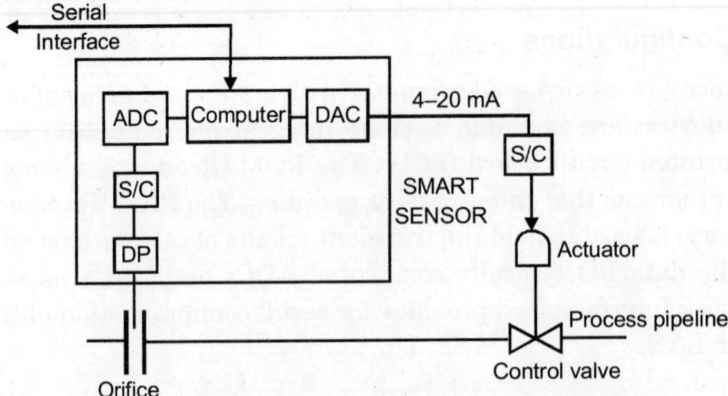

Fig 7.3 Smart sensor based single loop controller

7.2.1.2 Multiple-Loop Controllers

A single computer can be used for controlling more number of process control loops instead of controlling only one loop as in single loop controller. This helps to take care of interactions between loops in a process. This is also advantageous for economy point of view. Such multi-loop control is feasible as the computers are fast enough to take care of process variations. Fig. 7.4 shows how a process could be placed under the control of a single computer. Multiplexers and demultiplexers are used to allow the computer to read from various sensors and direct outputs to the right control elements. The network interface allows the computer to communicate with other computers so that operating parameters of the plant can be updated. However, even with modern fast computers control of multiple-loops stretches the ability of the computer. With the lowered price of microprocessor-based computer it is fiscally and may be technically better to let each control loop be controlled by a single computer. With the introduction of field buses to serially carry information between computers, sensors, and feedback elements the scenario has completely changed nowadays.

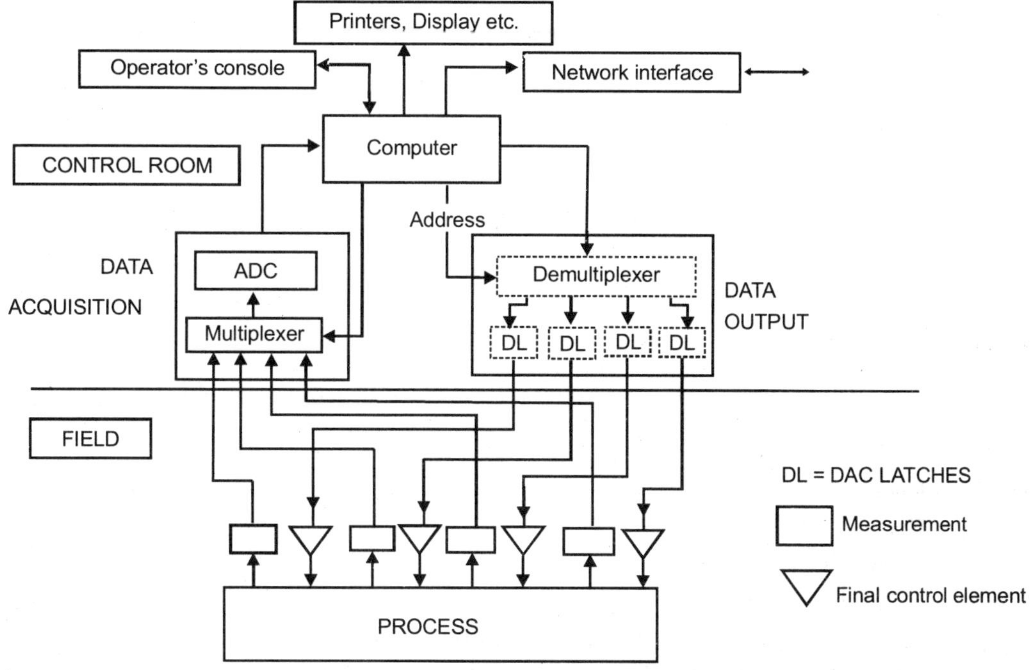

Fig. 7.4 Multiple-loop controllers

7.2.2 Software Requirements

When a computer is used as the controller, the computer must be able to solve the control equations introduced in chapter-2. The needed software is available as a 'control package' when the computer control is implemented on a general purpose computer. In the case of smart sensors and other dedicated control computers the control equations are built into the embedded computer. External commands can be used to select the desired mode (P, PI or PID etc.) and the gains for each mode. In the following sections we will see how the control equations can be modified to suit the digital environment and algorithms developed to implement the control equations.

7.2.2.1 Error

The computer accepts an input of the controlled variable from an ADC or over the bus from the sensor, encoded as a binary number. In describing the algorithms we assume the measurement range of the controlled variable is known, b_{min} to b_{max}. In chapter-2 we expressed error as percentage of span as reproduced below:

$$e_P = \frac{r-b}{b_{max}-b_{min}} \times 100 \qquad ...(2.4)$$

For the purposes of algorithm description we will assume the variable has been converted from a binary encoding to its actual value as a floating-point variable (temperature, pressure, etc.) in the control program. In the program the error will be used as a fractional quantity rather than a percent. Furthermore, we note that the variable value, and hence error, are only available as samples taken every Δt seconds. Thus the error will be expressed as,

$$e_i = \frac{r-b_i}{b_{max}-b_{min}} \qquad ...(7.1)$$

Again, we assume that when the binary number is brought into the computer it is passed to a floating-point processor (i.e., b_i is a base 10, floating-point number). This is typical of modern computers. With these assumptions about the input value and expressing the error sample as fraction of range, let us consider the three modes of control: proportional, integral, and derivative. The equations developed will provide a fractional number (0 to 1) representing what fraction of the controlling variable range should be sent to the final control element.

7.2.2.2 Proportional Mode

The proportional mode controller action is defined by a term that is directly proportional to the error. We have already seen the equation in chapter-2.

$$p = K_p e_p + p_o \qquad ...(2.14)$$

Where K_p = Proportional gain
e_p = Error
p_o = Controller output with no error
p = Controller output

This mode is easily implemented by the computer in the form of an algorithm that simply calculates equation (2.14) directly. The proportional mode is provided through the software by an equation that is entirely like the analog equation. Because we are expressing the error as a fraction of range, what is calculated is the fraction of the maximum output

$$\text{P} = \text{PO} + \text{KP} * \text{DE} \qquad ...(7.2)$$

$$\text{POUT} = \text{P} * \text{ROUT} \qquad ...(7.3)$$

Where
PO = Fraction of output with no error
KP = Proportional gain (%/%)
P = Fraction of output with error
ROUT = Maximum output
POUT = Output

DE = Error from equation (7.1) = DSP – DV (Set point/Reference – Input)

Fig. 7.5 shows a general flow chart for the proportional mode from which software can be developed.

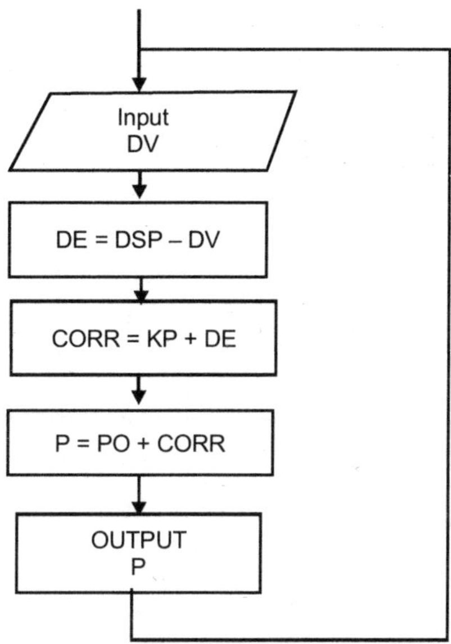

Fig. 7.5 Flow chart for proportional mode

7.2.2.3 Integral Mode

The integral or reset mode calculates a controller output that depends on the history of the controlled variable error. In a mathematical sense, history is measured by an integral of the error as given in equation (2.16).

$$p = K_I \int_0^t e_p \, dt + p(0) \qquad \text{...(2.16)}$$

To use this mode in computer control, we need a way of evaluating the integral of error. Many algorithms have been developed to do this, all of them only approximate, as only samples of the error in time are available. The simplest is called 'rectangular' and is often accurate enough to use in process control. The integral in equation (2.16) is merely the net area of the e_p curre from o to t as shown in Fig.7.6 (a).

$$\int_0^t e_p \, dt = \text{net area} = (\text{area of } e_p > 0) - (\text{area of } e_p < 0)$$

In rectangular integration, we simply use the periodic samples of e_p to construct a series of rectangles of height equal to the sample error and of width equal to the time between samples. The integral (or area) is then approximately equal to the sum of the rectangle areas. [Shown in Fig. 7.6 (b)]. In an equation, rectangular equation specifies that

$$\int_0^t e_p \, dt \simeq [S + e_{pi}]\Delta t$$

Where

Δt = Time between samples

S = $e_{p1} + e_{p2} + \ldots$ (Sum of errors calculated from previous variable samples)

e_{pi} = Last sample taken at time t specified in the integral.

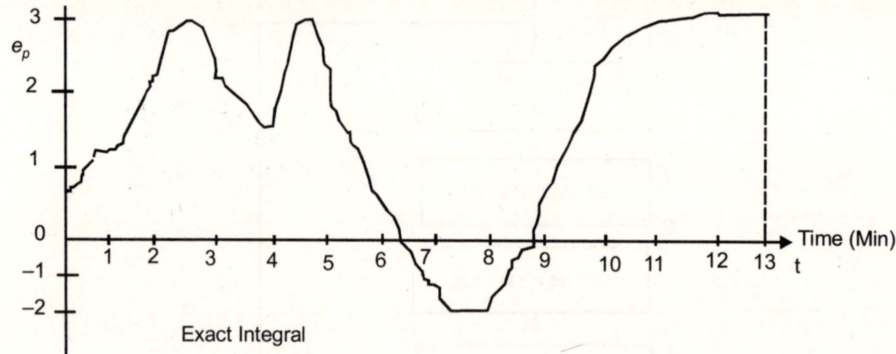

(a) $\int_0^t e_p \, dt$ = net area = area above – area below

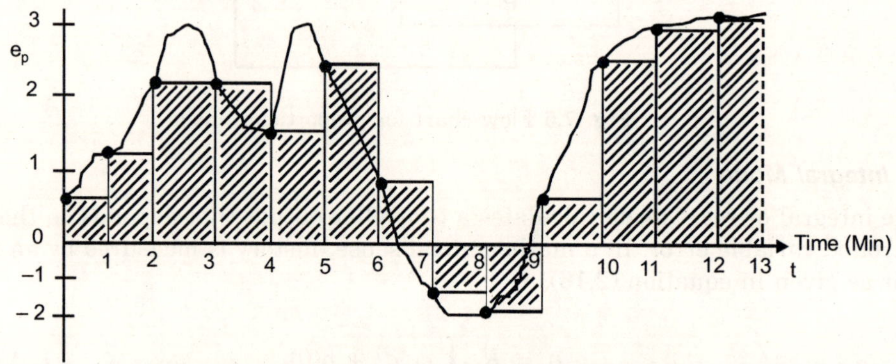

(b) Approximate integral = Sum of shaded rectangular areas

Fig. 7.6 The rectangular integration algorithm

It is clear that the smaller the time between samples, the more closely the approximate answer will approach the actual integral. Implementation of this mode in software involves the following basic equations:

$$\text{SUM} = \text{SUM} + \text{DE} \quad \ldots(7.4)$$

$$\text{PI} = \text{KI} * \text{DT} * \text{SUM} \quad \ldots(7.5)$$

$$\text{POUT} = \text{PI} * \text{ROUT}$$

Where

SUM = A running sum of errors

KI = The integral gain

DT = Time between samples

POUT = Fraction of maximum output

The flow chart shown in Fig. 7.7 illustrates how such a mode can be programmed. The time-delay routine must be built in to provide the required time between samples, because time appears as part of the mode equation [equation (7.5)] and must therefore be known. Another important point is that the units of KI and DT must be same.

7.2.2.4 Derivative Mode

The derivative controller mode, also called rate, derives a controller output that depends on the instantaneous rate of change of the error.

$$p = K_D \frac{de_p}{dt} \quad ...(2.17)$$

Where K_D = Derivative gain

$\frac{de_p}{dt}$ = Rate of error change

The gain expresses the percent controller output for each percent/second change in error. This mode is implemented in computer control by calculating an approximate derivative of the error from the data samples. A derivative is defined as the rate at which a quantity is changing at an instant in time. We can calculate only the rate at which it is changing over the sample period Δt, which is therefore only an approximation. In terms of an equation, we can express

$$\frac{de_{pi}}{dt} \simeq \frac{e_{pi} - e_{pi-1}}{\Delta t}$$

Where e_{pi} = Present error sample

e_{pi-1} = Previous error sample

Δt = Time between samples

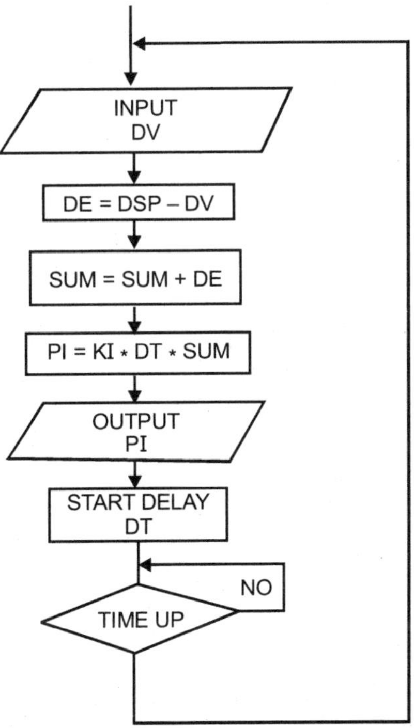

Fig 7.7 Flow chart for integral mode

Fig. 7.8 shows that this process results in a derivative that is not the actual derivative. As the time between samples is made smaller, the error will become less.

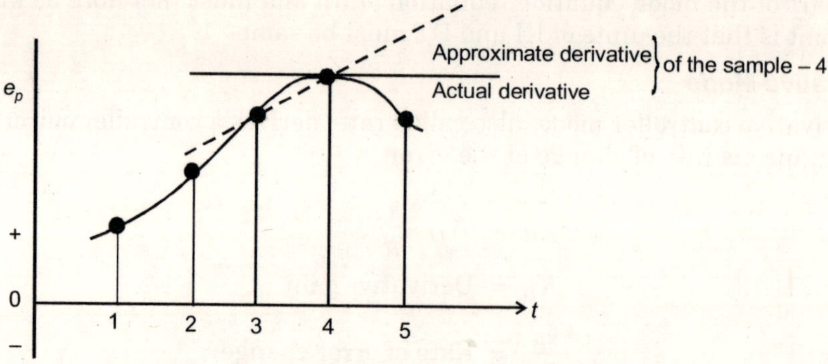

Fig. 7.8 Approximate derivative calculation

The set of equations for the derivative output can be developed directly from the definitions. We find

$$\text{DDE} = \text{DE} - \text{DEO} \qquad ...(7.6)$$

$$\text{DEO} = \text{DE} \qquad ...(7.7)$$

$$\text{PD} = \text{KD} * \text{DDE/DT} \qquad ...(7.8)$$

The flow chart for this mode is presented in Fig. 7.9.

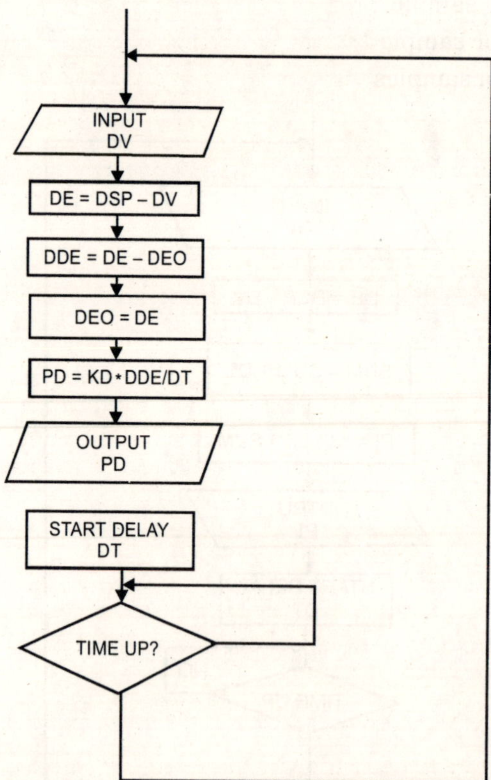

Fig. 7.9 Flow chart for derivative mode

7.2.2.5 PID Control Mode

The optimum control mode is a composite of the three modes namely proportional (P), integral (I), and derivative (D). With computer based control, a composite mode is developed by simply combining the three mode equations into the computation of the fractional output. According to the principles of PID control, the proportional gain should multiply all three forms. The control equations can be written as below:

$$\text{DDE} = \text{DE} - \text{DEO}$$

$$\text{DEO} = \text{DE}$$

$$\text{SUM} = \text{SUM} + \text{DE}$$

$$\text{PI} = \text{KP* KI* DT* SUM} \qquad ...(7.9)$$

$$\text{PD} = \text{KP* KD* DDE/DT} \qquad ...(7.10)$$

$$\text{P} = \text{KP* DE} + \text{PI} + \text{PD} \qquad ...(7.11)$$

$$\text{POUT} = \text{P* ROUT}$$

Where all the terms have been previously defined. These equations are then programmed into the control software for determination of the required output.

An alternative expression for the PID output can be constructed by using errors to provide corrections to the current output. To develop this, let us adopt a convention that a subscript will denote a particular sample. Thus, DE_i is the ith sample, and P_i is the fractional output for that sample. The output for the P_{i-1} sample, according to equation (7.11), can be written in the form

$$\text{P}_{i-1} = \text{KP*DE}_{i-1}\text{+KP*KI*DT*[SUM +DE}_{i-1}] + \text{KP*KD* [DE}_{i-1} - \text{DE}_{i-2}]/\text{DT}$$

The result for P_i will be

$$P_i = \text{KP*DE}_i \text{ +KP*KI*DT* [SUM} + \text{DE}_{i-1} + \text{DE}_i] + \text{KP*KD*[DE}_i - \text{DE}_{i-1}]/\text{DT}$$

Let us take the difference between these two expressions. This will give the correction to the previous output because of the present sample error. The result will be

$$\text{P}_i - \text{P}_{i-1} = \text{KP*[DE}_i - \text{DE}_{i-1}] + \text{KP*KI*DT*DE}_i + \text{KP*KD* [DE}_i - 2\text{DE}_{i-1} + \text{DE}_{i-2}]/\text{DT}$$

This equation can be simplified to give

$$\text{P}_i = \text{P}_{i-1} + \text{A*DE}_i - \text{B*DE}_{i-1} + \text{C*DE}_{i-2} \qquad ...(7.12)$$

Where

$$\text{A} = \text{KP} + \text{KP*KI*DT} + \text{KP*KD/DT}$$

$$\text{B} = \text{KP} + 2\text{*KP*KD/DT}$$

$$\text{C} = \text{KP*KD/DT}$$

Then the result of equation (7.12) is used to determine the output from

$$\text{POUT} = \text{P}_i\text{*ROUT} \qquad ...(7.13)$$

7.3 SUPERVISORY CONTROL AND DATA ACQUISITION (SCADA) SYSTEMS

7.3.1 Introduction

The advent of microprocessor has changed the field of process control completely. The tasks which were performed by complex and costly minicomputers are now easily programmed using microcomputers. In the past computer was not directly connected to the process but was used for supervision of analog controllers. The analog controllers were interfaced to the process directly as well as through specialized control for dedicated functions (Refer Fig. 7.10). The analog controllers and specialized controllers were called level 2 and level 1 control respectively.

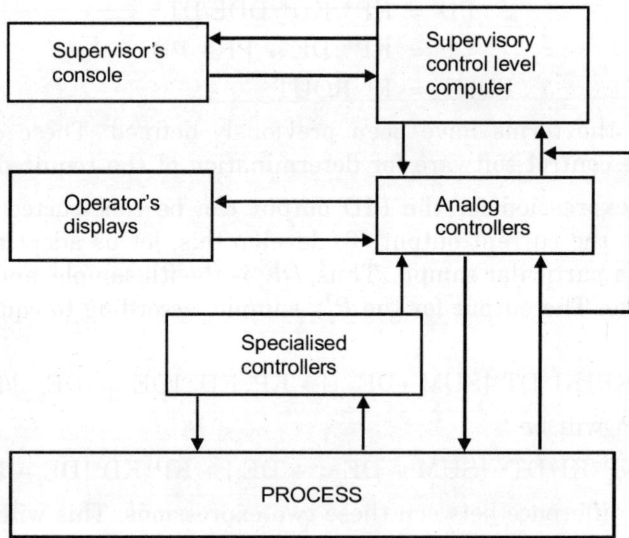

Fig. 7.10 Supervisory computer control

The emergence of economical and fast microprocessor has made analog controllers completely outdated, as the same functions can be performed by digital computers in more efficient and cost effective way. Also computers are used for many purposes in a process control facility beyond acting as the controller of loops. They also serve engineering and design functions, financial functions, and plant operations management functions. Data logging and Supervisory control applications support the engineering analysis and plant operations.

7.3.2 Data Logging (Data Acquisition System)

The efficient operation of a manufacturing process may involve the interplay of many factors, such as production rates, materials costs, and efficiencies of control. When the process requires implementation of many process control loops, then the interaction of one stage of the system with another often can be analysed in terms of the controlled variables of the loops. An example of this is the rate of production of one loop, expressed as a flow rate, which serves as a determining factor in the production rate of a following control system. An understanding of this type of interaction requires analysis from the variations of various process parameters during a production run. With the development of high-speed digital computers with mass digital storage, it became possible to record such data continuously and automatically, display the data on command and perform calculations on the data to reduce it to a form suitable for evaluation by appropriate technical individuals. Fig. 7.11 shows the block schematic of data acquisition system.

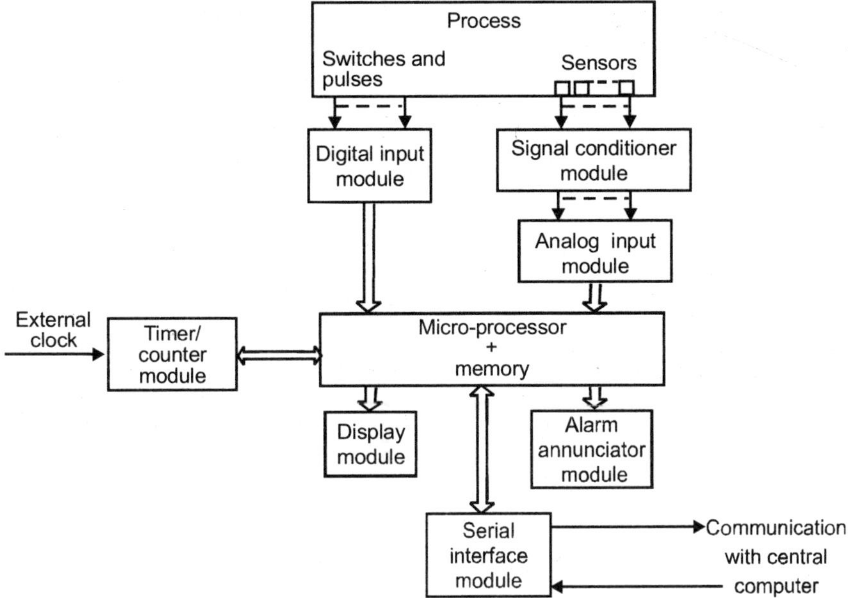

Fig. 7.11 Basic data acquisition system

The basic functions carried out by DAS are:

1. Channel scanning
2. Conversion into Engineering units
3. Data processing

Though we have discussed DAS briefly in chapter-6 the above functions will be dealt here with more details.

7.3.2.1 Channel Scanning

The microprocessor scans the channels to read the data, and this process is called 'polling'. In polling, the action of selecting a channel and addressing it is the responsibility of processor. The channel selection may be sequential or in any particular order decided by the designer. It is also possible to assign priority to some channels over others, i.e., some channels can be scanned more frequently than others. It is also possible to offer this facility of selecting the order of channel addressing and channel priorities to the operator level, i.e., make these facilities as dynamic.

The channel scanning and reading of data requires the following actions to be taken:

- Sending channel address to the multiplexer
- Sending start convert pulse to ADC
- Reading the digital data

For reading the digital data at ADC output, the end of conversion signal of ADC chip can be read by processor and when it is 'ON', the digital data can be read. Alternatively, the microprocessor can execute a group of instructions (which do not require this data) for the times which are equal to or greater than conversion time of ADC and then read ADC output. Another modification of this approach involves connecting the end of conversion line to one of the interrupt request-bins of the processor. In this case the interrupt service routine reads the ADC output and stores at predetermined memory location.

The channels can be polled sequentially, in which case the channel address in first step above increases by one every time or they may be scanned in some other order. In the later case, a channel Scan Array can be maintained in memory as shown in Fig. 7.12. The scan array contains the address of the channels in the order in which they should be addressed. The ASCN array in Fig. 7.12, has 9, 10, 1, 2......as entries in sequence. Thus, the first channel to be scanned will be channel 9, followed by 10, 1, 2,.... As the pointer reaches the last entry in the array, the first entry is again taken up (i.e., channel 9 is scanned).

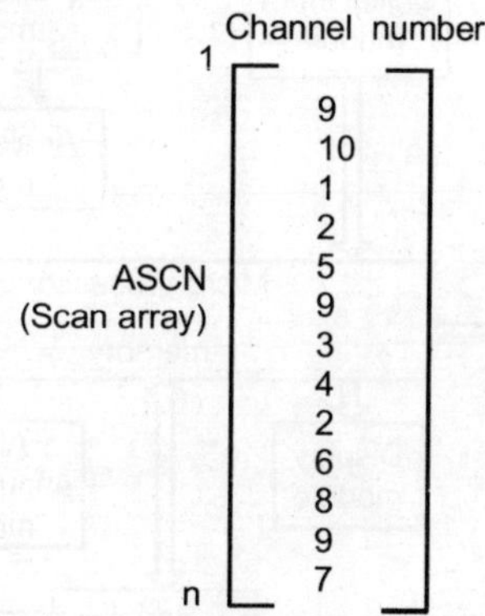

Fig. 7.12 Channel scan array

If a channel number is repeated in the array, then that particular channel will be scanned repeatedly. Thus it is possible to scan some channels more frequently than others. This gives them higher priority over others. In Fig. 7.12 the channel 9 is scanned 3 times, channel 2 is scanned 2 times while other channels are scanned once during a cycle. Fig. 7.13 shows the flow chart for one scan cycle.

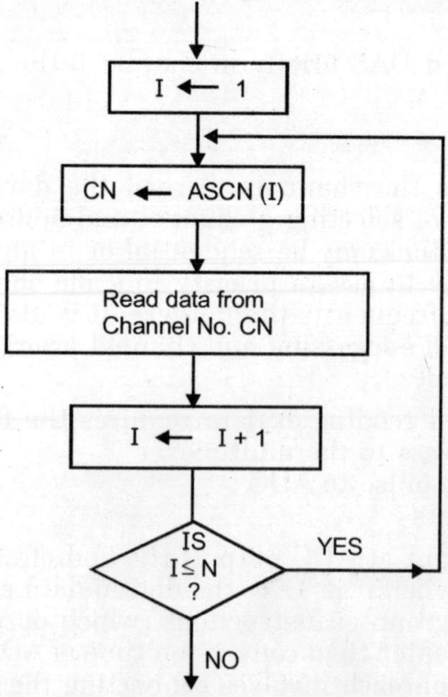

Fig. 7.13 Scan cycle flow chart

The processor may scan the channels continuously in the particular order illustrated by the flowchart or the channels may be scanned after every fixed time period. The second approach requires a timer/counter circuit whose output is connected to interrupt request input. The scan routine for one channel is incorporated in 'Interrupt Service Routine'. It is also possible to make the time gap between two channels as variable. This would require a n × 2 dimension scan array as shown in Fig. 7.14. The Interrupt Service Routine fetches the time gap value for next channel, loads the time/counter with the value and initiates the timer/counter before returning to main program.

	Channel number	Time gap
1	9	FF (H)
	10	OF (H)
	1	FO (H)
	2	
	5	
	9	
	3	
	4	
	2	
	6	
	8	
	9	
n	7	

Fig. 7.14 Scan array with time

The scan array may be decided at the design stage itself and fused permanently in ROM. Thus the channels are always scanned in that particular order. However, it may be desirable to offer the facility of changing the sequence at the operator level. The operator may like to take this action depending on the condition of the plant being monitored. This facility may be provided through a key switch which may be connected to interrupt request input of processor. The Interrupt Servicing Routine will accept new scan array and store in place of the old one.

Interrupt Scanning

Another way of scanning the channels may be to provide some primitive facility after transducer to check for violation of limits. It sends interrupt request signal to processor when the analog signal from transducer is not within High and Low limits boundary set by Analog High and Analog low signals. This is also called 'scanning by exception'. The limit checking circuit for one channel is shown in Fig. 7.15. Two analog comparators check whether the input signal is within high and low limits. The output is ORed and the final output is used as interrupt request to microprocessor.

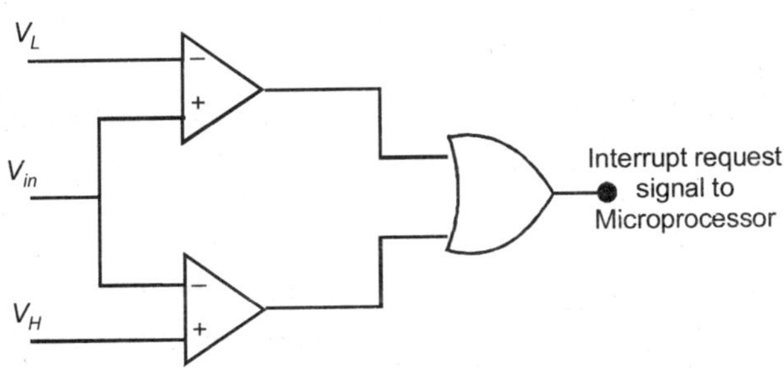

Fig. 7.15 Interrupt request generation on limit violation

7.3.2.2 Conversion to Engineering Units

The data read from the output of ADC should be converted to the equivalent engineering units before any analysis is done or the data is sent for display or printing. For an 8-bit ADC working in unipolar mode the output ranges between 0 and 255. An ADC output value will correspond to a particular engineering value based on the following parameters.

1. Calibration of transmitters
2. ADC mode and digital output lines.

Assume that the transmitter output is in the range of 0 to 5 *V*. Depending on the input range of measurand value for transmitter, a calibration factor is determined. If a transmitter is capable of measuring parameter within the input range X_1 and X_2 and provides 0-5 *V* signal at output then calibration factor is

$$1\,\text{volt} = \frac{X_2 - X_1}{5}\,\text{units}$$

If we are converting this signal to digital through an 8-bit ADC (input range 0-5 *V*) in unipolar mode then

$$5\text{ V} = 255 \text{ and } 0\text{ V} = 0$$

i.e.,

$$1\text{ volt} = \frac{255}{5}$$

Thus the conversion factor is

$$\text{ADC output}\,\frac{255}{5} = \frac{X_2 - X_1}{5}\,\text{units}$$

$$\text{ADC output } I = \frac{X_2 - X_1}{255}\,\text{engineering units}$$

If the ADC output is Y then the corresponding value in engineering units will be $\frac{Y(X_2 - X_1)}{255}$. Conversion factor is therefore $\frac{X_2 - X_1}{255}$.

The conversion of ADC output to engineering units, therefore, involves multiplication by conversion factor. The conversion factor is based on the ADC type, mode and the transmitter range. This multiplication may be achieved by shift and add method in case of 8-bit microprocessor. For 16-bit microprocessor, a single multiplication instruction will do the job.

7.3.2.3 Data Processing

The data read from the ADC output for various channels is processed by the microprocessor to carryout limit checking and performance analysis. For limit checking the 'Highest' and 'Lowest' limits for each channel are stored in an array as shown in Fig. 7.16. When any of the two limits is violated for any channel, appropriate action like alarm generation, printing, etc. is initiated. The limit array shown in Fig. 7.16, simplifies the limit checking routine. Through this, the facility to dynamically change the limits for any channel may also be provided, on the lines similar to scan array described in section 7.3.2.1.

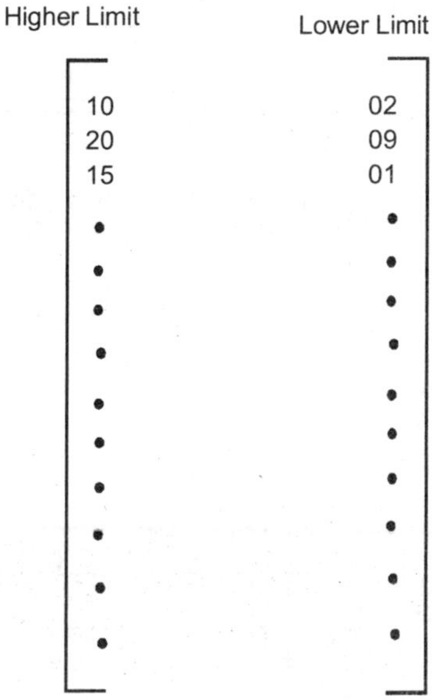

Higher Limit	Lower Limit
10	02
20	09
15	01
•	•
•	•
•	•
•	•
•	•
•	•
•	•
•	•
•	•
•	•

Fig. 7.16 Limit array

In addition to limit checking, the system performance may also be analysed and report could be generated for the manger level. This report will enable the managers to visualise the problems in the system and to take decisions regarding system modification or alternate operational strategy to increase the system performance. The analysis may include histogram generation, standard deviation calculation, plotting one parameter with respect to another, and so on. The software can be written depending on the type of analysis required. The analysis and report generation programs will be application dependent and will have to be written separately for different applications.

7.3.2.4 Distributed Data Acquisition System

In any application, if the number of channels are quite large then in order to interface these to processor, one has to use multiplexers at different levels. Fig. 7.17 shows the interfacing of 256 channels, using 17 multiplexer of 16 channels each. The 8 address lines are used to address 256 channels. Out of the 8 address lines, upper four are used to select a particular multiplexer and lower four are used to select a particular channel in the multiplexer.

For the process plants where the structure of Fig. 7.17 does not suit, the only alternative is to use more than one Data Acquisition system and distribute the channels among them. But, for performance analysis on the process plant, it is mandatory that the data from various channels should reach a central location where it can be consolidated and analysed to generate the reports on plant performance. A suitable configuration may be selected to interface the DA systems with central computer. The concept of local area networks or microprocessor interconnections can be used very effectively. Distributed DA system is the ultimate solution for complex process plant monitoring.

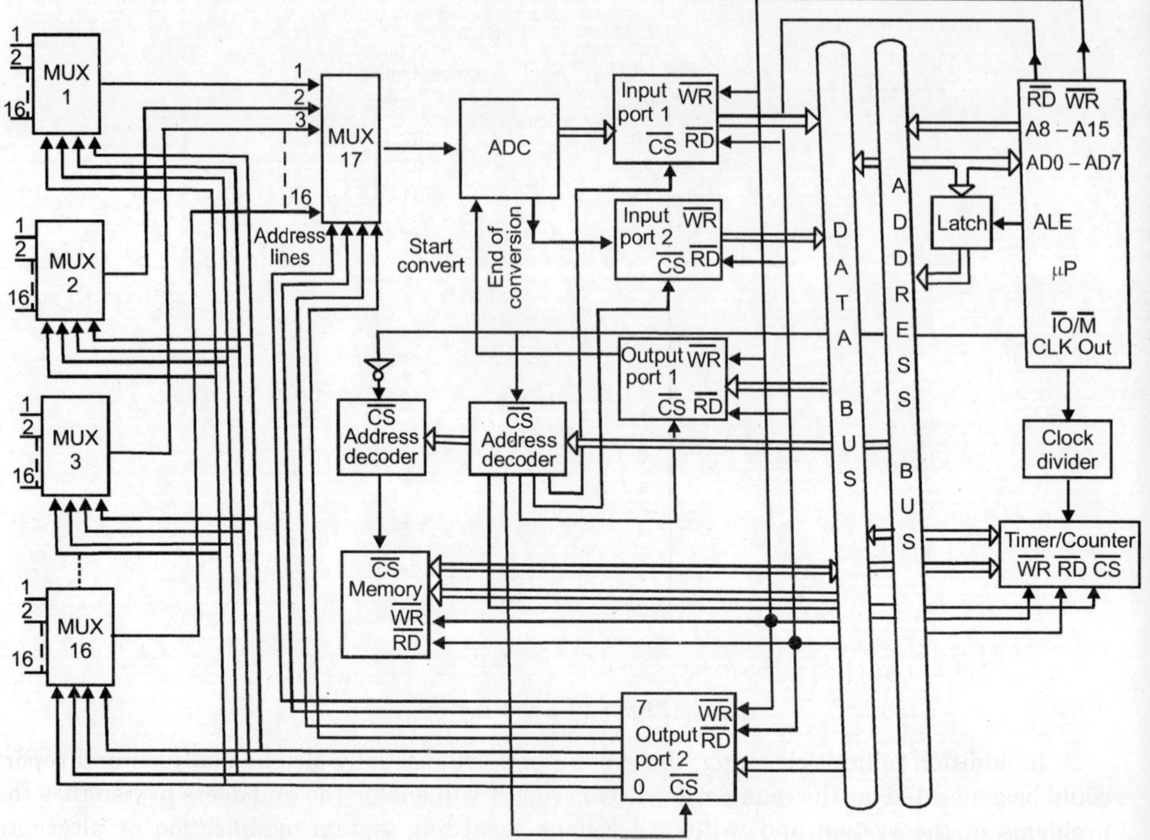

Fig. 7.17 256 channel DA with single microprocessor

7.3.3 Supervisory Control

A natural extension of a computer data-logging/data-acquisition system elaborated under section 7.3.2 involves computer feedback on the process through automatic adjustment of loop set points. As various loads in a process change, it is often advantageous to alter set points in certain loops to increase efficiency or to maintain the operation within certain pre calculated limits. In general, the choice of set point is a function of many other parameters in the process. In fact, a decision to alter one set point may necessitate the alteration of many other loop set points as interactive effects are taken into account. Given the number of loops, interactions and calculations required in such decisions, it is more natural and expedient to let a computer perform these operations under program control.

One such system is shown in Fig. 7.18, where the effect is shown by the addition of a Data-Output System (DOS) to Data Acquisition (DA) system. Such a system assumes the controllers of analog loops have been designed to accept set point values as some properly scaled voltage. By proper switch addressing, the computer then outputs a signal through the multiplexer and DACs, representing a new set point to a controller connected to that output line.

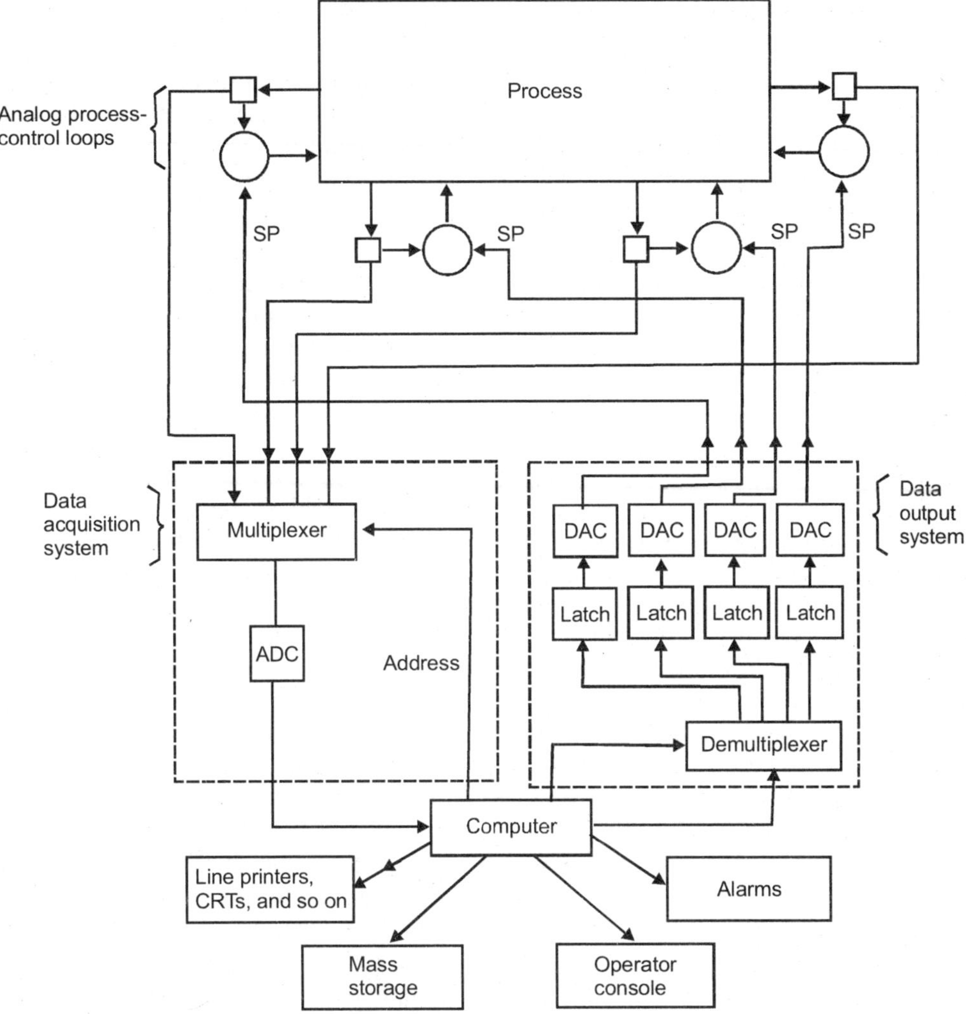

Fig. 7.18 Supervisory control and data acquisition system

7.4 CONTROL SYSTEM NETWORKS

Process control and control systems basically started with pneumatic measurements and controls with 0.1 to 1 kg/cm^2 (5 to 15 psi) as standard signal. The media used for communication is air (called in process control language as instrumentation air). Air is used as medium of communication mainly because of safety reasons. To avoid power carrying cables in hazardous and accident prone areas of petro chemical and chemical industries the air media was used. As time passed on, electronic sensors and instrumentation was introduced to avoid overvoltage entering the prohibited zone causing sparks and hence accidents. When electronics ruled the roast, number of standards like 0-50 mA, 0-100 mA, 0-250 mA, 0-10 V, 0-20 V etc. were followed by different companies. Then came the universally accepted 4 to 20 mA (or 1 to 5 V) as the standard signal for process control instrumentation. This again required lot of long cables running between field (plant) and centralized control room as because for every parameter one cable for carrying signal from field to controller and another pair of wires from controller to the final control element had to run. During this time analog controllers were used mostly.

There have been two major revolutions in the field of process control in the last 20 years. The first was the replacement of analog controllers by digital controllers using embedded computers to perform the control function. We have already studied this technology and showed how controller action is accomplished by software in the computer. Even we have gone up to digital Data Acquisition system and supervisory control of plant parameters in the previous sections.

The second revolution, which is happening even now, is replacement of the standard 4-20 mA analog signals for communication throughout a process plant by serial digital communication via a network. For control system data communication, we call the serial communication system a 'Field bus' because of the kind of network it employs. In this section the basic features of field bus serial communication systems will be presented.

7.4.1 Process Control Networks

When the first revolution was taking place in process control field, systems were developed to allow computers to communicate serially over a pair of wires in the computer industry. In this case, the data expressed in binary was sent serially as a stream of 1's and 0's from one computer to another. As the technology of serial communication matured, methods were developed to allow several computers to use a single pair of wires by sharing time slots on the wires so that one pair of computers communicated and then another pair, and so on, but so fast that it seemed that all computers were communicating at once over single pair of wires. This kind of multiple-user communication required development of protocols about how a computer could gain access to time on the line for communication and how computers could be identified so that communication could be between specific computers. This eventually gave rise to a myriad of open standard technologies including the Ethernet, local-area networks (LANs), TCP/IP addressing and transmission protocols, wide-area networks (WANs), and of course the World Wide Web (WWW) as a super network. The use of open standards for protocols and communication allows many manufacturers to design and market equipment and software to further develop this networking technology.

In the supervisory control (Section 7.3.3), process signals were predominantly carried by analog 4 to 20 mA current. The operations of controller computers were managed over a serial network, including changing set points, modes of controller action, and gains. Eventually there was a move, in the second revolution, to eliminate 4 to 20 mA data communication standard and let actual process data be carried from sensors to controllers and controllers to final control elements via digital serial networks.

Figure 7.19 shows the various configurations of the networks such as stars, rings, and a bus. The 'bus' uses straight runs of parallel wires with devices simply connected in parallel across the wires. Traditionally in a manufacturing facility the plant outside the control room is referred to as a field, so we have a 'fieldbus'. Many types of fieldbuses were put forward by process equipment manufacturers, many of which are not open. Finally several open systems were developed and, at the present time, there is competition between these standards for adoption by the process industries. The advantage of an open standard is that a process plant or manufacturing facility can buy equipment (sensors, actuators, controllers, etc.) from a variety of manufacturers and connect them all together seamlessly on the network. Fig. 7.20 shows how the serial field bus connects field hardware to computers.

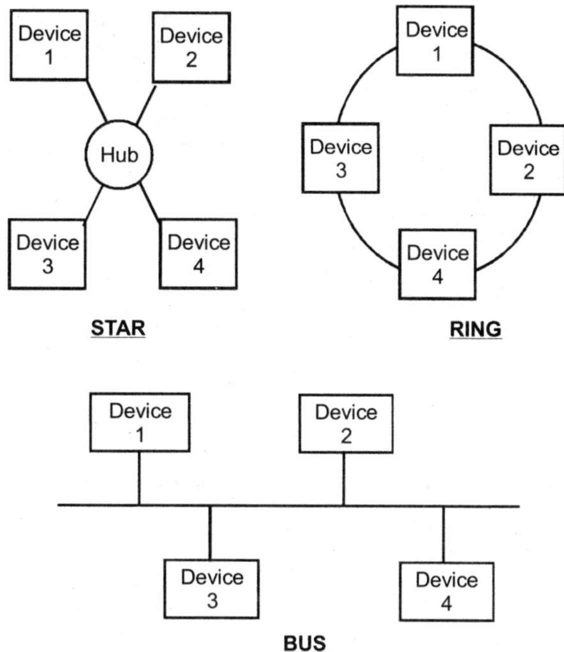

Fig. 7.19 Configurations of networks

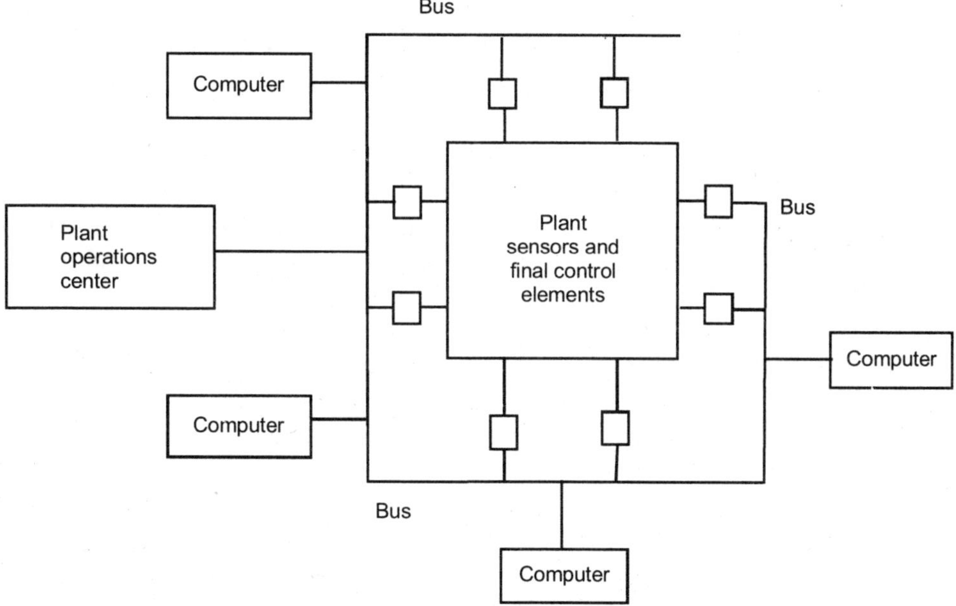

Fig. 7.20 Connection of serial field bus to field hardware and computers

7.4.2 Field Operations

To understand about field bus operations we may consider a process (heating liquid in a tank) as example. Among many other parameters we consider two specific parameter loops as shown in Fig. 7.21. First loop consists of a level sensor, computer and a final control element to adjust the liquid flow as per computer output. The second loop consists of a temperature sensor, computer and a control circuit with heater. Every device will have a unique network address. Each device also has the basic intelligence to communicate over the serial bus. The control process might go as follows, with all communication over the serial bus as digital bit streams:

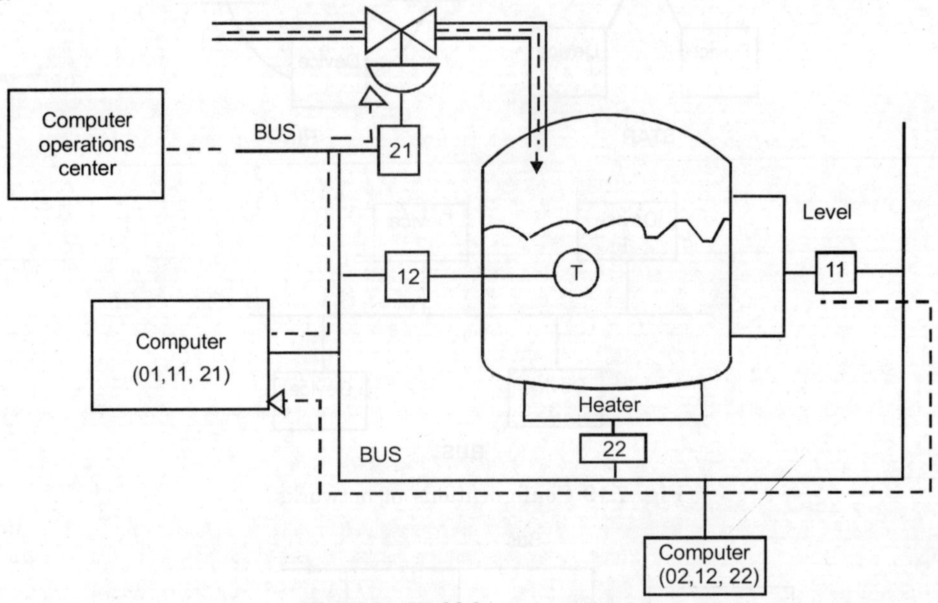

Fig 7.21 Field bus operations

1. Computer 01 sends a request to level sensor 11 asking for the actual value reading. To do this the computer forms a packet with its address (01), the sensor's address (11) and a code to request a reading.
2. The level sensor sends level data back to computer. The level sensor must have the intelligence to form a packet with the computer address, its address and the digitally encoded level.
3. After solving the control equations, computer 01 sends a valve-setting signal to valve 21. To do this the computer forms a packet with its address, the valve address (21), and a digitally encoded valve setting.
4. Valve 21 drives the valve to the new setting and, in some cases, may send an acknowledgement back to computer 01.

The cycle continues and the level is maintained at the required set point.

A similar measurement, evaluation, and feedback process may be going on at virtually the same time with the temperature control system with the help of computer (02), sensor(12), and the final control element (heater) (22). Because of their unique addresses, the two loops will concentrate on their own communications ignoring others. By sharing time on the serial line at very high speeds the two loops seem to use the bus at the same time. Meanwhile the operations center computer can also be requesting readings from the sensors to monitor the plant operations. The operations center computer can also request reports from the control computers. One of the big advantages of this system is that the large numbers of wire pairs necessary with the 4-20 mA current communication are eliminated.

7.4.3 Field Bus Characteristics

Many types of field buses are used in the process industries, each with their own features. The set of certain common characteristics called the protocol of the bus or network can be used to compare the buses. All these carry serial digital data. If we look at the signal on a bus wire, we would see a time series of pulses representing ones and zeros being propagated across the network.

Addresses

If many computers are to employ same bus for communication, there must be an addressing protocol so each device connected to the network can be uniquely identified. The number of bits allocated to the address determines the maximum number of devices that can be connected to the bus.

Data Packets

A data packet consists of the device address, often the address of the sender, the data itself and auxiliary bits for error correction and other functions. If a computer is using the bus to send an updated signal to its final control element device, it must package the new data along with the address of the device. The final control element recognizes its own address and will therefore accept the data and perform the update. Other devices on the bus will ignore the information.

Media

The actual nature of the medium that carries data varies greatly between buses. Typical media include twisted pair copper wire, coaxial cable, fiber-optic cable, and radio frequency propagation (wireless).

Speed

The speed by which data can be propagated across the network is typically given as the maximum number of bits per second (bps) that the bus can carry without serious degradation of the data. Speed is a function of not only the physical medium of the bus but also the number of devices connected to the bus and the complexity of the data being sent.

Cycle Time

An important aspect of the bus is the time required for a data update to propagate through the system. This is determined by the speed of the bus and also by how many other devices are on the bus and competing to send data.

Interoperability

An important feature of a bus is the ease by which computers and devices can be added to the bus and configured to operate as part of the plant. The manufacturer of a process-control computer can provide software that makes it compatible with a variety of bus protocols. However, manufactures of control valves, sensors, and other control system devices generally, have to build the protocol into the device. Thus they are interoperable on a particular bus only but not on other.

7.4.4 Field Bus Types

Many types of field buses have been developed and are in use in the manufacturing and process industries throughout the world. There are two open standards with wide applications and greatest support: Foundation Field bus and Profibus (Process field bus). Although both have their custom protocols they are often being allied with the Ethernet since this technology is so well established.

7.4.4.1 OSI 7–Layer Protocol

The protocol of a field bus refers to how the bus system packages data to be transmitted. In 1978, the International Standards Organisation (ISO), faced with a proliferation of closed network systems, i.e., manufacturer- specific (proprietary) networks with no possibility of multi-vendor participation, defined a 'Reference model for communication between open systems', the so called 'Open Systems Interconnection' or (OSI model). The OSI model applies to all communication systems, from the mainframe and personal computers operating at the administrative level to the data exchange between masters and slaves on a field bus. This is called the 7-layer OSI model which defines 7 functional layers as shown in Fig. 7.22. However, generally following three layers are used in process control applications: physical layer, data link layer and application layer. In fact it is not necessary for a communication system to employ all seven layers. Many field bus systems employ only the above three, about which we will discuss further.

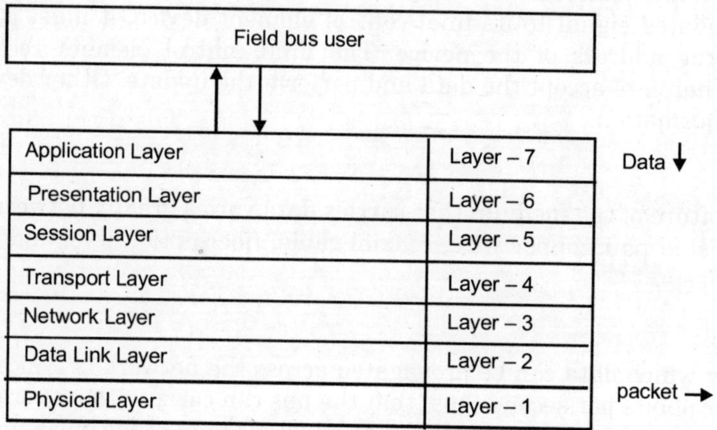

Fig. 7.22 OSI–7 Layer model architecture

Physical Layer

Physical layer defines how data are to be physically transported from device to device. It includes specifications of electrical signals, cabling, connectors, network topology, and transfer speed. The cabling or transmissions media is the backbone of all networks. The following three types of cables are used for this purpose:

1. Twisted pair of copper wires: For transmission rates from 3.75 KBPS (up to 300 meters) to 32 KBPS (up to 1500 meters).
2. Coaxial cable: More expensive than twisted pairs. Used for transmission rates from 5 MBPS (over 1000 meters over carrier) to 10 MBPS (over several kilometers over broad band).
3. Fiber optic cable: Has greater transmission capacity (more than 5 times that of coaxial cable) and high transmission rates (in the range of GBPS).

Data Link Layer

Data link layer ensures that data can be exchanged between devices. This layer governs not only network access and data format, but also mechanism to ensure data security. With several communication devices are present on a single line, clear rules must be given as to which are allowed to transmit. In principle, there are two methods of regulating access, central and de-central. In central bus control, a fixed master assigns the right to one station at a time. If master fails, then the entire communications network breaks down. For this reason, de-centralised bus control with flying masters has been developed. The following three de-centralised models have found a wide acceptance all over the world.

1. CSMA/CD model: Carrier sense multiple access with collision detection model
2. ETHERNET model as per IEEE 802.3
3. Token passing model as per IEEE 802.2.

Application Layer

Application layer defines the services which are supported by the network, for instance, read and write commands, program management, up-and-down loading of data, and virtual device images. User interfaces with this layer for programming the network. All the other layers refer to how the actual data, called the application layer (number 7), is packaged with control information such as the address of the sender, address of the receiver, size of the data, error correction etc.

7.4.4.2 Field Bus Category

The communication buses are classified into three categories depending upon how much data the system must carry.

1. Sensor or Bit Level: This type of bus is intended for application where single bits are the primary information carrier. This would principally include inputs with on/off conditions as from switches and alarms. Outputs would also be on/off such as lights, solenoid actuators, quick-acting valves, and so forth. Often basic PLC installations can use this type of bus.

2. Device or Byte Level: This is an intermediate application where one to several bytes (8 bits) are the primary information carrier. This type of bus is used when the state of some device requires more than one bit. This system can also be used to send and receive analog data that has been digitally encoded.

3. Field Bus or Message Level: Finally, we come to the actual field bus for which the information may be hundreds of bits; when combined with the encapsulation, it may be a thousand bits. This system includes error checking, correcting, and complex addresses as well. This type of bus can be used for any level but is most suited to operating control loops over the bus and the passing of large amounts of data. Data rates are often slower than sensor or device buses because of the large packet size.

7.4.4.3 Important Features of a Field Bus

A field bus is used to communicate control signals in a process plant among the sensors, controllers, and final control elements. Certain important characteristics of field bus are:

1. Physical Media Transmission: This refers to layer-1 of the OSI model and defines the actual communication carrier, such as twisted wire pairs, coaxial cable and fiber-optic cable.

2. Number of Devices: This refers to the fact that device addresses are carried in the information packet. Therefore, there will be a limit to the number of devices that can be addressed.

3. Distance Over which the Bus can Extend: The longer the bus, the weaker and more distorted the signals become. In some cases special repeater circuits can extend the basic bus distance.

4. Speed of Transmission: The serial data is like a square wave in the communication media where the state changes of the square wave represent bits. The bus speed refers to the frequency of the possible bit changes which is better described by the number of bits per second (bps) that the serial system can handle.

5. Bus Powered: This is a feature carried over from 4 to 20 mA analog systems. Some field buses have the capability that the data measurement serial bit stream rides on top of DC current providing power to the devices on the bus.

7.4.4.4 Comparison of Field Buses

Table 7.1 shows some of the field buses in use in the process industry and gives their approximate properties as previously defined. Some of the buses use the Ethernet for layer-1 but it is possible to simply use the Ethernet directly without the other protocol. The advantage to this is that Ethernet is an old and established system and its widespread use for LANs and office automation in general makes it a very familiar technology. Two other field buses which are in wide use throughout the world are (1) Foundation field bus and (2) Profibus (Process field bus)

Table 7.1 Comparison of field buses

Bua	Category	Physical media	Number of devices	Distance	Speed	Power from bus
ASI	Sensor	Twisted pair	31	100 m	167 kbps	Yes
Seriplex	Sensor	4-wire shielded	500	150 m	200 Mbps	No
CAN	Sensor	Twisted pair	127 nodes	25 m to 1 km (speed dependent)	10 kbps to 1 Mbps	No
Device net	Device	Twisted pair	64	500 m (6 km with repeater)	125 kbps to 500 kbps	No
LON works	All	Twisted pair fiber power line	32,000 per domain	2 km @ 78 kbps	1.25 Mbps	Yes
Profibus DP/PA	Field bus	Twisted pair fiber	127 nodes	100 m twisted pair 24 km fiber	DP: 500 kbps PA: 31.25 kbps	PA: Yes
Foundation (H1)	Field bus	Twisted pair fiber	240/segment 65 k segments	1900 m	31.25 kbps	Yes
Foundation (HSE)	Field bus	Twisted pair fiber	Unlimited	100 m twisted pair 2 km fiber	100 Mbps	No
Industrial ethernet	Field bus	Twisted pair fiber coax	Unlimited with routers	100 m twisted pair 2.5 km fiber	10 Mpbs 100 Mbps	No

7.5 PROBLEMS AND SOLUTIONS

Problem 7.1

Pressure in kg/cm² is measured and converted to a voltage by a sensor according to the relationship

$$V = 2.2\,(p + 12)^{1/2} - 10$$

The pressure range is 0 to 60 kg/cm² and the set point is 40 kg/cm². This voltage is provided as input to an 8-bit unipolar ADC with a 10 V reference and the resulting binary is provided as input to a control computer.

(*a*) *Develop the equations used to find the pressure from the binary input and then the error.*

(*b*) *Contrast the actual error with the computer sample error for a pressure of 42 kg/cm².*

Solution:

(*a*) Let us call the sample from the ADC as *Ni*, which is the base 10 equivalent of the binary output of the ADC. We can then find the voltage corresponding to this sample (within ΔV of the ADC) as,

$$V_i = \frac{10}{256} Ni \qquad \text{[Ref. equation 6.37]}$$

Now the pressure sample can be determined by using the given equation relating pressure and voltage.

$$p_i = \left(\frac{V_i + 10}{2.2}\right)^2 - 12$$

Then, the error as a fraction of range is found from equation (7.1)

$$e_i = \frac{r - b_i}{b_{max} - b_{min}} = \frac{40 - p_i}{60}$$

(*b*) To find the actual error for 42 kg/cm² we should use the previous equation with p_i = 42 kg/cm²

$$e = \frac{40 - 42}{60} = -0.033$$

To find the sample error we must take into account the loss in information due to the ADC. Thus, we calculate just as the computer will.

The voltage of the sensor is: $V_{42} = 2.2\,(42 + 12)^{1/2} - 10$
$= 6.17$ V

The output of the ADC will be

$$Ni = \text{Int}\left[\frac{6.17}{10} \times 256\right] = 157$$

Where the ADC truncated the fractional part. Thus in the computer, the voltage will be computed as

$$V_i = \frac{10}{256} \times 157 = 6.13 \text{ V}$$

and the pressure will be calculated as

$$p_i = \left(\frac{6.13+10}{2.2}\right)^2 - 12$$

$$= 41.73 \text{ kg/cm}^2$$

Therefore, the sample error will be

$$e_i = \frac{40 - 41.73}{60} = -0.029$$

Thus, because of truncation by the ADC there is difference in error representation of about 0.004. (i.e. 0.033 – 0.029 = 0.004) or about $\frac{0.004}{0.033} \times 100 = 12\%$

Problem 7.2

A proportional-mode controller has Kp = 3.2, input range of 255, and set point of 150. The output maximum is 200, and the output fraction with no error is 0.6.

(a) *Develop the control equations.*

(b) *What is the output for no error?*

(c) *Find the output for an input of 160.*

Solution:

(a) The equations are found simply from equations (7.1), (7.2) and (7.3).

$$\text{DE} = \frac{\text{DV} - \text{DSP}}{\text{Range}} = \frac{\text{DV} - 150}{255}$$

$$\text{P} = 0.6 + 3.2 * \text{DE}$$

$$\text{POUT} = \text{P} * 200$$

(b) Output when there is no error,

$$\text{DE} = 0$$

$$\text{POUT} = 0.6 * 200 = 120$$

(c) For an input 160, we get an error fraction of

$$\text{DE} = \frac{(160 - 150)}{255} = 0.039$$

$$\text{P} = 0.6 + 3.2 * (+0.039) = 0.6 + 0.1248$$

$$= 0.7248$$

$$\text{POUT} = \text{P} * 200 = 0.7248 * 200 = 144.96$$

Problem 7.3

Find the approximate integral of e_p in Fig. 7.6 from 0 to 12 min for the sample time (a) 1 min and (b) 2 min. (c) What percentage change in the value of the integral results from the difference in sample time?

Solution:

(a) We find the integral from the rectangular integral procedure with sample time of 1 min:

$$A_{1\text{ min}} = 1\,(0.6 + 1.2 + 2.1 + 2.1 + 1.5 + 2.2 + 0.8 - 1.2 - 1.8 + 0.5 + 2.8 + 2.9)$$

$$= (14.7 - 3) = 11.7\%\text{ min}$$

(b) With same procedure for sample time of 2 min:

$$A_{2\text{ min}} = 2\,[0.6 + 2.1 + 1.5 + 0.8 - 1.8 + 2.8]$$

$$= 2[6.0] = 12\%\text{ min}$$

(c) Percentage change in the value between two approaches

$$= \frac{12-11.7}{12}\times 100 = 2.5\%$$

Problem 7.4

Determine the approximate value of the derivative of e_p from Fig. 7.6, using samples every 2 min and every 1 min and compare the results

(a) At a time of 5 min.

(b) At a time of 12 min.

Solution:

(a) For 2 min samples, we get the derivative by using sample at 3 min and at 5 min.

$$\text{Derivative } e_{2\text{ min}} = \frac{2.2-2.1}{2} = 0.05\%\text{min}$$

$$\text{Derivative } e_{1\text{ min}} = \frac{\text{Sample at 5 min} - \text{Sample at 4 min}}{1}$$

$$= \frac{2.2-1.5}{1} = \frac{0.7}{1} = 0.7\%\text{ min}$$

This means there is a difference of about $\frac{0.7-0.05}{0.7}\times 100 = 93\%$

(b)

$$\text{Derivative } e_{2\text{ min}} = \frac{\text{Sample at 12min} - \text{Sample at 10min}}{2}$$

$$= \frac{3.1-2.8}{2} = 0.15\%\text{ min}$$

$$\text{Derivative } e_{1\text{ min}} = \frac{\text{Sample at 12 min} - \text{Sample at 11 min}}{1}$$

$$= \frac{3.1-2.9}{1} = 0.2\%\text{ min}$$

There is a difference at this point of about $\frac{0.2-0.15}{0.2}\times 100 = 25\%$.

Problem 7.5

A digital controller is to be developed with the following specifications: KP = 8%/%, KI = 0.6%/ (%-min), KD = 0.12%/(%/min), time between samples = 3s, input range 0 to 255, set point = 150. The output range is 0-255. Set up the control equations for PID control:

(a) By simply combining the three mode equations.

(b) By using errors to provide corrections to the current output.

Solution:

(a)

$$\text{DE} = (\text{DSP} - \text{DV})/255 - 0$$
$$\text{DDE} = \text{DE} - \text{DEO}$$
$$\text{DEO} = \text{DE}$$
$$\text{SUM} = \text{SUM} + \text{DE}$$

To compute the gains, we need to get the units the same

$$\text{KP*KI*DT} = (8)\ (0.6\ \text{min}^{-1})\ (3\text{s})\ (1\ \text{min}/60\text{s}) = 0.24$$
$$\text{KP*KD/DT} = (8)\ (0.12\ \text{min})/\ (5\text{s})\ (1\ \text{min}/60\text{s}) = 11.52$$

Thus the remaining equations become

$$\text{PI} = 0.24 * \text{SUM}$$
$$\text{PD} = 11.52 * \text{DDE}$$
$$\text{P} = 8 * \text{DE} + \text{PI} + \text{PD}$$
$$\text{POUT} = \text{P} * 255$$

(b) We need to evaluate the three constants of the equation (7.12)

$$\text{A} = \text{KP} + \text{KP} * \text{KI} * \text{DT} + \text{KP} * \text{KD/DT} = 8 + 0.24 + 11.52 = 19.76$$
$$\text{B} = \text{KP} + 2 * \text{KP} * \text{KD/DT} = 8 + 2 * 11.52 = 31.04$$
$$\text{C} = \text{KP} * \text{KD/DT} = 11.52$$

Then the equation (7.12) can be written

$$\text{P} = \text{P1} + 19.76 * \text{DE} - 31.04 * \text{DE1} + 11.52 * \text{DE2}$$
$$\text{DE2} = \text{DE1}$$
$$\text{DE1} = \text{DE}$$

The last two equations are necessary because the next sample DE1 and DE2 contains the previous two samples. Then the previous output must be updated.

Problem 7.6

A proportional control system with a 25% PB controls water flow rate by using tank level measurement. Specifications are as follows:

1. *Level range: 50 to 100 cms*
2. *Level set point : 75 cm*
3. *Signal conditioner converts 0 to 100 cm to a voltage scaled at 0.05 V/cm that is, 0 to 5V*
4. *An 8-bit ADC converts an input of 0 to 5V to OOH to FFH, where 5V input just produces FFH*
5. *Water flow rate is regulated by a final control element that operates directly from the output of an 8-bit DAC. There are 256 flow rates selected according to the following: OOH is 'O' flow rate, FFH is 100% flow rate (Maximum flow rate) and 80 H is the flow rate corresponding to the set point of 75 cm that is, zero error.*

Find the digital controller equations and express all numbers in their hex form.

Solution:

[The process assumed here is a hypothetical liquid tank which evaporates the liquid and sends out at the rate proportional to the level in the tank]

The proportional-mode equations are given by equations (7.2) and (7.3)

$$L_{min} = 50 \text{ cms, so}$$

$$D_{min} = (50/100)\ 256 = 128 \rightarrow 80\text{ H}$$

$$L_{max} = 100 \text{ cms}$$

$$D_{Max} = \left(\frac{100}{100}\right) 256 = 256 \rightarrow \text{FFH}$$

$$L_{SP} = 75 \text{ cm}$$

$$DSP = \left(\frac{75}{100}\right) 256 = 192 \rightarrow \text{COH}$$

The error equation becomes

$$\text{DE} = (\text{COH} - \text{DL}) / (\text{FFH} - 80\text{ H})$$

$$\text{DE} = (\text{COH} - \text{DL}) / 80\text{ H}$$

Because the zero-error output is given to be 80H, PO = 80H. A 25% PB means KP = 4%/%. Both the error and output are given as fractions of range so that this value of KP can be used directly. Thus,

$$\text{P} = 80\text{ H} + 4 * \text{DE}$$

This is the fraction of full-scale output. The final output will be

$$\text{POUT} = \text{P} * \text{FFH}$$

Actually, multiplying a 8-bit number by FFH is effectively equivalent to a shift of 8 bits, so that in terms of numbers,

$$\text{POUT} = \text{P}$$

7.6 PROBLEMS AND QUESTIONS

1. A data-logging system must take samples of 50 variables at 250 samples per second each. What is the maximum signal-acquisition and processing time in microseconds?
2. What are called smart transmitters/sensors?
3. Briefly discuss the features of distributed computer control systems with block diagram.
4. Describe with neat sketch the ISO reference model for communication.
5. Discuss about the various topologies used for data transmission.
6. What are the various field buses available for use in industries?
7. Briefly discuss about the various important features of field buses.
8. A computer must sequentially sample 100 process parameters. It requires 14 instructions at 5.2 μ_s/instruction for the computer to address and process one line of data. The multiplexer switching time is 2 μ_s, and the ADC conversion. time is 35 μ_s. Find the maximum sampling rate for a line.
9. What is meant by the protocol of the bus or network?
10. What is SCADA? Explain with a neat sketch.
11. A process is to operate under PID with a 60%. PB, 1.2 min integration time, and 0.05 min derivative time. If the error is available as per cent of span, develop the control equations and show a flow chart of computer controller action with all constants evaluated. The sample time is 0.8 mins.
12. Measurement of position L in mm is provided in terms of voltage by the relation $V = (e^{0.02L} - 1)/2$, if fed directly into an 8-bit ADC with a 5 V reference. Develop the control equations for the following controller specifications:

$$K_P = 2, K_I = 0.6\,\text{min}^{-1}, K_D = 0.25, \text{sample time 2s, set point of 80 mm, 8-bit output.}$$

13. Discuss the advantages and disadvantages of
    (*a*) Single loop digital controllers
    (*b*) Multi loop digital controllers
14. Name the three basic functions carried out by DAS.
15. Summarise the development of process control system since Pneumatic era through Electronic to Digital.

BIBLIOGRAPHY

1. B.G. Liptak, *'Process Control'–Instrument Engineers'* Handbook, Third edition, Butterworth and Heinemann, 1995.
2. Donald P. Eckman, *'Automatic Process Control'*, Wiley Eatern Ltd., New Delhi,1993.
3. George Stephanopoulos, *'Chemical Process Control'*, Prentice Hall of India, New Delhi, 2003.
4. Curtis D. Johnson, *'Process Control Instrumentation Technology'*, Prentice-Hall of India, New Delhi, 2007.
5. Peter Harriott, *'Process Control'*, Tata McGraw Hill Edition, New Delhi, 2004.
6. Dr. S. Sundaram and Dr. T. K. Radhakrishnan,*'Process Dynamics and Control'*, Delhi, 2003.
7. Werner G. Holzbock, *'Instruments for Measurement and Control'*, East-West Press Pvt. Ltd., New Delhi.
8. Sam G. Dukelow, *'The Control of Boilers'*, ISA Press, USA.
9. Siemens Pocket Hand book for *'Industrial Instrumentation and Process Control'*.
10. D. M. Considine, *'Process Instruments and Control Handbook'*, McGraw-Hill Publishing Company, 1985.
11. S. C. Thiagarajan, *'Sizing of pipes and control valves'*, SAIL, CET, Ranchi.
12. Faculty Development Programme on *'Process Control'*, Department of EIE, Kongu Engineering College, Erode-638 052.

INDEX

D

Q

R

S